Recent Developments in Cancer Systems Biology

Recent Developments in Cancer Systems Biology

Editors

Raghu Sinha
Kazim Yalcin Arga

MDPI • Basel • Beijing • Wuhan • Barcelona • Belgrade • Manchester • Tokyo • Cluj • Tianjin

Editors
Raghu Sinha
Biochemistry and Molecular
Biology
Penn State University College of
Medicine
Hershey
United States

Kazim Yalcin Arga
Bioengineering
Marmara University
Istanbul
Turkey

Editorial Office
MDPI
St. Alban-Anlage 66
4052 Basel, Switzerland

This is a reprint of articles from the Special Issue published online in the open access journal *Journal of Personalized Medicine* (ISSN 2075-4426) (available at: www.mdpi.com/journal/jpm/special_issues/Cancer_Biology).

For citation purposes, cite each article independently as indicated on the article page online and as indicated below:

LastName, A.A.; LastName, B.B.; LastName, C.C. Article Title. *Journal Name* **Year**, *Volume Number*, Page Range.

ISBN 978-3-0365-1472-7 (Hbk)
ISBN 978-3-0365-1471-0 (PDF)

Contents

About the Editors

Raghu Sinha

Raghu Sinha is an Associate Professor of Biochemistry and Molecular Biology, Penn State College of Medicine, Hershey, Pennsylvania. He is an expert in cancer chemoprevention, cancer therapy, and proteomics. His research focuses on studying mechanisms of breast, prostate, and pancreatic tumor growth inhibition by organo-selenium compounds, impact of dietary sulfur amino acids in animal cancer models and healthy humans, effects of tobacco products on lung epithelial cells, cancer systems biology, protein-protein interactions, and multi-omics approaches for identifying diagnostic and prognostic cancer biomarkers, as well as drug repurposing strategies. He has published his research in high impact journals and has mentored high school, undergraduate and graduate students, and post-doctoral fellows. Dr. Sinha received his BS and MS in Biophysics from Panjab University and PhD in Immunopathology from PGIMER in Chandigarh, India.

Kazim Yalcin Arga

Kazim Yalcin Arga is a Professor of Bioengineering and expert in systems biology, bioinformatics, and multi-omics research. His research centers on statistical and mathematical modeling, translation of Big Data to applications in systems biology, pharmaceutical innovation, drug repositioning, health care, and clinical practice. He is a frequent contributor to recognized international systems biology forums and articles in high-impact journals. His works span the continuum of metabolic systems engineering in bacteria to drug repurposing and diagnostics for complex human diseases. Prof. Arga has mentored several students who have gone on to pursue productive careers as independent researchers. Prof. Arga received his BS and PhD in Chemical Engineering from Bogazici University in Istanbul. He is currently Professor of Bioengineering at the Department of Bioengineering at Marmara University and at the Health Institutes of Turkey, both in Istanbul, Turkey.

Preface to "Recent Developments in Cancer Systems Biology"

The ebook is a compilation of literature reviews and original articles describing the tools used in cancer systems biology, and how and why these are important to move forward the field in the direction of personalized medicine. The editors had invited researchers from all over the world to share their findings and create a Special Issue on recent developments in cancer systems biology. The information provided within the articles will encourage young scientists to appreciate the value of cancer systems biology in tackling the various cancer types from the transcriptomics to proteomics tools for diagnosis, as well as drug repurposing and prognosis. Assistance from Ms. Esther Cao is specially acknowledged in completion of the ebook.

Raghu Sinha, Kazim Yalcin Arga
Editors

Journal of
Personalized
Medicine

Editorial

Recent Developments in Cancer Systems Biology: Lessons Learned and Future Directions

Kazim Y. Arga [1,*] and Raghu Sinha [2,*]

1   Department of Bioengineering, Marmara University, Istanbul 34722, Turkey
2   Department of Biochemistry and Molecular Biology, Penn State College of Medicine, Hershey, PA 17033, USA
*   Correspondence: kazim.arga@marmara.edu.tr (K.Y.A.); rsinha@pennstatehealth.psu.edu (R.S.)

check for
updates

Citation: Arga, K.Y.; Sinha, R. Recent
Developments in Cancer Systems
Biology: Lessons Learned and Future
Directions. *J. Pers. Med.* **2021**, *11*, 271.
https://doi.org/10.3390/jpm11040271

Received: 30 March 2021
Accepted: 2 April 2021
Published: 4 April 2021

Publisher's Note: MDPI stays neutral
with regard to jurisdictional claims in
published maps and institutional affiliations.

Cancer is a complex disease involving multiple mechanisms and critical players, at broad genomic, transcriptional, translational and/or biochemical levels. One could envision discovering new biomarkers for early detection by understanding the behavior of cancer development and progression, but to date, there are few biomarkers approved for use in the clinical setting. Therefore, there is a critical need to improve strategies and methods by using novel state-of-the-art tools and strategies to identify and validate newer biomarkers. In addition to biomarkers, there is also a demand for effective methods to identify new targets to inhibit tumor growth. Technically, there is a growing requirement to find new targets using personalized approaches in a targeted and much more effective manner, as existing drugs often become resistant over time in cancer patients. Opportunities to improve this strategy would, therefore, be to find better druggable targets and provide options for drug combinations and/or drug repurposing. More importantly, the ultimate goal of an oncologist and the desire of the cancer patient is to improve overall survival and this could be achieved in part through better prognostic models. Cancer systems biology has undoubtedly emerged as an integrative tool to achieve such advances.

This Special Issue on "recent developments in cancer systems biology" has compiled several novel approaches that use cutting-edge technologies to build a strong foundation of systems biology in cancer research. The issue includes eight original research articles and four literature reviews on recent efforts that use a variety of in silico tools along with experimental approaches to discover novel biomarker candidates for diagnosis and prognosis and to identify drugs and their targets for treatments that could be used in thirteen cancers and their subtypes.

Several "omics" investigations, including genomics, proteomics, metabolomics, glycomics and metagenomics, provide potential candidate biomarkers that can be measured in plasma, tissue and saliva in several lethal cancer types including Pancreatic Cancer [1]. Integrative analysis of these "omics" data would likely discover novel biomarkers for diagnosis and prognosis as well as targets for effective therapy. Moreover, distinguishing clinically similar cancers can be challenging and focusing on genomic and transcriptomic variations may prove beneficial, this issue describes details on various methods available for ovarian and breast cancers [2] and types of lung cancer [3,4] and renal cell carcinoma [5] for identifying key genes and pathways that might assist in proposing diagnostic and prognostic predictions. In addition, integrating multi-omics is important particularly in the use of patient-derived experimental models [6] that can be used in the clinical setting to provide personalized treatment options. Another genome-level advancement that surpasses next-generation sequencing is the identification of somatic structural variants (SVs) that influence functional and cancer-related genes [7]. This optical genome mapping and SVs analysis can be applied to a variety of solid tumors for better cancer prognosis and treatment.

Discovering new targets in cancers provide opportunities especially for recurrences since the drug resistance is proving to be challenging to treat. Several drug targets have been identified using transcriptomics and biological networks in different cancer types including miR-1246 targeting several genes [4] and Sestrin-2 [8] in lung adenocarcinomas, ELK1 [9] and ETS [10] genes in glioma. Additionally, drug repurposing strategies are not only extensively used to discover new uses for already approved drugs, but also provide opportunities for potentially treatment of drug resistance in various cancers. In another article [11], drug repurposing efforts were reviewed in triple-negative breast cancer, an aggressive breast cancer subtype that has a high rate of recurrence and metastasis. These authors compared different repurposing strategies, including structure-based, transcription signature-based, biological network-based and data mining-based drug repositioning. In another study, seven distinct gene programs representing different biological processes involved in drug-induced changes in AML were identified [12]. Furthermore, a data-driven dynamic model of acquired resistance to combined drugs was constructed by these authors and revealed several interventions that can specifically disrupt portions of the system-wide drug response, which could allow co-targeting and lead to synergistic treatments that can overcome resistance and prevent potential recurrence.

In conclusion, all of the articles published in this Special Issue cover recent developments with attractive approaches to a wide range of topics encompassing the Cancer Systems Biology. These articles and reviews propose a variety of biomarkers for clinical diagnosis, prognosis and therapeutic strategies including "drug repurposing" for various cancers that pose a major health challenge with significant socioeconomic consequences. We would like to make an appeal to researchers around the world to join forces and contribute to the development of a common platform for personalized medicine using a combination of the different biomarkers proposed in this Special Issue in a diagnostic and/or prognostic setting, allowing the identification of patients at risk, which would facilitate the early initiation of personalized treatments. This Special Issue also highlights the various predictive models and the use of integrated biological network analysis to identify target genes and correlate them with prognosis. It is of utmost importance that all predictive models must undergo extensive validation.

Funding: The research received no external funding.

Acknowledgments: We are very grateful to all the authors who have provided excellent contributions to this Special Issue. Moreover, we would like to thank *Journal of Personalized Medicine* for offering us the opportunity to make this Special Issue a reality and in particular, Esther Cao for her availability, professionalism, help and constant presence, support and kindness. Finally, we would like to acknowledge the excellent and efficient work of the expert reviewers who reviewed submissions in a timely, fair and constructive manner.

Conflicts of Interest: The authors declare no conflict of interest.

References

1. Turanli, B.; Yildirim, E.; Gulfidan, G.; Arga, K.Y.; Sinha, R. Current State of "Omics" Biomarkers in Pancreatic Cancer. *J. Pers. Med.* **2021**, *11*, 127. [CrossRef] [PubMed]
2. Khella, C.A.; Mehta, G.A.; Mehta, R.N.; Gatza, M.L. Recent Advances in Integrative Multi-Omics Research in Breast and Ovarian Cancer. *J. Pers. Med.* **2021**, *11*, 149. [CrossRef] [PubMed]
3. Zengin, T.; Önal-Süzek, T. Comprehensive Profiling of Genomic and Transcriptomic Differences between Risk Groups of Lung Adenocarcinoma and Lung Squamous Cell Carcinoma. *J. Pers. Med.* **2021**, *11*, 154. [CrossRef] [PubMed]
4. Huang, S.; Wei, Y.-K.; Kaliamurthi, S.; Cao, Y.; Nangraj, A.S.; Sui, X.; Chu, D.; Wang, H.; Wei, D.-Q.; Peslherbe, G.H.; et al. Circulating miR-1246 Targeting UBE2C, TNNI3, TRAIP, UCHL1 Genes and Key Pathways as a Potential Biomarker for Lung Adenocarcinoma: Integrated Biological Network Analysis. *J. Pers. Med.* **2020**, *10*, 162. [CrossRef] [PubMed]
5. Caliskan, A.; Gulfidan, G.; Sinha, R.; Arga, K.Y. Differential Interactome Proposes Subtype-Specific Biomarkers and Potential Therapeutics in Renal Cell Carcinomas. *J. Pers. Med.* **2021**, *11*, 158. [CrossRef] [PubMed]
6. Yalcin, G.D.; Danisik, N.; Baygin, R.C.; Acar, A. Systems Biology and Experimental Model Systems of Cancer. *J. Pers. Med.* **2020**, *10*, 180. [CrossRef] [PubMed]
7. Goldrich, D.Y.; LaBarge, B.; Chartrand, S.; Zhang, L.; Sadowski, H.B.; Zhang, Y.; Pham, K.; Way, H.; Lai, C.-Y.J.; Pang, A.W.C.; et al. Identification of Somatic Structural Variants in Solid Tumors By Optical Genome Mapping. *J. Pers. Med.* **2021**, *11*, 142. [CrossRef] [PubMed]

8. Chae, H.S.; Gil, M.; Saha, S.K.; Kwak, H.J.; Park, H.-W.; Vellingiri, B.; Cho, S.-G. Sestrin2 Expression Has Regulatory Properties and Prognostic Value in Lung Cancer. *J. Pers. Med.* **2020**, *10*, 109. [CrossRef]
9. Sogut, M.S.; Venugopal, C.; Kandemir, B.; Dag, U.; Mahendram, S.; Singh, S.; Gulfidan, G.; Arga, K.Y.; Yilmaz, B.; Aksan Kurnaz, I.A. ETS-Domain Transcription Factor Elk-1 Regulates Stemness Genes in Brain Tumors and CD133+ BrainTumor-Initiating Cells. *J. Pers. Med.* **2021**, *11*, 125. [CrossRef] [PubMed]
10. Babal, Y.K.; Kandemir, B.; Kurnaz, I.A. Gene Regulatory Network of ETS Domain Transcription Factors in Different Stages of Glioma. *J. Pers. Med.* **2021**, *11*, 138. [CrossRef] [PubMed]
11. Ávalos-Moreno, M.; López-Tejada, A.; Blaya-Cánovas, J.L.; Cara-Lupiañez, F.E.; González-González, A.; Lorente, J.A.; Sánchez-Rovira, P.; Granados-Principal, S. Drug Repurposing for Triple-Negative Breast Cancer. *J. Pers. Med.* **2020**, *10*, 200. [CrossRef] [PubMed]
12. Wooten, D.J.; Gebru, M.; Wang, H.G.; Albert, R. Data-Driven Math Model of FLT3-ITD Acute Myeloid Leukemia Reveals Potential Therapeutic Targets. *J. Pers. Med.* **2021**, *11*, 127. [CrossRef]

Journal of
*Personalized
Medicine*

MDPI

Article

Data-Driven Math Model of FLT3-ITD Acute Myeloid Leukemia Reveals Potential Therapeutic Targets

David J. Wooten [1], Melat Gebru [2], Hong-Gang Wang [2] and Réka Albert [1,*]

[1] Department of Physics, Pennsylvania State University, University Park, PA 16802, USA; dzw347@psu.edu
[2] Department of Pediatrics, Penn State College of Medicine, Hershey, PA 17033, USA; mtg194@psu.edu (M.G.); hwang3@pennstatehealth.psu.edu (H.-G.W.)
* Correspondence: rza1@psu.edu

Abstract: *FLT3*-mutant acute myeloid leukemia (AML) is an aggressive form of leukemia with poor prognosis. Treatment with *FLT3* inhibitors frequently produces a clinical response, but the disease nevertheless often recurs. Recent studies have revealed system-wide gene expression changes in *FLT3*-mutant AML cell lines in response to drug treatment. Here we sought a systems-level understanding of how these cells mediate these drug-induced changes. Using RNAseq data from AML cells with an internal tandem duplication *FLT3* mutation (*FLT3*-ITD) under six drug treatment conditions including quizartinib and dexamethasone, we identified seven distinct gene programs representing diverse biological processes involved in AML drug-induced changes. Based on the literature knowledge about genes from these modules, along with public gene regulatory network databases, we constructed a network of *FLT3*-ITD AML. Applying the BooleaBayes algorithm to this network and the RNAseq data, we created a probabilistic, data-driven dynamical model of acquired resistance to these drugs. Analysis of this model reveals several interventions that may disrupt targeted parts of the system-wide drug response. We anticipate co-targeting these points may result in synergistic treatments that can overcome resistance and prevent eventual recurrence.

Keywords: acute myeloid leukemia; Boolean model; drug resistance; network

check for
updates

Citation: Wooten, D.J.; Gebru, M.; Wang, H.-G.; Albert, R. Data-Driven Math Model of FLT3-ITD Acute Myeloid Leukemia Reveals Potential Therapeutic Targets. *J. Pers. Med.* **2021**, *11*, 193. https://doi.org/10.3390/jpm11030193

Academic Editor: Raghu Sinha

Received: 28 January 2021
Accepted: 7 March 2021
Published: 11 March 2021

1. Introduction

Acute myeloid leukemia (AML), characterized by the pathological accumulation of myeloblast cells in blood or bone marrow, is a heterogeneous and aggressive form of leukemia. About 30% of AML cases carry a mutation in the *FLT3* gene, which encodes a receptor critical for normal hematopoiesis [1]. By far the most common mutation is an internal tandem duplication (*FLT3*-ITD), which occurs in about 25% of all AML cases [1], a mutation placing patients in a poor prognosis category [2]. Highly specific *FLT3* inhibitors are therapeutically promising [1,2], though the disease often recurs.

Recent experimental results have suggested that while *FLT3*-inhibition can kill *FLT3*-ITD cells, some cells survive and become drug tolerant persisters (DTPs) [3,4]. Targeting the therapeutic vulnerabilities of drug-tolerant FLT3 mutant AML cells can enhance the anti-leukemic efficacy of FLT3 inhibitors to eliminate minimal residual disease, mutational drug resistance and relapse. The mechanisms underlying this phenotypic change are not fully understood. A recent study found that DTPs exhibit the upregulation of inflammation pathways, and combination treatment with quizartinib (a *FLT3* inhibitor) and dexamethasone (a glucocorticoid that reduces inflammation) was synergistic [4]. This is an example of reprogramming therapy, in which the phenotypes or gene expression patterns induced by one drug are countered by another simultaneous intervention.

The idea of reprogramming cancer cells into drug-sensitive states [5–9] or even non-malignant states [10,11] has become increasingly promising. Reprogramming drug-sensitivity follows from the hypothesis that drug treatment induces reversible, system-wide gene expression and epigenetic changes, causing cells to achieve a resistant or tolerant subtype [12,13]. Targeting these changes and reverting them may then reprogram the cells. With this view, we seek to gain a systems-level understanding of the gene expression and phenotypic changes of *FLT3*-ITD AML cells in response to drug treatment with quizartinib and dexamethasone, and their evolution into DTPs.

To this end, we identified several modules of co-expressed genes that correspond to different treatment conditions with quizartinib, dexamethasone, or their combination. Based on genes within these modules, we built a network model of *FLT3*-ITD AML drug response. Using data-driven tools, we derived a probabilistic, dynamical gene regulatory model that recapitulates the expression changes of AML cells following these drug treatments and can be used to predict the effects of perturbations and interventions in the cells. We focused on identifying interventions that downregulate modules associated with drug resistance, and upregulate modules associated with cell death. The interventions we identified represent promising strategies to improve response to *FLT3* inhibitors in *FLT3*-ITD AML.

2. Materials and Methods

2.1. Data Acquisition and Processing

RNAseq data of MV4-11 cells were collected by M. Gebru, as described in [4], and previously made publicly available on GEO (GSE116432). Data consisted of triplicate measurements, each of (1) 10 nM quizartinib treatment for 48 h, (2) 10 nM quizartinib treatment for five days, (3) 100 nM dexamethasone treatment for 48 h, (4) combination 10 nM quizartinib + 100 nM dexamethasone for 48 h (we refer to this combination as Quiz + Dex), (5) Quiz + Dex for five days (quizartinib for five days and dexamethasone added on day 3 because the combination for 5 days would kill almost all cells), and (6) DMSO (GEO: GSE116432). Data were transformed as log(1 + FPKM). Only transcripts with a matched HGNC symbol were kept.

2.2. Weighted Gene Co-Expression Network Analysis

We used v1.69 of the WGCNA package in R v4.0.2. We used the pckSoftThresold function with a "signed" network type to identify power = 10 as the smallest power that achieved a scale-free R^2 value >= 0.9 (Figure S1). We built a topological overlap matrix using a signed adjacency matrix obtained from power = 10. Genes were hierarchically clustered using the "average" method, and genes were assigned to co-expression modules using WGCNA's cutreeDynamic function with deepSplit = 2, pamRespectsDendro = FALSE, and minClusterSize = 100. This analysis resulted in seven modules of co-expressed genes (Figure 1A and Figure S1). Following WGCNA convention, the modules are denoted by color: turquoise (7219 genes), blue, yellow, brown, green, black (164 genes).

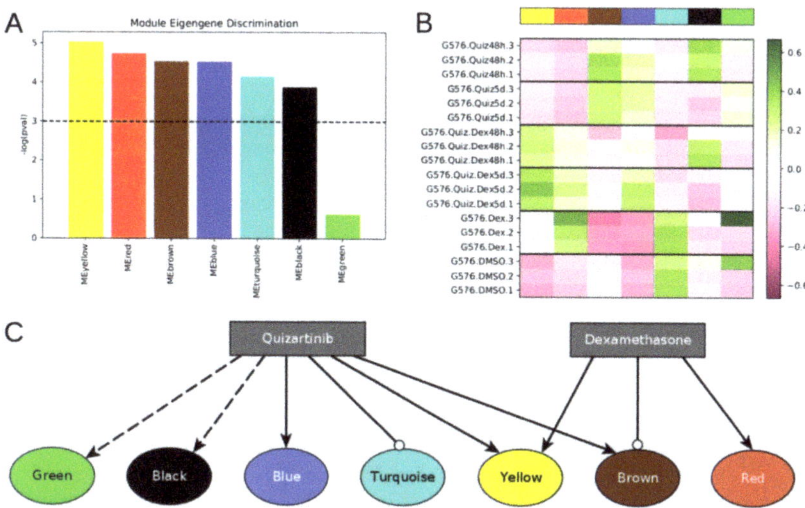

Figure 1. Differentially expressed modules respond differently to different treatment conditions. (**A**) WGCNA identified seven gene co-expression modules from the DMSO and drug treated *FLT3*-ITD AML expression dataset, six of which are differentially expressed across the six different treatment conditions. (**B**) Heatmap showing module eigengene expression for each module in each sample. High module eigengene expression reflects high average expression of genes within that module. (**C**) Qualitative model showing the effect of each drug on the expression of genes within each module. Arrow-tipped edges indicate activation, while circle-tipped edges indicate repression. The dotted edge from quizartinib to the black module reflects the observation that black module genes are upregulated at 48 h by quizartinib, but become downregulated again by five days of treatment. The dotted edge from quizartinib to the green module reflects that the green module is not upregulated after 48 h, but is after 5 days.

2.3. Molecular Biology of the Cell (MBCO) Ontology Analysis

MBCO analysis was completed using the source code from https://github.com/SBCNY/Molecular-Biology-of-the-Cell/commit/9ff6c87 (accessed on 15 March 2020). The background gene set consisted of all genes from the RNAseq dataset, and ontology analysis was performed independently for each WGCNA gene module. Enrichment results are given in Figures S2–S8 and File S2.

2.4. Network Construction

To build the network, we integrated interactions from multiple databases that aggregated literature-based or predicted interactions, SIGNOR [14], TRRUST [15], and RegNetwork [16], as well as published networks related to AML [14,17], NFKappaB signaling [17], NOTCH signaling [18], tumor promoting inflammation [19,20], and apoptosis [20].

Many of these network resources have minor variations in gene names or use different aliases for different genes. We applied two methods to transform gene names from different sources into a common space so that all interactions with a given gene may be identified, even if the different sources use different names for that gene. First, we considered that many sources use different capitalization, or interchangeably use ".", "-", or "_" characters. To address this, we capitalized all characters in each gene name, and removed all ".", "-", and "_" characters. Second, to match gene aliases across different network sources, we used three separate gene name alias data sources, including Entrez Homo_sapiens gene info (https://ftp.ncbi.nih.gov/gene/DATA/GENE_INFO/Mammalia/Homo_sapiens.gene_info.gz (accessed on 25 October 2020)), BioMart from Ensemble (https://useast.ensembl.org/biomart (accessed on 25 October 2020)), and HGNC (https://www.genenames.org/ (accessed on 25 October 2020)). Each source includes multiple aliases for each gene name. We constructed a gene name alias graph whose nodes represent gene names, and in which each edge represents that two nodes are aliases for the same gene from one of those resources. Within this alias graph, if there exists a path from one node to another, it indicates they refer to the same gene.

There were several properties we wanted the final AML network topology to have, and the strategy we used to build the network was refined until we reached a network that satisfied these properties. First, we wanted the network to be large enough to capture enough regulatory details (e.g., more than about four nodes per module, or about 30 nodes total), but not too large to be able to model or simulate well (e.g., fewer than about 200 nodes). Second, we wanted all seven gene co-expression modules to be similarly represented, even though some modules are much larger than others (turquoise and blue have thousands of genes each but green and black only have a few hundred genes each). Third, we wanted the in-degree of nodes to not be too large (e.g., more than about 7 incoming edges). This is because a Boolean regulatory function with N inputs has 2^N possible input conditions for which an output value must be specified. When inferring Boolean functions of 7 or more variables using the BooleaBayes algorithm, the probability that any given sample constrains a given input condition becomes extremely small, and the resulting Boolean function becomes nearly completely stochastic.

The process we used to build the final network is shown in Figure S9. First, we merged all the network sources (e.g., SIGNOR) into a single large network, wherein nodes that were aliases of one another from different sources were merged into a single node. This network contained 8614 nodes and 35,710 edges. Of the nodes, 2374 had non-zero out-degrees and represented a gene from the RNAseq dataset. We then extracted subgraphs consisting of only genes from the brown, red, green, yellow, and black modules. We focused on these first as they are smaller modules than blue and turquoise, and we wanted to include as many of these nodes as possible to ensure they are well represented in the final network. We merged these five subgraphs together, which resulted in a disconnected graph. This graph contained only two components with five or more nodes, one of which consisted of 18 green nodes, the other consisted of 53 brown nodes. We hypothesized that nodes from the red, yellow, or black modules may be connected into these components through paths (successions of edges) containing nodes not in the brown, red, green, yellow, or black modules. For example, no blue or turquoise nodes had been included at this point. We searched for paths of no more than four nodes that could connect nodes from the red, yellow, or black modules into the above-mentioned components (Figure S9B). Anytime multiple paths were found, we only added the shortest path. If there were multiple equally short paths, all were added.

We removed all sink nodes because they do not feed back into the dynamics of the network, and thus cannot be drivers. The resulting graph contained 186 nodes and 888 edges. Of the nodes, 52 belonged to the brown module, 34 to turquoise, 23 blue, 21 yellow, 20 red, 15 green, and 9 black, while the others belonged to no module. This satisfied our goals of having approximately equal representation of the different modules, and not too few nor too many nodes. However, many nodes in the resulting graph had extremely high in-degree. For example, RELA had 43 in-edges, TP53 had 37, and FOXO3 had 36. The Boolean regulatory update function for RELA would then have $2^{43} \sim= 10^{12}$ conditions that must be specified, which would be impractical, and impossible given available data.

To avoid such excessively high in-degree nodes in the network, we calculated an edge score that we used to retain only the most confident edges. We set a threshold that must be exceeded to include an edge, and made this threshold increase as more in-edges are added to a node. This process preserves a node's regulators if it has low in-degree, but provides an increasingly strict criterion for edges to be included as the in-degree becomes larger.

The edge score was based on the following factors: (1) whether or not the source node is a transcription factor (TF), (2) the number of references supporting the edge, (3) the number of different databases (e.g., SIGNOR) or literature-based networks that included the edge, and (4) the edge confidence given by the network resource, including "belief" (networks from Indra) or "score" (SIGNOR, TRUUST). Regarding point (1), if the source node is a TF, the edge score is multiplied by $TF_{MUL} = 2$. If not, $TF_{MUL} = 1$. Regarding point (4), for network resources that did not provide edge confidence, the confidence was assumed to be 0. With these metrics, the edge score was calculated as:

$$score = \left(N_{references} + N_{resources} + 10 \cdot \frac{confidence}{N_{edges}} \right) \cdot TF_{MUL}$$

The minimum possible score was 2, as $N_{references}$ and $N_{resources}$ were at least each 1 for every edge. For each node, all its incoming edges were scored and ordered from highest to lowest. The (up to) three edges with the three highest scores were always included. Following these, each subsequent edge was included if $score > N_{IN-EDGES} - 1$. For example, given five incoming edges with scores (5, 4.5, 4, 3, 2), the first three edges (scores 5, 4.5, and 3) are automatically included. The next edge has score = 3, which is greater than $N_{IN\ EDGES} - 1 = 3 - 1 = 2$, so it is included. The next edge has score = 2, which is no longer greater than $N_{IN\ EDGES} - 1 = 4 - 1 = 3$, so it, and any lower score edges, would not be included.

Finally, once again, all sink nodes, or nodes that do not belong to a component of at least size = 4, were removed. The final network had 106 nodes and 270 edges.

2.5. Regulatory Function Inference using the BooleaBayes Algorithm

Using the transcription data from the RNAseq dataset, the node activation data constructed as described in the next section, and the network topology, we inferred probabilistic Boolean regulatory functions using the BooleaBayes algorithm as described in [6]. Briefly, BooleaBayes tries to find Boolean logic functions consistent with steady-state gene expression data and a network topology. As BooleaBayes needs normalized expression, RNAseq data for each gene were normalized between 0 and 1 by setting all values less than the 20th percentile to 0, all values above the 80% to 1, and all values in-between were linearly interpolated between 0 and 1.

BooleaBayes infers a probabilistic Boolean regulatory function for each node in the network. For each function, all input regulators are assigned a significance value by BooleaBayes, defined as the maximum possible (absolute value) difference in output value the regulator can make if it switches from OFF to ON. We set a minimum threshold of 0.1 for this value. With this threshold, each regulator must, in at least one condition, mean the difference between a 0.45 or less output, and a 0.55 or greater output.

When fitting the function for a node, if at least one regulator did not exceed this threshold, the regulator with the lowest significance was removed, and the function was inferred again using only the remaining regulators. This process was repeated until either all regulators exceeded the minimum significance threshold, or no regulators remained. In the latter case, the target node becomes a source node for later analyses.

2.6. Extension of BooleaBayes to Post-Translational Regulation

Unlike previous work with BooleaBayes, which focused purely on transcriptional regulation, the AML network includes post-translational modifications. However, the expression data only include transcription quantification. To apply the BooleaBayes algorithm, we must separate the probability of a node being transcribed from the probability of a node being active. For instance, if node A regulates node B, node A may be transcribed but not active, in which case the input value of node A into node B's Boolean function should be OFF.

To this end, we first distinguished for each edge whether it represented transcriptional regulation or post-translational regulation. An edge whose source node is a transcription factor according to [21] was considered to be a transcriptional edge. All other edges were considered as post-translational. Post-translational edges were assigned as positive (activating) or negative (de-activating) based on edge annotations from the source network. For example, SIGNOR and Indra edges indicate whether the regulator up-regulates or down-regulates the target. For edges with no consistent database annotation, edge weights were obtained from literature search when possible, or assumed to be positive if no specific supporting information could be found.

Any node that is only transcriptionally regulated is assumed to be active as long as it is transcribed. All nodes that are post-translationally regulated (such as a node named "X") were split into transcript (X_T) and active protein (X_A) forms. Any outgoing edges (regulatory effects) from nodes that have _T and _A forms are assumed to come only from the _A form.

To fit BooleaBayes functions, the values from the RNAseq data are used directly for X_T. Values for X_A for each sample must be determined prior to applying BooleaBayes, so that the target nodes of X use X_A for their training data, instead of X_T. We assumed that protein post-translational activation follows an inhibitory dominant form. For example, if X_A is activated by nodes J and K, and deactivated by node M, we say

X_A = X_T and (J or K) and not M

(or J_A, K_A, or M_A, if any of those regulators require an activated form). As X_T, J, K, and M are not strictly Boolean variables, but rather probabilities, we transform this into a sloppy logic form by replacing "or" with "+", "and" with "*", and "not" with "1-". Further, each term is strictly held within 0 and 1. Thus

X_A = X_T * min(J+K, 1) * max(1-M, 0)

More generally, as long as X has at least one activator we say

X_A = X_T * min(sum(ACTIVATORS_X), 1) * max(1-sum(INHIBITORS_X), 0)

while, if X has no activators, we say

X_A = X_T * max(1-sum(INHIBITORS_X), 0)

This distinction prevents nodes that have no activators from always being inactive—they are assumed to be active unless deactivated. We constructed such an equation for every node that must be activated. These equations formed a system of nonlinear algebraic equations which we solved numerically using the scipy.optimize.fsolve() function in Python v3.8, with an initial guess for each node of X_A = X_T. The resulting values of X_A were then added to the gene expression dataset to be used for inferring BooleaBayes functions for any node regulated by an _A form of a regulator.

2.7. Identification of Pseudo-Attractors

Pseudo-attractors of a probabilistic discrete system are states, or collections of states, that the system keeps revisiting. Expressed more technically, pseudo-attractors are collections of states for which transitions into them are more likely than transitions out, along every axis. The sum of forward and backward transition probabilities between two BooleaBayes states always adds up to 1. Therefore, if a transition is more likely into a state than out of a state, the out-transition will be less than 0.5. Thus, pseudo-attractors of a BooleaBayes-inferred system will correspond to the attracting strongly connected components of the state transition system, for which all transitions with probability less than 0.5 are removed. This corresponds exactly to the attractors of the closest approximating deterministic Boolean system, obtained by rounding all probabilities to the nearest 0 or 1—all transitions in the probabilistic system with probability less than 0.5 are absent.

Thus, to identify pseudo-attractors of the probabilistic AML drug network, we approximated each BooleaBayes-inferred update function to its closest deterministic function. We used the AttractorRepertoire module from the StableMotifs [22] python package to find attractors of the deterministic system. The system has a very large number of source nodes (nodes with no regulators), which allows many attractors. To isolate the attractors most relevant to AML drug response, we determined the Boolean state of these source nodes for each of the six experimental conditions by averaging their probability to be ON or OFF from the data. For each node, if it was more likely to be ON across the three replicates, we plugged in the value ON to the deterministic system and propagated its value through the Boolean update functions, and likewise for OFF.

2.8. Node Interventions

We sought to understand how interventions that target specific nodes influence the stability of WGCNA gene modules. We considered two types of interventions: holding a node in the OFF (0) state, akin to knockout (KO) and holding a node in the ON (1) state, akin to constitutive activation (CA). We assumed that any intervention targeting a gene that was separated into transcribed and active protein forms applied to both forms. During simulations, the states of controlled nodes were held constant, and other nodes were updated as in the WT system.

2.9. Definition and Calculation of Influence Index

Systematic in silico intervention experiments require a significant number of computational resources, thus we wanted to prioritize the most likely candidates for up- or down-regulating a target module. To this end, we calculated an "influence index" for each node-intervention-module tuple, for example the tuple "GSK3B, KO, blue module". The influence index is designed to estimate how likely it is that the influence of a node-intervention on the node's direct targets aligns with the up- or down-regulation goal of a specified module.

The influence index is based on the concepts of necessary and sufficient regulation. If node A = ON is necessary for node B = ON, this means that A = OFF implies B = OFF. Conversely, if node A = ON is sufficient for node B = ON, this means that A = ON implies B = ON [23]. For each edge we developed scores quantifying the likelihood that the edge represents necessary regulation or sufficient regulation. In total we calculated four scores for each edge: (1) the source is necessary for the target to be ON (called N_{ON}), (2) the source is sufficient for the target to be ON (called S_{ON}), (3) the source is necessary for the target to be OFF (called N_{OFF}), and (4) the source is sufficient for the target to be OFF (called S_{OFF}). These scores are based on the average value of the probabilistic function output when the node at the source of the edge is ON (avg_{ON}), or the source node is OFF (avg_{OFF}). For example, consider a node C whose regulatory function is $f(A, B)$. For the edge A→B,
$$avg_{ON} = \frac{f(1,0)+f(1,1)}{2} \text{ while } avg_{ON} = \frac{f(0,0)+f(0,1)}{2}.$$

Using this definition of avg_{ON} and avg_{OFF}, N_{ON}, S_{ON}, N_{OFF}, and S_{OFF} were calculated as follows:

If an edge represents overall positive regulation (meaning that switching the source node from OFF to ON increases the likelihood that the target turns on)

$$N_{ON} = 1 - avg_{OFF}$$
$$S_{ON} = avg_{ON}$$
$$N_{OFF} = 0$$
$$S_{OFF} = 0$$

Conversely, if an edge represents overall negative regulation (meaning that switching the source node from OFF to ON decreases the likelihood that the target turns on)

$$N_{ON} = 0$$
$$S_{ON} = 0$$
$$N_{OFF} = avg_{OFF}$$
$$S_{OFF} = 1 - avg_{ON}$$

To illustrate these definitions, consider a node D with deterministic Boolean update function f(A,B,C) = A or (B and C). This function means that node D will turn on if A is ON or if B and C are simultaneously ON. For the edge A → D, we can calculate

$$avg_{ON} = \frac{f(1,0,0)+f(1,0,0)+f(1,1,0)+f(1,1,1)}{4} = 1$$
$$avg_{OFF} = \frac{f(0,0,0)+f(0,0,0)+f(0,1,0)+f(0,1,1)}{4} = 0.25$$

A is a positive regulator of D, so $N_{ON} = 1 - 0.25 = 0.75$, $S_{ON} = 1$, and $N_{OFF} = S_{OFF} = 0$. This means that in 75% of input conditions A would be necessary to turn D ON (only when B=C=1 does D turn ON without A). Conversely, A is sufficient to turn D ON in all input conditions. Finally, A is never sufficient nor necessary to turn D OFF, as A is a positive regulator of D.

With this, we define the influence index of each node-intervention-module tuple using one of the following formulas:

Source node intervention: KO; Target module goal: DOWN

$$InfluenceIndex = \sum[(N_{ON} - N_{OFF}) + 0.5 \cdot (S_{ON} - S_{OFF})]$$

Source node intervention: KO; Target module goal: UP

$$InfluenceIndex = \sum[(N_{OFF} - N_{ON}) + 0.5 \cdot (S_{OFF} - S_{ON})]$$

Source node intervention: CA; Target module goal: DOWN

$$InfluenceIndex = \sum[(S_{OFF} - S_{ON}) + 0.5 \cdot (N_{OFF} - N_{ON})]$$

Source node intervention: CA; Target module goal: UP

$$InfluenceIndex = \sum[(S_{ON} - S_{OFF}) + 0.5 \cdot (N_{ON} - N_{OFF})]$$

where in each case the sum is over all target nodes of the perturbed node that are in the target gene module. The higher weight on necessary edges in KO interventions reflects the fact that turning OFF a necessary regulator is sufficient to control its output. The higher weight on sufficient edges in CA interventions reflects the fact that turning ON a sufficient regulator is sufficient to control its output.

2.10. Analyzing the Effect of Node Interventions

In contrast to attractors of deterministic systems, our stochastic model can evolve away from pseudo-attractors (i.e., pseudo-attractors are not trap spaces). We start simulations from a system state that corresponds to the average state of all pseudo-attractors associated to a given experimental condition, and examine how many steps are required for a given module's overall expression to increase or decrease relative to its start state.

To accomplish this, we quantify a module's "activation" as the fraction of nodes in the module that are ON. For the purpose of this calculation, we exclude all source nodes, as those nodes cannot be activated or silenced based on interventions of other nodes, and are, therefore, insensitive to any perturbation. During simulations we very rarely observed a module achieve more than $3/4$ of non-source nodes becoming ON. We thus considered switches between states that have low module activation (fewer than $1/4$ non-source nodes are ON) and intermediate module activation (between $1/4$ and $3/4$ non-source nodes are ON).

We simulated the dynamics of the WT system by starting from a pseudo-attractor and updating a single, randomly selected node at each time step [6]. For modules that start in the low activation state, we counted how many steps were required for the module to switch to the intermediate state for the first time. For modules that start in the intermediate state, we instead counted how many steps were required to switch to the low activation state for the first time. We repeated these simulations, restarting from the start state, 100 times. For each simulation, we updated the system 5000 times. If a module did not switch within that time, we assigned a value 5001. For subsequent statistical analyses, we used a non-parametric ordinal test, so in most cases it does not practically matter how much above 5001 it really would have been.

We then chose a set of interventions to test, based on analysis of the network and influence index of various nodes. We considered single node KO or CA, or combinations of multiple nodes individually controlled. As in the WT system, we performed 100 iterations of 5000 steps, counting how many steps were required for a module to switch for the first time. We used a two-sided Mann–Whitney U test to test whether the average number of steps from the intervention simulations was statistically different from the WT. All p-values were FDR-corrected using the Benjamini–Hochberg (BH) method, and the threshold for significance was defined as BH-adjusted $p < 0.05$.

Following intervention, if a module requires more steps before it switches from low to intermediate activation, or vice versa, compared to WT, then the module's original state was stabilized by the intervention. If the module began in the low state, we then classified the intervention as down-regulating. If the module began in the high state, we classified the intervention as up-regulating. Conversely, if an intervention makes a module require fewer steps to switch, then the module's original state was destabilized by the intervention. If the module began in the intermediate state, we classified the intervention as down-regulating. If the module began in the low state, we classified the intervention as upregulating.

3. Results

3.1. Identification of Gene Co-Expression Modules Associated with Distinct Treatments

We analyzed an RNAseq dataset [4] consisting of MV4-11 cells (a *FLT3*-ITD AML cell line) exposed to six different treatment conditions. These included triplicate measurements each of (1) 10 nM quizartinib treatment for 48 h, (2) 10 nM quizartinib treatment for five days, (3) 100 nM dexamethasone treatment for 48 h, (4) combination 10 nM quizartinib + 100 nM dexamethasone for 48 h (we refer to this combination as Quiz + Dex), (5) Quiz+Dex for five days (quizartinib for five days and dexamethasone added on day 3), and (6) DMSO (GEO: GSE116432). Previous work found that dexamethasone and quizartinib in combination were synergistic in *FLT3*-ITD cells [4].

Applying weighted gene co-expression network analysis (WGCNA) [24] to this gene expression dataset, we identified seven modules of co-expressed genes (Figure 1A and Figure S1). WGCNA assigns color names to each module. The modules we identified ranged in size from 164 genes (black module) up to 7219 genes (turquoise module). The genes in each module are given in File S1.

Given a WGCNA gene module, the module's eigengene (defined as the first principal component) is commonly used as a single metric capturing the overall expression of all genes within that module. Based on module eigengene expression, we found that six modules were statistically differentially expressed across treatment conditions (Figure 1A, Kruskal–Wallis test, BH-adjusted p-value < 0.05): the yellow, red, brown, blue, turquoise, and black modules.

Of these modules, we find that the yellow module is upregulated (relative to DMSO) by all treatments, most significantly by the combination of dexamethasone + quizartinib (Figure 1B). The red module is upregulated by dexamethasone, with or without the addition of quizartinib, while we detected no response of this module to quizartinib alone. The blue module is upregulated by quizartinib, with or without the addition of dexamethasone, while we detected no response of this module to dexamethasone alone. Opposite to blue, the turquoise module is downregulated by all treatments, including quizartinib. Finally, the black module is upregulated after 48 h of treatment with quizartinib (with or without dexamethasone) but returns to DMSO levels after 5 days of treatment. Though the green module was not significantly differentially expressed, we noticed that, within the triplicate measures in both DMSO and dexamethasone treatment, one sample of each appears to be a clear outlier in the green module eigengene expression (Figure 1B). Without those samples, the green module is upregulated following quizartinib treatment for 5 days. Thus, the green module may still be relevant to understanding AML drug response, and we consider its possible role later. These results are summarized as interactions between the drugs and modules in Figure 1C, which shows that dexamethasone reverses quizartinib-induced upregulation of the brown module but does not reverse other modules that are affected by quizartinib.

3.2. Ontology Analysis Reveals Biological Processes Unique to Each Module

To uncover the biological character of each gene module, we performed ontology enrichment analysis using the Molecular Biology of the Cell Ontology (MBCO) method [25]. This analysis searches not only for enriched sub-cellular processes, but enriched relationships between processes. The results of MBCO analysis are reported in Figures S2–S8, and in File S2. These analyses revealed several biological processes and pathways that become activated by different drug treatments, that may play critical roles in mediating AML drug response.

We found the yellow module was enriched for cell–cell communication, especially via NOTCH signaling, as well as extracellular matrix homeostasis. The red module was strongly enriched for extracellular matrix homeostasis, including collagen biosynthesis and crosslinking. The brown module was enriched for immune response activation and actin and lamellipodium structure. Collectively, these three modules thus may be responsible for mediating the tumor microenvironment response through cell–cell communication, structural changes, and immune activation.

The blue module, which is upregulated by quizartinib, was highly enriched in drug export and cellular detoxification, indicating a potentially essential role in mediating cell survival following treatment with quizartinib. The turquoise module was highly enriched in cell-cycle progression, suggesting a possible role in mediating proliferation. Treatment with quizartinib downregulates the turquoise module, which is consistent with the prior hypothesis that DTPs slow down their growth in the presence of drug [3,26,27]. Quizartinib's simultaneous activation of the blue module, and downregulation of the turquoise module may be able to quickly lead to the emergence of DTPs, allowing cells to then acquire other changes, or change their environment, leading to more favorable cancer-cell survival. Dexamethasone treatment did not reverse the effect of quizartinib on these DTP-associated modules (Figure 1C).

The black module was found to be enriched in transcription and translation. The black module is upregulated by quizartinib in the short term, but returns to untreated levels by 5 days of treatment. This suggests that, early on in treatment, cells may quickly activate several gene transcription programs, but the activation of new programs may relax by day five as cells reach a new equilibrium.

Finally, though it was not significantly differentially expressed across subtypes, the green module was highly enriched in the regulation of apoptosis. As noted above, excluding two outlier samples, the green module is upregulated following quizartinib treatment for 5 days.

3.3. Network Analysis

3.3.1. Construction of Gene Regulatory Network Governing AML Drug Response

To understand how cells mediate these drug-induced gene expression changes, we constructed a gene regulatory network model of highly differentially expressed genes from within each module. The full details of network construction are presented in Methods Sections 2.4–2.6 and an overview is in Figure S9. Briefly, we aggregated interactions from the public databases SIGNOR [14], TRRUST [15], and RegNetwork [16]. Based on the ontology analysis that implicated cell–cell communication, inflammation, and apoptosis, we also integrated published networks related to AML [14,17], NFKappaB signaling [17], NOTCH signaling [18], tumor promoting inflammation [19,20], and apoptosis [20]. The final network (Figure 2 and Figure S10) was constructed with the aim of avoiding over- or under-representation of any single module.

3.3.2. Inference of Predictive Dynamic AML Drug Resistance Network Model and Drug-Induced Pseudo-Attractors

We next sought to understand how the genes in the AML drug response network interact. To this end, we applied the BooleaBayes algorithm [6] to infer probabilistic regulatory functions for each node in the network (File S3). Briefly, BooleaBayes tries to find Boolean logic functions consistent with steady-state gene expression data and a network topology. In our case, the gene expression data are derived from normalization of the 18 AML RNAseq samples (see Methods Section 2.5) and the network topology is that of Figure 2. The inferred logic functions use binary values. For example, f(nodeA, nodeB) = nodeA AND nodeB, where nodeA and nodeB have binary (ON or OFF) values. BooleaBayes produces probabilistic functions indicating the probability that the target node will be ON or OFF, depending on the ON/OFF status of its regulators, where a value 0 indicates 100% confidence the target node is OFF, a value 1 indicates 100% confidence the target node is ON, and a value 0.5 indicates equal chance of being ON or OFF. See Figure 3 for an example showing how BooleaBayes finds these values.

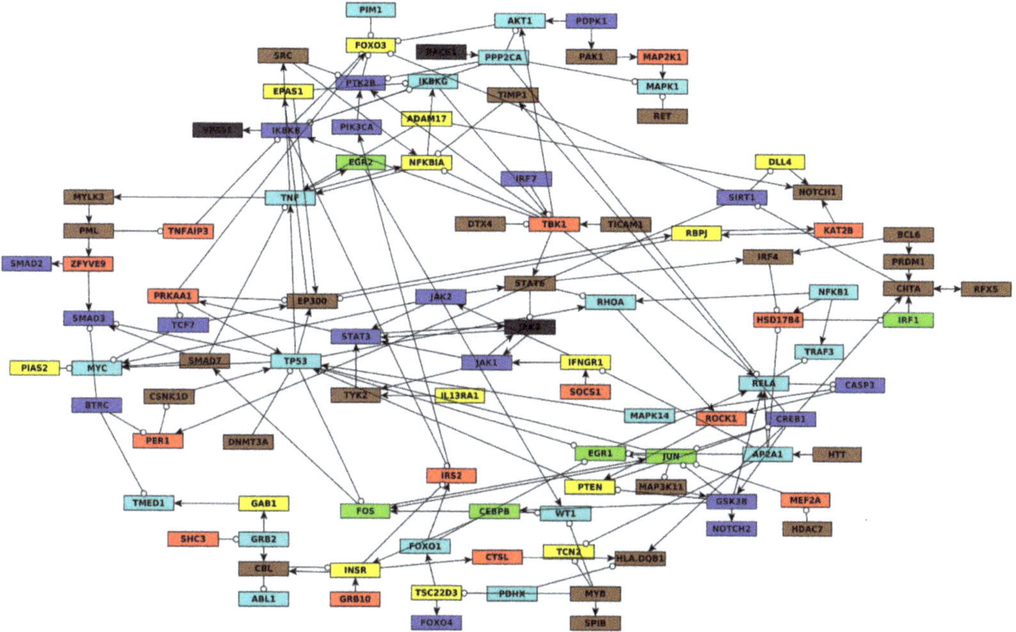

Figure 2. Gene regulatory network of *FLT*-ITD AML quizartinib and dexamethasone response. Nodes are genes and are colored by the WGCNA module to which they belong. Edges ending in arrows represent net positive regulation, while edges ending in circles indicate net negative regulation. Edge regulation sign is determined using the BooleaBayes algorithm (see Methods Sections 2.5 and 2.6). The nodes in this network combine the transcript and the protein encoded by the same gene. For example, the node JUN has transcriptional regulators (FOS, CREB1, and MEF2A) and a posttranslational regulator (GSK3B). For nodes like this with post-translational modifications, the full network, shown in Figure S10, has separate nodes corresponding to their transcript and active protein.

Unlike previous work with BooleaBayes, which focused exclusively on transcription factors, the AML network also includes post-translational regulation. For nodes with post-translational regulation, we distinguish between the transcription of the gene, and the activation of the protein product. Protein activation is assumed to follow inhibitory dominant Boolean rules, which means that at least one activator is required, but any inhibitor is enough to prevent activation. Full details of how this was implemented into BooleaBayes are available in Section 2.6.

With a deterministic Boolean model, one may search for its attractors, which represent long-term stable behaviors of the system. Once the system reaches an attractor, it can no longer escape it without an external intervention. As BooleaBayes is a probabilistic system, it has no inescapable attractors. Nevertheless, there are states which the system is more likely to enter than to leave, termed pseudo-attractors.

We asked whether the AML network has pseudo-attractors corresponding to the drug-treatment conditions. To find these, we approximated the probabilistic BooleaBayes regulatory functions by finding their closest-matching deterministic Boolean functions (File S4). The network has 65 source nodes (nodes without regulators), which can be 0 or 1 with no constraints, indicating that there will be at least $2^{65} \sim= 10^{19}$ possible attractors (at least one per source node configuration), and even more pseudo-attractors. Not all of these pseudo-attractors necessarily correspond to true attractors of the system, but may instead reflect uncertainty of BooleaBayes functions far from the observed data. To specifically find pseudo-attractors associated with the drug-response, we plugged into the source nodes their respective observed values in each of the drug conditions, and propagated those substitutions to find simplified systems for each drug state. Attractors of these simplified systems were found using the StableMotif [22] Python package (Figure 4). These attractors of the simplified deterministic system correspond to pseudo-attractors of the probabilistic BooleaBayes functions. Pseudo-attractors for the 5-day Quiz+Dex treatment have three oscillating nodes: ABL1_active, CBL_active, and INSR_active, driven by a negative feedback loop between CBL_active and INSR_active. All other pseudo-attractors are steady states. Many modules have clear consensus of the activity of their genes, in which almost all of them are ON or almost all are OFF. For example, almost all yellow nodes and almost all red nodes are OFF in pseudo-attractors corresponding to quizartinib treatment (either for two days or for five days). This agrees with the low module eigengene expression (pink color) on Figure 1B. Other modules are more split. For example, nearly half of the brown and green module nodes are ON and nearly half are OFF in pseudo-attractors corresponding to combination treatment for five days. These also agree with the module eigengene expressions (white color). Overall, there is a good agreement between all modules' average activation in pseudo-attractors and their eigengene expression shown in Figure 1B.

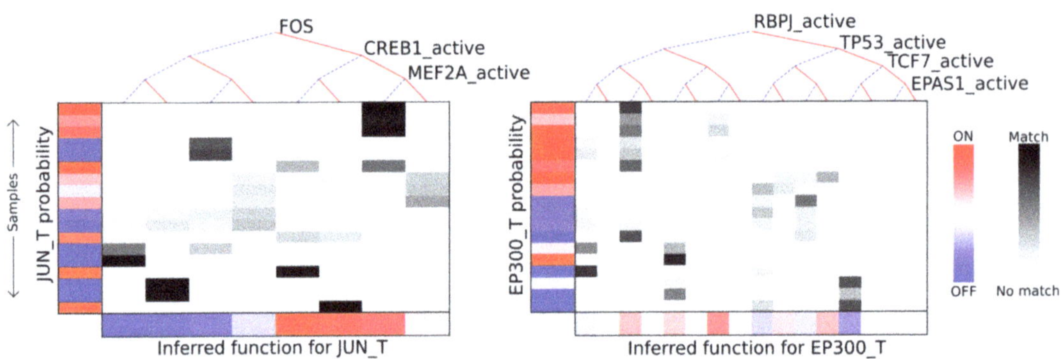

Figure 3. Examples demonstrating BooleaBayes regulatory function inference. Left: the inferred function for JUN transcription. JUN_T has three regulators: FOS, CREB1_active, and MEF2A_active. A Boolean function of 3 regulators has $2^3 = 8$ possible input configurations (e.g., FOS = 0, CREB1_active = 0, MEF2A_active = 0). Each column of the figure corresponds to one of the possible regulator configurations, from all regulators being OFF (left-most column) to all regulators being ON (rightmost column). Each row corresponds to one of the 18 AML samples. The red and blue colors along the far left show whether JUN_T (the target node) is ON or OFF in each sample. The white-black color scale shows how closely each sample (row) corresponds to a given input configuration (column). For example, the top three samples (rows) are most likely to correspond to FOS = 1, CREB1_active = 1, MEF2A_active = 0, as is shown by the black and dark grey cells in the first three rows of that column. In both samples, JUN_T is likely to be OFF (indicated by the blue color). Thus, the inferred regulatory function for JUN_T (bottom row) says that if FOS = CREB1_active = MEF2A_active = 0, JUN_T is very likely to turn OFF. Right: inferred function for EP300 transcription. Unlike JUN_T, there are many conditions for which there was no observed data, such as RBPJ_active = 0, TP53_active = 0, TCF7_active = 0, and EPAS1_active = 1 (second condition from the left). In these cases, the inferred rule has a near 50% chance for EP300_T to turn ON or OFF, as there are no data indicating what should happen in these cases.

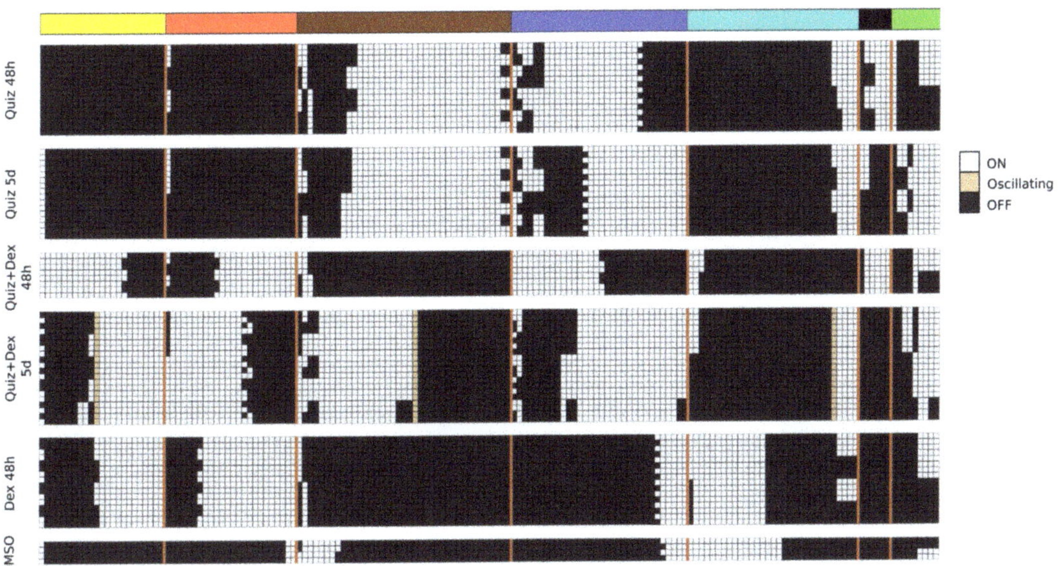

Figure 4. Pseudo-attractors corresponding to DMSO and drug-treated AML, which are also the attractors of the deterministic approximation of the BooleaBayes network. The values of the network's 65 source nodes are fixed to match one of the six treatment conditions, including DMSO, and attractors of the reduced systems are shown above. Each row is one attractor, and columns are nodes of the network. Columns are grouped and colored based on the module the node belongs in. The vertical orange lines delimit the nodes of each module.

3.3.3. Identification of Intervention Targets That Disrupt AML Drug Resistance Modules

With the dynamic model of AML drug response, there are many possible questions one could pursue. We focus on identifying interventions that we hypothesize may be able to reprogram DTP cells into drug susceptible states. Specifically, we previously showed that dexamethasone increases cell death of quizartinib-induced DTPs in FLT3-ITD AML [4]. However, there remain gene expression modules that dexamethasone does not reverse (Figures 1 and 4), including some that are natural markers of DTPs. To identify additional targets that may be able to improve combination quizartinib and dexamethasone treatment, we thus focus on the pseudo-attractors corresponding to combination treatment with Quiz + Dex for 5 days, compared to the attractors in DMSO. As discussed above, several differentially expressed modules are enriched in biological functions that may be responsible for mediating drug resistance. Of greatest interest, the blue module is enriched in detoxification and drug export, the green module is enriched in regulated cell death and apoptosis, and the turquoise module is enriched in cell cycle progression. We hypothesize that downregulating the blue module may prevent the emergence of resistance mechanisms. Activating the turquoise module may enhance proliferation, preventing cells from entering the DTP state. Activating the green module may enhance apoptosis. It is also of note that the yellow and red modules are more highly expressed following Quiz + Dex treatment than DMSO. We hypothesize that reverting these modules to the DMSO state may improve therapy response. Collectively, these changes may extend the efficacy of combination Quiz + Dex treatment.

To this end, we picked control objectives of downregulating the blue, red, and yellow modules, and upregulating the turquoise or green modules. Upregulation of the green module was chosen due to its enrichment in apoptosis, even though this would push the green module further away from the DMSO state. We first quantified the stability of the gene expression modules near the Quiz + Dex 5-day pseudo-attractors. To accomplish this, we simulated 100 random walks of 5000 steps (see Methods Section 2.10) from an initial state determined by the average of the Quiz + Dex 5-day pseudo-attractors. For each step along the walk, we quantified the fraction of nodes from each module that are ON to get an overall module activation score. We characterized modules with fewer than 1/4 active nodes to be in a low state, between 1/4 and 3/4 to be in an intermediate state, and above 3/4 to be in a high state. These ranges were chosen to ensure each module began sufficiently far from the boundary. With these definitions, all modules began in either the low or intermediate states. We then quantified how long it took for each random-walk simulation to cross from low to intermediate activation, or from intermediate to low activation, for the first time (we only rarely observed a cross from intermediate to high activation, so this transition was excluded). The distributions of the crossing times in Figure 5A capture the baseline stability of each module.

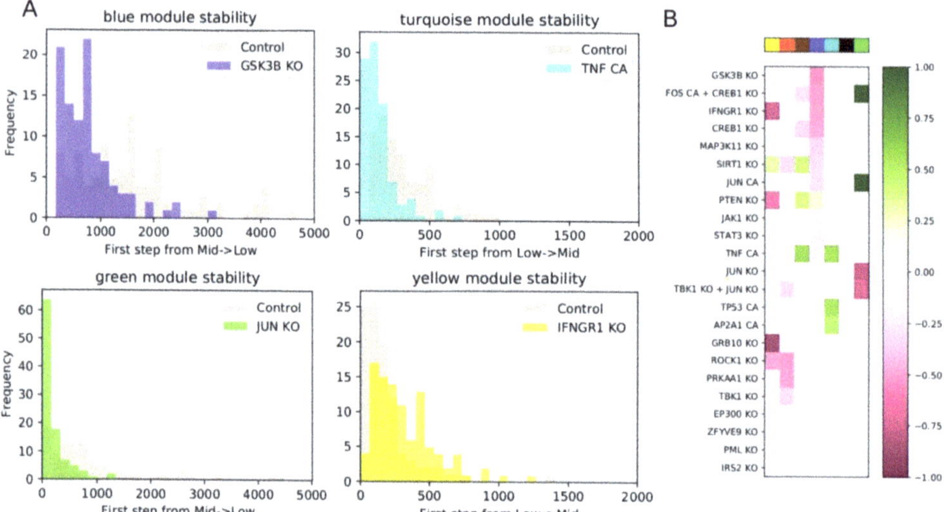

Figure 5. Targeted interventions of driver nodes cause up- or down-regulation of gene modules. (**A**) Distribution of the number of steps required for the module to switch between low and intermediate activation. The distributions in beige show the dynamics of the system with no manipulations, while the colored distributions show the manipulated systems. For modules transitioning from low to intermediate activation, an intervention shifting the distribution to a longer time to switch (rightward shift) maintains the module in the downregulated state, while a leftward shift upregulates the module. For modules transitioning from intermediate to low, a rightward shift indicates the module is maintained in an upregulated state, while a leftward shift indicates downregulation. (**B**) Heatmap showing the up- or down-regulation of statistically significant interventions (two-sided Mann–Whitney U Test, FDR-adjusted p-value < 0.05) compared to control for each module. Only significant interventions are colored. Colors are scaled so that a value of $+/-1$ indicates a 100% relative shift in mean transition time compared to control.

We then asked which interventions shift the module distributions to lower or higher numbers of steps. If a module starts in the intermediate state and transitions into the low state, then interventions shifting the distribution to a higher number of steps stabilize the more active state of the module. Conversely, interventions shifting the distribution to a lower number of steps downregulate the module. The opposite interpretation holds for modules starting in the low state and transitioning to the intermediate state: a shift to a lower number of steps indicates upregulating the module, while a shift to a higher number of steps indicates maintaining the module in the low state.

Nodes targeted by in silico intervention were fixed as either ON or OFF, and not allowed to update during the simulation. We prioritized nodes to target by (1) analysis of the regulatory paths in the network (Figure 2 and Figure S10), and (2) calculation of an influence index for each possible node interventions and each module (see Methods Section 2.9 for details). Briefly, the influence index considers the most likely effect an intervention of a given node will have on the nodes it targets. A positive influence index indicates that those effects are likely to align with our control objectives, while a negative influence index indicates that those effects are likely to contradict our objectives. Influence indices for each intervention are given in File S5.

We tested the interventions shown in Figure 5 by simulation. For each intervention, we quantified how the distribution of steps required to cross the low-intermediate threshold shifts relative to the baseline control. We determined significant upregulating or downregulating shifts using a two-sided Mann–Whitney U test (see Methods Section 2.10). The most significant regulators for each module are shown in Figure 5B.

We predict several interventions that may lead to the downregulation of the blue module, which is enriched in genes related to drug resistance. The most significant are knockout of GSK3B, IFNGR1, CREB1, SIRT1, or MAP3K11. Investigating these further, GSK3B inhibition has previously been proposed as a differentiation-inducing therapy for AML [28,29]. Nevertheless, it has also been found in a CRISPR screen that GSK3B KO leads to the reactivation of FGF/Ras/ERK and Wnt signaling that can confer resistance to quizartinib monotherapy in *FLT3*-ITD AML [30]. CREB1 overexpression has been associated with poor outcome in AML patients [31], and SIRT1 activation has been previously associated with drug resistance of *FLT3*-ITD AML stem cells [32].

For the green module, which is enriched in apoptosis regulation, we find JUN constitutive activation leads to activation of the green module, while JUN knockout inhibits it. Previous work has found that JUN KO increased apoptosis in AML cells [33]. JUN is a master regulator of apoptosis, but also involved in AML cell survival via inflammatory pathways, indicating it may have dual roles. Expanding the network to include relevant downstream JUN activity may better elucidate how these competing effects may be activated or controlled.

For the turquoise module, which is enriched in cell cycle progression, we found activation of TNF, TP53, or AP2A1 support upregulation of the turquoise module. TNF-alpha is highly upregulated in AML patients, and has been shown to induce proliferation of leukemic blasts [34,35]

We additionally tested combination interventions to simultaneously control multiple gene modules. The combined knockout of TBK1 and JUN and found that it leads to downregulation of both the red and yellow modules. Simultaneous constitutive activation of FOS with knockout of CREB1 led to downregulation of the brown and blue modules, and stabilization of the green module.

Focusing on modules other than blue, green, and turquoise, we predict that GRB10 KO strongly downregulates the yellow module, which is upregulated by both quizartinib and dexamethasone treatment, and GRB10 overexpression has previously been associated with aggressive phenotypes in *FLT3*-ITD AML [36]. ROCK1 KO downregulates both the yellow and red modules, and ROCK inhibition has been shown to inhibit cell growth in *FLT3* mutant AML patient-derived blasts [37]. We predict that TBK1 KO downregulates the red module; it has previously been suggested as a therapeutic target in AML due to its activation of MYC-dependent survival pathways [38].

Four interventions had relatively high influence indices but did not lead to a statistically significant shift: KO of EP300, ZFYVE9, PML, or IRS2. Of these, ZFYVE9 and IRS2 KO have a large effect on their direct targets, but that effect clearly will not propagate through the network. For instance, ZFYVE9 is a necessary regulator for two blue module genes: SMAD2 and SMAD3, which BooleaBayes did not detect significant regulatory functions for in this network, and thus they became sink nodes. IRS2 is necessary for PIK3CA to activate, and PIK3CA_A only regulates a single target node, PTK2B_T. PTK2B_T is also regulated by SRC, and SRC has a much stronger regulatory influence than PIK3CA (File S3). The remaining two interventions, EP300 and PML KO, have multiple downstream paths, but nevertheless our full simulations of the network did not detect upregulating or downregulating influences of these interventions on any module.

The concordance of our identified interventions with previous work from the literature supports the validity of our findings. Furthermore, we make several additional novel predictions. For example, we predict the blue module can be downregulated by the KO of IFNGR1 or MAP3K11. The turquoise module can be upregulated by CA of AP2A1 or TP53. The validation of these novel predictions is needed as a next step in establishing the predictive value of this model and is the subject of future work in our labs. Collectively, we anticipate that these interventions would synergize with combination quizartinib and dexamethasone treatment in patients with *FLT3*-ITD AML.

4. Discussion

Here we constructed a dynamic model of a gene regulatory network relevant to *FLT3*-mutant acute myeloid leukemia. The model integrates multiple types of information: RNAseq data consisting of MV4-11 cells exposed to drug treatment and several databases of signal transduction and gene regulation. Model development included multiple state of the art analysis methodologies: weighted gene co-expression analysis, ontology analysis, inference of regulatory relationships using BooleaBayes, attractor analysis, and control theory. We also developed new capabilities for BooleaBayes, and new confidence scores to prioritize interactions to be included in the network and new influence scores to prioritize interventions. Overall, this work illustrates the challenges and capabilities of computational systems biology analysis in cancer research and the potential for this type of analysis to advance personalized medicine.

The model attractors recapitulate the activation of the modules (compare Figure 4 to Figure 1B), and the most significant predicted model interventions match well with literature reports on drivers of proliferation, survival, and drug resistance (Figure 5). Collectively, these results strongly support the model's validity. Nevertheless, there are several possible avenues for further model improvement. This model was derived from data in MV4-11 cells treated with quizartinib and dexamethasone. We previously showed that the gene expression profile of MV4-11 cells was predictive of sensitivity of multiple *FLT3*-ITD cell lines and patient cells to treatment with quizartinib and dexamethasone [4]. Nevertheless, including data collected from other cell lines, or cells treated with other drugs, such as other *FLT3* inhibitors or glucocorticoids may reveal alternative pathways and processes involved in mediating drug resistance. Finding common resistance mechanisms, as well as system-specific resistance mechanisms, may lead to a more generalizable model. Furthermore, during network construction we removed sink nodes to focus on nodes that contribute feedback into the network dynamics. Nevertheless, those sink nodes may be valuable phenotypic markers, or could be regulators of other nodes we may include in the future. Additionally, the large number of source nodes (65) should eventually be decreased. Many of these became source nodes because BooleaBayes was not able to determine a significant role for their regulators, and so those edges were removed. Additional expression datasets, or literature knowledge, may elucidate functional forms of those interactions. Additionally, more nodes may be added by including more AML-specific literature knowledge (e.g., MCL as downstream target of GSK3B, downstream targets of JUN to further elucidate the dual effect of its inhibition on apoptosis and inflammation).

In Figure 5 we showed four interventions that had high influence indices, but this did not translate into significant up- or down-regulation of any modules. In at least two cases, we determined that these interventions led to sink nodes, or nodes with weak influence, explaining why the influence index was not predictive of overall impact. To address this, the influence index of a node may be extended to consider the influence index of its downstream targets. Further, nodes can have conflicting downstream effects, and resolving these may improve the predictive value of influence index.

The dynamic model may eventually be used to answer other fundamental questions, such as how does drug treatments lead to the resistant state. To this end, the network could be extended by integrating known drug targets, though in practice drugs often have multiple off-target effects. One possible way to overcome this would be to prioritize adding drug targets that can induce the changes between the status of the source nodes in the untreated and drug-treated conditions. Future work is focused on expanding and improving the network model by incorporating information about drug targets, additional cell lines, and additional drug perturbation datasets. We are also working to validate the model's novel predictions, such as combining Quiz+Dex treatment with KO of IFNG1 or MAP3K11, or CA of AP2A1.

Finally, we anticipate that data-driven predictive modeling, as demonstrated in this work, may eventually help accelerate patient-specific precision treatments. The dynamics of the AML model emerged from the expression data we used to train it, thus incorporating patient-specific data may help reveal patient-specific drug resistance pathways or targets.

Supplementary Materials: The following are available online at https://www.mdpi.com/2075-4426/11/3/193/s1, Figure S1: WGCNA module identification, Figure S2: Black module enrichment, Figure S3: Blue module enrichment, Figure S4: Brown module enrichment, Figure S5: Green module enrichment, Figure S6: Red module enrichment, Figure S7: Turquoise module enrichment, Figure S8: Yellow module enrichment, Figure S9: Network construction, Figure S10: Full network, File S1.csv: WGCNA modules, File S2.xlsx: MBCO enrichment, File S3.txt: BooleaBayes fit probabilistic update functions, File S4.txt: Deterministic Boolean approximation, File S5.csv: Influence indices.

Author Contributions: Conceptualization, D.J.W. and R.A.; methodology, D.J.W. and R.A.; formal analysis, D.J.W. and R.A.; resources, D.J.W., M.G., H.-G.W. and R.A.; writing—original draft preparation, D.J.W. and R.A.; writing—review and editing, D.J.W., R.A., M.G., H.-G.W.; visualization, D.J.W.; supervision, R.A.; project administration, R.A.; funding acquisition, R.A. All authors have read and agreed to the published version of the manuscript.

Funding: This research was funded by National Science Foundation (NSF), grant numbers PHY 1545832, MCB-1715826, and IIS-1814405 to R.A.

Data Availability Statement: No new data were created or analyzed in this study. Data sharing is not applicable to this article.

Conflicts of Interest: The authors declare no conflict of interest. The funders had no role in the design of the study; in the collection, analyses, or interpretation of data; in the writing of the manuscript, or in the decision to publish the results.

References

1. Daver, N.; Schlenk, R.F.; Russell, N.H.; Levis, M.J. Targeting FLT3 mutations in AML: Review of current knowledge and evidence. *Leukemia* **2019**, *33*, 299–312. [CrossRef]
2. O'Donnell, M.R.; Tallman, M.S.; Abboud, C.N.; Altman, J.K.; Appelbaum, F.R.; Arber, D.A.; Bhatt, V.; Bixby, D.; Blum, W.; Coutre, S.E.; et al. Acute Myeloid Leukemia, Version 3.2017, NCCN Clinical Practice Guidelines in Oncology. *J. Natl. Compr. Cancer Netw.* **2017**, *15*, 926–957. [CrossRef]
3. Gebru, M.T.; Wang, H.-G. Therapeutic targeting of FLT3 and associated drug resistance in acute myeloid leukemia. *J. Hematol. Oncol.* **2020**, *13*, 155. [CrossRef] [PubMed]
4. Gebru, M.T.; Atkinson, J.M.; Young, M.M.; Zhang, L.; Tang, Z.; Liu, Z.; Lu, P.; Dower, C.M.; Chen, L.; Annageldiyev, C.; et al. Glucocorticoids enhance the antileukemic activity of FLT3 inhibitors in FLT3-mutant acute myeloid leukemia. *Blood* **2020**, *136*, 1067–1079. [CrossRef] [PubMed]
5. Wooten, D.J.; Quaranta, V. Mathematical models of cell phenotype regulation and reprogramming: Make cancer cells sensitive again! *Biochim. Biophys. Acta Rev. Cancer* **2017**, *1867*, 167–175. [CrossRef]
6. Wooten, D.J.; Groves, S.M.; Tyson, D.R.; Liu, Q.; Lim, J.S.; Albert, R.; Lopez, C.F.; Sage, J.; Quaranta, V. Systems-level network modeling of Small Cell Lung Cancer subtypes identifies master regulators and destabilizers. *PLoS Comput. Biol.* **2019**, *15*, e1007343. [CrossRef]
7. Issa, M.E.; Takhsha, F.S.; Chirumamilla, C.S.; Perez-Novo, C.; Vanden Berghe, W.; Cuendet, M. Epigenetic strategies to reverse drug resistance in heterogeneous multiple myeloma. *Clin. Epigenetics* **2017**, *9*, 17. [CrossRef]
8. Gong, L.; Yan, Q.; Zhang, Y.; Fang, X.; Liu, B.; Guan, X. Cancer cell reprogramming: A promising therapy converting malignancy to benignity. *Cancer Commun.* **2019**, *39*, 48. [CrossRef] [PubMed]
9. Gómez Tejeda Zañudo, J.; Scaltriti, M.; Albert, R. A network modeling approach to elucidate drug resistance mechanisms and predict combinatorial drug treatments in breast cancer. *Cancer Converg.* **2017**, *1*, 5. [CrossRef]

10. Huang, S.; Kauffman, S. How to escape the cancer attractor: Rationale and limitations of multi-target drugs. *Semin. Cancer Biol.* **2013**, *23*, 270–278. [CrossRef] [PubMed]
11. Zhou, S.; Abdouh, M.; Arena, V.; Arena, M.; Arena, G.O. Reprogramming Malignant Cancer Cells toward a Benign Phenotype following Exposure to Human Embryonic Stem Cell Microenvironment. *PLoS ONE* **2017**, *12*, e0169899. [CrossRef]
12. Shaffer, S.M.; Dunagin, M.C.; Torborg, S.R.; Torre, E.A.; Emert, B.; Krepler, C.; Beqiri, M.; Sproesser, K.; Brafford, P.A.; Xiao, M.; et al. Rare cell variability and drug-induced reprogramming as a mode of cancer drug resistance. *Nature* **2017**, *546*, 431–435. [CrossRef]
13. Pisco, A.O.; Huang, S. Non-genetic cancer cell plasticity and therapy-induced stemness in tumour relapse: 'What does not kill me strengthens me'. *Br. J. Cancer* **2015**, *112*, 1725–1732. [CrossRef] [PubMed]
14. Licata, L.; Lo Surdo, P.; Iannuccelli, M.; Palma, A.; Micarelli, E.; Perfetto, L.; Peluso, D.; Calderone, A.; Castagnoli, L.; Cesareni, G. SIGNOR 2.0, the SIGnaling Network Open Resource 2.0: 2019 update. *Nucleic Acids Res.* **2020**, *48*, D504–D510. [CrossRef]
15. Han, H.; Cho, J.-W.; Lee, S.; Yun, A.; Kim, H.; Bae, D.; Yang, S.; Kim, C.Y.; Lee, M.; Kim, E.; et al. TRRUST v2: An expanded reference database of human and mouse transcriptional regulatory interactions. *Nucleic Acids Res.* **2018**, *46*, D380–D386. [CrossRef]
16. Liu, Z.-P.; Wu, C.; Miao, H.; Wu, H. RegNetwork: An integrated database of transcriptional and post-transcriptional regulatory networks in human and mouse. *Database* **2015**, *2015*. [CrossRef]
17. Gyori, B.M.; Bachman, J.A.; Subramanian, K.; Muhlich, J.L.; Galescu, L.; Sorger, P.K. From word models to executable models of signaling networks using automated assembly. *Mol. Syst. Biol.* **2017**, *13*, 954. [CrossRef]
18. Kandasamy, K.; Mohan, S.S.; Raju, R.; Keerthikumar, S.; Kumar, G.S.S.; Venugopal, A.K.; Teilikicherla, D.; Navarro, D.J.; Mathivanan, S.; Pecquet, C.; et al. NetPath: A public resource of curated signal transduction pathways. *Genome Biol.* **2010**, *11*, R3. [CrossRef]
19. Hanahan, D.; Weinberg, R.A. Hallmarks of cancer: The next generation. *Cell* **2011**, *144*, 646–674. [CrossRef]
20. Martens, M.; Verbruggen, T.; Nymark, P.; Grafström, R.; Burgoon, L.D.; Aladjov, H.; Andón, F.T.; Evelo, C.T.; Willighagen, E.L. Introducing WikiPathways as a Data-Source to Support Adverse Outcome Pathways for Regulatory Risk Assessment of Chemicals and Nanomaterials. *Front. Genet.* **2018**, *9*, 661. [CrossRef]
21. Lambert, S.A.; Jolma, A.; Campitelli, L.F.; Das, P.K.; Yin, Y.; Albu, M.; Chen, X.; Taipale, J.; Hughes, T.R.; Weirauch, M.T. The Human Transcription Factors. *Cell* **2018**, *172*, 650–665. [CrossRef]
22. Rozum, J.C.; Zañudo, J.G.T.; Gan, X.; Albert, R. Parity and Time-Reversal Elucidate Decisions in High-Dimensional State Space—Application to Attractor Scaling in Critical Boolean Networks. 2020. Available online: http://arxiv.org/abs/2009.05526 (accessed on 30 October 2020).
23. Maheshwari, P.; Albert, R. A framework to find the logic backbone of a biological network. *BMC Syst. Biol.* **2017**, *11*, 122. [CrossRef]
24. Langfelder, P.; Horvath, S. WGCNA: An R package for weighted correlation network analysis. *BMC Bioinform.* **2008**, *9*. [CrossRef]
25. Hansen, J.; Meretzky, D.; Woldesenbet, S.; Stolovitzky, G.; Iyengar, R. A flexible ontology for inference of emergent whole cell function from relationships between subcellular processes. *Sci. Rep.* **2017**, *7*. [CrossRef]
26. Sharma, S.V.; Lee, D.Y.; Li, B.; Quinlan, M.P.; Takahashi, F.; Maheswaran, S.; McDermott, U.; Azizian, N.; Zou, L.; Fischbach, M.A.; et al. A chromatin-mediated reversible drug-tolerant state in cancer cell subpopulations. *Cell* **2010**, *141*, 69–80. [CrossRef]
27. Vallette, F.M.; Olivier, C.; Lézot, F.; Oliver, L.; Cochonneau, D.; Lalier, L.; Cartron, P.-F.; Heymann, D. Dormant, quiescent, tolerant and persister cells: Four synonyms for the same target in cancer. *Biochem. Pharmacol.* **2019**, *162*, 169–176. [CrossRef]
28. Gupta, K.; Stefan, T.; Ignatz-Hoover, J.; Moreton, S.; Parizher, G.; Saunthararajah, Y.; Wald, D.N. GSK-3 Inhibition Sensitizes Acute Myeloid Leukemia Cells to 1,25D-Mediated Differentiation. *Cancer Res.* **2016**, *76*, 2743–2753. [CrossRef]
29. Hu, S.; Ueda, M.; Stetson, L.; Ignatz-Hoover, J.; Moreton, S.; Chakrabarti, A.; Xia, Z.; Karan, G.; De Lima, M.; Agrawal, M.K.; et al. A Novel Glycogen Synthase Kinase-3 Inhibitor Optimized for Acute Myeloid Leukemia Differentiation Activity. *Mol. Cancer Ther.* **2016**, *15*, 1485–1494. [CrossRef]
30. Hou, P.; Wu, C.; Wang, Y.; Qi, R.; Bhavanasi, D.; Zuo, Z.; Dos Santos, C.; Chen, S.; Chen, Y.; Zheng, H.; et al. A Genome-Wide CRISPR Screen Identifies Genes Critical for Resistance to FLT3 Inhibitor AC220. *Cancer Res.* **2017**, *77*, 4402–4413. [CrossRef]
31. Cho, E.-C.; Mitton, B.; Sakamoto, K.M. CREB and leukemogenesis. *Crit. Rev. Oncog.* **2011**, *16*, 37–46. [CrossRef]
32. Li, L.; Osdal, T.; Ho, Y.; Chun, S.; McDonald, T.; Agarwal, P.; Lin, A.; Chu, S.; Qi, J.; Hsieh, Y.-T.; et al. SIRT1 activation by a c-MYC oncogenic network promotes the maintenance and drug resistance of human FLT3-ITD acute myeloid leukemia stem cells. *Cell Stem Cell* **2014**, *15*, 431–446. [CrossRef] [PubMed]
33. Zhou, C.; Martinez, E.; Di Marcantonio, D.; Solanki-Patel, N.; Aghayev, T.; Peri, S.; Ferraro, F.; Skorski, T.; Scholl, C.; Fröhling, S.; et al. JUN is a key transcriptional regulator of the unfolded protein response in acute myeloid leukemia. *Leukemia* **2017**, *31*, 1196–1205. [CrossRef]
34. Oster, W.; Cicco, N.A.; Klein, H.; Hirano, T.; Kishimoto, T.; Lindemann, A.; Mertelsmann, R.H.; Herrmann, F. Participation of the cytokines interleukin 6, tumor necrosis factor-alpha, and interleukin 1-beta secreted by acute myelogenous leukemia blasts in autocrine and paracrine leukemia growth control. *J Clin Investig.* **1989**, *84*, 451–457. [CrossRef]
35. Kagoya, Y.; Yoshimi, A.; Kataoka, K.; Nakagawa, M.; Kumano, K.; Arai, S.; Kobayashi, H.; Saito, T.; Iwakura, Y.; Kurokawa, M. Positive feedback between NF-κB and TNF-α promotes leukemia-initiating cell capacity. *J Clin Investig.* **2014**, *124*, 528–542. [CrossRef]

36. Kazi, J.U.; Rönnstrand, L. FLT3 signals via the adapter protein Grb10 and overexpression of Grb10 leads to aberrant cell proliferation in acute myeloid leukemia. *Mol. Oncol.* **2013**, *7*, 402–418. [CrossRef]
37. Mali, R.S.; Ramdas, B.; Ma, P.; Shi, J.; Munugalavadla, V.; Sims, E.; Wei, L.; Vemula, S.; Nabinger, S.C.; Goodwin, C.B.; et al. Rho kinase regulates the survival and transformation of cells bearing oncogenic forms of KIT, FLT3, and BCR-ABL. *Cancer Cell* **2011**, *20*, 357–369. [CrossRef] [PubMed]
38. Liu, S.; Marneth, A.E.; Alexe, G.; Walker, S.R.; Gandler, H.I.; Ye, D.Q.; Labella, K.; Mathur, R.; Toniolo, P.A.; Tillgren, M.; et al. The kinases IKBKE and TBK1 regulate MYC-dependent survival pathways through YB-1 in AML and are targets for therapy. *Blood Adv.* **2018**, *2*, 3428–3442. [CrossRef] [PubMed]

Journal of
*Personalized
Medicine*

Article

Comprehensive Profiling of Genomic and Transcriptomic Differences between Risk Groups of Lung Adenocarcinoma and Lung Squamous Cell Carcinoma

Talip Zengin [1,2] and **Tuğba Önal-Süzek** [2,3,*]

1 Department of Molecular Biology and Genetics, Muğla Sıtkı Koçman University, 48000 Muğla, Turkey; talipzengin@mu.edu.tr
2 Department of Bioinformatics, Muğla Sıtkı Koçman University, 48000 Muğla, Turkey
3 Department of Computer Engineering, Muğla Sıtkı Koçman University, 48000 Muğla, Turkey
* Correspondence: tugbasuzek@mu.edu.tr

Abstract: Lung cancer is the second most frequently diagnosed cancer type and responsible for the highest number of cancer deaths worldwide. Lung adenocarcinoma (LUAD) and lung squamous cell carcinoma (LUSC) are subtypes of non-small-cell lung cancer which has the highest frequency of lung cancer cases. We aimed to analyze genomic and transcriptomic variations including simple nucleotide variations (SNVs), copy number variations (CNVs) and differential expressed genes (DEGs) in order to find key genes and pathways for diagnostic and prognostic prediction for lung adenocarcinoma and lung squamous cell carcinoma. We performed a univariate Cox model and then lasso-regularized Cox model with leave-one-out cross-validation using The Cancer Genome Atlas (TCGA) gene expression data in tumor samples. We generated 35- and 33-gene signatures for prognostic risk prediction based on the overall survival time of the patients with LUAD and LUSC, respectively. When we clustered patients into high- and low-risk groups, the survival analysis showed highly significant results with high prediction power for both training and test datasets. Then, we characterized the differences including significant SNVs, CNVs, DEGs, active subnetworks, and the pathways. We described the results for the risk groups and cancer subtypes separately to identify specific genomic alterations between both high-risk groups and cancer subtypes. Both LUAD and LUSC high-risk groups have more downregulated immune pathways and upregulated metabolic pathways. On the other hand, low-risk groups have both up- and downregulated genes on cancer-related pathways. Both LUAD and LUSC have important gene alterations such as CDKN2A and CDKN2B deletions with different frequencies. SOX2 amplification occurs in LUSC and PSMD4 amplification in LUAD. EGFR and KRAS mutations are mutually exclusive in LUAD samples. EGFR, MGA, SMARCA4, ATM, RBM10, and KDM5C genes are mutated only in LUAD but not in LUSC. CDKN2A, PTEN, and HRAS genes are mutated only in LUSC samples. The low-risk groups of both LUAD and LUSC tend to have a higher number of SNVs, CNVs, and DEGs. The signature genes and altered genes have the potential to be used as diagnostic and prognostic biomarkers for personalized oncology.

Keywords: TCGA; non-small-cell lung cancer; lung adenocarcinoma (LUAD); lung squamous cell carcinoma (LUSC); differential expression; SNV; CNV; risk group; signature; survival

---

check for
updates

Citation: Zengin, T.; Önal-Süzek, T. Comprehensive Profiling of Genomic and Transcriptomic Differences between Risk Groups of Lung Adenocarcinoma and Lung Squamous Cell Carcinoma. *J. Pers. Med.* **2021**, *11*, 154. https://doi.org/10.3390/jpm11020154

Academic Editor: Raghu Sinha

Received: 30 December 2020
Accepted: 19 February 2021
Published: 23 February 2021

Publisher's Note: MDPI stays neutral with regard to jurisdictional claims in published maps and institutional affiliations.

1. Introduction

Lung cancer is the second most frequently diagnosed cancer type and the leading cause of cancer-related mortality worldwide [1]. Lung cancer treatments used in the clinic are surgery, radiotherapy, chemotherapy, targeted therapy, and emerging immunotherapy. The clinical treatment decisions are made based on tumor stage, histology, genetic alterations of a few driver oncogenes for targeted therapies, and patient's condition [2]. However, most of the patients are diagnosed at an advanced and metastatic stage, with high mortality and poor benefit from therapies [3]. Although the targeted therapeutics and immunotherapeutics including immune-checkpoint inhibitors are introduced for patients at an advanced stage, these options are beneficial only for limited subsets of patients and these patients still can develop resistance [4]. Therefore, the majority of patients with advanced-stage lung cancer die within 5 years of diagnosis [5].

Histologically there are four major types of lung cancer, including small-cell carcinoma (SCLC), and adenocarcinoma, squamous cell carcinoma, large cell carcinoma as grouped non-small-cell carcinoma (NSCLC). Lung adenocarcinoma (LUAD) and lung squamous cell carcinoma (LUSC) account for 50% and 23% of all lung cancers, respectively [6]. Lung cancer is both histologically and molecularly heterogeneous disease and characterizing the genomics and transcriptomics of its nature is very important for effective therapies. Lung cancer has many subtypes with distinct genetic characteristics, resulting in intra-tumoral heterogeneity [7].

The Cancer Genome Atlas (TCGA) database serves different types of data such as transcriptome profiling, simple nucleotide variation, copy number variation, DNA methylation, clinical and biospecimen data of 84,392 cancer patients with 68 primary sites [8]. The Cancer Genome Atlas Research Network reported molecular profiling of 230 lung adenocarcinoma samples using mRNA, microRNA and DNA sequencing integrated with copy number, methylation and proteomic analyses. They identified 18 significantly mutated genes, including TP53, KRAS which is mutually exclusive with EGFR, BRAF, PIK3CA, MET, STK11, KEAP1, NF1, RB1, CDKN2A, GTPase gene RIT1, including activating mutations and MGA including loss-of-function mutations. DNA and mRNA sequence from the same tumor highlighted splicing alterations including exon 14 skipping in MET mRNA in 4% of cases. They also showed DNA hyper-methylation of several key genes: CDKN2A, GATA2, GATA4, GATA5, HIC1, HOXA9, HOXD13, RASSF1, SFRP1, SOX17, WIF1, and MYC over-expression was significantly associated with the hyper-methylation phenotype as well [9].

The Cancer Genome Atlas Research Network also profiled 178 lung squamous cell carcinomas and detected mutations in 11 genes, including mutations in TP53 (81%), CDKN2A, PTEN, PIK3CA, KEAP1, MLL2, HLA-A, NFE2L2, RB1, NOTCH1 including truncating mutations and loss-of-function mutations in the HLA-A class I major histocompatibility gene. They identified altered pathways such as NFE2L2 and KEAP1 in 34%, squamous differentiation genes in 44%, PI3K pathway genes in 47%, and CDKN2A and RB1 in 72% of tumors. CNV analysis revealed the amplification of NFE2L2, MYC, CDK6, MDM2, BCL2L1 and EYS, and deletions of FOXP1, PTEN and NF1 genes with previously identified CNV genes, SOX2, PDGFRA, KIT, EGFR, FGFR1, WHSC1L1, CCND1, and CDKN2A. They identified overexpression and amplification of SOX2 and TP63, loss-of-function mutations in NOTCH1, NOTCH2 and ASCL4 and focal deletions in FOXP1 which have known roles in squamous cell differentiation. CDKN2A is downregulated in over 70% of samples through epigenetic silencing by methylation (21%), inactivating mutation (18%), exon 1β skipping (4%), or homozygous deletion (29%) [10].

Recently, many studies have been published on gene expression signatures predicting the survival risk of patients with lung adenocarcinoma. These recent studies have been mostly using TCGA data, but their methods generated different gene signatures. Seven-gene expression signature including ASPM, KIF15, NCAPG, FGFR1OP, RAD51AP1, DLGAP5 and ADAM10 genes, was obtained for early stage cases from seven published lung adenocarcinoma cohorts and the signature showed high hazard rations in Cox regression analysis [11]. Shukla et al. developed TCGA RNAseq data-based prognostic signature including four protein-coding genes RHOV, CD109, FRRS1, and the lncRNA gene LINC00941, which showed high hazard ratios for stage I, EGFR wild-type, and EGFR mutant groups [12]. A prognostic signature that was independent of other clinical factors, was developed and validated based on the TCGA data. Patients were grouped into risk groups using signature genes, and patients with high-risk scores tended to have poor survival rate at 1-, 3- and 5-year follow-up. The developed eight-gene signature including

TTK, HMMR, ASPM, CDCA8, KIF2C, CCNA2, CCNB2, and MKI67 were highly expressed in A549 and PC-9 cells [13].

Twelve-gene signature (RPL22, VEGFA, G0S2, NES, TNFRSF25, DKFZP586P0123, COL8A2, ZNF3, RIPK5, RNFT2, ARHGEF12 and PTPN20A/B) was established by using published microarray dataset from 129 patients and the signature was independently prognostic for lung squamous carcinoma but not for lung adenocarcinoma [14]. A four-gene clustering model in 14-Genes (DPPA, TTTY16, TRIM58, HKDC1, ZNF589, ALDH7A1, LINC01426, IL19, LOC101928358, TMEM92, HRASLS, JPH1, LOC100288778, GCGR) was established and these genes plays role in positive regulation of ERK1 and ERK2 cascade, angiogenesis, platelet degranulation, cell–matrix adhesion, extracellular matrix organization and macrophage activation [15].

Lu et.al. identified differentially expressed genes between lung adenocarcinoma and lung squamous cell carcinoma by using microarray data from the Gene Expression Omnibus database. They identified 95 upregulated and 241 downregulated DEGs in lung adenocarcinoma samples, and 204 upregulated and 285 downregulated DEGs in lung squamous cell carcinoma samples, compared to the normal lung tissue samples. The genes play role in cell-cycle, DNA replication and mismatch repair. The top five genes from global network, HSP90AA1, BCL2, CDK2, KIT and HDAC2 have differential expression profiles between lung adenocarcinoma and lung squamous cell carcinoma [16]. Recently, Wu et.al. identified diagnostic and prognostic genes for lung adenocarcinoma and squamous cell carcinoma by using weighted gene expression profiles. The five-gene diagnostic signature including KRT5, MUC1, TREM1, C3 and TMPRSS2 and the five-gene prognostic signature including ADH1C, AZGP1, CLU, CDK1 and PEG10 obtained a log-rank P-value of 0.03 and a C-index of 0.622 on the test set [17].

A considerable number of genetic and transcriptomic alterations have been identified in mostly LUAD and poorly in LUSC. Although many gene expression signatures have been identified in LUAD recently, there is less work on LUSC expression signatures. Additionally, the molecular differences between risk groups of LUAD and LUSC have not yet been systematically described. In this study, we aimed to identify the genomic and transcriptomic differences between risk groups of lung adenocarcinoma and lung squamous cell carcinoma. We performed a univariate Cox model and then Lasso-Regularized Cox Model with Leave-One-Out Cross-Validation (LOOCV) by using TCGA gene expression data in tumor samples, and identified best gene signatures to cluster patients into low- and high-risk groups. We generated 35- and 33-gene signatures for prognostic risk prediction based on the overall survival time of the patients with LUAD and LUSC. When we clustered patients into high- and low-risk groups, the survival analysis showed highly significant results for both training and test datasets. Then, we characterized the differences including significant SNVs, CNVs, DEGs and active subnetwork DEGs between risk groups in LUAD and LUSC.

2. Materials and Methods

2.1. Data

Simple Nucleotide Variation (SNV), Transcriptome Profiling, Copy Number Variation (CNV) and Clinical data of patients who have all of these data types in LUAD and LUSC projects, was downloaded separately using *TCGAbiolinks* R package [18]. Using the same package and the reference of hg38; Simple Nucleotide Variations (SNVs) and Copy Number Variations (CNVs); and transcriptomic variations were processed to identify the genomic alterations of the LUAD and LUSC patients (Table 1). The method described below can be found as flowchart in Figure S1.

Table 1. Summary of clinical variables of train and test group of patients with LUAD and LUSC analyzed in the study.

	LUAD		LUSC	
Category	**Train Group (n: 436)**	**Test Group (n: 56)**	**Train Group (n: 431)**	**Test Group (n: 47)**
Age at diagnosis (median; range)	66; 33–88	66.5; 42–86	68; 39–90	69; 45–85
Gender				
Female	232	33	112	14
Male	204	23	319	33
Tumor stage				
I	241	28	211	25
II	106	13	138	16
III	68	13	76	5
IV	23	2	6	1
Vital status				
Alive	284	30	275	18
Dead	152	26	156	29
Smoked years (median; range)	33; 2–61	31.5; 4–64	40; 8–62	40; 10–60
Smoked packs per year (median; range)	40; 0.15–154	48; 5–94.5	50; 1–240	50; 2–157.5

2.2. Gene Expression Signature Analysis

Clinical data and Gene Expression Quantification data (HTSeq counts) of patients with unpaired RNAseq data (tumor samples without normal samples) was downloaded from the TCGA database using the *TCGAbiolinks* R package. Raw HTSeq counts of tumor samples were normalized by TMM (trimmed mean of M values) method and Log_2 transformed after filtering to remove genes that consistently have zero or low counts. Univariate Cox Proportional Hazards Regression analysis was performed using *survival* R package [19] to identify survival-related genes. For these survival-related potential biomarker genes ($p \leq 0.05$), Lasso-Regularized Cox Model (by using minimum lambda calculated in the model) with Leave-One-Out Cross-Validation (LOOCV) was performed to determine a gene expression signature using *glmnet* R package [20]. Multivariate Cox Regression for the signature genes was performed and the predictive performance of the model was scored using *riskRegression* R package [21]. The risk score of each patient was predicted based on multivariate Cox regression model using the *survival* R package. Patients were clustered into high-risk and the low-risk group based on the best cutoff value for ROC, calculated by *cutoff* R package [22].

For the validation of the gene signature, HTSeq counts belonging to the tumor samples of patients who have paired RNAseq data (tumor samples with the paired adjacent normal samples) were downloaded from the TCGA database, filtered, normalized by TMM method and Log_2 transformed. Multivariate Cox Regression for the signature genes was performed and the predictive performance of the model was scored. The risk score of every patient in the validation group was predicted based on multivariate Cox regression model and each patient was assigned to the high- or low-risk group using the best cutoff value for ROC. These analyses were performed for LUAD and LUSC patients separately.

2.3. Differential Expression Analysis

Gene Expression Quantification data (HTSeq counts) of both the primary tumor (TP) and the paired normal tissue adjacent to the tumor (NT) was downloaded from the TCGA database. Raw HTSeq counts of both tumor and normal samples were normalized by TMM method after filtering to remove genes which have zero or low counts. Differentially expressed ($q < 0.01$) genes were determined using *limma* [23] and *edgeR* [24] R packages by limma-voom method with duplicate-correlation function. HUGO symbols and NCBI Gene identifiers of the differentially expressed genes were downloaded using the *biomaRt* R package. This analysis was performed for high- and low-risk group patients of LUAD and LUSC, separately.

2.4. Active Subnetwork Analysis

Active subnetworks of the differentially expressed genes were determined using *DEsubs* R package [25]. *DEsubs* package accepts the differentially expressed genes output of the *limma* package along with their FDR adjusted p values (q values). *DEsubs* package both computes and plots the active subnetworks. All the plots and computations were generated for the high- and low-risk group patients of the LUAD and LUSC projects, separately.

2.5. Copy Number Variation Analysis

The Copy Number Variation data of the primary tumor samples of patients was downloaded using *TCGAbiolinks* package (Masked Copy Number Segment as data type). The chromosomal regions which are significantly aberrant in tumor samples were determined and plotted by *gaia* R package [26]. Gene enrichment from genomic regions which have significant differential copy number was performed using *GenomicRanges* [27] and *biomaRt* R packages. R codes used in this analysis were modified from the codes presented at "TCGA Workflow" article [28]. All the computations and the plots were generated for the high- and low-risk groups of LUAD and LUSC projects, separately.

2.6. Simple Nucleotide Variations Analysis

The masked Mutation Annotation Format (maf) files of the TCGA mutect2 pipeline in tumor samples were downloaded to obtain the somatic mutations. The maf files are filtered using the *maftools* [29] to obtain the subset of the mutations corresponding to the patient barcodes. Summary plot and oncoplot were generated to summarize the mutation data using *maftools* R package. Somatic mutations were filtered and assigned to either oncogene (OG) or tumor suppressor gene (TSG) groups along with a significance score ($q < 0.05$) using the *SomInaClust* R package [30]. *SomInaClust* computes a background mutation value to identify the hot spots using the known set of somatic mutations in "COSMIC" and the "Cancer Gene Census" (v92) datasets of COSMIC database for GRCh38 [31]. SNV analysis was performed for high- and low-risk group patients of LUAD and LUSC projects, separately.

2.7. Visualization

Scatter plots showing risk score and survival time of patients were generated by *ggrisk* R package [32] and Kaplan–Meier (KM) survival curves were plotted by *survminer* R package [33] displaying the overall survival difference between the risk groups stratified on the proposed gene signature. ROC curves were plotted for the risk scores based on each gene signature using *survivalROC* R package [34]. Univariate and multivariate Cox regression analyses were performed and forest plots were generated for risk score with clinical variables using *survival* and *forestmodel* [35] R packages.

Gene and pathway enrichment analyses were performed by *biomaRt* [36] and *clusterProfiler* [37] R packages and plotted by *enrichplot* R package [38]. Heatmap plots were generated using *ComplexHeatmap* R package [39]. Mosaic plots to compare the categorical variables were generated using the *vcd* R package [40,41].

OncoPrint showing CNVs among patient samples was generated using *Complex-Heatmap* R package. OncoPlot for significant mutated genes was drawn using *maftools*, and oncoPrint showing SNVs and CNVs together was generated using *ComplexHeatmap* R package. Circos plot showing all non-synonymous SNVs in original data of risk groups and significant CNVs at genome-scale were generated using *circlize* R package [42].

All possible relations between DEGs; active subnetwork DEGs; CNV genes; SNV genes of LUAD and LUSC risk groups were identified by using *VennDiagram* R package [43].

3. Results

3.1. Gene Expression Signature Analysis of LUAD and LUSC Patients

In order to identify gene expression prognosis risk model, clinical data and gene expression quantification data of tumor samples of patients with LUAD and h LUSC with unpaired RNAseq data as two separate training groups (Table 1) were downloaded from the TCGA database. A 35-gene expression signature for LUAD and a 33-gene expression signature for LUSC were identified by Lasso-Regularized Cox Model with LOOCV after univariate Cox regression analysis. The risk scores of each patient in training groups and test groups were predicted using signature genes, then patients were clustered into high- and low-risk groups based on the cutoff values.

The genes of the LUAD expression signature model identified are AC005077.4, AC113404.3, ADAMTS15, AL365181.2, ANGPTL4, ASB2, ASCL2, CCDC181, CCL20, CD200R1, CPXM2, DKK1, ENPP5, EPHX1, GNPNAT1, GRIK2, IRX2, LDHA, LDLRAD3, LINC00539, LINC00578, MS4A1, OGFRP1, RAB9B, RGS20, RHOQ, SAMD13, SLC52A1, STAP1, TLE1, U91328.1, WBP2NL, ZNF571-AS1, ZNF682, ZNF835. Twenty-seven of them are protein-coding genes while two of them are long intergenic non-protein coding RNA (LINC00539, LINC00578), one is antisense RNA (ZNF571-AS1), three of them are pseudogenes (AC005077.4, AC113404.3, OGFRP1) and two of them are novel transcripts (AL365181.2, U91328.1) (Table S1). Pathway enrichment analysis by using *clusterProfiler* R package did not give any results for this 35-gene list; therefore, enrichment analysis was performed manually using the online KEGG Mapper tool. The genes play role in metabolic pathways, cancer and immune system-related pathways such as Central carbon metabolism in cancer, Glycolysis, Cholesterol metabolism, Amino sugar and Nucleotide sugar metabolism, HIF-1 signaling pathway, TNF signaling pathway, IL-17 signaling pathway, Chemokine signaling pathway and Wnt signaling pathway (Table S2). Multivariate Cox regression analysis was performed for the signature genes and the predictive performance of the model was scored. The AUC was 0.963 ($p = 1.1 \times 10^{-15}$) for LUAD training group. The risk score of each patient was predicted and patients were clustered into high- and low-risk groups based on the cutoff value. Low- and high-risk groups have different expression patterns of the signature genes and significantly different survival probabilities ($p < 0.0001$). The prediction power of the risk score is around 0.78 (AUC) for 1, 3, 5 and 8 years for LUAD training group (Figure S2). Risk group clustering is independent from tumor stages because risk groups have also significantly different survival probability for each tumor stage (Figure S3). Vital status is highly correlated with risk groups that high-risk group is positively correlated with death ($p = 1.5 \times 10^{-13}$), while only tumor stage IA and III are associated with risk groups (Figure S4). The risk score has highly significant prognostic ability (HR:2.59, $p < 0.001$) when multivariate Cox regression analysis was performed with other clinical variables (Figures S5 and S6).

In order to validate the gene expression signature, gene expression quantification data of tumor samples of patients with LUAD who have paired RNAseq data were downloaded from the TCGA database. The risk scores of each patient in test group were predicted using the gene signature lists and patients were clustered into high- and low-risk groups based on the best cutoff values for ROC. Risk groups have differential signature gene expression patterns; high-risk group has lower survival time and higher number of deaths resulting a significantly different survival probability ($p < 0.0001$). The risk score has high prediction powers, 0.97, 0.92, 0.93 and 0.92 (AUC) for 1, 3, 5 and 8 years, respectively, for LUAD test group (Figure 1).

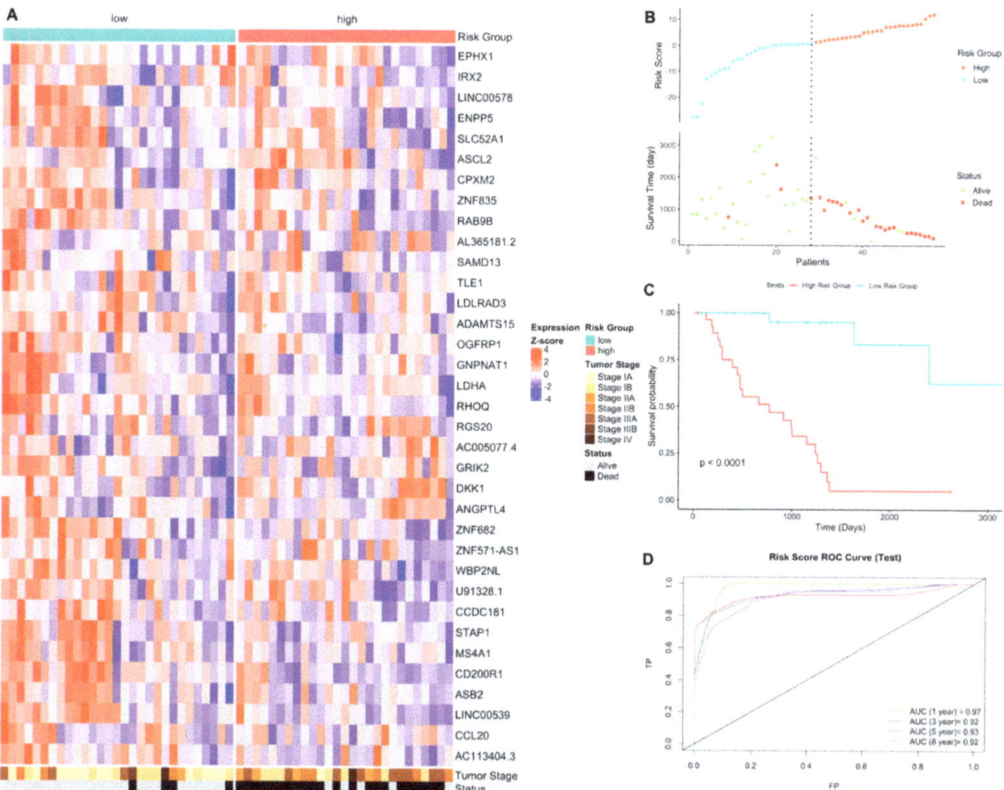

Figure 1. Gene expression signature and risk clustering of LUAD test dataset. Test dataset patients were clustered into high- and low-risk groups based on risk scores of patients calculated by predicting the effect of the signature genes of the signature genes expression on overall survival. (**A**) Expression heatmap of the signature genes in tumor samples of LUAD patients in the test dataset. (**B**) Scatter plot showing risk scores, survival time and separation point of the patients into risk groups. (**C**) KM survival plot showing the overall survival probability between risk groups. (**D**) ROC curve showing prediction power of risk score in the test dataset for 1, 3, 5 and 8 years.

Risk groups have significantly different survival probability for each tumor stage in LUAD test group as well (Figure S7). Vital status is highly correlated with risk groups. The high-risk group is positively correlated with death ($p = 3.87 \times 10^{-7}$), while only tumor stage I is positively associated with low-risk group ($p = 0.016$) (Figure S8). The risk score has highly significant prognostic ability (HR:2.79, $p < 0.001$) as the result of multivariate Cox regression analysis was performed with other clinical variables (Figure S9).

Expression signature model identified for LUSC includes these genes: AC078883.1, AC096677.1, AC106786.1, ADAMTS17, ALDH7A1, ALK, COL28A1, EDN1, FABP6, HKDC1, IGSF1, ITIH3, JHY, KBTBD11, LINC01426, LINC01748, LPAL2, NOS1, PLAAT1, PNMA8B, RGMA, RPL37P6, S100A5, SLC9A9, SNX32, SRP14-AS1, STK24, UBB, UGGT2, WASH8P, Y_RNA, ZNF160, ZNF703. Twenty-three of them are protein coding genes while two of them are long intergenic non-protein coding RNA (LINC01748, LINC01426), one is antisense RNA (SRP14-AS1), three of them are pseudo-genes (LPAL2, RPL37P6, WASH8P), three of them are novel transcripts (AC106786.1, AC096677.1, AC078883.1) and one is Y RNA (Table S3). They play role in mostly in metabolic pathways, cancer and immunity related pathways such as Arginine and proline metabolism, Glycolysis/Gluconeogenesis, HIF-1 signaling pathway, Non-small-cell lung cancer, PD-L1 expression and PD-1 checkpoint pathway in cancer and TGF-beta signaling pathway (Table S4).

The predictive performance score of the signature model is 80.8 (AUC) ($p = 1.3 \times 10^{-6}$) in multivariate Cox regression analysis for LUSC training group. The risk score of each patient was predicted and patients were clustered into high- and low-risk groups based on the cutoff value. Low- and high-risk groups have different expression patterns of the signature genes and significant difference of survival probability ($p < 0.0001$). The AUC values showing prediction power of the risk score are 0.76, 0.82, 0.87 and 0.92 for 1, 3, 5 and 8 years, respectively, for LUSC training group (Figure S10). Risk groups have also significantly different survival probability for tumor stages I, II and III (Figure S11). Risk groups are highly correlated with vital status. The high-risk group has highly significant positive correlation with death ($p = 8.5 \times 10^{-15}$), while low-risk group is negatively correlated. Tumor stages did not show any association with risk groups (Figure S12). The risk score has highly significant prognostic ability (HR:2.85, $p < 0.001$) when multivariate Cox regression analysis was performed with other clinical variables (Figure S13).

In order to validate the gene expression signature for LUSC, gene expression quantification data of tumor samples of patients with LUSC who have paired RNAseq data were downloaded. The risk scores of each patient in LUSC test group were predicted using gene signature lists and patients were clustered into high- and low-risk groups based on the best cutoff values for ROC. Risk groups have differential signature gene expression pattern; high-risk group has lower survival time and higher number of deaths. Risk groups have significantly different survival probability ($p < 0.0001$). The risk score has high prediction powers, 0.93, 0.95, 0.96 and 0.97 (AUC) for 1, 3, 5 and 8 years, respectively, for LUSC test group (Figure 2).

Risk groups have also significantly different survival probability for tumor stages in test group (Figure S14). Vital status is not correlated with risk groups of LUSC test group that number of deaths is higher for high-risk group insignificantly ($p = 0.07$). Tumor stages are not associated with risk groups (Figure S15). The risk score has highly significant prognostic ability (HR:2.66, $p < 0.001$) while other clinical variables have no effect on overall survival in multivariate Cox regression analysis (Figure S16).

The expression gene signatures of LUAD and LUSC do not have any common gene, however they share eight common pathways which are mostly metabolic pathways: Central carbon metabolism in cancer, Glycolysis/Gluconeogenesis, HIF-1 signaling pathway, Pyruvate metabolism, PPAR signaling pathway, Amino sugar and nucleotide sugar metabolism, TNF signaling pathway and Pathways of neurodegeneration—multiple diseases.

3.2. Differential Expression and Active Subnetwork Analysis of Risk Groups

Gene expression quantification data of both primary tumor and adjacent normal tissues of patients who have paired RNAseq data (test groups) in LUAD and LUSC projects were downloaded from the TCGA database. Differentially expressed ($q < 0.01$) genes (DEGs) were determined in tumor samples according to normal samples for high- and low-risk patient groups in test sets of LUAD and LUSC, separately. Then, active subnetworks of DEGs in tumor samples were determined using the DEGs with their q values.

In tumor samples of the LUAD low-risk group, the number of the genes which are dysregulated significantly ($q < 0.01$) more than 2-fold is 3615 (2439 down-, 1176 upregulated) while 3610 genes (2239 down-, 1371 upregulated) are dysregulated for the LUAD high-risk group. LUAD low- and high-risk groups have 2745 common differentially expressed genes (Figure S17). The top 20 significant DEGs highlighted as purple at volcano plot in Figure 3A,B are different between LUAD risk groups as dysregulation pattern is different between risk groups albeit the shared 2745 DEGs.

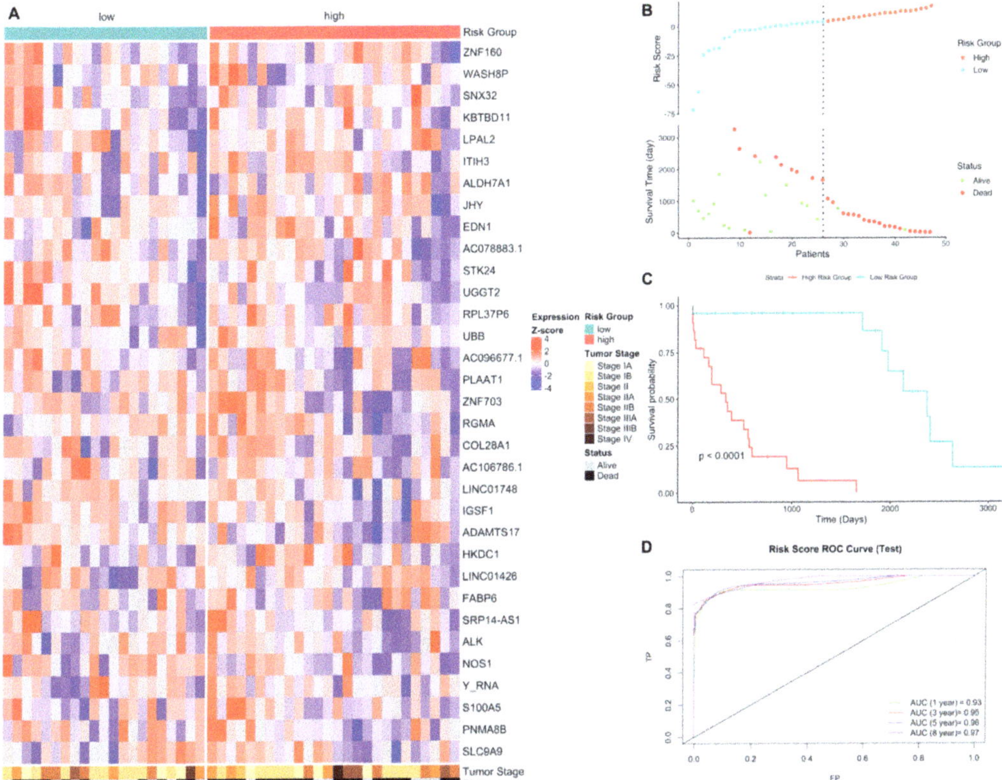

Figure 2. Gene expression signature and risk clustering of LUSC test dataset. Test dataset patients were clustered into high- and low-risk groups based on risk scores of patients calculated by predicting the effect of the signature genes' expression on overall survival. (**A**) Expression heatmap of the signature genes in tumor samples of LUSC patients in the test dataset. (**B**) Scatter plot showing risk scores, survival time and separation point of the patients into risk groups. (**C**) KM survival plot showing the overall survival probability between risk groups. (**D**) ROC curve showing prediction power of risk score in the test dataset for 1, 3, 5, and 8 years.

Seven of the signature genes (GNPNAT1, CCDC181, LDHA, ADAMTS15, IRX2, LINC00578, AC005077.4) are dysregulated in both risk groups. ANGPTL4 is upregulated in the high-risk group while MS4A1, GRIK2, and OGFRP1 are upregulated in the low-risk group.

Risk groups of LUAD share dysregulated pathways (Figure 3C,D), highly related to cancer, such as Cell cycle, Biosynthesis of amino acids and Protein digestion and absorption which are upregulated for both risk groups (Figure S18), on the other hand, they also share ECM–receptor interaction, Cell adhesion molecules pathways with immune system-related pathways such as Complement and coagulation cascades and Cytokine-cytokine receptor interaction which are downregulated for both risk groups (Figure S18). However, the high-risk group has more dysregulated immune system-related pathways such as Allograft rejection, Graft-versus-host disease, Inflammatory bowel disease, Intestinal immune network for IgA production, Rheumatoid arthritis, Staphylococcus aureus infection (Figure 3C,D), which are downregulated pathways in LUAD high-risk group (Figure S18).

Active subnetworks of differentially expressed genes in tumor samples of the LUAD risk groups were identified and low-risk group has 191 genes while high-risk group has 168 genes including 112 common genes, which are acting on active subnetworks (Figure S17).

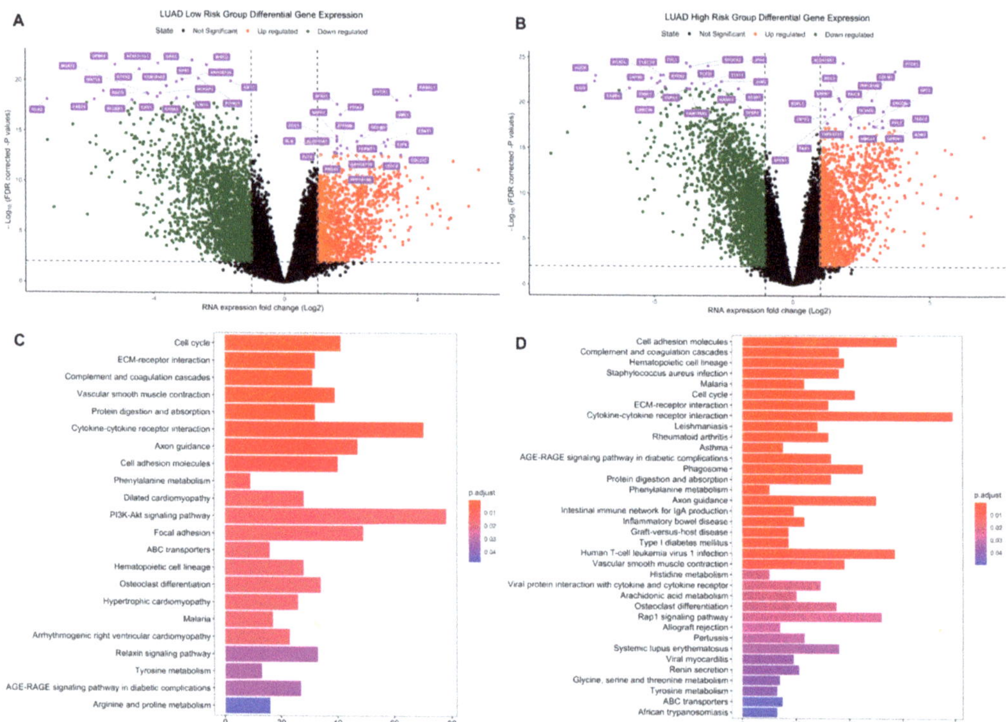

Figure 3. Differential expression analysis of the LUAD risk groups. LUAD test dataset patients were clustered into high- and low-risk groups based on risk scores of patients and differentially expressed genes in tumor samples were determined based on expressions in normal tissues. (**A**) Volcano plot showing differentially expressed genes more than 2-fold (Log$_2$ =1) for LUAD low-risk group. The top 20 significant downregulated and upregulated genes are highlighted as purple. FDR corrected p-values threshold is 0.01 (-Log$_{10}$ = 2). Red: Upregulated, Green: Downregulated, Black: Not significant or low than 2-fold. (**B**) Volcano plot showing differentially expressed genes more than two-fold (Log$_2$ = 1) for the LUAD high-risk group. The top 20 significant downregulated and upregulated genes are highlighted as purple. FDR corrected *p*-values threshold is 0.01 (-Log$_{10}$ = 2). Red: Upregulated, Green: Downregulated, Black: Not significant or low than 2-fold. (**C**) Dysregulated pathways of differentially expressed genes for LUAD low-risk group. (**D**) Dysregulated pathways of differentially expressed genes for LUAD high-risk group.

Pathway enrichment of DEGs at active subnetworks shows that the genes playing role in active subnetworks are much more related to cancer pathways such as PI3K-Akt signaling pathway, Ras signaling pathway, Small-cell lung cancer, Breast cancer, Gastric cancer, Proteoglycans in cancer and Rap1 signaling pathway (Figure 4). LUAD risk groups have mostly similar cancer-related active pathways, however only low-risk group has FoxO signaling pathway and TNF signaling pathway while high-risk group has Estrogen signaling pathway, Growth hormone synthesis, secretion, and action with immune system pathways such as Antigen processing and presentation, Intestinal immune network for IgA production and Leukocyte trans-endothelial migration.

The number of dysregulated genes expressed significantly ($q < 0.01$) more than 2-fold in tumor samples of the LUSC low-risk group is 5596 (3394 downregulated, 2202 upregulated) while 5403 genes (3338 downregulated, 2065 upregulated) are dysregulated for LUSC high-risk group. LUSC low- and high-risk groups have 4562 common differentially expressed genes (Figure S17). The top 20 significant DEGs highlighted at volcano plot in Figure 5A,B include common genes and dysregulation pattern is similar between risk groups.

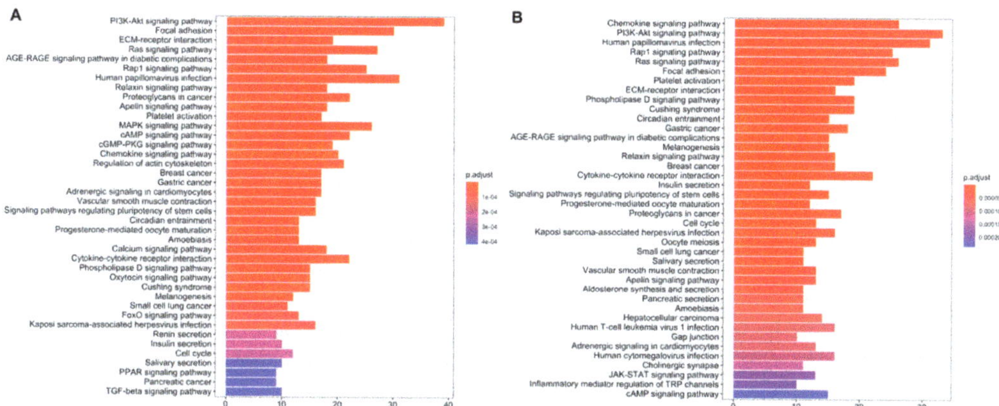

Figure 4. Pathway enrichment of differentially expressed genes at active subnetworks of the LUAD risk groups. Active subnetworks were determined by using differential expression analysis results and pathway enrichment analysis was performed for the genes at subnetworks. (**A**) Pathways of differentially expressed genes in active subnetworks for LUAD low-risk group. (**B**) Pathways of differentially expressed genes in active subnetworks for LUAD high-risk group.

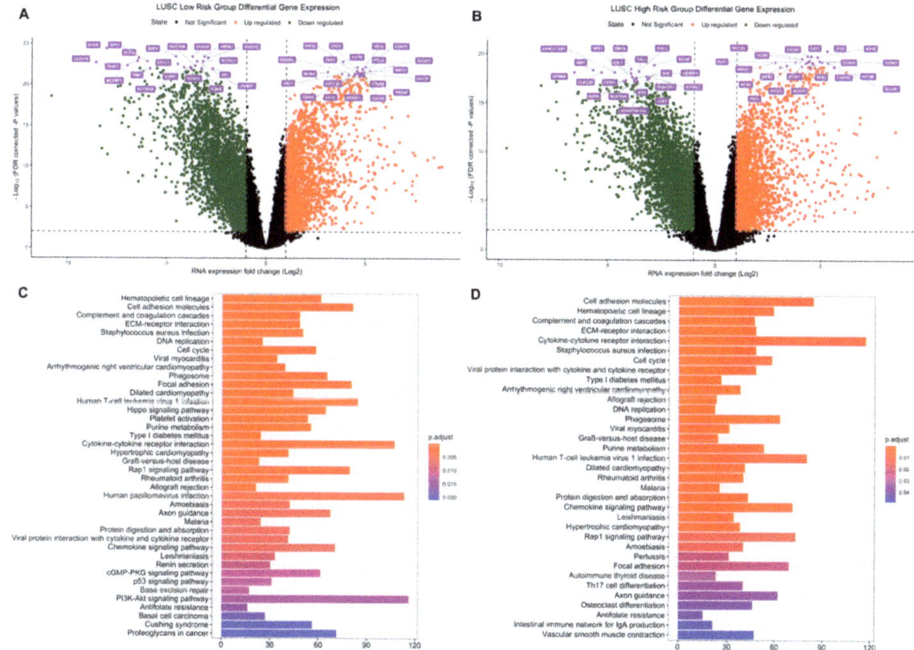

Figure 5. Differential expression analysis of the LUSC risk groups. LUSC test dataset patients were clustered into high- and low-risk groups based on risk scores of patients and differentially expressed genes in tumor samples were determined based on expressions in normal tissues. (**A**) Volcano plot showing differentially expressed genes more than 2-fold ($Log_2 = 1$) for LUSC low-risk group. The top 20 significant downregulated and upregulated genes are highlighted as purple. FDR corrected p-values threshold is 0.01 (-$Log_{10} = 2$). Red: Upregulated, Green: Downregulated, Black: Not significant or low than 2-fold. (**B**) Volcano plot showing differentially expressed genes more than two-fold ($Log_2 = 1$) for LUSC high-risk group. The top 20 significant downregulated and upregulated genes are highlighted as purple. FDR corrected p-values threshold is 0.01 (-$Log_{10} = 2$). Red: Upregulated, Green: Downregulated, Black: Not significant or low than 2-fold. (**C**) Dysregulated pathways of differentially expressed genes for LUSC low-risk group. (**D**) Dysregulated pathways of differentially expressed genes for LUSC high-risk group.

LUSC signature genes have 10 common genes (EDN1, JHY, PLAAT1, HKDC1, ITIH3, KBTBD11, RGMA, ZNF703, S100A5, LPAL2) with DEGs of both risk groups. Three of the signature genes, ADAMTS17, IGSF1, and LINC01426, are upregulated in the low-risk group; others, NOS1 and SRP14-AS1 are downregulated while Y_RNA is upregulated in the high-risk group.

Risk groups of LUSC have common dysregulated pathways (Figure 5C,D), which are highly related to cancer, such as Cell cycle, DNA replication, Base excision repair, p53 signaling pathway which are upregulated at both risk groups (Figure S19), on the other hand, they also share ECM–receptor interaction, Cell adhesion molecules, Focal adhesion pathways with immune system-related pathways such as Chemokine signaling pathway, Complement and coagulation cascades, Cytokine–cytokine receptor interaction, which are downregulated at both risk groups (Figure S19). However, the high-risk group has more upregulated metabolic pathways such as Central carbon metabolism in cancer, Protein digestion and absorption, Alanine, aspartate and glutamate metabolism, Arginine and proline metabolism, Cysteine and methionine metabolism, Glutathione metabolism, Ribosome biogenesis in eukaryotes; and downregulated immune-related pathways such as JAK-STAT signaling pathway, TNF signaling pathway, Primary immunodeficiency, T cell receptor signaling pathway distinctly from low-risk group (Figure S19). LUSC low-risk group has downregulated PI3K-Akt signaling pathway, Phenylalanine metabolism, Tyrosine metabolism, Phospholipase D signaling pathway, Proteoglycans in cancer and Tight junction pathways with upregulated Hippo signaling pathway and Small-cell lung cancer distinctly from high-risk group (Figure S19).

Active subnetworks of differentially expressed genes in tumor samples of the LUSC risk groups has 357 genes for the low-risk group while 350 genes for high-risk group including 245 common genes (Figure S17). Active pathways of the LUSC risk groups, are highly related to cancer pathways such as PI3K-Akt signaling pathway, Ras signaling pathway, Small-cell lung cancer, Proteoglycans in cancer and Rap1 signaling pathway (Figure 6A,B). LUSC risk groups have mostly similar cancer-related active pathways, however only low-risk group has Nucleotide excision repair, Adherens junction and Alpha-Linolenic acid metabolism pathways, while high-risk group has cancer and metabolism-related pathways such as Basal cell carcinoma, Prolactin signaling pathway, Apoptosis, Mitophagy, Choline metabolism in cancer, Insulin signaling pathway, Carbohydrate digestion and absorption, Central carbon metabolism in cancer with immune system-related Measles and Influenza A pathways.

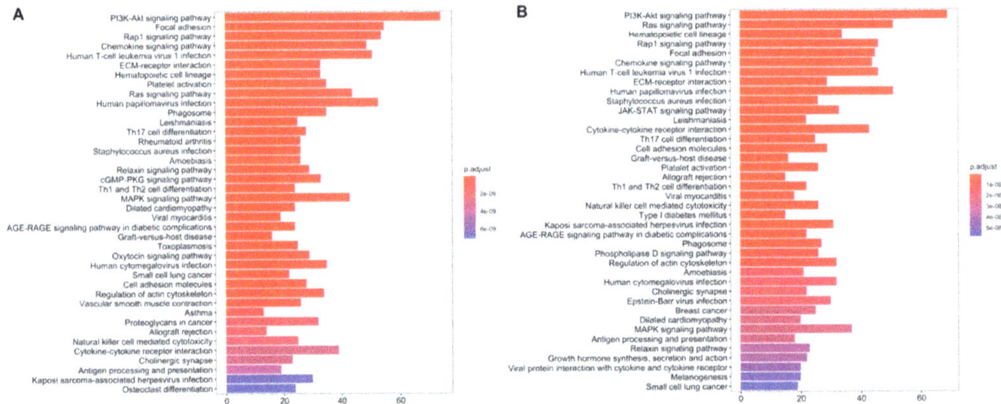

Figure 6. Pathway enrichment of differentially expressed genes at active subnetworks of the LUSC risk groups. Active subnetworks were determined by using differential expression analysis results and pathway enrichment analysis was performed for the genes at subnetworks. (**A**) Active pathways of differentially expressed genes for LUSC low-risk group. (**B**) Active pathways of differentially expressed genes for LUSC high-risk group.

3.3. Copy Number Variations Analysis

The significant aberrant genomic regions in tumor samples of patients were determined and then gene enrichment from genomic regions which have differential copy number was performed. Pathway enrichment analysis of genes which have CNVs was performed and plotted. LUAD low- and high-risk groups have different CNV profiles as seen at CNV plots showing amplified or deleted genomic regions on chromosomes. Chromosomes 1, 6, 7, 10, 13, 16, 17, 28 and 20 have different significant aberrant genomic regions (q < 0.01) between risk groups (Figure 7A,B). The highest frequencies of the amplified genes are 45%, 49% and the deleted genes are 31%, 45% in the low- and high-risk groups, respectively. The top 10 the highest frequently amplified or deleted genes in tumor samples of risk groups are different and patients in the same group may have different aberration patterns (Figure 7C,D). The numbers of the deleted genes and the amplified genes are 10,144 and 10,412, respectively, in tumor samples of the LUAD low-risk group. LUAD high-risk group has 5379 deleted and 8442 amplified genes in tumor samples. Risk groups have 4921 deleted and 6559 amplified genes in common (Figure S22).

Pathways of CNV genes are different between LUAD risk groups; mostly immune system pathways such as Allograft rejection, Graft-versus-host disease, Antigen processing and presentation, Complement and coagulation cascades, Inflammatory bowel disease and Viral carcinogenesis pathways have amplified CNVs in the low-risk group (Figure S20) while Herpes simplex virus 1, Cytosolic DNA sensing pathway, Natural killer cell mediated cytotoxicity and Nod-like receptor signaling pathways have deleted CNVs (Figure S20) in the high-risk group (Figure 7). Complement and coagulation cascades pathway has amplified genes in both risk groups while Natural killer cell mediated cytotoxicity and Nod-like receptor signaling pathways have deleted genes in both risk groups (Figure S20). The low-risk group patients have immune system pathways with amplified genes whereas high-risk group patients have immune system pathways with deleted genes. On the other hand, high-risk group has amplified genes in metabolic pathways such as Gastric acid secretion and Insulin secretion (Figure S20).

LUSC risk groups have different significant aberrant genomic regions obviously on chromosomes 5, 6, 8 and X (Figure 8A,B). The highest frequencies of amplified genes are 84%, 77% and of the deleted genes are 55%, 51% in the low- and high-risk groups, respectively. LUSC risk groups have higher frequency of amplified genes than deleted genes. Risk groups have common genes from top 25 the highest frequently amplified genes such as SOX2, GHSR, TNFSF10 and miRNAs, miR-7977 and miR-569, with variable frequencies. Risk groups have also common deleted genes such as CDK inhibitors, CDKN2A and CDKN2B, and miR-1284 (Figure 8C,D). LUSC low-risk group has 10,720 deleted and 10,264 amplified genes while LUSC high-risk group has 9477 deleted and 10,250 amplified genes in tumor samples. Risk groups have 7820 deleted and 8659 amplified genes in common (Figure S22).

Pathways of CNV genes highly overlap between LUSC risk groups and they share cancer-related pathways such as PI3K-Akt signaling pathway, JAK-STAT signaling pathway, Ras signaling pathway, Gastric cancer (Figure 8E,F). However, some pathways differ between risk groups, low-risk group has CNVs at mTOR signaling pathway, VEGF signaling pathways and Central carbon metabolism in cancer, while high-risk group has CNVs at Chemical carcinogenesis, Drug metabolism—cytochrome P450, Carbohydrate digestion and absorption pathways (Figure 8E,F). Steroid hormone biosynthesis and Bile secretion pathways have multiple amplified genes while NOD-like receptor signaling pathway has deleted genes, in both risk groups. Only low-risk group has multiple amplified genes at Growth hormone synthesis, secretion and action, and Complement and coagulation cascades pathways. Only high-risk group has amplified genes at Chemical carcinogenesis and Drug metabolism pathways while has deleted genes at Cytokine-cytokine receptor interaction and Fatty acid biosynthesis pathways (Figure S21).

1518/1158 to 10s with median 167 and 172.5 for low- and high-risk groups, respectively. Missense mutation is the highest frequent mutation type, and C > A and C > T substitutions are the most frequent ones for both risk groups. LUAD risk groups have a similar set of mutated genes with varying frequencies. TP53 is the highest frequently mutated gene with 45% and 53% for low- and high-risk groups, and the following ones are MUC16 (39%, 40%) and CSMD3 (38%, 35%) for both groups (Figure S23). SomInaClust analysis was performed to determine driver genes, and 39 genes and 19 genes are strong candidate driver genes for the low-risk group and high-risk group, respectively (Tables S5 and S6). Interestingly, LUAD risk groups share 18 of these driver genes (Figure S27). SomInaClust calculates TSG and OG scores based on background mutation rate and hot spots, then classifies the genes based on TSG/OG scores and cancer gene census data (Figure S25). The driver genes determined in LUAD low-risk group are KRAS, TP53, EGFR, BRAF, STK11, MGA, NF1, RB1, PIK3CA, ATM, RBM10, SETD2, ARID1A, CTNNB1, CMTR2, SF3B1, CSMD3, ATF7IP, KEAP1, HMCN1, EPHA5, ARID2, TTK, SMAD4, KDM5C, SMARCA4, APC, NFE2L2, RIT1, DDX10, LTN1, CDH10, SPTA1, LRP1B, COL11A1, MAP3K12, USH2A, AKAP6 and RASA1. The driver genes determined in LUAD high-risk group are KRAS, TP53, STK11, EGFR, BRAF, RBM10, PIK3CA, SETD2, ARID2, NF1, RB1, MGA, KEAP1, CSMD3, SMARCA4, CTNNB1, KDM5C, IDH1 and ATM (Figure S25; Tables S5 and S6). TP53 and CSMD3 genes are the most frequently mutated genes with 47%, 56% and 41%, 37% frequencies, respectively for low- and high-risk groups (Figure 9A,B). More than half of the genes are mutated in less than 12% of patients. For common genes, LUAD high-risk group has mostly higher frequencies. TP53 has differential mutation types, while KRAS has mostly missense mutations. CSMD3 has more multi-hits (multiple mutations in one patient) in the low-risk group than the high-risk group. EGFR has in frame deletions in both risk groups and other common genes have similar mutation type pattern between risk groups (Figure 9A,B). Pathways of driver mutated genes are highly lung cancer-related pathways such as Non-small-cell lung cancer, EGFR tyrosine kinase inhibitor resistance, Platinum drug resistance, MAPK signaling, mTOR signaling, Ras signaling pathway, PI3K-Akt signaling (Figure 9C,D) and other immunologic and metabolic pathways such as Signaling pathways regulating pluripotency of stem cells, FoxO signaling pathway, Rap1 signaling pathway, Central carbon metabolism in cancer, Proteoglycans in cancer, Human T-cell leukemia virus 1 infection, PD-L1 expression and PD-1 checkpoint pathway in cancer and Natural killer cell mediated cytotoxicity pathways, for both risk groups. Many common pathways are enriched because these mutated driver genes play role in many crucial important pathways. However, Wnt signaling pathway and Hippo signaling pathways are mutated only in the low-risk group, while Gap junction, GnRH signaling pathway, C-type lectin receptor signaling pathway, T cell receptor signaling pathway, HIF-1 signaling pathway, Growth hormone synthesis, secretion and action and AMPK signaling pathways are mutated only in the high-risk group (Figure 9C,D).

LUSC low-risk group has 14,038 mutated genes, while LUSC low-risk group has 14,616 mutated genes, and 11,947 genes are common (Figure S27). LUSC patients have a range of mutation numbers from 2300/1488 to 10s with median 201 for low- and high-risk groups, respectively. Missense mutation is the highest frequent mutation type, and C > A and C > T substitutions are the most frequent ones for both risk groups. LUSC risk groups have overlapping list of mutated genes with varying frequencies. TP53 is the highest frequently mutated gene with 80% and 78% for low- and high-risk groups, and the following ones are CSMD3 (42%, 42%) and MUC16 (39%, 40%) for both groups (Figure S24). As candidate driver genes, 30 genes and 19 genes were identified for the low-risk group and the high-risk group, respectively (Tables S7 and S8). LUSC risk groups share 14 of these driver genes (Figure S27). The driver genes determined in LUSC low-risk group are TP53, KMT2D, NFE2L2, PIK3CA, CDKN2A, PTEN, RB1, FAT1, ARID1A, NF1, RASA1, CUL3, KDM6A, NRAS, KRT5, ZNF750, EP300, FGFR3, TAOK1, CSMD3, NSD1, HRAS, SI, PDS5B, KRAS, KEAP1, API5, HNRNPUL1, SLC16A1, FBXW7. The driver genes determined in LUSC high-risk group are TP53, NFE2L2, PIK3CA, KMT2D, FAT1, CDKN2A, RB1, PTEN, NOTCH1,

ARID1A, RASA1, NF1, KMT2C, BRAF, PIK3R1, CSMD3, STK11, HRAS, KEAP1 (Figure S26; Tables S7 and S8). TP53 (83%, 82%), CSMD3 (44%, 44%) and KMT2D (25%, 23%) are most frequent mutated genes for low- and high-risk groups (Figure 10A,B). For common genes, risk groups have similar frequencies. TP53 and KMT2D genes have differential mutation types, while CSMD3 has mostly missense and multi-hit mutations. CDKN2A has mostly truncating mutations in both risk groups and other common genes have similar mutation type pattern between risk groups (Figure 10A,B). Pathways of driver mutated genes are highly lung cancer-related pathways such as Non-small-cell lung cancer, EGFR tyrosine kinase inhibitor resistance, Platinum drug resistance, MAPK signaling and Ras signaling (Figure 10C,D) and other immunologic and metabolic pathways such as FoxO signaling pathway, Central carbon metabolism in cancer, Proteoglycans in cancer, Hepatitis B, Hepatitis C, PD-L1 expression and PD-1 checkpoint pathway in cancer for both risk groups. Many common pathways are enriched because these mutated driver genes play role in many crucial important pathways. However, Gap junction and Ubiquitin mediated proteolysis pathways are mutated only in the low-risk group, while HIF-1 signaling and TNF signaling pathways are mutated only in the high-risk group (Figure 10C,D).

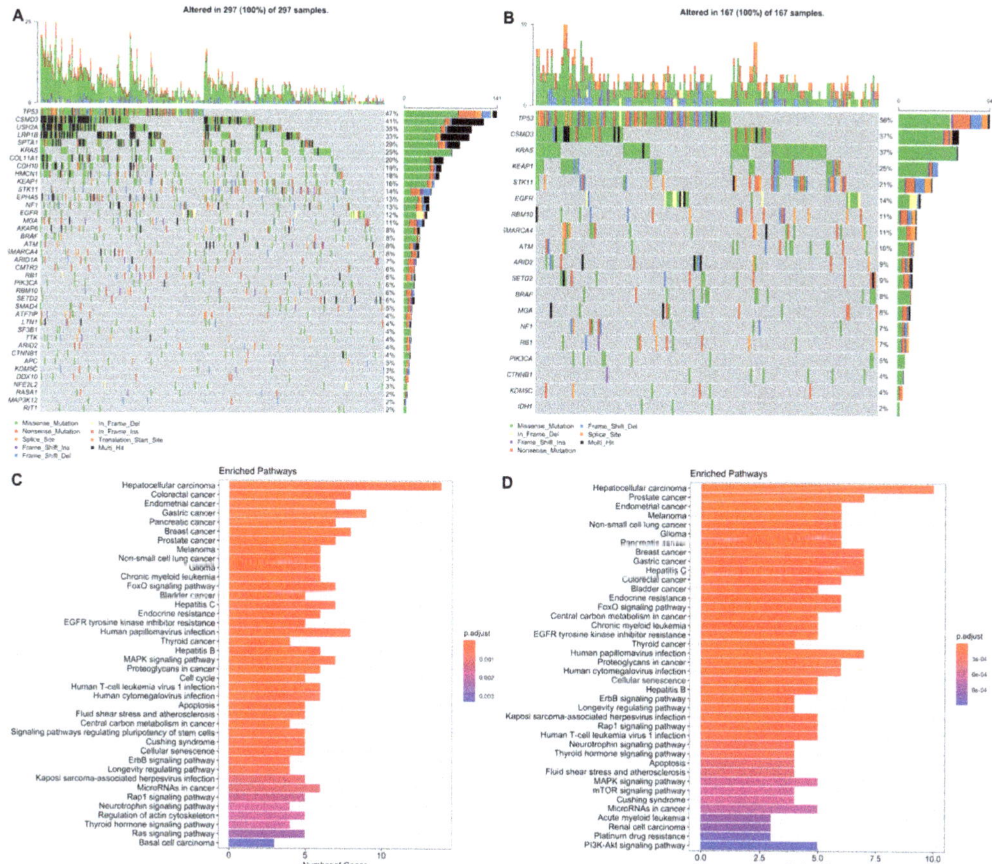

Figure 9. Oncoplot of potential driver genes containing significant SNVs of the LUAD risk groups. (**A**) Oncoplot showing significant SNV genes in tumor samples of the LUAD low-risk group patients. (**B**) Oncoplot showing significant SNV genes in tumor samples of the LUAD high-risk group patients. (**C**) Pathway enrichment of the significant SNV genes of the LUAD low-risk group. (**D**) Pathway enrichment of the significant SNV genes of the LUAD high-risk group.

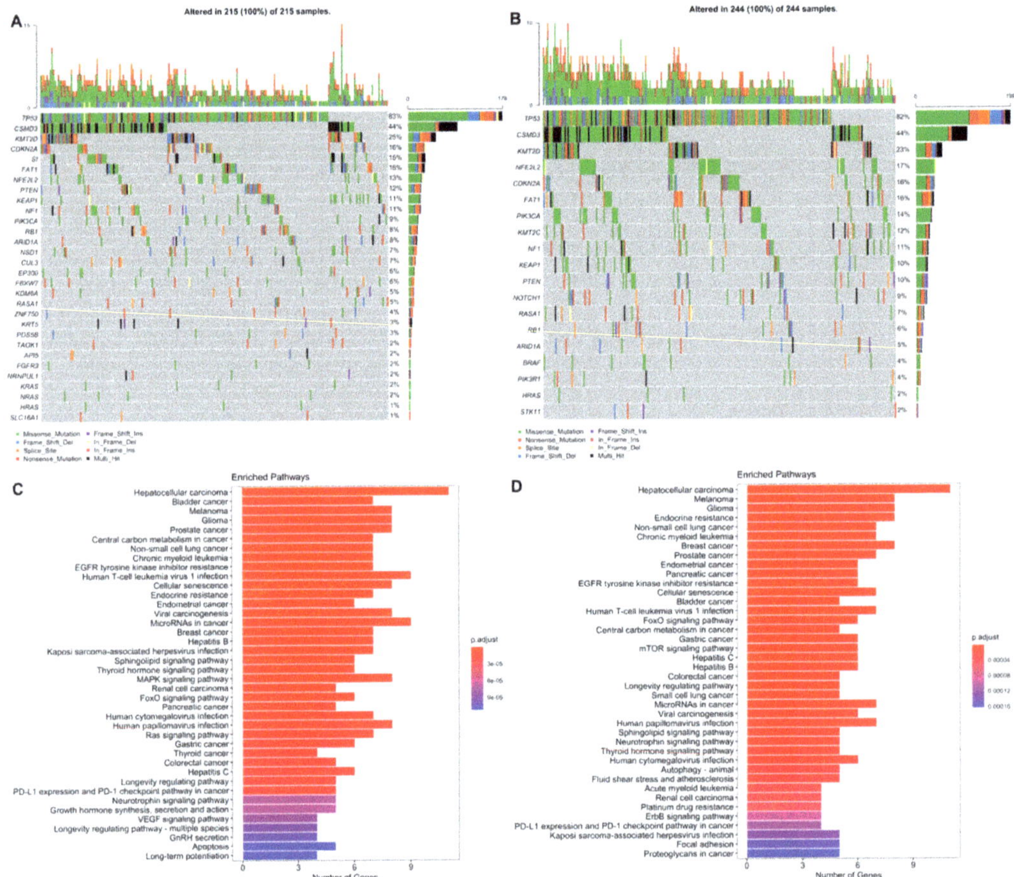

Figure 10. Oncoplot of potential driver genes containing significant SNVs of the LUSC risk groups. (**A**) Oncoplot showing significant SNV genes in tumor samples of the LUSC low-risk group patients. (**B**) Oncoplot showing significant SNV genes in tumor samples of the LUSC high-risk group patients. (**C**) Pathway enrichment of the significant SNV genes of the LUSC low-risk group. (**D**) Pathway enrichment of the significant SNV genes of the LUSC high-risk group.

When venn diagram is drawn by using all driver genes, all cancer and risk groups have TP53, CSMD3, KEAP1, NF1, RB1 and PIK3CA mutations. KRAS, STK11, BRAF, ARID1A, NFE2L2 and RASA1 genes are shared by 3 different groups. LUAD high-risk group has only IDH1 oncogene as different from LUAD low-risk group while LUSC high-risk group has KMT2C, NOTCH1 and PIK3R1 tumor suppressor genes as different from LUSC low-risk group. EGFR, MGA and SMARCA4 are not driver genes in LUSC while CDKN2A, PTEN, HRAS and FAT1 are not driver genes in LUAD groups (Figure 11).

Significant SNVs and CNVs on driver genes are co-displayed as OncoPrint. Although there exist some genes with both SNVs and significant CNVs while others have only SNVs. Moreover, some patients have only SNVs or only CNVs or both for a particular driver gene.

TP53, STK11, KEAP1, SMARCA4 and MGA genes have deletions while CSMD3 and PIK3CA genes have amplification beside SNVs in both LUAD risk group. KRAS and EGFR genes have amplification in the high-risk group; however, they do not have significant CNVs in the low-risk group. Oncogenes tend to have amplifications while tumor suppressor genes tend to have deletions in both risk groups with exceptions (CSMD3, CDH10, HMCN1, AKAP6 and CTNNB1) (Figure 12).

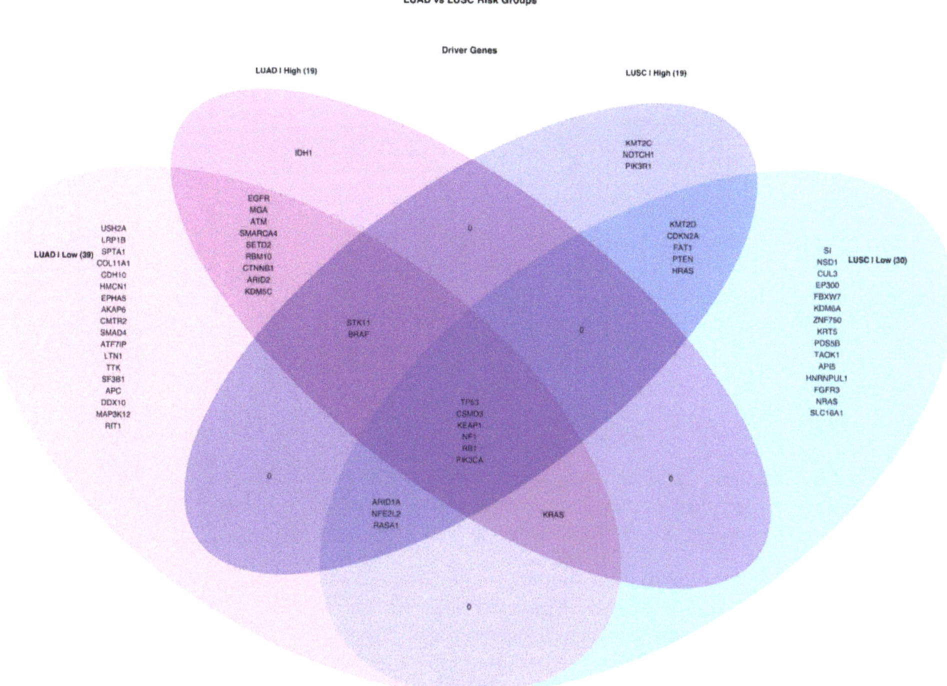

Figure 11. Venn diagram of driver genes containing Simple Nucleotide Variation (SNV) in tumor samples of LUAD and LUSC risk groups.

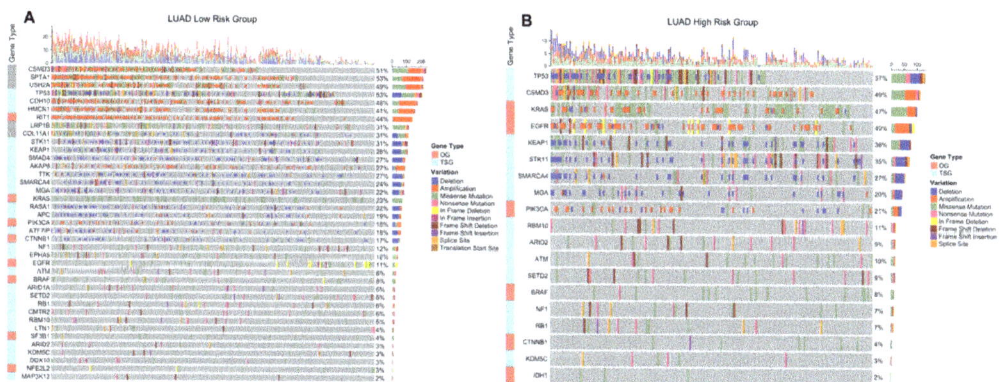

Figure 12. OncoPrint of the driver genes containing significant SNVs and CNVs in LUAD risk groups. Significant SNVs and CNVs are plotted together on potential driver genes in tumor samples of the LUAD risk groups. (**A**) OncoPrint of the driver genes in LUAD low-risk group. (**B**) OncoPrint of the driver genes in LUAD high-risk group.

OncoPrints in Figure 13 show that TP53, CDKN2A, FAT1, RASA1, ARID1A and HRAS genes have deletions while only PIK3CA gene has amplification beside SNVs in both LUSC risk groups. PIK3R1, KEAP1 and STK11 genes have deletions only in the high-risk group while SI, CSMD3, ZNF750, KRAS genes have amplification and NSD1, FGFR3, PTEN, SLC16A1, NRAS and CUL3 have deletion only in the low-risk group. Oncogenes tend to have amplifications while tumor suppressor genes tend to have deletions in both risk groups with exceptions (CSMD3, FGFR3, ZNF750, NRAS, HRAS, KEAP1) (Figure 13).

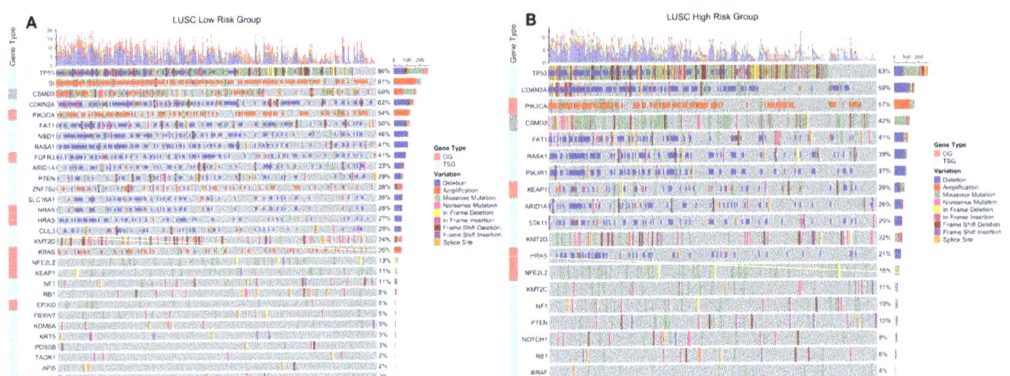

Figure 13. OncoPrint of the driver genes containing significant SNVs and CNVs in LUSC risk groups. Significant SNVs and CNVs are plotted together on potential driver genes in tumor samples of the LUSC risk groups. (**A**) OncoPrint of the driver genes in LUSC low-risk group. (**B**) OncoPrint of the driver genes in LUSC high-risk group.

Circos plots showing all non-synonymous SNVs in original data of risk groups and significant CNVs at genomic scale on chromosomes were drawn to show the genomic alterations between risk groups of LUAD and LUSC.

LUAD low-risk group has more genome-wide CNVs and SNVs than the high-risk group. The low-risk group has more genomics regions containing missense, nonsense and frame-shift insertions/deletions mutations. Moreover, low-risk group has extra deletions on chromosomes 1, 3, 5, 6, 12, 15 and X with extra amplifications on chromosomes 6, 10, 14, and 20. The high-risk group has extra amplifications on chromosomes 7, 11, 12, and 17. The CNVs of high-risk group are localized mostly on 1, 3, 5, 6, 7, 8 and 17 whereas low-risk group has CNVs on more chromosomes (Figure 14).

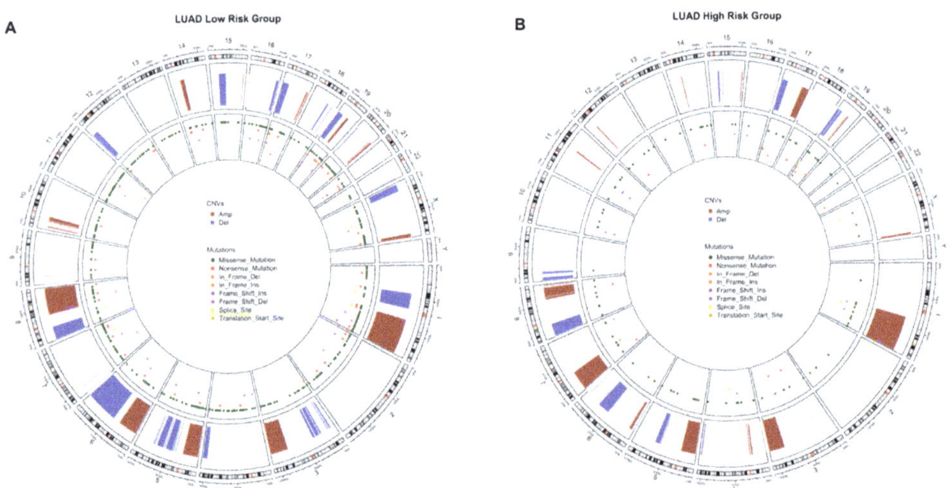

Figure 14. Circos plot of chromosome regions containing all SNVs and CNVs in LUAD risk groups. Significant CNVs ($q < 0.01$) and all SNVs in original data are plotted together on chromosome regions in tumor samples of the LUAD risk groups. (**A**) Circos plot of the LUAD low-risk group. (**B**) Circos plot of the LUAD high-risk group.

LUSC high-risk group has more genomic regions containing missense and nonsense mutations than the low-risk group. However, they have similar amount of CNVs although with different localizations. The high-risk group has extra amplifications on chromosomes 4, 6 and 11; has extra deletions on chromosomes 15, 19 and X. The low-risk group has only extra deletions on chromosomes 1, 5, 6, 11 and 16 (Figure 15).

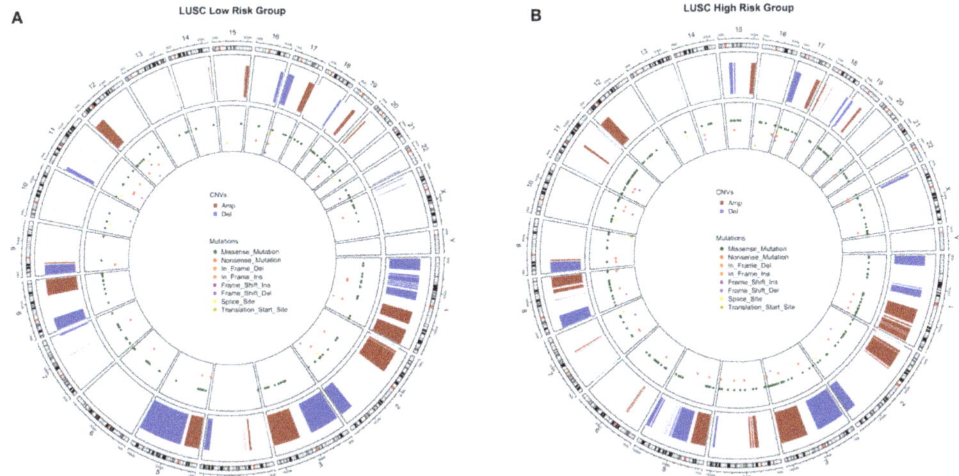

Figure 15. Circos plot of chromosome regions containing all SNVs and CNVs in LUSC risk groups. Significant CNVs ($q < 0.01$) and all SNVs in original data are plotted together on chromosome regions in tumor samples of the LUSC risk groups. (**A**) Circos plot of the LUSC low-risk group. (**B**) Circos plot of the LUSC high-risk group.

4. Discussion

In order to profile the genetic differences between risk groups of LUAD and LUSC, gene expression signatures were generated and the patients were clustered into low- and high-risk groups and then significant DEGs, DEGs at active subnetworks, CNVs and SNVs were identified in each risk group. The biological alterations for these data types were compared between risk groups and between lung cancer subtypes.

Expression signature for LUAD consists of 35 gene which 27 of are protein-coding genes while two are long intergenic non-protein coding RNA, one is antisense RNA, three are pseudogenes and two are novel transcripts. Many of the coding genes are lung cancer or other cancer types related such as ADAMTS15 [44], ASB2 [45] and EPHX1 [46] with potential tumor suppressor roles; ANGPTL4 [47], ASCL2 [48], CCL20 [49], DKK1 [50], GRIK2 [51], LDHA [52], RGS20 [53], RHOQ [54], TLE1 [55] and WBP2 [56] with potential oncogenic roles; and CD200 [57], CD200R1 [57], CCDC181 [58], GNPNAT1 [59], IRX2 [60], LDLRAD3 [61], STAP1 [62], LINC00578 [63] with prognostic potential. Moreover, MS4A1 is dysregulated in asbestos-related lung squamous carcinoma [64], RAB9B is a target of miR-15/16 which are highly related to lung cancer [65], LINC00539 is related to tumor immune response [66] while long non-coding RNA, OGFRP1, regulates non-small-cell lung cancer progression [67]. The remaining signature genes, CPXM2, ENPP5, SAMD13, SLC52A1, ZNF682, ZNF835, ZNF571-AS1 and U91328.1, have not been related to carcinoma, yet. However, they showed highly prognostic power through risk score to distinguish low- and high-risk of overall survival in LUAD.

LUSC gene expression signature including 33 genes of which ALDH7A1 [68], ALK [69], EDN1 [70], FABP6 [71], HKDC1 [72], IGSF1 [73], KBTBD11 [74], NOS1 [75], SLC9A9 [76], STK24 [77], UBB [78], ZNF703 [79] have been shown with oncogenic relations while RGMA [80] is candidate tumor suppressors. ITIH3 [81] and S100A5 [82] has been related to prognostic biomarker potentials. Other cancer-related genes are ADAMTS17 [83],

LINC01748 [84], LPAL2 [85], SRP14-AS1 [86] and WASH8P [87]. Long intergenic non-protein coding RNA, LINC01426, promotes cancer progression via AZGP1 and predicts poor prognosis in patients with LUAD [88]. COL28A1 has prognostic values in glioblastoma [89]. Many of the genes such as JHY, PLAAT1, PNMA8B, RPL37P6, SNX32, UGGT2 and Y_RNA have not been related to any cancer, yet.

Gene expression signatures of LUAD and LUSC share eight pathways which are mostly metabolic pathways. LUAD signature plays role in immune-related pathways as different from those in LUSC. However, pathway enrichment shows us that risk prediction works on metabolic pathways, therefore if we put a name to important mutations as driver mutations, in this case we can say that reprogramming of energy metabolism is the alternative fuel of the cancer [90–92]. The differential expression on them with immune system effect in count can hold the passage of cancer.

High-risk groups of both LUAD and LUSC have more immune pathways including downregulated genes and metabolic pathways including upregulated genes. On the other hand, low-risk groups have both upregulated and downregulated genes on cancer-related pathways. Although LUAD and LUSC seem to have similar characteristics of risk groups, close signature gene pathways and similar differential expression pathways sharing 2106 DEGs in total, they are displayed separately in PCA, especially at analysis of test groups.

At CNV level both risk groups and cancer subtypes have huge number of genes with amplifications or deletions which can cause genomic instability and uncontrolled regulation. Both LUAD and LUSC risk groups have important gene alterations such as CDKN2A and CDKN2B deletions which are associated with NSCLC [93] and promotes KRAS and EGFR mutant tumorigenesis [94,95] while SOX2 oncogene amplification in LUSC which is a common event in squamous cell carcinomas [96,97] and amplification of PSMD4 in LUAD, with oncogenic roles in breast, hepatocellular, colorectal and prostate cancer cells [98–101]. CNVs also play role in metabolic and immune-related pathways which can differ between risk groups and cancer subtypes. If we look from a higher perspective, the LUAD low-risk group has much more CNVs and SNVs on its genome than the high-risk group. On the other hand, the LUSC high-risk group has more SNVs than the low-risk group while CNVs do not vary too much.

SNV analysis gives similar results with literature for example EGFR and KRAS mutations are mutually exclusive in LUAD samples that is confirmed again [9]. Additionally, EGFR [102], MGA [103], SMARCA4 [104], ATM [105], RBM10 [106] and KDM5C [107] which are lung cancer related genes are mutated only in LUAD but not in LUSC. On the other hand, CDKN2A [108], PTEN [109] and HRAS [110] genes are mutated only in LUSC. In general, low-risk groups have more mutated genes for both LUAD and LUSC samples. When SNV and CNV genes are plotted together, it can be seen that LUAD high-risk group has obvious oncogene amplifications and tumor suppressor deletions, while LUAD low-risk group has both tumor suppressor deletions and tumor suppressor amplifications with a few oncogene amplifications. This SNV and copy number differential pattern can cause differential gene expression profiles and characteristics of tumor. LUSC patients have mostly deletions on driver genes with only PIK3CA [111] and KRAS [111] oncogene amplifications. Both LUSC risk groups have obvious TP53 [111] and CDKN2A tumor suppressor gene deletions, but amplification of CSMD3, which has differential roles in lung cancer [112,113], does not occur in LUSC high-risk group. Again, only these driver genes which have differential alterations and frequencies can create the risk difference based on gene expression levels.

5. Conclusions

This study has been performed to profile the genomic and transcriptomic differences not only between LUAD and LUSC but also between risk groups to understand the driving differences between them. Treatment options can vary between cancer subtypes and risk groups because of differential targetable mutation patterns. Nowadays, many groups and government institutions are working on the integration of the drug bioactivity and molecular data to investigate more effective molecularly targeting therapeutics for individual patients for the personalized therapy.

Supplementary Materials: The supplementary data are available online at https://www.mdpi.com/2075-4426/11/2/154/s1; Figure S1: Flowchart of method and used R packages in this study. The other R packages not written in this flowchart can be found at Materials and Method part of the article; Figure S2: Gene expression signature and risk clustering of LUAD training dataset; Figure S3: Survival analysis of risk groups clustered by using signature gene expression at different tumor stages in LUAD training dataset; Figure S4: Mosaic plots showing association analysis of categorical variables for LUAD training dataset. Pearson residuals show the positive (blue) or negative (red) association between levels of categories; Figure S5: Multivariate Cox Regression results of clinical variables and risk score in LUAD training dataset. Only risk score has significant result when all clinical variables are included into multivariate analysis; Figure S6: Multivariate Cox Regression results of selected clinical variables (which have significant results in univariate Cox analysis) and risk score in LUAD training dataset. Risk score, t, n, m stages and history of prior malignancy have significant effects on survival. When pathologic tumor stage is used instead of t, n, m stages, only risk score and history of prior malignancy show significant effect on survival; Figure S7: Survival analysis of risk groups clustered by using signature gene expression at different tumor stages in LUAD test dataset; Figure S8: Mosaic plots showing association analysis of categorical variables for LUAD test dataset; Figure S9: Multivariate Cox Regression results of selected clinical variables (which have significant results in univariate Cox analysis) and risk score in LUAD test dataset. Risk score and n stages have significant effect on survival. When pathologic tumor stage is used instead of t, n, m stages, only risk score shows significant effect on survival; Figure S10: Gene expression signature and risk clustering of LUSC training dataset; Figure S11: Survival analysis of risk groups clustered by using signature gene expression at different tumor stages in LUSC training dataset; Figure S12: Mosaic plots showing association analysis of categorical variables for LUSC training dataset. Pearson residuals show the positive (blue) or negative (red) association between levels of categories; Figure S13: Multivariate Cox Regression results of selected clinical variables (which have significant results in univariate Cox analysis) and risk score in LUSC training dataset. Risk score, tissue or organ of origin, t and n stages and history of prior malignancy have significant effects on survival. When pathologic tumor stage is used instead of t, n, m stages, tissue or organ of origin, risk score and history of prior malignancy show significant effect on survival; Figure S14: Survival analysis of risk groups clustered by using signature gene expression at different tumor stages in LUSC test dataset; Figure S15: Mosaic plots showing association analysis of categorical variables for LUSC test dataset. Pearson residuals show the positive (blue) or negative (red) association between levels of categories; Figure S16: Multivariate Cox Regression results of selected clinical variables (which have significant results in univariate Cox analysis) and risk score in LUSC test dataset. Only risk score has significant effect on survival either t, n, m stages or pathologic tumor stage is used instead of t, n, m stages; Figure S17. Venn diagram of differentially expressed genes in tumor samples of risk groups for LUAD and LUSC test groups; Figure S18: Pathway enrichment of DEGs of LUAD risk groups; Figure S19: Pathway enrichment of DEGs of LUSC risk groups; Figure S20: Pathway enrichment of CNV genes of LUAD risk groups; Figure S21: Pathway enrichment of CNV genes of LUSC risk groups; Figure S22: Venn diagram of genes which have significant copy number alterations in tumor samples of LUAD and LUSC risk groups; Figure S23: Summary of SNVs in LUAD risk groups; Figure S24: Summary of SNVs in LUSC risk groups; Figure S25: SomInaClust result of potential driver genes containing significant SNVs in LUAD risk groups. SomInaClust calculates oncogene (OG) score and tumor suppressor gene (TSG) score for each significant gene and classifies the gene according to the score threshold (20) and reference database; Figure S26: SomInaClust result of potential driver genes containing significant SNVs in LUSC risk groups. SomInaClust calculates oncogene (OG) score and tumor suppressor gene (TSG) score for each significant gene and classifies the gene according to the score threshold (20) and reference database; Figure S27: Venn diagram of all genes and potential driver genes containing SNVs of LUAD and LUSC risk groups, Table S1: Gene list of expression signature in LUAD. Ensemble Gene IDs were used in signature analysis and then enriched by using BioMart database; Table S2: KEGG pathway enrichment of expression signature gene list in LUAD by using KEGG Mapper tool; Table S3: Gene list of expression signature in LUSC. Ensemble Gene IDs were used in signature analysis and then enriched by using BioMart database; Table S4: KEGG pathway enrichment of expression signature gene list in LUSC by using *clusterProfiler*

R package; Table S5: SomInaClust result of SNV data in tumor samples of LUAD low-risk group; Table S6: SomInaClust result of SNV data in tumor samples of LUAD high-risk group; Table S7: SomInaClust result of SNV data in tumor samples of LUSC low-risk group; Table S8: SomInaClust result of SNV data in tumor samples of LUSC high-risk group.

Author Contributions: Methodology, T.Z.; formal analysis, T.Z.; resources, T.Z., T.Ö.-S.; data curation, T.Z.; writing—original draft preparation, T.Z.; writing—review and editing, T.Ö.-S.; visualization, T.Z.; project administration, T.Ö.-S. All authors have read and agreed to the published version of the manuscript.

Funding: T.Z. and T.Ö.-S. were partially funded by Turkish National Institutes of Health (TÜSEB) grant number 4583.

Informed Consent Statement: Not applicable.

Data Availability Statement: The datasets supporting the conclusions of this article are publicly available and can be downloaded from TCGA data portal (https://portal.gdc.cancer.gov) or by using *TCGAbiolinks* R package [18]. The R code used in this study is available upon request.

Conflicts of Interest: The authors declare no conflict of interest.

References

1. GLOBOCAN 2020: Cancer Today. Available online: https://gco.iarc.fr/today/home (accessed on 29 December 2020).
2. Alexander, M.; Kim, S.Y.; Cheng, H. Update 2020: Management of Non-Small Cell Lung Cancer. *Lung* **2020**, *198*, 897–907. [CrossRef] [PubMed]
3. Chansky, K.; Detterbeck, F.C.; Nicholson, A.G.; Rusch, V.W.; Vallières, E.; Groome, P.; Kennedy, C.; Krasnik, M.; Peake, M.; Shemanski, L.; et al. The IASLC Lung Cancer Staging Project: External Validation of the Revision of the TNM Stage Groupings in the Eighth Edition of the TNM Classification of Lung Cancer. *J. Thorac. Oncol.* **2017**, *12*, 1109–1121. [CrossRef]
4. Camidge, D.R.; Doebele, R.C.; Kerr, K.M. Comparing and contrasting predictive biomarkers for immunotherapy and targeted therapy of NSCLC. *Nat. Rev. Clin. Oncol.* **2019**, *16*, 341–355. [CrossRef] [PubMed]
5. Wang, B.-Y.; Huang, J.-Y.; Chen, H.-C.; Lin, C.-H.; Lin, S.-H.; Hung, W.-H.; Cheng, Y.-F. The comparison between adenocarcinoma and squamous cell carcinoma in lung cancer patients. *J. Cancer Res. Clin. Oncol.* **2019**, *146*, 43–52. [CrossRef] [PubMed]
6. Travis, W.D. Lung Cancer Pathology. *Clin. Chest Med.* **2020**, *41*, 67–85. [CrossRef] [PubMed]
7. Zhang, J.; Fujimoto, J.; Wedge, D.C.; Song, X.; Seth, S.; Chow, C.-W.; Cao, Y.; Gumbs, C.; Gold, K.A.; Kalhor, N.; et al. Intratumor heterogeneity in localized lung adenocarcinomas delineated by multiregion sequencing. *Science* **2014**, *346*, 256–259. [CrossRef] [PubMed]
8. The Cancer Genome Atlas Research Network; Weinstein, J.N.; Collisson, E.A.; Mills, G.B.; Shaw, K.R.M.; Ozenberger, B.A.; Ellrott, K.; Shmulevich, I.; Sander, C.; Stuart, J.M. The Cancer Genome Atlas Pan-Cancer analysis project. *Nat. Genet.* **2013**, *45*, 1113–1120. [CrossRef] [PubMed]
9. The Cancer Genome Atlas Research Network Comprehensive molecular profiling of lung adenocarcinoma. *Nat. Cell Biol.* **2014**, *511*, 543–550. [CrossRef]
10. The Cancer Genome Atlas Research Network Comprehensive genomic characterization of squamous cell lung cancers. *Nat. Cell Biol.* **2012**, *489*, 519–525. [CrossRef]
11. Krzystanek, M.; Moldvay, J.; Szüts, D.; Szallasi, Z.; Eklund, A.C. A robust prognostic gene expression signature for early stage lung adenocarcinoma. *Biomark. Res.* **2016**, *4*, 1–7. [CrossRef] [PubMed]
12. Shukla, S.; Evans, J.R.; Malik, R.; Feng, F.Y.; Dhanasekaran, S.M.; Cao, X.; Chen, G.; Beer, D.G.; Jiang, H.; Chinnaiyan, A.M. Development of a RNA-Seq Based Prognostic Signature in Lung Adenocarcinoma. *J. Natl. Cancer Inst.* **2017**, *109*, 200. [CrossRef] [PubMed]
13. Li, Z.; Qi, F.; Li, F. Establishment of a Gene Signature to Predict Prognosis for Patients with Lung Adenocarcinoma. *Int. J. Mol. Sci.* **2020**, *21*, 8479. [CrossRef] [PubMed]
14. Zhu, C.-Q.; Strumpf, D.; Li, C.-Y.; Li, Q.; Liu, N.; Der, S.; Shepherd, F.A.; Tsao, M.-S.; Jurisica, I. Prognostic Gene Expression Signature for Squamous Cell Carcinoma of Lung. *Clin. Cancer Res.* **2010**, *16*, 5038–5047. [CrossRef] [PubMed]
15. Li, J.; Wang, J.; Chen, Y.; Yang, L.; Chen, S. A prognostic 4-gene expression signature for squamous cell lung carcinoma. *J. Cell. Physiol.* **2017**, *232*, 3702–3713. [CrossRef] [PubMed]
16. Lu, C.; Chen, H.; Shan, Z.; Yang, L. Identification of differentially expressed genes between lung adenocarcinoma and lung squamous cell carcinoma by gene expression profiling. *Mol. Med. Rep.* **2016**, *14*, 1483–1490. [CrossRef]
17. Wu, X.; Wang, L.; Feng, F.; Tian, S. Weighted gene expression profiles identify diagnostic and prognostic genes for lung adenocarcinoma and squamous cell carcinoma. *J. Int. Med. Res.* **2020**, *48*, 0300060519893837. [CrossRef]
18. Colaprico, A.; Silva, T.C.; Olsen, C.; Garofano, L.; Cava, C.; Garolini, D.; Sabedot, T.S.; Malta, T.M.; Pagnotta, S.M.; Castiglioni, I.; et al. TCGAbiolinks: An R/Bioconductor package for integrative analysis of TCGA data. *Nucleic Acids Res.* **2016**, *44*, e71. [CrossRef] [PubMed]

19. Therneau, T. *A Package for Survival Analysis in R. R Package Version 3.2-7.* 2020. Available online: https://cran.r-project.org/package=survival (accessed on 21 May 2020).
20. Simon, N.; Friedman, J.H.; Hastie, T.; Tibshirani, R. Regularization Paths for Cox's Proportional Hazards Model via Coordinate Descent. *J. Stat. Softw.* **2011**, *39*, 1–13. [CrossRef]
21. Gerds, T.A.; Ozenne, B. *RiskRegression: Risk Regression Models and Prediction Scores for Survival Analysis with Competing Risks. R Package Version 2020.12.08.* 2020. Available online: https://cran.r-project.org/package=riskRegression (accessed on 21 May 2020).
22. Zhang, J.; Jin, Z. *Cutoff: Seek the Significant Cutoff Value. R Package Version 1.3.* 2019. Available online: https://cran.r-project.org/package=cutoff (accessed on 21 May 2020).
23. Ritchie, M.E.; Phipson, B.; Wu, D.; Hu, Y.; Law, C.W.; Shi, W.; Smyth, G.K. Limma powers differential expression analyses for RNA-sequencing and microarray studies. *Nucleic Acids Res.* **2015**, *43*, e47. [CrossRef] [PubMed]
24. McCarthy, D.J.; Chen, Y.; Smyth, G.K. Differential expression analysis of multifactor RNA-Seq experiments with respect to biological variation. *Nucleic Acids Res.* **2012**, *40*, 4288–4297. [CrossRef]
25. Vrahatis, A.G.; Balomenos, P.; Tsakalidis, A.K.; Bezerianos, A. DEsubs: An R package for flexible identification of differentially expressed subpathways using RNA-seq experiments. *Bioinformatics* **2016**, *32*, 3844–3846. [CrossRef]
26. Morganella, S.; Pagnotta, S.M.; Ceccarelli, M. *GAIA: An R Package for Genomic Analysis of Significant Chromosomal Aberrations. R Package Version 2.32.0.* 2020. Available online: https://bioconductor.org/packages/gaia (accessed on 21 May 2020).
27. Lawrence, M.; Huber, W.; Pagès, H.; Aboyoun, P.; Carlson, M.; Gentleman, R.; Morgan, M.T.; Carey, V.J. Software for Computing and Annotating Genomic Ranges. *PLoS Comput. Biol.* **2013**, *9*, e1003118. [CrossRef]
28. Silva, T.C.; Colaprico, A.; Olsen, C.; D'Angelo, F.; Bontempi, G.; Ceccarelli, M.; Noushmehr, H. TCGA Workflow: Analyze cancer genomics and epigenomics data using Bioconductor packages. *F1000Research* **2016**, *5*, 1542. [CrossRef]
29. Mayakonda, A.; Lin, D.-C.; Assenov, Y.; Plass, C.; Koeffler, H.P. Maftools: Efficient and comprehensive analysis of somatic variants in cancer. *Genome Res.* **2018**, *28*, 1747–1756. [CrossRef] [PubMed]
30. Eynden, J.V.D.; Fierro, A.C.; Verbeke, L.P.C.; Marchal, K. SomInaClust: Detection of cancer genes based on somatic mutation patterns of inactivation and clustering. *BMC Bioinform.* **2015**, *16*, 1–12. [CrossRef] [PubMed]
31. Tate, J.G.; Bamford, S.; Jubb, H.C.; Sondka, Z.; Beare, D.M.; Bindal, N.; Boutselakis, H.; Cole, C.G.; Creatore, C.; Dawson, E.; et al. COSMIC: The Catalogue Of Somatic Mutations In Cancer. *Nucleic Acids Res.* **2018**, *47*, D941–D947. [CrossRef]
32. Zhang, J.; Jin, Z. *Ggrisk: Risk Score Plot for Cox Regression. R Package Version 1.2.* 2020. Available online: https://cran.r-project.org/package=ggrisk (accessed on 21 May 2020).
33. Kassambara, A.; Kosinski, M.; Biecek, P. *Survminer: Drawing Survival Curves Using "ggplot2". R Package Version 0.4.8.* 2020. Available online: https://cran.r-project.org/package=survminer (accessed on 21 May 2020).
34. Heagerty, P.J.; Saha-Chaudhuri, P. *survivalROC: Time-Dependent ROC Curve Estimation from Censored Survival Data. R Package Version 1.0.3.* 2013. Available online: https://cran.r-project.org/package=survivalROC (accessed on 21 May 2020).
35. Kennedy, N. *Forestmodel: Forest Plots from Regression Models. R Package Version 0.6.2.* 2020. Available online: https://cran.r-project.org/package=forestmodel (accessed on 21 May 2020).
36. Durinck, S.; Spellman, P.T.; Birney, E.; Huber, W. Mapping identifiers for the integration of genomic datasets with the R/Bioconductor package biomaRt. *Nat. Protoc.* **2009**, *4*, 1184–1191. [CrossRef]
37. Yu, G.; Wang, L.-G.; Han, Y.; He, Q.-Y. clusterProfiler: An R Package for Comparing Biological Themes Among Gene Clusters. *OMICS J. Integr. Biol.* **2012**, *16*, 284–287. [CrossRef]
38. Yu, G. *Enrichplot: Visualization of Functional Enrichment Result. R Package Version 1.8.1.* 2020. Available online: https://github.com/GuangchuangYu/enrichplot (accessed on 21 May 2020).
39. Gu, Z.; Eils, R.; Schlesner, M. Complex heatmaps reveal patterns and correlations in multidimensional genomic data. *Bioinformatics* **2016**, *32*, 2847–2849. [CrossRef] [PubMed]
40. Meyer, D.; Zeileis, A.; Hornik, K. *Vcd: Visualizing Categorical Data. R Package Version 1.4-8.* 2020. Available online: https://cran.r-project.org/package=vcd (accessed on 21 May 2020).
41. Meyer, D.; Zeileis, A.; Hornik, K. The Strucplot Framework: Visualizing Multi-way Contingency Tables with vcd. *J. Stat. Softw.* **2006**, *17*, 1–48. [CrossRef]
42. Gu, Z.; Gu, L.; Eils, R.; Schlesner, M.; Brors, B. circlize implements and enhances circular visualization in R. *Bioinformatics* **2014**, *30*, 2811–2812. [CrossRef] [PubMed]
43. Chen, H. *VennDiagram: Generate High-Resolution Venn and Euler Plots. R Package Version 1.6.20.* 2018. Available online: https://cran.r-project.org/package=VennDiagram (accessed on 21 May 2020).
44. Kumar, S.; Rao, N.; Ge, R. Emerging Roles of ADAMTSs in Angiogenesis and Cancer. *Cancers* **2012**, *4*, 1252. [CrossRef]
45. Li, Z.; Weng, H.; Su, R.; Weng, X.; Zuo, Z.; Li, C.; Huang, H.; Nachtergaele, S.; Dong, L.; Hu, C.; et al. FTO Plays an Oncogenic Role in Acute Myeloid Leukemia as a N 6 -Methyladenosine RNA Demethylase. *Cancer Cell* **2017**, *31*, 127–141. [CrossRef]
46. Li, X.; Hu, Z.; Qu, X.; Zhu, J.; Li, L.; Ring, B.Z.; Su, L. Putative EPHX1 Enzyme Activity Is Related with Risk of Lung and Upper Aerodigestive Tract Cancers: A Comprehensive Meta-Analysis. *PLoS ONE* **2011**, *6*, e14749. [CrossRef]
47. Zhu, X.; Guo, X.; Wu, S.; Wei, L. ANGPTL4 Correlates with NSCLC Progression and Regulates Epithelial-Mesenchymal Transition via ERK Pathway. *Lung* **2016**, *194*, 637–646. [CrossRef] [PubMed]
48. Hu, X.-G.; Chen, L.; Wang, Q.-L.; Zhao, X.-L.; Tan, J.; Cui, Y.-H.; Liu, X.-D.; Zhang, X.; Bian, X.-W. Elevated expression of ASCL2 is an independent prognostic indicator in lung squamous cell carcinoma. *J. Clin. Pathol.* **2015**, *69*, 313–318. [CrossRef]

49. Kadomoto, S.; Izumi, K.; Mizokami, A. The CCL20-CCR6 Axis in Cancer Progression. *Int. J. Mol. Sci.* **2020**, *21*, 5186. [CrossRef]
50. Zhang, J.; Zhang, X.; Zhao, X.; Jiang, M.; Gu, M.; Wang, Z.; Yue, W. DKK1 promotes migration and invasion of non–small cell lung cancer via β-catenin signaling pathway. *Tumor Biol.* **2017**, *39*. [CrossRef] [PubMed]
51. Inoue, R.; Hirohashi, Y.; Kitamura, H.; Nishida, S.; Murai, A.; Takaya, A.; Yamamoto, E.; Matsuki, M.; Tanaka, T.; Kubo, T.; et al. GRIK2 has a role in the maintenance of urothelial carcinoma stem-like cells, and its expression is associated with poorer prognosis. *Oncotarget* **2017**, *8*, 28826–28839. [CrossRef] [PubMed]
52. Yu, C.; Hou, L.; Cui, H.; Zhang, L.; Tan, X.; Leng, X.; Li, Y. LDHA upregulation independently predicts poor survival in lung adenocarcinoma, but not in lung squamous cell carcinoma. *Futur. Oncol.* **2018**, *14*, 2483–2492. [CrossRef]
53. Yang, L.; Lee, M.M.; Leung, M.M.; Wong, Y.H. Regulator of G protein signaling 20 enhances cancer cell aggregation, migration, invasion and adhesion. *Cell. Signal.* **2016**, *28*, 1663–1672. [CrossRef]
54. Han, S.-W.; Kim, H.-P.; Shin, J.-Y.; Jeong, E.-G.; Lee, W.-C.; Kim, K.Y.; Park, S.Y.; Lee, D.-W.; Won, J.-K.; Jeong, S.-Y.; et al. RNA editing in RHOQ promotes invasion potential in colorectal cancer. *J. Exp. Med.* **2014**, *211*, 613–621. [CrossRef]
55. Yuan, D.; Yang, X.; Yuan, Z.; Zhao, Y.; Guo, J. TLE1 function and therapeutic potential in cancer. *Oncotarget* **2016**, *8*, 15971–15976. [CrossRef]
56. Tabatabaeian, H.; Rao, A.; Ramos, A.; Chu, T.; Sudol, M.; Lim, Y.P. The emerging roles of WBP2 oncogene in human cancers. *Oncogene* **2020**, *39*, 4621–4635. [CrossRef] [PubMed]
57. Yoshimura, K.; Suzuki, Y.; Inoue, Y.; Tsuchiya, K.; Karayama, M.; Iwashita, Y.; Kahyo, T.; Kawase, A.; Tanahashi, M.; Ogawa, H.; et al. CD200 and CD200R1 are differentially expressed and have differential prognostic roles in non-small cell lung cancer. *OncoImmunology* **2020**, *9*, 1746554. [CrossRef]
58. Gao, C.; Zhuang, J.; Li, H.; Liu, C.; Zhou, C.; Liu, L.; Sun, C. Exploration of methylation-driven genes for monitoring and prognosis of patients with lung adenocarcinoma. *Cancer Cell Int.* **2018**, *18*, 1–11. [CrossRef] [PubMed]
59. Zheng, X.; Li, Y.; Ma, C.; Zhang, J.; Zhang, Y.; Fu, Z.; Luo, H. Independent Prognostic Potential of GNPNAT1 in Lung Adenocarcinoma. *BioMed Res. Int.* **2020**, *2020*, 1–16. [CrossRef] [PubMed]
60. Wang, Q.; Qiu, X. Comprehensive Analysis of the Expression and Prognosis for IRXs in Non-small Cell Lung Cancer. *Res. Sq.* **2020**. [CrossRef]
61. Puderecki, M.; Szumiło, J.; Marzec-Kotarska, B. Novel prognostic molecular markers in lung cancer (Review). *Oncol. Lett.* **2020**, *20*, 9–18. [CrossRef]
62. Zhao, R.; Ding, D.; Yu, W.; Zhu, C.; Ding, Y. The Lung Adenocarcinoma Microenvironment Mining and Its Prognostic Merit. *Technol. Cancer Res. Treat.* **2020**, *19*. [CrossRef]
63. Wang, L.; Zhao, H.; Xu, Y.; Li, J.; Deng, C.; Deng, Y.; Bai, J.; Li, X.; Xiao, Y.; Zhang, Y. Systematic identification of lincRNA-based prognostic biomarkers by integrating lincRNA expression and copy number variation in lung adenocarcinoma. *Int. J. Cancer* **2019**, *144*, 1723–1734. [CrossRef] [PubMed]
64. Wright, C.M.; Francis, S.M.S.; Tan, M.E.; Martins, M.U.; Winterford, C.; Davidson, M.R.; Duhig, E.E.; Clarke, B.E.; Hayward, N.K.; Yang, I.A.; et al. MS4A1 Dysregulation in Asbestos-Related Lung Squamous Cell Carcinoma Is Due to CD20 Stromal Lymphocyte Expression. *PLoS ONE* **2012**, *7*, e34943. [CrossRef] [PubMed]
65. Qi, J.; Mu, D. MicroRNAs and lung cancers: From pathogenesis to clinical implications. *Front. Med.* **2012**, *6*, 134–155. [CrossRef]
66. Sage, A.P.; Ng, K.W.; Marshall, E.A.; Stewart, G.L.; Minatel, B.C.; Enfield, K.S.S.; Martin, S.D.; Brown, C.J.; Abraham, N.; Lam, W.L. Assessment of long non-coding RNA expression reveals novel mediators of the lung tumour immune response. *Sci. Rep.* **2020**, *10*, 1–13. [CrossRef]
67. Tang, L.-X.; Chen, G.-H.; Li, H.; He, P.; Zhang, Y.; Xu, X.-W. Long non-coding RNA OGFRP1 regulates LYPD3 expression by sponging miR-124-3p and promotes non-small cell lung cancer progression. *Biochem. Biophys. Res. Commun.* **2018**, *505*, 578–585. [CrossRef] [PubMed]
68. Giacalone, N.J.; Den, R.B.; Eisenberg, R.; Chen, H.; Olson, S.J.; Massion, P.P.; Carbone, D.P.; Lu, B. ALDH7A1 expression is associated with recurrence in patients with surgically resected non-small-cell lung carcinoma. *Futur. Oncol.* **2013**, *9*, 737–745. [CrossRef] [PubMed]
69. Wang, J.; Shen, Q.; Shi, Q.; Yu, B.; Wang, X.; Cheng, K.; Lu, G.; Zhou, X. Detection of ALK protein expression in lung squamous cell carcinomas by immunohistochemistry. *J. Exp. Clin. Cancer Res.* **2014**, *33*, 1–7. [CrossRef] [PubMed]
70. Boldrini, L.; Gisfredi, S.; Ursino, S.; Faviana, P.; Lucchi, M.; Melfi, F.; Mussi, A.; Basolo, F.; Fontanini, G. Expression of endothelin-1 is related to poor prognosis in non-small cell lung carcinoma. *Eur. J. Cancer* **2005**, *41*, 2828–2835. [CrossRef] [PubMed]
71. Zhang, Y.; Zhao, X.; Deng, L.; Li, X.; Wang, G.; Li, Y.; Chen, M. High expression of FABP4 and FABP6 in patients with colorectal cancer. *World J. Surg. Oncol.* **2019**, *17*, 1–13. [CrossRef]
72. Wang, X.; Shi, B.; Zhao, Y.; Lu, Q.; Fei, X.; Lu, C.; Li, C.; Chen, H. HKDC1 promotes the tumorigenesis and glycolysis in lung adenocarcinoma via regulating AMPK/mTOR signaling pathway. *Cancer Cell Int.* **2020**, *20*, 1–12. [CrossRef]
73. Guan, Y.; Wang, Y.; Bhandari, A.; Xia, E.; Wang, O. IGSF1: A novel oncogene regulates the thyroid cancer progression. *Cell Biochem. Funct.* **2019**, *37*, 516–524. [CrossRef]
74. Gong, J.; Tian, J.; Lou, J.; Wang, X.; Ke, J.; Li, J.; Yang, Y.; Gong, Y.; Zhu, Y.; Zou, D.; et al. A polymorphic MYC response element in KBTBD11 influences colorectal cancer risk, especially in interaction with an MYC-regulated SNP rs6983267. *Ann. Oncol.* **2017**, *29*, 632–639. [CrossRef]

75. Zou, Z.; Li, X.; Sun, Y.; Li, L.; Zhang, Q.; Zhu, L.; Zhong, Z.; Wang, M.; Wang, Q.; Liu, Z.; et al. NOS1 expression promotes proliferation and invasion and enhances chemoresistance in ovarian cancer. *Oncol. Lett.* **2020**, *19*, 2989–2995. [CrossRef]

76. Ueda, M.; Iguchi, T.; Masuda, T.; Komatsu, H.; Nambara, S.; Sakimura, S.; Hirata, H.; Uchi, R.; Eguchi, H.; Ito, S.; et al. Up-regulation of SLC9A9 Promotes Cancer Progression and Is Involved in Poor Prognosis in Colorectal Cancer. *Anticancer Res.* **2017**, *37*, 2255–2263. [CrossRef]

77. Huang, N.; Lin, W.; Shi, X.; Tao, T. STK24 expression is modulated by DNA copy number/methylation in lung adenocarcinoma and predicts poor survival. *Futur. Oncol.* **2018**, *14*, 2253–2263. [CrossRef]

78. Tang, Y.; Geng, Y.; Luo, J.; Shen, W.; Zhu, W.; Meng, C.; Li, M.; Zhou, X.; Zhang, S.; Cao, J. Downregulation of ubiquitin inhibits the proliferation and radioresistance of non-small cell lung cancer cells in vitro and in vivo. *Sci. Rep.* **2015**, *5*, 1–12. [CrossRef]

79. Baykara, O.; Dalay, N.; Kaynak, K.; Buyru, N. ZNF703 Overexpression may act as an oncogene in non-small cell lung cancer. *Cancer Med.* **2016**, *5*, 2873–2878. [CrossRef]

80. Li, J.; Ye, L.; Mansel, R.E.; Jiang, W.G. Potential prognostic value of repulsive guidance molecules in breast cancer. *Anticancer Res.* **2011**, *31*, 1703–1711. [PubMed]

81. Chong, P.K.; Lee, H.; Zhou, J.; Liu, S.-C.; Loh, M.C.S.; Wang, T.T.; Chan, S.P.; Smoot, D.T.; Ashktorab, H.; So, J.B.Y.; et al. ITIH3 Is a Potential Biomarker for Early Detection of Gastric Cancer. *J. Proteome Res.* **2010**, *9*, 3671–3679. [CrossRef] [PubMed]

82. Liu, Y.; Cui, J.; Tang, Y.-L.; Huang, L.; Zhou, C.-Y.; Xu, J.-X. Prognostic Roles of mRNA Expression of S100 in Non-Small-Cell Lung Cancer. *BioMed Res. Int.* **2018**, *2018*, 1–11. [CrossRef] [PubMed]

83. Jia, Z.; Gao, S.; M'Rabet, N.; De Geyter, C.; Zhang, H. Sp1 Is Necessary for Gene Activation of Adamts17 by Estrogen. *J. Cell. Biochem.* **2014**, *115*, 1829–1839. [CrossRef]

84. Li, R.; Yang, Y.-E.; Jin, J.; Zhang, M.-Y.; Liu, X.-X.; Yin, Y.-H.; Qu, Y.-Q. Identification of lncRNA biomarkers in lung squamous cell carcinoma using comprehensive analysis of lncRNA mediated ceRNA network. *Artif. Cells Nanomed. Biotechnol.* **2019**, *47*, 3246–3258. [CrossRef] [PubMed]

85. Han, B.-W.; Ye, H.; Wei, P.-P.; He, B.; Han, C.; Chen, Z.-H.; Chen, Y.-Q.; Wang, W.-T. Global identification and characterization of lncRNAs that control inflammation in malignant cholangiocytes. *BMC Genom.* **2018**, *19*, 1–13. [CrossRef] [PubMed]

86. Rao, Y.; Liu, H.; Yan, X.; Wang, J. In Silico Analysis Identifies Differently Expressed lncRNAs as Novel Biomarkers for the Prognosis of Thyroid Cancer. *Comput. Math. Methods Med.* **2020**, *2020*, 1–10. [CrossRef] [PubMed]

87. Zhang, W.; Ye, Y.J.; Ren, X.W.; Huang, J.; Shen, Z.L. Detection of preoperative chemoradiotherapy sensitivity molecular characteristics of rectal cancer by transcriptome second generation sequencing. *J. Peking Univ. Health Sci.* **2019**, *51*, 542–547. [CrossRef]

88. Tian, B.; Han, X.; Li, G.; Jiang, H.; Qi, J.; Li, J.; Tian, Y.; Wang, C. A Long Intergenic Non-coding RNA, LINC01426, Promotes Cancer Progression via AZGP1 and Predicts Poor Prognosis in Patients with LUAD. *Mol. Ther. Methods Clin. Dev.* **2020**, *18*, 765–780. [CrossRef]

89. Yang, H.; Jin, L.; Sun, X. A thirteen-gene set efficiently predicts the prognosis of glioblastoma. *Mol. Med. Rep.* **2019**, *19*, 1613–1621. [CrossRef]

90. Hanahan, D.; Weinberg, R.A. Hallmarks of Cancer: The Next Generation. *Cell* **2011**, *144*, 646–674. [CrossRef] [PubMed]

91. Phan, L.M.; Yeung, S.J.; Lee, M.-H. Cancer metabolic reprogramming: Importance, main features, and potentials for precise targeted anti-cancer therapies. *Cancer Biol. Med.* **2014**, *11*, 1–19.

92. Keenan, M.M.; Chi, J.-T. Alternative Fuels for Cancer Cells. *Cancer J.* **2015**, *21*, 49–55. [CrossRef] [PubMed]

93. Hamada, K.; Kohno, T.; Kawanishi, M.; Ohwada, S.; Yokota, J. Association of CDKN2A (p16)/CDKN2B (p15) alterations and homozygous chromosome arm 9p deletions in human lung carcinoma. Genes, Chromosom. *Cancer* **1998**, *22*, 232–240. [CrossRef]

94. Schuster, K.; Venkateswaran, N.; Rabellino, A.; Girard, L.; Peña-Llopis, S.; Scaglioni, P.P. Nullifying the CDKN2AB Locus Promotes Mutant K-ras Lung Tumorigenesis. *Mol. Cancer Res.* **2014**, *12*, 912–923. [CrossRef]

95. Jiang, J.; Gu, Y.; Liu, J.; Wu, R.; Fu, L.; Zhao, J.; Guan, Y. Coexistence of p16/CDKN2A homozygous deletions and activating EGFR mutations in lung adenocarcinoma patients signifies a poor response to EGFR-TKIs. *Lung Cancer* **2016**, *102*, 101–107. [CrossRef] [PubMed]

96. Bass, A.J.; Watanabe, H.; Mermel, C.H.; Yu, S.; Perner, S.; Verhaak, R.G.; Kim, S.Y.; Wardwell, L.; Tamayo, P.; Gat-Viks, I.; et al. SOX2 is an amplified lineage-survival oncogene in lung and esophageal squamous cell carcinomas. *Nat. Genet.* **2009**, *41*, 1238–1242. [CrossRef]

97. Maier, S.; Wilbertz, T.; Braun, M.; Scheble, V.; Reischl, M.; Mikut, R.; Menon, R.; Nikolov, P.; Petersen, K.; Beschorner, C.; et al. SOX2 amplification is a common event in squamous cell carcinomas of different organ sites. *Hum. Pathol.* **2011**, *42*, 1078–1088. [CrossRef]

98. Fejzo, M.S.; Anderson, L.; Chen, H.-W.; Guandique, E.; Kalous, O.; Conklin, D.; Slamon, D.J. Proteasome ubiquitin receptor PSMD4 is an amplification target in breast cancer and may predict sensitivity to PARPi. Genes, Chromosom. *Cancer* **2017**, *56*, 589–597. [CrossRef]

99. Cai, M.-J.; Cui, Y.; Fang, M.; Wang, Q.; Zhang, A.-J.; Kuai, J.-H.; Pang, F.; Cui, X.-D. Inhibition of PSMD4 blocks the tumorigenesis of hepatocellular carcinoma. *Gene* **2019**, *702*, 66–74. [CrossRef]

100. Cheng, Y.-M.; Lin, P.-L.; Wu, D.-W.; Wang, L.; Huang, C.-C.; Lee, H. PSMD4 is a novel therapeutic target in chemoresistant colorectal cancer activated by cytoplasmic localization of Nrf2. *Oncotarget* **2018**, *9*, 26342–26352. [CrossRef] [PubMed]

101. Türkoğlu, S.A.; Dayi, G.; Köçkar, F. Upregulation of PSMD4 Gene By Hypoxia in Prostate Cancer Cells. *Turk. J. Boil.* **2020**, *44*, 275–283. [CrossRef]
102. O'Leary, C.; Gasper, H.; Sahin, K.B.; Tang, M.; Kulasinghe, A.; Adams, M.N.; Richard, D.J.; O'Byrne, K.J. Epidermal Growth Factor Receptor (EGFR)-Mutated Non-Small-Cell Lung Cancer (NSCLC). *Pharmaceuticals* **2020**, *13*, 273. [CrossRef]
103. Mathsyaraja, H.; Catchpole, J.; Eastwood, E.; Babaeva, E.; Geuenich, M.; Cheng, P.F.; Freie, B.; Ayers, J.; Yu, M.; Wu, N.; et al. Loss of MGA mediated Polycomb repression promotes tumor progression and invasiveness. *bioRxiv* **2020**. [CrossRef]
104. Xue, Y.; Meehan, B.; Fu, Z.; Wang, X.Q.D.; Fiset, P.O.; Rieker, R.; Levins, C.; Kong, T.; Zhu, X.; Morin, G.; et al. SMARCA4 loss is synthetic lethal with CDK4/6 inhibition in non-small cell lung cancer. *Nat. Commun.* **2019**, *10*, 1–13. [CrossRef]
105. Xu, Y.; Gao, P.; Lv, X.; Zhang, L.; Zhang, J. The role of the ataxia telangiectasia mutated gene in lung cancer: Recent advances in research. *Ther. Adv. Respir. Dis.* **2017**, *11*, 375–380. [CrossRef] [PubMed]
106. Sun, X.; Jia, M.; Sun, W.; Feng, L.; Gu, C.; Wu, T. Functional role of RBM10 in lung adenocarcinoma proliferation. *Int. J. Oncol.* **2018**, *54*, 467–478. [CrossRef]
107. Chang, S.; Yim, S.; Park, H. The cancer driver genes IDH1/2, JARID1C/ KDM5C, and UTX/ KDM6A: Crosstalk between histone demethylation and hypoxic reprogramming in cancer metabolism. *Exp. Mol. Med.* **2019**, *51*, 1–17. [CrossRef] [PubMed]
108. Tam, K.W.; Zhang, W.; Soh, J.; Stastny, V.; Chen, M.; Sun, H.; Thu, K.; Rios, J.J.; Yang, C.; Marconett, C.N.; et al. CDKN2A/p16 Inactivation Mechanisms and Their Relationship to Smoke Exposure and Molecular Features in Non–Small-Cell Lung Cancer. *J. Thorac. Oncol.* **2013**, *8*, 1378–1388. [CrossRef]
109. Gkountakos, A.; Sartori, G.; Falcone, I.; Piro, G.; Ciuffreda, L.; Carbone, C.; Tortora, G.; Scarpa, A.; Bria, E.; Milella, M.; et al. PTEN in Lung Cancer: Dealing with the Problem, Building on New Knowledge and Turning the Game Around. *Cancers* **2019**, *11*, 1141. [CrossRef] [PubMed]
110. Pązik, M.; Michalska, K.; Żebrowska-Nawrocka, M.; Zawadzka, I.; Łochowski, M.; Balcerczak, E. Clinical significance of HRAS and KRAS genes expression in patients with non–small-cell lung cancer—Preliminary Findings. *BMC Cancer* **2021**, *21*, 1–13. [CrossRef] [PubMed]
111. Zhao, J.; Han, Y.; Li, J.; Chai, R.; Bai, C. Prognostic value of KRAS/TP53/PIK3CA in non-small cell lung cancer. *Oncol. Lett.* **2019**, *17*, 3233–3240. [CrossRef] [PubMed]
112. Liu, P.; Morrison, C.; Wang, L.; Xiong, D.; Vedell, P.; Cui, P.; Hua, X.; Ding, F.; Lu, Y.; James, M.; et al. Identification of somatic mutations in non-small cell lung carcinomas using whole-exome sequencing. *Carcinogenesis* **2012**, *33*, 1270–1276. [CrossRef] [PubMed]
113. La Fleur, L.; Falk-Sörqvist, E.; Smeds, P.; Berglund, A.; Sundström, M.; Mattsson, J.S.; Brandén, E.; Koyi, H.; Isaksson, J.; Brunnström, H.; et al. Mutation patterns in a population-based non-small cell lung cancer cohort and prognostic impact of concomitant mutations in KRAS and TP53 or STK11. *Lung Cancer* **2019**, *130*, 50–58. [CrossRef] [PubMed]

Journal of
*Personalized
Medicine*

MDPI

Article

Differential Interactome Proposes Subtype-Specific Biomarkers and Potential Therapeutics in Renal Cell Carcinomas

Aysegul Caliskan [1,2,†], Gizem Gulfidan [1,†], Raghu Sinha [3,*] and Kazim Yalcin Arga [1,*]

[1] Department of Bioengineering, Marmara University, Istanbul 34722, Turkey;
    gulaysecaliskan@hotmail.com (A.C.); gizemgulfidn@gmail.com (G.G.)
[2] Faculty of Pharmacy, Istinye University, Istanbul 34010, Turkey
[3] Department of Biochemistry and Molecular Biology, Penn State College of Medicine, Hershey, PA 17033, USA
* Correspondence: rsinha@pennstatehealth.psu.edu (R.S.); kazim.arga@marmara.edu.tr (K.Y.A.)
† These authors contributed equally to this work.

Abstract: Although many studies have been conducted on single gene therapies in cancer patients, the reality is that tumor arises from different coordinating protein groups. Unveiling perturbations in protein interactome related to the tumor formation may contribute to the development of effective diagnosis, treatment strategies, and prognosis. In this study, considering the clinical and transcriptome data of three Renal Cell Carcinoma (RCC) subtypes (ccRCC, pRCC, and chRCC) retrieved from The Cancer Genome Atlas (TCGA) and the human protein interactome, the differential protein–protein interactions were identified in each RCC subtype. The approach enabled the identification of differentially interacting proteins (DIPs) indicating prominent changes in their interaction patterns during tumor formation. Further, diagnostic and prognostic performances were generated by taking into account DIP clusters which are specific to the relevant subtypes. Furthermore, considering the mesenchymal epithelial transition (MET) receptor tyrosine kinase (PDB ID: 3DKF) as a potential drug target specific to pRCC, twenty-one lead compounds were identified through virtual screening of ZINC molecules. In this study, we presented remarkable findings in terms of early diagnosis, prognosis, and effective treatment strategies, that deserve further experimental and clinical efforts.

Keywords: renal cancers; protein interactome; diagnostic biomarker; prognostic biomarker; virtual screening; docking

check for
updates

Citation: Caliskan, A.; Gulfidan, G.; Sinha, R.; Arga, K.Y. Differential Interactome Proposes Subtype-Specific Biomarkers and Potential Therapeutics in Renal Cell Carcinomas. *J. Pers. Med.* **2021**, *11*, 158. https://doi.org/10.3390/jpm11020158

Academic Editor: Susan M. Bailey

Received: 28 December 2020
Accepted: 18 February 2021
Published: 23 February 2021

Publisher's Note: MDPI stays neutral with regard to jurisdictional claims in published maps and institutional affiliations.

1. Introduction

Kidney cancer is among the 10 most common cancers in adults and renal cell carcinoma (RCC) shows a steady increase in prevalence [1]. RCC is known to be the most common type of kidney cancer and is responsible for up to 85% of cases; it is more common in males than in females (ratio, 1.7:1), and most of the patients are at an older age (average age of 64 years) [1]. Primarily, RCC is categorized into subtypes according to histological classification under a microscope, including clear cell (ccRCC, also known as KIRC), papillary (pRCC, also known as KIRP), chromophobe (chRCC, also known as KICH), and some other, less common subtypes such as collecting duct, medullary RCC, and unclassified RCC [2]. The most prevalent one among kidney cancers is ccRCC which represents 75–80% of RCC [3] and derives its name from its clear cytoplasm on the pathologic analysis [4]. The rest are papillary (10–15%), chromophobe (5%), and rare kidney cancers. Although improvement of the state-of-the-art treatment technologies, the overall prognosis is still poor in RCCs, particularly for patients who present with the advanced-stage disease [1]. Therefore, early diagnosis and successful urological procedures with partial or total nephrectomy can be life-saving. However, only about 10% of RCC patients present with urological problems or other known clinical symptoms. More than sixty percent of patients are incidentally noticed at imaging investigations [5], and metastasis has already begun in nearly 20–30% of the patients when diagnosed [6]. In this context, biomarker identification from secretion fluids is extremely important for early diagnosis. Furthermore, biomarkers are becoming increasingly significant to facilitate the discovery of anti-cancer agents, to distinguish cancer cells from the other cells, to understand drug action mechanisms, to predict prognosis, to design personalized medication, and to understand the mechanisms underlying response to therapy. All types of kidney cancers are different in many respects including tumor location within the kidney, the cell type from which they originate, and alterations on their genotype, making it even more crucial to characterize the pathology of each type and to identify specific proteins as druggable targets.

Biomarkers play an important role in the implementation of personalized medicine in clinics with respect to defining subtype phenotypes, predicting clinical course and prognosis, and determining the appropriate therapeutic approach. In this respect, a comprehensive pool of molecular markers from different biological levels (hub proteins, receptors, miRNAs, mRNAs, reporter TFs, and metabolites) were presented from a systematic integrative biology perspective with the potential to provide in-depth knowledge into the disease mechanisms in RCC subtypes [7]. On the other hand, the limited diagnostic and prognostic performance of a molecular biomarker revealed the need for system biomarkers to be obtained with approaches that consider interactions between critical molecules such as the differential protein interactome [8,9].

The differential interactome methodology is based on the idea that significant alterations occur in the protein–protein interactions (PPIs) among phenotypes. The success of this approach has been effectively demonstrated in various cancers and their subtypes [8–10]. The differential interactome approach made it possible to estimate the probability distributions for any possible co-expression profile of gene pairs (encoding proteins that interact with each other) across phenotypes and to determine the uncertainty of whether a PPI is meeting the corresponding phenotype.

The Cancer Genome Atlas (TCGA) is one of the comprehensive cancer genomics datasets available. The availability of TCGA allows researchers to uncover the molecular profiling of tumors through the application of genome analysis technologies, including large-scale genome sequencing. In our present study, we investigated the TCGA transcriptome data from 892 individuals and used the differential interactome methodology [8] that integrates transcriptome data with the human protein interactome network to analyze and compare the differential protein–protein interactions among healthy and tumor groups. Three common subtypes (ccRCC, pRCC, and chRCC) of RCC were investigated and compared in terms of the differential interactome profiles. These analyses allowed us to identify differentially interacting proteins (DIPs) that represent significant changes in their interaction patterns during the transition from "normal" to "tumor" phenotypes and are therefore differently related to the corresponding tumor [9]. We also determined candidate protein panels with high diagnostic and/or prognostic performance, which might allow us to develop novel drug candidates and to diagnose patients in the early stage. Furthermore, we offer drug candidates that showed an inhibitory effect on mesenchymal epithelial transition (MET) receptor tyrosine kinase which is one of the DIPs that have activated interactions in the case of pRCC.

2. Materials and Methods

2.1. Collecting of Gene Expression Data

The transcriptome datasets consisting of three different subtypes of kidney cancer (chRCC, ccRCC, and pRCC) were acquired from the TCGA database [11] to analyze their gene expression profiles. The number of the primary tumor and the matched normal tissue samples were 538 and 72 for ccRCC, 289 and 32 for pRCC, and 65 and 24 for chRCC, respectively.

2.2. Obtaining Protein–Protein Interactions Data

Physical PPI data experimentally detected in humans was obtained from the BioGRID database using the latest version (v. 4.0.189) [12]. The data contained 51,745 PPIs among

10,177 human proteins. After filtering the PPI data for proteins encoded by genes having transcriptome data in TCGA datasets, a network was reconstructed with 34,604 PPIs among 8322 proteins.

2.3. Identification of Differential Interactome and Differentially Interacting Proteins

The gene expression profiles of RCC subtypes were analyzed together with the obtained PPI data through the differential interactome algorithm revised in the study of Gulfidan et al. [8] using R (version 3.6.1). This algorithm presents the differential PPIs (dPPIs) between the tumor phenotype and normal phenotype, taking into account the relative observation frequencies (q-value) of each PPI as described earlier [8,9]. The criteria of the algorithm for obtaining significant dPPIs were set as q-value < 0.10 (significantly repressed in tumor phenotype), q-value > 0.90 (significantly activated in tumor phenotype), and a normalized observation frequency either in normal or tumor phenotype > 20%.

DIPs, the proteins having differential interactions, were classified into two groups according to their interaction patterns: (i) DIPs having repressed interactions under tumor condition, and (ii) DIPs having activated interactions under tumor condition. DIPs that were specific to the RCC subtypes and were common in all subtypes were detected for further analyses. The networks consisting of dPPIs and DIPs were visualized through the Cytoscape 3.4.0 [13].

2.4. Evaluation of the Secretion Levels of Subtype-Specific DIPs in Body Fluids

The secretion levels (ppm) of subtype-specific DIPs in plasma, serum, urine, and saliva were investigated through protein expression data which is accessible in the GeneCards [14] database curating the proteomics databases; ProteomicsDB [15], MaxQB [16], and MOPED [17].

2.5. Analysis of Diagnostic Performance and Prognostic Power

Principal component analyses (PCA) were carried out for the assessment of the diagnostic potential of subtype-specific DIPs using the expression values of genes encoding the DIPs which had the secretion levels in body fluids. The simulations were performed using the gene expression data of tumor samples of ccRCC, pRCC, and chRCC datasets for each subtype-specific DIPs, separately.

To explore the prognostic performance of each subtype-specific DIP, survival analyses were carried out through stratification of patients into high- and low-risk groups based on their prognostic index (PI), which is the linear component of the Cox model ($PI = \beta_1 x_1 + \beta_2 x_2 + \ldots + \beta_p x_p$, where β_i is coefficient acquired from the Cox fitting, x_i is the expression value of each gene). Analyses were implemented through the SurvExpress tool [18] utilizing two RNA-Seq originated datasets of ccRCC with 415 samples, and pRCC with 278 samples including clinical data. In addition, RNA–Seq originated chRCC dataset with 9 samples with clinical data retrieved from TCGA [11] was analyzed separately through the pipeline established in our previous study [8] due to the absence of any dataset related to the chRCC subtype in the SurvExpress database. The signatures of survival in each risk group were estimated by Kaplan–Meier curves and Hazard Ratios (HR). Statistical significance of each plot was evaluated by the cut-off for log-rank p-value < 0.05. Hazard ratio ($HR = O_1/E_1/O_2/E_2$) was calculated to discover the significance of the survival curves based on the ratio between the relative death rate in group 1 (O_1/E_1) and the relative death rate in group 2 (O_2/E_2), where O denotes the observed number of deaths, and E denotes the expected number of deaths.

2.6. Identification of Candidate Drugs through Virtual Screening

We set the following criteria to determine the potential drug target protein among DIPs in docking studies: (i) its interactions should be activated in the disease state, and (ii) it should have at least 5 interactions. Among DIP proteins of pRCC, MET protein satisfied all the criteria and came to the forefront as a potential drug target. Through virtual screening, potential molecules targeting MET were determined. To have an insight into the ligand-receptor interactions, the available X-ray crystal structures of MET were fetched from the Protein Data Bank (PDB) (www.rcsb.org) [19]. PDB entry 3DKF was chosen for all the docking studies according to the resolution, Ramchandran outliers, and structural similarity between the screened ligands and the co-crystallized ligands. Virtual Screening binding analysis was carried out on the assigned binding site of the X-ray crystal structure of MET [20] exploiting ZINC molecules described by the publicly available ZINC15 library [21]. Molecular docking studies were executed for 703 substances retrieved from the ZINC15 library through AutoDock Vina [22] in the PyRx virtual screening tool (v. 0.8) [23].

3. Results

3.1. Differential Interactome Estimation in Subtypes of RCC

RNA-seq transcriptome data of three RCC subtypes were retrieved from TCGA to apply differential interactome methodology [8] for prediction of highly probable PPIs in each state and identification of differential PPIs. To this end, we examined transcriptomic data for three common subtypes of RCC with an adequate number of samples ($n > 24$) in both normal and tumor groups (see Materials and Methods section). The scale-free topology of the differential interactome network brings out the presence of hubs called DIPs indicating substantial changes in their interaction patterns during the transition from "normal" to "tumor" phenotypes [8,9]. We determined 628 DIPs for chRCC, 50 DIPs for ccRCC, and 29 DIPs for pRCC as subtype-specific DIPs, whereas 33 DIPs were common in all subtypes (Supplementary Table S1). The tumor-specificity of DIPs varied according to the subtype (Figure 1).

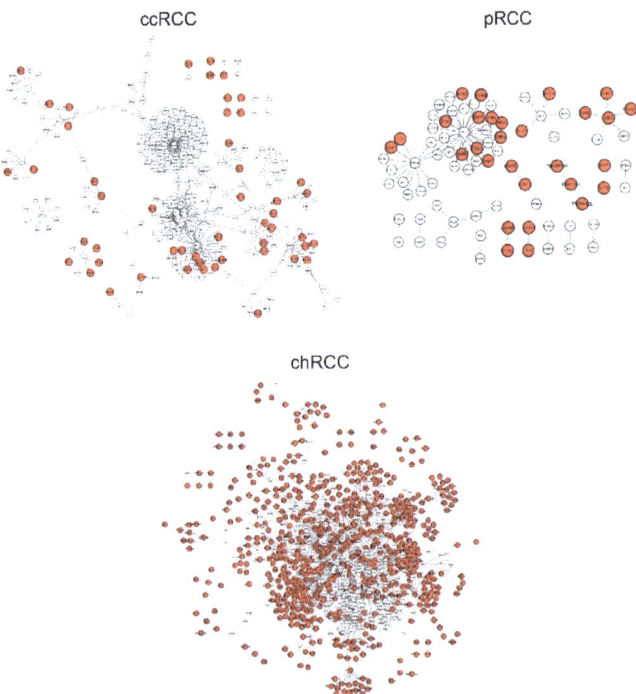

Figure 1. Differential interactome networks reconstructed with differential protein–protein interactions (dPPIs) around differentially interacting proteins (DIPs) in three Renal Cell Carcinoma RCC subtypes. Red nodes represent DIPs specific to the subtype of interest. ccRCC: Clear Cell Renal Carcinoma; pRCC: Papillary Renal Cell Carcinoma; chRCC: Chromophobe Renal Cell Carcinoma.

Further analyses (i.e., determination of prognostic power, diagnostic performance, and druggability) were implemented by taking into account 50 DIPs specific to ccRCC, 29 DIPs specific to pRCC, and the top 50 DIPs having the most interactions specific to chRCC (Table 1). We considered those DIPs as a cluster for each subtype and suggested them as potential systems biomarkers for the development of effective diagnosis, prognosis, and treatment strategies.

Table 1. Differentially interacting proteins (DIPs) specific to RCC subtypes.

Specificity	s-DIPs [1]	Non s-DIPs [2]
ccRCC-specific	ABCC2, B2M, BST2, CALU, CCDC106, CENPA, CYB5R3, DDX3X, DKC1, DNAJB4, DTNBP1, GABBR1, GIT2, HLA-B, HSPBP1, IMMT, MAPK3, NRP1, PDIA4, PEA15, PFDN2, PFKM, PPIB, PRKCD, RGCC, RPS6KA3, SDHA, UBQLN1, TNIP1	AZIN1, CDT1, ELF4, FBXW8, GPS2, IL32, IRF1, LDOC1, MCM7, MCM9, MTF1, MTOR, P4HA2, PHLPP1, RSL1D1, SCD, TAF1, TAPBP, TOMM20, USP2, ZNF668
pRCC-specific	CS, CUL3, DFFA, DHFR, EIF4A2, FLOT2, G6PD, GSTA2, IGBP1, ITGA3, MET, MME, MVP, PARP4, PGM2, PNPT1, PPM1A, TRAPPC1	GSTA4, HGF, LBH, LGALS8, MMGT1, RANBP9, SF3A3, SOCS1, TRAPPC12, TRAPPC2L, UNG
chRCC-specific	ANXA5, AQP1, ARF1, BAD, CHMP4B, CYLD, ECH1, EEF1B2, FLOT1, FUS, HADHA, HADHB, HSD17B10, JUP, KRT18, MAPRE1, PARK7, PFN1, PHB, PHB2, PPP1CB, PRDX1, PRDX2, PRDX3, PRDX5, PSMB4, PSMB6, PSME1, PTGES3, PTMA, RAB1A, RAB7A, S100A10, TGOLN2, TXN, UBB, UBE3A, YWHAB, YWHAE	ABL1, AMFR, ARAF, CDK9, FOS, JUND, MCL1, MORF4L2, SF3B5, STAU1, TRIM8

[1] Protein expression was observed at least in one of the following body fluids: serum, plasma, saliva, urine; [2] Protein expression was not observed in any of the following body fluids: serum, plasma, saliva, urine.

Then, we filtered DIPs by considering whether they are secreted in body fluids and renamed secreted proteins as "s-DIPs" (Table 1, Figure 2). s-DIPs represent proteins that were expressed at least in one of the following media: serum, plasma, saliva, or urine (www.genecards.org) [14]. The importance of secretion in body fluids that can be accessed without surgery is that it might provide serious convenience for early diagnosis. While s-DIPs were used for diagnosis analysis, all DIPs (s-DIPs and non-s-DIPs) were considered in prognosis and druggability (virtual screening) analyses.

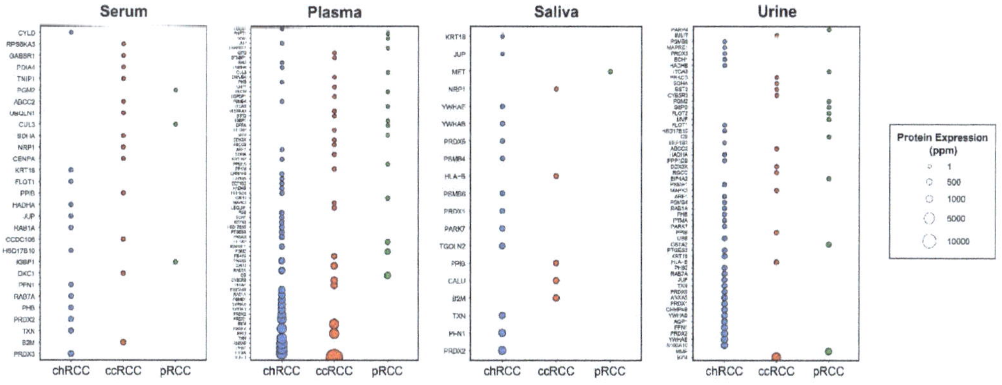

Figure 2. Bubble plots indicating protein expression levels of DIPs specific to three subtypes in different body fluids including serum, plasma, saliva, and urine. The x-axis indicates subtypes while the y-axis indicates protein symbols.

3.2. Prognostic and Diagnostic Capabilities of DIPs Clusters

We considered the clusters of DIPs as potential systems biomarkers for each RCC subtype and analyzed their diagnostic performance and prognostic power.

The diagnostic analysis was performed via PCA using s–DIPs (Table 1). All s–DIP clusters exhibited significantly high diagnostic performance for relevant subtypes (Figure 3A).

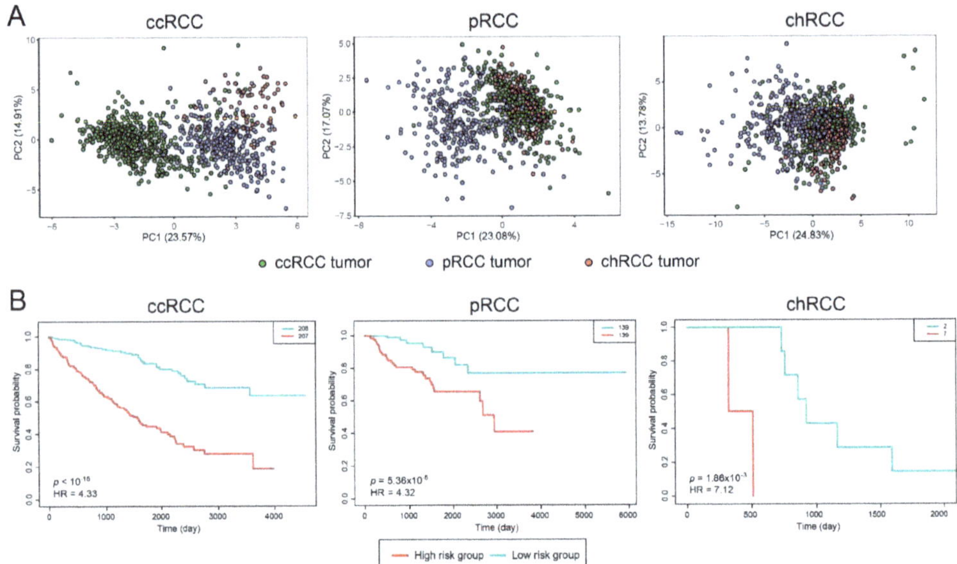

Figure 3. Diagnostic and prognostic performance analysis results for Renal Cell Carcinoma (RCC) subtypes. (**A**) Principal component analyses (PCA) plots, visualized by considering s-DIPs, indicating the individual differences in the gene expression profiles in tumor samples among the subtypes. (**B**) Kaplan–Meier curves estimating patients' survival for three subtypes based on categorization of patients into high- and low-risk groups via prognostic index. ccRCC: Clear Cell Renal Carcinoma; pRCC: Papillary Renal Cell Carcinoma; chRCC: Chromophobe Renal Cell Carcinoma; HR: Hazard Ratio; PC: Principal component.

Prognostic capabilities of gene clusters were quantified through log-rank p-values and visualized by Kaplan–Meier curves (Figure 3B). Cox (proportional hazards) regression was also engaged to estimate HRs. These analyses were carried out utilizing TCGA clinical datasets (see Materials and Methods section). Gene clusters were significantly predictive in terms of patient survival risk assessment for the respective subtype (Figure 3B, ccRCC $p < 1 \times 10^{-15}$, pRCC $p = 5.36 \times 10^{-5}$, chRCC $p = 1.86 \times 10^{-3}$). Through Cox-proportional hazard analysis, HR values were estimated as 4.33, 4.32, and 7.12 for ccRCC, pRCC, and chRCC, respectively.

3.3. Discovery of Drug Candidates through Virtual Screening Analyses

In silico simulation techniques have become an indispensable tool for modern-day drug discovery programs. Molecular docking currently offers the best alternative to quickly estimate the binding conformations of ligands that are energy-efficient to interact with a pharmacological receptor site. It has become more popular as it is time and cost effective in the pipeline of drug discovery and development. Interactions of some DIP proteins within the module were activated during the tumorigenesis, while some were found to be repressed. We hypothesized that, if we manage to break through the interactions that are activated, we might model a strategy to cure the disease. For this purpose, we considered DIPs with activated interactions in the tumor state as potential drug targets.

For instance, among DIP proteins of the pRCC subtype, MET protein came into prominence as a potential drug target. Candidate molecules targeting MET were determined via virtual screening of the ZINC15 library via the available crystal structures of MET from PDB. All available X-Ray crystal structures of MET (PDB IDs: 3DKF, 2RFN, 3EFJ, 3U6H, 4EEV) and their bound ligands were superposed, and potential binding sites were determined to identify the binding site location on the receptor (Figure 4A). Virtual Screening binding analysis was accomplished on the assigned binding site of the X-ray structure of MET (PDB ID: 3DKF) utilizing ZINC molecules which were described by the ZINC15 library. The virtual screening analysis revealed twenty-one ZINC molecules with high binding affinities ($\Delta G^0 \leq -12$, LE > 0.35) (Table 2; Figure 4B).

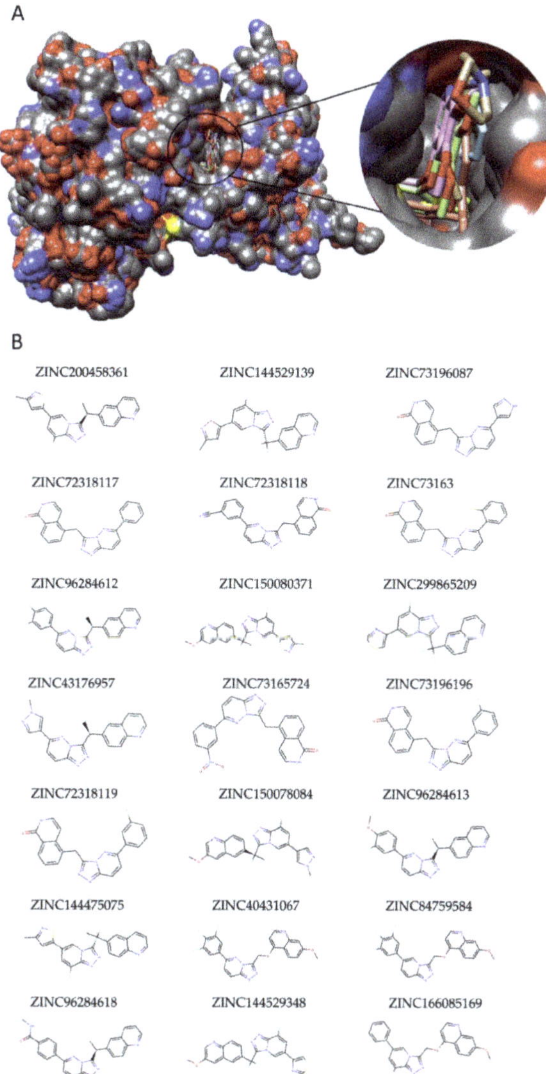

Figure 4. Virtual screening to identify potential hit drug candidates for pRCC. (**A**) Superposition of X-ray crystal structures of MET retrieved from RCSB for the validation of docking protocol. (**B**) 2D structures of ZINC molecules that showed high binding affinities to MET protein in virtual screening.

Table 2. The ZINC molecules presented the best binding affinities to MET.

Ligand ZINC15 ID	Vina Binding Affinity (kcal/mol)	Ligand Efficiency (LE)
ZINC200458361	−12.7	0.41
ZINC144529139	−12.6	0.39
ZINC73196087	−12.6	0.45
ZINC72318117	−12.5	0.44
ZINC72318118	−12.5	0.41
ZINC73163075	−12.5	0.42
ZINC96284612	−12.5	0.41
ZINC150080371	−12.4	0.38
ZINC299865209	−12.4	0.42
ZINC43176957	−12.4	0.43
ZINC73165724	−12.4	0.39
ZINC73196196	−12.4	0.43
ZINC72318119	−12.3	0.41
ZINC150078084	−12.2	0.37
ZINC96284613	−12.2	0.39
ZINC144475075	−12.1	0.4
ZINC40431067	−12.1	0.37
ZINC84759584	−12.1	0.36
ZINC96284618	−12.1	0.37
ZINC144529348	−12	0.41
ZINC166085169	−12	0.38

4. Discussion

Dysregulations in various biochemical pathways play an important role in cancer formation and development. Genetic studies have identified numerous molecular defects in cancer cells and suggested multiple potential targets for therapeutic intervention. Conventional drug design has mainly focused on the inhibition of a single protein, usually an enzyme or receptor; however, this strategy has not been successful enough, as the development and progression of cancers are mostly due to the coordinated action of a group of biological entities rather than a single molecule dysfunction [24]. Hereby, PPIs have become highly promising targets that cover many therapeutic areas and potential intervention points for the development of anticancer agents. Until now, significant progress has been made in identifying small molecule inhibitors of various protein–protein systems in the field of oncology, and powerful and selective drug-like molecules that inhibit many interactions such as p53-MDM2 interaction have been discovered [25]. Furthermore, a number of these small-molecule inhibitors, such as Siremadlin, AMG-232, and APG-115 have progressed to early phase clinical trials [26].

Our study reports the generation of the dPPI networks in RCC subtypes through the implementation of high throughput transcriptome and protein interactome data. The integration of respective RNA-seq datasets and differential interactome approach allowed the identification of dPPIs in different conditions (tumor/normal) in RCC subtypes. The study unveils and compares the dPPIs for each subtype and identifies DIPs through a differential interactome. Further analyses on DIPs may be useful in understanding the tumor mechanisms. For instance, our findings revealed that HspB1 protein is one of the common DIPs for three subtypes. The correlation between HspB1 expression in RCC subtypes and metastasis process has been revealed in previous studies and HspB1 is known to facilitate metastasis by suppressing anti-cancer response such as apoptosis and senescence [7,27].

DIP clusters were used for diagnostic and prognostic analyses for each subtype. Despite the improvements in the state of the art treatment technologies, the overall prognosis is still poor in RCCs and more than 50% of RCCs are diagnosed incidentally [28]. Even the detection of the early asymptomatic stage during routine examination could have a profound impact on clinical outcome. Therefore, an effective, clinically useful test for early detection of RCC subtypes should be measurable in readily accessible body fluids, such as plasma, serum, urine, or saliva. For this purpose, we filtered DIPs by considering whether they are expressed in those body fluids at the protein level and defined the s-DIP concept here for the first time in literature. s-DIP clusters characterize patients well in terms of the diagnostic group (subtype) to which they belong. Hence, we offer that s-DIPs might be used for the diagnosis of candidate RCC patients after further experimental and clinical validations.

Saliva is one of the complex and important multi-constituent body fluids that reflects a wide variety of physiological knowledge due to its contents extensively supplied by the blood. Moreover, a saliva-based diagnosis has been drawing attention in the diagnosis of systemic diseases such as renal cancers, due to the source, composition, function, and interaction of saliva with the substances that make up the plasma [29,30]. In the present study, besides blood components and urine, we also demonstrated the potential of saliva as a non-invasive potential media for RCC diagnosis, especially in chRCC.

The three basic elements for the art of medicine are diagnosis, therapeutics, and prognosis. Therefore, after making the correct and early diagnosis, determining the optimal treatment strategies would be important and as a follow-up, one could provide up-to-date information on the patient's prognosis. Our present investigation also aimed to provide new targets for the design of novel therapies in RCC subtypes and putative biomarkers with prognostic significance. In this study, DIP clusters appear to be strong putative candidates for the prognostic marker in each related subtype. Survival analyses through stratification of patients according to clinicopathological variables such as tumor stage or grade would demonstrate the prognostic power of the potential biomarkers better. However, despite the presence of comprehensive gene expression profiling efforts such as TCGA, transcriptome data with available clinical information is still limited for RCCs, even for the most common subtypes.

Additionally, to shed light on the further experimental studies, we identified MET protein as an ideal potential drug target in pRCC and showed the high potential of twenty-one Zinc molecules (Table 2) as candidate therapeutics for future preclinical studies. The integration of the transcriptome and protein interactome data with the drug knowledge helped to uncover 21 in silico validated potential drug candidates for pRCC. These in-silico findings can be used further to design and synthesize novel MET inhibitors. Furthermore, ZINC73196087, ZINC72318117, ZINC72318118, ZINC73163075, ZINC73165724, ZINC73196196, and ZINC72318119 have been shown to demonstrate effective anti-proliferative activity against a panel of c-Met-amplified gastric cancer cell lines [31]. We propose that these ZINC compounds should also be evaluated with experimental studies for RCC cell lines and we conclude that these molecules might be potential therapeutics for the management of the pRCC. Further in vitro/in vivo pharmacological evaluations and clinical validations are needed for approval of these candidate drugs.

The major limitation of the study is the lack of experimental validations of the identified ZINC compounds on the RCC samples or cell lines. Future in vitro studies need to be conducted to evaluate the effects of ZINC compounds identified on cell viability, proliferation, and migration. Moreover, the mechanism of actions of these molecules need to be investigated in detail to elucidate their effect on molecular pathways such as apoptosis and cell cycle. Rather than being considered as a single agent, these compounds can also be regarded as adjuvant therapy to the baseline therapeutics, then, the critical extension of this work would be to learn whether the observations of in vitro studies can be recapitulated by in vivo studies and eventually in clinical trials. Another point that has a crucial role in translation to the clinic is sampling where body fluids are favorable for the detection of the biomarkers. Proteomics studies also need be verified for the proteins exhibiting significantly high diagnostic and prognostic performance for relevant subtypes. Moreover, these biomarkers could also assist oncologists to assist in optimal diagnosis and prognosis management.

Supplementary Materials: The following are available online at https://www.mdpi.com/2075-4426/11/2/158/s1, Table S1: List of Differentially Interacting Proteins (DIPs) in chRCC, ccRCC, pRCC.

Author Contributions: Conceptualization, K.Y.A., R.S., and A.C.; formal analysis, A.C. and G.G.; investigation, A.C.; writing—original draft preparation, A.C. and G.G.; writing—review and editing, K.Y.A. and R.S.; visualization, G.G. and A.C.; supervision, K.Y.A. and R.S.; project administration, K.Y.A. and, R.S. All authors have read and agreed to the published version of the manuscript.

Funding: This research received no external funding.

Data Availability Statement: Publicly available datasets were analyzed in this study. Transcirptome data can be found here: [https://portal.gdc.cancer.gov/, Primary site: Kidney]. Protein interactome data is available here: [https://thebiogrid.org, File repository: BIOGRID-4.0.189].

Conflicts of Interest: The authors declare no conflict of interest.

Abbreviations

ccRCC	Clear Cell Renal Cell Carcinoma
chRCC	Chromophobe Renal Cell Carcinoma
DIP	Differentially interacting protein
dPPI	Differential protein–protein interaction
HR	Hazard ratio
KICH	Kidney Chromophobe
KIRC	Kidney Renal Clear Cell Carcinoma
KIRP	Kidney Renal Papillary Cell Carcinoma
LE	Ligand efficiency
ns-DIP	Non-secreted DIP
PC	Principle component
PCA	Principal component analysis
PDB	Protein data bank
PPI	Protein–protein interaction
pRCC	Papillary Renal Cell Carcinoma
RCC	Renal cell carcinoma
s-DIP	Secreted DIP
TCGA	The Cancer Genome Atlas
ZINC	ZINC Is Not Commercial

References

1. Barata, P.C.; Rini, B.I. Treatment of renal cell carcinoma: Current status and future directions. *CA Cancer J. Clin.* **2017**, *67*, 507–524. [CrossRef]
2. Linehan, W.M.; Ricketts, C.J. The Cancer Genome Atlas of renal cell carcinoma: Findings and clinical implications. *Nat. Rev. Urol.* **2019**, *16*, 539–552. [CrossRef] [PubMed]
3. Priolo, C.; Khabibullin, D.; Reznik, E.; Filippakis, H.; Ogórek, B.; Kavanagh, T.R.; Nijmeh, J.; Herbert, Z.T.; Asara, J.M.; Kwiatkowski, D.J. Impairment of gamma-glutamyl transferase 1 activity in the metabolic pathogenesis of chromophobe renal cell carcinoma. *Proc. Natl. Acad. Sci. USA* **2018**, *115*, E6274–E6282. [CrossRef]
4. Cairns, P. Renal cell carcinoma. *Cancer Biomark.* **2011**, *9*, 461–473. [CrossRef]
5. Petejova, N.; Martinek, A. Renal cell carcinoma: Review of etiology, pathophysiology and risk factors. *Biomed. Pap. Med. Fac. Palacky Univ. Olomouc* **2016**, *160*, 183–194. [CrossRef]
6. Wang, Y.; Su, J.; Fu, D.; Wang, Y.; Chen, Y.; Chen, R.; Qin, G.; Zuo, J.; Yue, D. The role of YB1 in renal cell carcinoma cell adhesion. *Int. J. Med. Sci.* **2018**, *15*, 1304. [CrossRef]
7. Caliskan, A.; Andac, A.C.; Arga, K.Y. Novel molecular signatures and potential therapeutics in renal cell carcinomas: Insights from a comparative analysis of subtypes. *Genomics* **2020**, *112*, 3166–3178. [CrossRef]
8. Gulfidan, G.; Turanli, B.; Beklen, H.; Sinha, R.; Arga, K.Y. Pan-cancer mapping of differential protein-protein interactions. *Sci. Rep.* **2020**, *10*, 3272. [CrossRef]
9. Ayyildiz, D.; Gov, E.; Sinha, R.; Arga, K.Y. Ovarian Cancer Differential Interactome and Network Entropy Analysis Reveal New Candidate Biomarkers. *Omi. A J. Integr. Biol.* **2017**, *21*, 285–294. [CrossRef]
10. Turanli, B.; Karagoz, K.; Bidkhori, G.; Sinha, R.; Gatza, M.L.; Uhlen, M.; Mardinoglu, A.; Arga, K.Y. Multi-omic data interpretation to repurpose subtype specific drug candidates for breast cancer. *Front. Genet.* **2019**, *10*, 420. [CrossRef]

11. Tomczak, K.; Czerwińska, P.; Wiznerowicz, M. The Cancer Genome Atlas (TCGA): An immeasurable source of knowledge. *Contemp. Oncol.* **2015**, *19*, A68–A77. [CrossRef]

12. Chatr-Aryamontri, A.; Oughtred, R.; Boucher, L.; Rust, J.; Chang, C.; Kolas, N.K.; O'Donnell, L.; Oster, S.; Theesfeld, C.; Sellam, A.; et al. The BioGRID interaction database: 2017 update. *Nucleic Acids Res.* **2017**, *45*, D369–D379. [CrossRef] [PubMed]

13. Shannon, P. Cytoscape: A Software Environment for Integrated Models of Biomolecular Interaction Networks. *Genome Res.* **2003**, *13*, 2498–2504. [CrossRef]

14. Stelzer, G.; Rosen, N.; Plaschkes, I.; Zimmerman, S.; Twik, M.; Fishilevich, S.; Iny Stein, T.; Nudel, R.; Lieder, I.; Mazor, Y.; et al. The GeneCards suite: From gene data mining to disease genome sequence analyses. *Curr. Protoc. Bioinform.* **2016**, *54*, 1.30.1–1.30.33. [CrossRef]

15. Samaras, P.; Schmidt, T.; Frejno, M.; Gessulat, S.; Reinecke, M.; Jarzab, A.; Zecha, J.; Mergner, J.; Giansanti, P.; Ehrlich, H.-C.; et al. ProteomicsDB: A multi-omics and multi-organism resource for life science research. *Nucleic Acids Res.* **2019**, *48*, D1153–D1163. [CrossRef] [PubMed]

16. Schaab, C.; Geiger, T.; Stoehr, G.; Cox, J.; Mann, M. Analysis of high accuracy, quantitative proteomics data in the MaxQB database. *Mol. Cell. Proteom.* **2012**, *11*, M111.014068. [CrossRef]

17. Montague, E.; Stanberry, L.; Higdon, R.; Janko, I.; Lee, E.; Anderson, N.; Choiniere, J.; Stewart, E.; Yandl, G.; Broomall, W.; et al. MOPED 2.5—An integrated multi-omics resource: Multi-omics profiling expression database now includes transcriptomics data. *Omics J. Integr. Biol.* **2014**, *18*, 335–343. [CrossRef] [PubMed]

18. Aguirre-Gamboa, R.; Gomez-Rueda, H.; Martínez-Ledesma, E.; Martínez-Torteya, A.; Chacolla-Huaringa, R.; Rodriguez-Barrientos, A.; Tamez-Peña, J.G.; Treviño, V. SurvExpress: An Online Biomarker Validation Tool and Database for Cancer Gene Expression Data Using Survival Analysis. *PLoS ONE* **2013**, *8*, e74250. [CrossRef] [PubMed]

19. Berman, H.M.; Westbrook, J.; Feng, Z.; Gilliland, G.; Bhat, T.N.; Weissig, H.; Shindyalov, I.N.; Bourne, P.E. The protein data bank. *Nucleic Acids Res.* **2000**, *28*, 235–242. [CrossRef] [PubMed]

20. Buchanan, S.G.; Hendle, J.; Lee, P.S.; Smith, C.R.; Bounaud, P.-Y.; Jessen, K.A.; Tang, C.M.; Huser, N.H.; Felce, J.D.; Froning, K.J. SGX523 is an exquisitely selective, ATP-competitive inhibitor of the MET receptor tyrosine kinase with antitumor activity in vivo. *Mol. Cancer Ther.* **2009**, *8*, 3181–3190. [CrossRef]

21. Sterling, T.; Irwin, J.J. ZINC 15–ligand discovery for everyone. *J. Chem. Inf. Model.* **2015**, *55*, 2324–2337. [CrossRef] [PubMed]

22. Trott, O.; Olson, A.J. AutoDock Vina: Improving the speed and accuracy of docking with a new scoring function, efficient optimization, and multithreading. *J. Comput. Chem.* **2009**, *31*, 455–461. [CrossRef]

23. Dallakyan, S.; Olson, A.J. Small-molecule library screening by docking with PyRx. *Chem. Biol.* **2015**, *1263*, 243–250.

24. Jonsson, P.F.; Bates, P.A. Global topological features of cancer proteins in the human interactome. *Bioinformatics* **2006**, *22*, 2291–2297. [CrossRef] [PubMed]

25. Shangary, S.; Wang, S. Small-molecule inhibitors of the MDM2-p53 protein-protein interaction to reactivate p53 function: A novel approach for cancer therapy. *Annu. Rev. Pharmacol. Toxicol.* **2009**, *49*, 223–241. [CrossRef]

26. Konopleva, M.; Martinelli, G.; Daver, N.; Papayannidis, C.; Wei, A.; Higgins, B.; Ott, M.; Mascarenhas, J.; Andreeff, M. MDM2 inhibition: An important step forward in cancer therapy. *Leukemia* **2020**, *34*, 2858–2874. [CrossRef]

27. Nagaraja, G.M.; Kaur, P.; Asea, A. Role of human and mouse HspB1 in metastasis. *Curr. Mol. Med.* **2012**, *12*, 1142–1150. [CrossRef]

28. Ljungberg, B.; Cowan, N.C.; Hanbury, D.C.; Hora, M.; Kuczyk, M.A.; Merseburger, A.S.; Patard, J.-J.; Mulders, P.F.A.; Sinescu, I.C. EAU guidelines on renal cell carcinoma: The 2010 update. *Eur. Urol.* **2010**, *58*, 398–406. [CrossRef] [PubMed]

29. Zhang, X.L.; Wu, Z.Z.; Xu, Y.; Wang, J.G.; Wang, Y.Q.; Cao, M.Q.; Wang, C.H. Saliva proteomic analysis reveals possible biomarkers of renal cell carcinoma. *Open Chem.* **2020**, *18*, 918–926. [CrossRef]

30. Dudek, A.; Appleyard, L.; O'Brien, T.S.; Chowdhury, S.; Champion, P.; Challacombe, B.; Kooiman, G.; Carpenter, G. Salivary markers in renal cell carcinoma. *J. Clin. Oncol.* **2014**, *32*, 489. [CrossRef]

31. Ryu, J.W.; Han, S.-Y.; Yun, J.I.; Choi, S.-U.; Jung, H.; Du Ha, J.; Cho, S.Y.; Lee, C.O.; Kang, N.S.; Koh, J.S. Design and synthesis of triazolopyridazines substituted with methylisoquinolinone as selective c-Met kinase inhibitors. *Bioorg. Med. Chem. Lett.* **2011**, *21*, 7185–7188. [CrossRef]

Journal of
Personalized
Medicine

MDPI

Review

Recent Advances in Integrative Multi-Omics Research in Breast and Ovarian Cancer

Christen A. Khella, Gaurav A. Mehta, Rushabh N. Mehta and Michael L. Gatza *

Rutgers Cancer Institute of New Jersey, New Brunswick, NJ 08903, USA; cak241@scarletmail.rutgers.edu (C.A.K.);
gam160@cinj.rutgers.edu (G.A.M.); rnm53@scarletmail.rutgers.edu (R.N.M.)
* Correspondence: michael.gatza@cinj.rutgers.edu; Tel.: +1-732-235-8751

Abstract: The underlying molecular heterogeneity of cancer is responsible for the dynamic clinical landscape of this disease. The combination of genomic and proteomic alterations, including both inherited and acquired mutations, promotes tumor diversity and accounts for variable disease progression, therapeutic response, and clinical outcome. Recent advances in high-throughput proteogenomic profiling of tumor samples have resulted in the identification of novel oncogenic drivers, tumor suppressors, and signaling networks; biomarkers for the prediction of drug sensitivity and disease progression; and have contributed to the development of novel and more effective treatment strategies. In this review, we will focus on the impact of historical and recent advances in single platform and integrative proteogenomic studies in breast and ovarian cancer, which constitute two of the most lethal forms of cancer for women, and discuss the molecular similarities of these diseases, the impact of these findings on our understanding of tumor biology as well as the clinical applicability of these discoveries.

Keywords: genomics; proteomics; breast; ovarian; cancer

Citation: Khella, C.A.; Mehta, G.A.;
Mehta, R.N.; Gatza, M.L. Recent
Advances in Integrative Multi-Omics
Research in Breast and Ovarian
Cancer. *J. Pers. Med.* **2021**, *11*, 149.
https://doi.org/10.3390/jpm11020149

Academic Editor: Raghu Sinha

Received: 14 January 2021
Accepted: 14 February 2021
Published: 19 February 2021

Publisher's Note: MDPI stays neutral
with regard to jurisdictional claims in
published maps and institutional affiliations.

1. Introduction

Each year more than 1.8 million people are diagnosed with cancer in the United States including more than 270,000 breast cancer patients and 21,000 ovarian cancer patients [1]. Despite advances in diagnostic tools, predictive biomarkers, and new therapies over the past 20 years which have led to declining mortality rates, more than 285,000 people will die each year in the US due to their disease, including more than 50,000 breast and 13,000 ovarian cancer patients [1]. Enormous clinical variability, including disease progression and response to therapy has been shown to exist for most forms of cancer. These observed differences are driven, in part, by underlying genetic, genomic, and proteomic alterations unique to each patient [2–5]. In essence, cancer is not a single disease but rather a collection of genetically driven malignancies affecting a given tissue. As a result, all tumors, even within a given tissue type, cannot be treated equally [6–12]. New tools, therapies, biomarkers, and treatment strategies are being developed or will need to be developed, to identify and target those mutations and/or signaling pathways essential for each tumor to improve clinical outcome and quality of life for each patient.

The underlying genetic heterogeneity within human cancers creates several challenges both clinically and from a basic science perspective. From a mechanistic standpoint, variability in patterns of genomic and proteomic alterations create a challenge in separating the key drivers of oncogenic signaling, tumor development, and progression from those mutations that are tumor-promoting but non-transforming or that do not directly contribute to tumorigenesis (i.e., passenger mutations). This is essential as not all mutated or aberrantly expressed genes are required for tumorigenesis nor do they equally contribute to therapeutic response [13,14]. As a result, there is a need to develop tools and approaches to understand the interplay between altered genes and to determine how these genes or proteins promote aberrant signaling, including the identification of novel signaling networks and cellular processes that contribute to tumor growth and progression. Finally, utilizing the compendium of alterations across a given tumor type, we must develop approaches to identify novel therapeutic targets and determine predictive biomarkers to recognize patients that are likely to benefit from specific therapeutic regimens.

Over the past 20 years, beginning with the sequencing of the human genome to the more recent development of next-generation sequencing (NGS), advances in genomics, proteomics, and systems biology have allowed us to begin to catalogue, visualize, compare and dissect patterns of DNA mutations and copy number alterations, mRNA and miRNA expression patterns, protein and phosphorylated protein expression and epigenetic alterations between individual patients, across specific forms of cancer and between malignancies affecting different tissues [2–5,15,16]. These studies, coupled with functional genomic studies, have begun to identify and provide insight into key drivers of oncogenic signaling, mediators of specific tumor characteristics, including response to therapy, and identify novel treatment strategies. In this review, we will examine historical and recent advances in genome and proteome-wide analyses in breast and ovarian cancer and discuss the impact of these findings on our understanding of tumor biology as well as the clinical applicability of these discoveries.

2. Clinical Characterization of Breast and Ovarian Cancer

Breast cancer is the most commonly diagnosed and the second leading cause of cancer-related mortality for women in the United States [1]. While it is estimated that approximately 50,000 women in the US and 522,000 women worldwide will die from this disease annually, survival rates have steadily increased by over 40% over the past 30 years [1,17]. Currently, more than 98% of patients diagnosed with early stage disease are expected to live for at least 10 years and the current 5-year survival rate is ~90% across all stages [1,17–19]. These improvements can be attributed, in part, to increased early detection from earlier screening and improved imaging technology as well as the development of novel therapeutic regimens incorporating chemotherapeutics, targeted therapies, radiation, surgery, and immunotherapy [20–22]. Despite these advances, the prognosis for patients with locally advanced and metastatic disease remains poor. Patients with advanced metastatic disease have a 5-year survival rate of less than 30% and a significant percentage of patients whose tumors are inoperable and/or refractory to current therapies will succumb to their disease within 5 years irrespective of tumor stage at diagnosis [18].

Part of the challenge in developing effective treatments for this disease lies in the molecular and clinical heterogeneity that exists between each patient's tumor. Clinically, breast tumors are classified based on morphological features with ~70% of tumors being classified as invasive ductal carcinomas (IDC), ~15% categorized as invasive lobular carcinoma (ILC), and the remaining tumors regarded as rare subtypes [18,23]. Prognosis and treatment strategies are largely dictated by classical histopathologic features including tumor size, histological grade and stage, lymph node status, and the expression of hormone receptors or HER2 (human epidermal growth factor receptor 2) status [18,19]. Among the histological subtypes, estrogen receptor (ER), progesterone receptor (PR), and HER2 status can be used to further delineate patients into ER+/PR+ (60–70% of patients), HER2+ (10–20%), and triple negative breast cancer (TNBC, 15–20%). However, differences in the prevalence of these histological subtypes are seen between women with different ancestries. Notably, women of African American decent have a higher incidence of TNBC when compared to American women of European ancestry (36.3% vs. 13.7%) [1,24,25]. Importantly, these biomarkers are used to direct current standard-of-care treatments with endocrine-based therapies comprising the core of therapeutic regimens to treat hormone receptor-positive (HR+) breast tumors; HER2-family inhibitors forming the foundation for therapies used to treat HER2+ patients [26], and multi-agent cytotoxic chemotherapies providing the basis for the treatment of TNBC patients [27,28]. While endocrine-based therapies result in remission in the majority of patients with HR+ tumors, approximately

30–50% of patients manifest primary or acquired resistance [29,30]. Recent studies have reported that the emergence of hormone therapy resistance in ER+ breast cancers can arise through four predominant mechanisms including ESR1 mutations (18%), altered MAPK signaling (13%), MYC or transcription factor activation (9%), and other/unknown factors (60%) [31]. The findings of multiple clinical trials have resulted in FDA and international approval for use of the mTOR inhibitor everolimus in conjunction with exemestane for the treatment of patients with advanced or metastatic ER+, PR+, HER2-negative, PIK3CA mutant tumors [32]. Likewise, the PI3K inhibitor alpelisib has been approved for the same patient population in combination with fulvestrant [33]. More recently, CDK4/6 inhibitors palbociclib, ribociclib, and abemaciclib have been approved and, in conjunction with hormone therapy, have become the primary treatment regimen for HR+/HER2- treatment naïve or hormone therapy treated metastatic breast cancer patients [34]. Finally, the treatment of TNBC tumors has begun to evolve to include immune checkpoint inhibitors while patients with BRCA mutations are treated with PARP inhibitors in conjunction with chemotherapy [28,35,36].

In contrast to breast cancer, ovarian cancer is a less frequently diagnosed malignancy with approximately 21,750 women in the US and 250,000 women worldwide being diagnosed with this disease annually [1]. Unfortunately, however ovarian cancer is the most lethal form of gynecological cancer with an estimated 13,940 women dying from this disease in the United States and more than 150,000 women dying worldwide in 2020 [1]. This translates to a death-to-case ratio of approximately 64%, far outpacing the lethality of breast cancer [1].

Similar to breast cancer, the clinical complexity of ovarian cancer is due, in part, to histological and molecular heterogeneity. Ovarian tumors are classified into four major classes: high (70%) and low (4.1%) grade serous, endometrioid (8.3%), clear cell (9.5%) and mucinous (3.2%) carcinoma [37–39]. Of note, a study by Beckmeyer-Borowko and colleagues showed that non-Hispanic Black ovarian cancer patients were more likely to be diagnosed with stage four HGSOC, clear cell or mucinous carcinomas when compared to non-Hispanic White patients [40]. Beyond these classifications, ovarian epithelial tumors have been divided into Type I and Type II tumors [41–44]. Type I tumors typically encompass low-grade and indolent tumors including low-grade serous, low-grade endometrioid, clear cell, and mucinous carcinomas that tend to present as stage I tumors while Type II tumors include more aggressive and high-grade tumors including high-grade serous, high-grade endometrioid, malignant mixed mesodermal tumors, and undifferentiated carcinomas [42]. Genetically, Type I and II tumors are characterized by specific mutations: mutations common to Type I tumors include KRAS, BRAF, ERBB2, CTNNB1, PTEN, PIK3CA, ARID1A, and PPP2R1A, while Type II tumors have a high frequency of TP53 mutations (>95%) as well as mutation or aberrant expression of BRCA1 or BRCA2 [3,42–44]. Importantly, these mutations appear to be largely confined to each subtype with Type II tumors rarely expressing Type I mutations and Type I tumors being largely wild-type for TP53, except for low-grade mucinous tumors (~25%) [45].

While more than 92% of Stage I ovarian cancer patients are successfully treated, only 15% of patients are diagnosed with the early stage disease [1]. High-grade serous ovarian cancer (HGSOC) is the most prominent form of ovarian cancer and accounts for 70% of ovarian cancer-related deaths [46,47]. Although most patients will initially respond favorably to standard-of-care cytoreduction surgery followed by platinum- and taxane-based treatment, approximately 80% will eventually relapse and develop resistance in late stage disease [39,48–51]. In addition, 25% of patients are inherently resistant to standard-of-care therapy and demonstrate disease progression within six months of treatment [52]. More recent studies have determined that HGSOC tumors are characterized by homologous recombination deficiencies (HRD) which render these tumors sensitive to PARP inhibition [53–56]. As such, PARP inhibitors (olaparib, rucaparib, niraparib) were FDA approved for treatment of platinum-sensitive recurrent, BRCA mutated, and HRD-positive epithelial ovarian cancer [34,56–61]. In addition, olaparib in combination with bevacizumab has been approved for the treatment of patients with advanced epithelial ovarian cancer. This combination treatment nearly doubled the progression-free survival in HRD-positive tumors when compared to bevacizumab alone [62]. Finally, clinical trials examining the impact of multiple novel combinatorial strategies, including VEGF inhibitors (VEGFi) in combination with PARP inhibitors (PARPi) as well as anti-PD-1 inhibitors, alone or in combination with VEGFi and/or PARPi, are ongoing [63]. While significant advances in the molecular characterization of ovarian cancer have led to a better understanding of this disease, the prognosis has not significantly improved over the past several decades; poor prognosis is attributed to lack of early detection and resistance (inherent and acquired) to platinum-/taxane-based therapies [3,46].

Despite the inherent differences in clinical manifestation between breast and ovarian cancer, a portion of these malignancies are intrinsically linked, as women with specific inherited germline mutations including BRCA1, BRCA2, PALB2, TP53, CDH1, and PTEN have an increased lifetime risk of developing either disease [64,65]. BRCA1 or BRCA2 mutations are the most prevalent cause of high penetrance inherited breast or ovarian cancers and have been shown to affect patients irrespective of race or ethnicity. Overall, the rate for germline BRCA1 or BRCA2 mutations is relatively low with 4–6% of breast and 8–15% of ovarian tumors expressing one of these mutations [66–73]. However, it is estimated that 39–63% of women with a BRCA1 mutation will develop ovarian cancer while 46–87% will develop breast cancer by age 70. Likewise, BRCA2 mutation carriers are strongly predisposed to develop ovarian (17–27%) or breast (38–84%) cancer [65,74–77]. Clinically, BRCA1/2-mutated breast tumors tend to be classified as TNBC invasive ductal carcinoma with high nuclear grade while BRCA1/2-mutated ovarian tumors are predominantly classified as HGSOC [78–80]. Although BRCA-mutated breast and ovarian tumors are often highly aggressive, a number of studies suggest that these patients may achieve a slightly better short-term therapeutic response (2–3 year overall survival) compared to patients with wild-type BRCA1 or BRACA2, as these tumors may be more responsive to DNA-damaging drugs; however, long-term survival and/or progression-free survival differences remain unclear [65,79–82].

While inherited breast and ovarian cancers have similar features, including response to specific inhibitors, as we will discuss below, non-familial ovarian and some subsets of breast tumors also demonstrate striking genome- and proteome-wide similarities including somatic mutations, patterns of copy number alterations, and expression of specific genes, proteins and signaling pathways. By utilizing this information, more recent treatment strategies for breast and ovarian cancers have begun to incorporate targeted therapies in conjunction with standard-of-care treatments [18,19,21,83]. While these novel regimens have improved clinical response and quality of life, as we have discussed, these treatments are often limited to patients with specific genomic alterations or clinical subtypes and not all patients will respond equally. These observations highlight the need to not only develop new, more effective therapies but also illustrate that it is necessary to develop a genome- or proteome-wide portrait of the underlying molecular heterogeneity of each of these diseases. Gaining a more complex view of the underlying biological mechanisms driving disease development, progression, and response to treatment will allow investigators to identify and develop biomarkers that will enable the design and evolution of treatment regimens based on the underlying biology of a given patient's tumor.

3. Molecular Classification and Characterization of Breast Cancer

Seminal studies by Perou and colleagues used microarray-based gene expression profiling and unsupervised hierarchical clustering to identify a 496 intrinsic gene list that defined five molecularly distinct subtypes of breast cancer [84,85]. These subtypes clustered largely along the estrogen receptor status with ER-positive tumors being classified into luminal A (LumA) or luminal B (LumB) subtypes while ER-negative tumors were classified as HER2 enriched (HER2E), basal like, or normal like [84–87] (Figure 1).

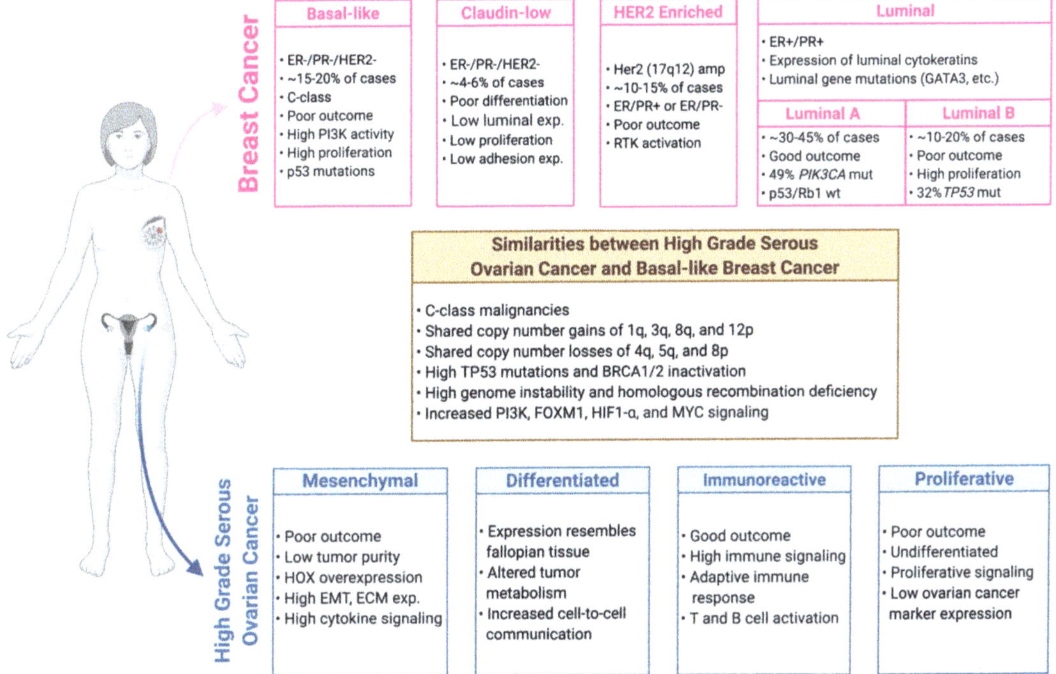

Figure 1. Gene expression-based classification of breast and ovarian cancers. The major molecular classifications of breast and ovarian cancers are depicted here. Further highlighted are the molecular similarities between high grade serous ovarian and basal-like breast cancer.

The ER-positive luminal tumors express luminal cytokeratins 8 and 18 and are enriched for genes expressed by breast luminal epithelial cells, including GATA3, FOXA1, ESR1, and MYB. Among luminal tumors, LumB tumors are defined by higher expression of proliferation-related genes, high genomic risk, and poorer clinical outcome than LumA tumors. HER2E tumors are predominantly ER negative, characterized by the amplification of the HER2 gene on chromosome 17q12, and are associated with poor prognosis and increased risk of metastasis. The basal-like subtype is largely synonymous with triple negative breast cancer (TNBC). These tumors express basal epithelial cell markers keratin 5/6 and are characterized by enrichment of the genes expressed by breast basal or myoepithelial cells [2]. Basal-like tumors represent the most diverse subtype of breast cancer and are associated with high proliferation rates, high mutational burden, higher risk of metastasis, and poor survival rates [85,86]. Finally, the normal-like breast cancer subtype has also been described and is typified by high expression of genes known to be expressed by basal epithelial cells and adipose cells. However, the biological relevance and clinical importance of this subtype remains unclear [84–87].

The association between molecular subtypes and disease-specific outcomes demonstrate that tumor cell response to treatment is not determined by anatomical prognostic factors but rather inherent molecular features, indicating the potential clinical value of these expression-based patient classifications [84,85]. However, the 'intrinsic' gene set used by Perou and group to experimentally categorize patients was not readily employable in the clinic due to its relatively large size [84–87]. Utilizing microarray data and several minimization methods, Parker et al. developed a reliable 50-gene signature to identify breast cancer intrinsic subtypes [88]. Combined with common histologic criteria, such as tumor grade and pathologic staging, the 50-gene signature (PAM50) provided significant prognostic and predictive value through classification and generating risk-of-relapse

(ROR) scores for all patients [88]. While the clinical implications of the PAM50 subtype predictor remain to be fully resolved, the Prosigna assay, which is derived from the initial intrinsic analyses, is used clinically to help predict risk of relapse and to guide therapeutic intervention [89–91].

The recent advances in gene expression profiling platforms have led to the identification of additional molecular subtypes, further defining the biological and clinical heterogeneity of breast cancer. The claudin-low subtype was identified to be predominantly triple negative and poorly differentiated subgroup of breast tumors which are enriched for cancer stem cell-like genomic signatures and immune response genes [92,93]. These tumors are characterized by low expression of luminal genes, proliferation genes, and genes involved in tight junctions and cell–cell adhesion [92,93]. More recent gene expression studies employed by Lehmann et al. initially categorized TNBC tumors into six molecular subtypes, including BL1 and BL2 (basal-like), immunomodulatory (IM), mesenchymal (M), mesenchymal stem-like (MSL), and luminal androgen receptor (LAR) [94]. This classification has since been further refined to include the four (TNBCtype-4) tumor-specific subtypes (BL1, BL2, M, and LAR) and exclude the IM and MSL subtypes due to the identification of transcripts from infiltrating lymphocytes and tumor-associated stromal cells, respectively [95]. The TNBC type-4 subtypes demonstrated significant differences in histopathology, grade, and local and distant disease progression [95]. These subtypes were characterized by unique identities of pathway activation which stimulated the use of known inhibitors and therapies to exploit signaling vulnerabilities, exhibiting early evidence of clinical applicability [94,95].

Decomposing vast amount of information from profiling studies represents a key step in developing patient-specific therapeutic regimens. In light of this, pathway signatures were developed as an underlying platform to provide a functional interpretation of the gene expression data within each subtype and further dissect the heterogeneity of breast cancer [96]. Integrated analysis using gene expression and pathway activation probabilities contributed to stratifying tumor subtypes and characterizing distinct clinical and biological features [96–100]. Along these lines, Gatza et al. utilized pathway activation probabilities that reflect in vivo activity levels to identify subgroups that reflect the status of important signaling pathways in breast tumors [97]. These subgroups corresponded to the intrinsic subtypes and exhibited distinct patterns of pathway activation, DNA copy number changes as well as clinical and biological characteristics [97,98].

While microarray-based gene expression profiling of breast tumors has been able to distinguish tumor subgroups and begin to define underlying biological and clinical diversity, these studies were limited in their ability to create true "molecular portraits" of breast cancer. Large-scale integration of multiple proteogenomic platforms through The Cancer Genome Atlas (TCGA) project provided a more comprehensive view of breast cancer heterogeneity and underlying biology. The TCGA project ($n = 1072$) used data from six different high-throughput technology platforms, including mRNA expression microarrays (and mRNA sequencing), DNA methylation, genomic DNA copy number arrays, microRNA sequencing, whole-exome sequencing, and reverse-phase protein array (RPPA) to examine specific genetic, epigenetic, and proteomic alterations in breast cancer and to link these alterations to clinical data and characteristics [2]. Intriguingly, while the overall patterns of proteogenomic alterations were found to be variable amongst patients, including between subtypes, intra-subtype variation was limited. Remarkably consistent patterns of genomic and proteomic alterations were found to be associated with each of the mRNA-based PAM50 subtypes.

Luminal tumors are characterized by an increased frequency and diversity of significantly mutated genes in addition to a lower frequency of copy number alterations [2,101,102]. These tumors exhibit increased mutations in luminal genes including GATA3 and FOXA1, as well as genes belonging to the p38-JNK pathway (MAP3K1 and MAP2K4), which were mutated in a mutually exclusive manner. PIK3CA, which is the most frequently mutated gene in breast cancer, was predominantly altered in luminal tumors and was mutated at a much higher frequency in LumA (45%) relative to LumB (29%) tumors. Despite the high frequency of activating PIK3CA mutations in LumA subtype tumors, the PI3K/AKT signaling axis has not been shown to be consistently upregulated in these tumors. In contrast to LumA tumors, LumB tumors are characterized by higher inactivation of the TP53 pathway associated with a higher rate of mutation in the TP53 gene, loss of ATM2, and MDM2 amplification [2]. More recent integrative analysis using 52 gene expression signatures that measure oncogenic signaling pathways identified a limited number of genes that are amplified and overexpressed in aggressive luminal subtype tumors. Among these genes, a subset (FGD5, METTL6, CPT1A, DTX3, MRPS23, EIF2S2, EIF6, and SLC2A10) was found to be essential for cell growth and, in some instances, correlated with clinical outcome [99,103–105]. This study further suggests that not only do LumA and LumB tumors express unique mutation profiles, but that these alterations result in distinct patterns of oncogenic signaling beyond differences in proliferation.

The HER2E subtype is characterized by high amplification of the HER2 amplicon (80%) on chromosome 17q12. These tumors can be either ER negative or positive and demonstrate increased expression of the HER2 oncogene as well as other genes on the 17q12-amplicon, including GRB7. However, not all clinically defined HER2-positive tumors are categorized into this subtype, as some ER+/HER2+ tumors demonstrate increased expression of specific luminal genes (i.e., GATA3, BCL2, and ESR1) and cluster largely into the LumB subtype. TP53 (72%) and PIK3CA (39%) mutations are highly enriched in this subtype and show significantly higher expression and activation of receptor tyrosine kinases such as FGFR4, EGFR, and HER2 [2].

Basal-like breast cancers represent the most heterogeneous subtype with a high frequency of TP53 mutations which are present in an overwhelming 80–90% of tumors. In addition to TP53 truncating mutations, these tumors are characterized by loss of RB1 and BRCA1 along with amplification and hyperactivation of the MYC and FOXM1 genes. Increased activation of PI3K/AKT signaling, relative to other subtypes is a distinguishing feature of basal-like tumors despite a low incidence of PIK3CA (9%) mutations. Expression of keratins 5, 6, and 17 and cell proliferation genes are significantly upregulated in these tumors owing to the increased expression of FOXM1 as a transcriptional driver of this gene signature [2].

Similar multiplatform analysis was also conducted to provide molecular context to invasive lobular breast cancer, which is the second most commonly diagnosed invasive breast cancer and comprise approximately 10–15% of all cases. Despite histological differences, invasive lobular carcinomas (ILC) and ER+ invasive ductal carcinomas (IDC) patients have historically been treated similarly, emphasizing the need to more robustly understand the molecular underpinnings of the disease for better therapeutic interventions [106]. Multiplatform studies carried out by the TCGA project and Desmedt et al. identified mutations in the E-cadherin (CDH1) gene (63% in ILC vs. 2% in IDC) which is the hallmark feature of ILCs. In addition to CDH1 loss, mutations in PTEN, TBX3, FOXA1, and ESR1 were enriched in ILC relative to IDC tumors. Mutations in PIK3CA were reported in 48% of ILC relative to 33% of IDC tumors which, along with loss of PTEN function, defines the significant upregulation of PI3K signaling in ILC tumors [23,106]. Transcriptomic analysis identified molecular ILC subtypes which were characterized by unique molecular profiles and clinical outcomes with more proliferative tumors demonstrating a worse clinical prognosis [23,106]. Overall, these multiplatform analyses not only better distinguished between lobular and ductal carcinomas but also identified clinically relevant heterogeneity that may help to better differentiate and treat these carcinomas. In addition to TCGA, the METABRIC (Molecular taxonomy of breast cancer international consortium) study used an integrated clustering approach to examine the genomic and transcriptomic architecture of 2000 breast tumors (along with clinical data) and classify them into 10 integrative clusters (IntClust 1–10) which demonstrate distinct alterations and clinical outcomes [9,107]. Importantly, this classification strategy demonstrated that incorporation of both mRNA and cDNA copy number data identified additional granularity within the PAM50 subtypes as well as molecularly distinct entities based on the underlying genetic alterations. These data, coupled with multi-platform orthogonal analyses performed by TCGA have provided enormous insight into the underlying genetic framework of breast cancer; however, these studies were limited in their ability to associate the genomic and transcriptomic features with the proteome and phosphoproteome that drives the phenotypic characteristics of a tumor. The RPPA platform used by TCGA for quantifying protein abundance and post-translational modifications is limited by antibody quality and an inability to detect mutant protein forms.

Consistent with this premise, analysis of the proteome and phosphoproteome was performed by the Clinical Proteomic Tumor Analysis Consortium (CPTAC) using mass spectrometry-based analyses to integrate and contextualize genome-scale alterations of 105 tumors and adjacent normal samples [5]. In breast cancer, these analyses resulted in the identification of an average of more than 11,000 proteins and 26,000 phosphosites per tumor significantly extending the previous work from TCGA where only 141 proteins and 31 phosphosites were captured [2,5,23]. Phosphoproteomic analysis informed the translational outcomes of PIK3CA mutations in breast cancer, which often are not correlated with the transcriptional signature of breast tumors. These analyses resulted in the identification of 62 different phosphosites in PIK3CA mutated breast tumors, including *RPS6KA5* and *EIF2AK4*, explaining the activation of the pathway and revealing possible druggable kinases in this pathway [5]. The CPTAC project highlights the need for integrating data across proteogenomic platforms to connect somatic mutations with the activation of various oncogenic signaling pathways in tumors for better therapeutic outcomes. In addition, CyTOF (Cytometry by Time of Flight), has been used for real-time high-dimensional analysis of breast cancer [108–110]. For example, a recent study by Ali et al. emphasized the significance of multiplatform analyses when coupled with multidimensional imaging mass cytometry in highlighting the tumor heterogeneity both on tumor-specific and tumor microenvironment levels which in turn affect the tumor evolution, ecosystem and clinical outcomes [109]. Similarly, imaging mass cytometry has been used to generate high-dimensional images of 281 human breast tumor samples in order to identify the spatial architecture, and to define heterogeneity between intra and inter-tumoral cell subpopulations [110].

4. Molecular Classification and Characterization of Ovarian Cancer

Similar to studies in breast cancer, studies by Tothill (n = 285) [111], the TCGA project (n = 489) [3], Helland (n = 939) [112], and Konecny (n = 174) [113] utilized K-means clustering and non-negative matrix factorization consensus clustering to classify HGSOC into four distinct gene expression-based subtypes. These four molecularly distinct subtypes (Figure 1) were termed immunoreactive, proliferative, mesenchymal, and differentiated based on molecular and clinical characteristics [3]. However, in contrast to molecular subtypes of breast cancer which have clear biological and clinical implications, these relationships do not appear to be as robust in HGSOC.

Mesenchymal tumors have been reported to have the worst clinical prognosis of the four HGSOC molecular subtypes [111,113,114]. These tumors are defined by low tumor purity and demonstrate increased desmoplasia and reactive stromal components, including CD3+ infiltrates [4,111,115]. Phenotypically, these tumors exhibit increased epithelial to mesenchymal transition (EMT), angiogenesis, extracellular matrix (ECM) remodeling, and proteolysis [113,115]. Consistent with these findings, mesenchymal subtype tumors demonstrate increased expression of HOX genes which contribute to development regulation as well as aberrant TGFβ, stromal-associated, wound response, and fos-jun signaling as demonstrated by gene expression signatures [116]. Global proteomic analyses by the CPTAC project further demonstrated that these tumors exhibit increased expression of ECM and cytokine signaling at the protein level [4].

Similar to mesenchymal subtype tumors, immunoreactive tumors are defined by low tumor purity [4]. However, immunoreactive subtype tumors are associated with a good clinical prognosis [111,113–115]. While these tumors do demonstrate infiltration of stromal cells, immunoreactive tumors appear to be defined by increased immune signaling, likely due to increased immune cell infiltration [117]. Gene and protein expression profiling studies have reported activation of the adaptive immune response as well as increased T and B cell activation markers, antigen presentation, and chemokine signaling [3,111,113,115]. Consistent with these findings, it was reported that mesenchymal and immunoreactive tumors are more closely related to each other, as compared to the proliferative or differentiated subtypes, despite differences in patterns of signaling network activity and clinical outcomes [113,118]. These similarities are likely due to the low tumor cell purity that is apparent in mesenchymal and immunoreactive tumors, while the distinction between these groups is driven by both underlying tumor biology as well as the composition of infiltrating cell populations in the tumor microenvironment. Consistent with these ideas, recent single-cell RNAseq studies have demonstrated that unique aspects of the tumor microenvironment may define signaling within these subtypes; immunoreactive tumors were shown to have immune-related cell clusters while mesenchymal tumors contained cell clusters enriched for cancer-associated fibroblast signaling [117].

Tumors classified in the proliferative subtype are associated with poor overall survival [106,107,109]. In contrast to immunoreactive or mesenchymal tumors, these tumors exhibit high tumor cellularity and low infiltration of CD3+ and CD45+ stromal cells [111,112]. Proliferative subtype tumors are defined by an undifferentiated phenotype and express pro-proliferative signaling including increased expression of developmental transcription factors, proliferation markers, ECM-related genes, and WNT/β-catenin signaling, as well as increased expression of proteins involved in DNA replication [3,4,111,113]. In addition, it has been noted that these tumors express low levels of ovarian cancer marker genes (MUC1, MUC16, KLK6, KLK7, and KLK8) and high expression of the developmental transcription factors HMGA2 and SOX1. These tumors were also associated with an increased expression of FANC genes and homologous recombination [113,115].

Finally, differentiated subtype tumors have been shown to most closely resemble normal fallopian tissue at the gene expression level [111,115]. At the genetic level, these tumors are defined by increased expression of MUC1, MUC16, SLP1 (secretary fallopian tube marker), epithelial cell differentiation markers, and folliculogenesis-related genes which are indicative of increased tumor cell differentiation [3,113,115]. Proteomic analyses from the CPTAC project were able to further dissect signaling networks activated in these tumors to identify enrichment of protein expression programs associated with altered tumor cell metabolism and increased cell-to-cell communication [4] providing additional insight into subtype-specific mechanisms driving tumor development and progression.

Although these seminal studies were able to identify four largely concordant subtypes based on gene expression profiling, a number of recent studies have suggested that these subtypes are not consistent across platforms and populations [113–115,118]. These more recent studies have observed that tumors were able to be more robustly classified into fewer groups and/or that alternative strategies may provide additional insight into the underlying biology of this disease. Notably, studies from the CPTAC project were able to utilize proteome-wide data from 9600 proteins and 6769 phosphoproteins from 174 tumor samples to identify altered signaling networks in the transcriptome based subtypes further refining and validating the distinct signaling networks in these tumors, as well as identifying signaling pathways correlated with homologous recombination deficiency phenotype and patient survival [4]. However, in this proteogenomic analysis of ovarian tumors, Zhang et al. also identified five distinct protein-based subtypes and were able to show that three of the five subgroups were largely concordant with the TCGA mRNA-based subtypes. The remaining two subgroups represented tumors defined by unique underlying biology that would not be apparent by assessing mRNA data alone. Consistent with this premise, a number of recent studies have attempted to move beyond mRNA- or protein-based approaches to incorporate phosphoproteomic or glycoproteomic profiling to investigate the heterogeneity of HGSOC tumors [119,120]. These studies have provided additional depth to our understanding of HGSOC tumorigenesis by identifying subgroups defined by unique patterns of active kinases and altered cell signaling which contributing to tumor development, progression, and clinical outcome. Likewise, recent work by Karagoz et al. [116] assessed patterns of oncogenic signaling using a panel of 62 gene expression-based signatures across the four TCGA subtypes in three unique datasets [3,111,121]. As noted above, these studies identified unique oncogenic and tumorigenic signaling pathways associated with each mRNA-based subtype. However, in contrast to similar analyses in breast tumors which demonstrated clear differences in pathway patterns between the PAM50 subtypes, the distinctions amongst ovarian subtypes appeared to be more subtle and included increased intra-subtype heterogeneity [99,116].

Collectively, these data reinforce the premise that ambiguity in HGSOC subtype assignment could be a result of shared common biological underpinnings, the existence of intermediate subtypes, or biased by tumor cellularity and/or composition. As such, it is apparent that further refinement of the molecular subtypes, potentially through the incorporation of multiple genomic or proteomic platforms, may be necessary for these classification schemes to be clinically relevant.

At the molecular level, HGSOC has been classified as a C-class malignancy (chromosomally unstable) that is defined by extensive structural variants [102]. Consistent with this classification, mutational profiling of HGSOC by the TCGA project using whole-exome sequencing has identified a limited number of significantly mutated genes that define this disease [3]. The most prominent among these is TP53 mutations which are evident in nearly all patients and are believed to arise early in the transformation process [3,44,122–124]. Beyond altered p53 signaling, transforming oncogenic mutations in PIK3CA, BRAF, KRAS and NRAS have been detected in HGSOC, albeit at low frequencies (<1%). Almost half of HGSOC tumors are characterized by homologous recombination (HR) deficiency through germline or somatic mutations in BRCA1/2 (20%), BRCA1 hypermethylation (11%), and/or dysregulation of other HR genes including PTEN, ATM or ATR, RAD51C, EMSY and Fanconi anemia genes [3]. While few significant mutations are apparent in HGSOC, DNA copy number alterations are more frequent in these tumors [3,102]. This includes amplification of MECOM, MYC, and CCNE1 which are among the most significant focal amplifications and found in more than 20% of HGSOC cases in addition to KRAS and MAPK1 which are found in more than 10% of cases.

Interestingly, while specific genes are mutated at a low frequency in HGSOC, pathway analyses incorporating orthogonal whole-exome sequencing and copy number data demonstrated that HGSOC tumors are characterized by aberrant RB1/E2F (67%), PI3K/RAS (45%), and NOTCH (22%) signaling as well as dysregulation of the FOXM1 transcription factor network (87%) [3]. Further pathway analysis, based on phosphoproteomic profiles of HGSOC tumors demonstrated differential expression of RhoA-regulatory, PDFRB, and integrin-linked kinase pathways between poor and good prognostic HGSOC patients [4].

Finally, a number of recent studies have used integrative analyses to identify novel oncogenes and tumor suppressors that promote HGSOC and biomarkers to predict therapeutic response and risk. These studies relied on integrative analyses of DNA copy number, methylation, and gene expression data to identify potential oncogenes and tumor suppressor proteins in HGSOC and clear cell carcinoma [125–127]. Similarly, studies from Karagoz et al. assessed orthogonal genomic and proteomic data from human HGSOC tumors from the TCGA and CPTAC studies in the context of a prognosis gene expression signature. These analyses, along with data from a genome-wide RNAi screen in ovarian cancer cell lines, identified ADNP as a novel oncogene in HGSOC and in vitro studies showed that this protein regulates cell survival through altered cell cycle checkpoints [116]. While the therapeutic potential of these genes remains unclear, studies by Kurimchak et al. incorporated kinome profiling of human tumors and PDX models to identify MRCKA as a potentially drug-able oncogene activated in a subset of HGSOC tumors. Subsequent loss-of-function studies demonstrated that this gene could regulate HGSOC tumorigenesis and could be pharmacologically inhibited suggesting it may have potential as a novel therapeutic target [128]. Finally, studies from Coscia et al. identified CT45 as a biomarker for platinum-sensitivity in HGSOC using global proteomic profiling and demonstrated that mRNA or protein expression was associated significantly with chemosensitivity and disease-free survival [129].

5. Genetic and Genomic Relationship between Breast and Ovarian Tumors

As discussed above, both familial and non-inherited breast and ovarian cancers have been shown to have similar genetic and genomic features (Figure 1). Beyond the previously discussed correlation between inherited mutations and the increased risk of breast or ovarian tumor development, analysis of human breast tumors demonstrated that HGSOC tumors also express a basal-like gene expression signature [2]. This relationship was further validated by multi-platform genomic analyses in which basal-like and HGSOC tumors were found to have a strong genomic association based on global mRNA profiling and to express a similar pattern of DNA copy number alterations [130]. While similar patterns of gene expression between these two diseases were noted by Hoadley and colleagues in a pan-cancer analysis of 12 tumor types and by the TCGA breast cancer paper, this association was not as clear when studied within the context of 33 tumor types potentially reflecting differences driven by tumor cell of origin, additional variability due to a more diverse tumor population, or other technical or biological factors [2,130,131]. Regardless, basal-like and high-grade serous ovarian tumors are classified as C-class malignancies and are characterized by predominant recurrent copy number alterations [102]. Specifically, these tumors share copy number gains of 1q, 3q, 8q, and 12p, and copy number losses of 4q, 5q, and 8p [2,100]. Among the commonly amplified genes are MYC (8p21.21), CCNE1 (19q13.2), MECOM (3q26.2), FGF3 (4p16.3), MCL1 (1q21.3) and ERBB3 (12q13.2) [43]. Additionally, basal-like and HGSOC tumors share RB1 loss in 20% and 10% of tumors, respectively [2,3].

Beyond copy number alterations, these tumor subtypes have been shown to express similar mutation profiles for a limited number of key oncogenes and tumor suppressor genes. Basal-like and high-grade serous ovarian tumors are enriched for BRCA1/2 inactivation and express TP53 mutations in 90–95% of tumors [2,3,5]. In addition, both tumor types exhibit an increased frequency of genome breakpoints as well as a loss of heterozygosity and allelic imbalance indicating genomic instability and homologous recombination deficiency [132–134]. More recent studies have indicated that these tumors demonstrate high homologous recombination deficiency (HRD) scores, accumulation of large-scale state transitions, increased loss of heterozygosity (LOH), and telomeric allelic imbalance scar signatures. Clinically, these alterations have been shown to be significantly correlated with pathologic complete response and minimal residual disease in TNBC patients treated with platinum-based therapies and with a better prognosis in HGSOC [36,135,136].

In addition to specific mutations and genomic alterations, basal-like breast and HGSOC tumors have been shown to express similar signaling networks including increased activation of PI3K signaling [2,3,102,137–139]. While PIK3CA mutations are relatively rare events in each tumor type, a number of unique alterations have emerged as contributing to aberrant signaling [2,3,5]. In HGSOC, DNA copy number gains in PIK3CA (18%), AKT1 or AKT2 (9% combined) and to a lesser extent, homozygous deletion of PTEN (7%) are the main drivers for this pathway [3,140,141]. In contrast, basal-like tumors are regulated by alterations in multiple genes (EGFR, IGFR1, AKT3) that occur at a low frequency (2–4%), as well as a loss of PTEN (35%) or INPP4B (30%), SOX4 amplification and overexpression, and MAGI3-AKT3 gene fusion [2,137,142,143]. Interestingly, these data indicate that while both tumor types are characterized by high PI3K signaling, the mutations activating signaling in each tumor type differed in prevalence and composition.

Similarly, on a pathway activity level, basal-like and HGSOC tumors share increased FOXM1, HIF1-α, and MYC signaling [2]. Basal-like breast cancers have increased altered cell cycle checkpoint regulation, DNA damage repair, MYC, and immune response signaling [5], while proteins associated with recurrent copy number alterations in HGSOC converge on cell migration/invasion and immune regulation pathways [4]. Consistent with common alterations between basal-like and high-grade serous ovarian tumors, Marcotte and colleagues [144] used a genome-wide pooled shRNA screen in 29 breast and 15 ovarian cancer cell lines to identify genes uniformly essential for cell viability as well as genes required within each disease type. While cell line-specific genes were identified, these analyses also identified 66 ovarian and 155 breast cancer-specific genes as well as 297 genes that were essential for viability in the majority of cell lines irrespective of tissue type. While the latter set of genes did not necessarily take into account distinctions between molecular subtypes, these studies further reinforce the shared underlying biology of these diseases.

6. Advances in Genomic Analyses of Breast and Ovarian Cancer

Single-platform genomic and proteomic analyses have allowed for the identification and cataloging of mutations; copy number alterations; and altered gene, miRNA, protein, or phosphoprotein expression profiles [2,3,7,12,23,145–152]. As we have outlined above, patterns of genomic and proteomic alterations can define tissue- and histological-specific differences in underlying biology and can be used to define molecularly distinct subtypes of cancer, including breast and ovarian cancer [2,3,11,23,97,118–120,136,153–155]. These genomic and proteomic patterns can identify oncogenic mechanisms that contribute to disease development, progression and in some instance can serve as therapeutic targets or markers of therapeutic response [10,62,63,99,102,113,126,128,129,135,137,142,156–172]. However, single platform analyses can be limited in their ability to visualize altered signaling networks and oncogenic processes. Given the complexity of mechanisms regulating these processes, multiplatform analyses, incorporating orthogonal genomic and proteomic data, enable the visualization of various types of alterations, in multiple key components within a given network to better define the state of signaling within specific tumors types and/or subtypes [102]. More importantly, integrative multiplatform analyses have led to the comprehensive identification of actionable alterations through reverse engineering of signaling pathways, while identifying upstream effectors and downstream targets using multiple omics platforms [156,173–176] (Figure 2). Ultimately, integrative analyses have resulted in the discovery of novel tumor-promoting mechanisms with higher confidence.

Breast and ovarian tumors are comprised of a complex collection of cell types including multiple populations of tumor cells, stroma, immune cells, fibroblasts, and other cells that encompass the tumor microenvironment [177,178]. As we have discussed, omics technologies that rely on analysis of the entire tumor (i.e., bulk analysis) have provided an enormous amount of insight into tumor biology; however, these approaches represent an averaged view of the tumor landscape and do not allow for fine resolution at the single cell level. Although a number of approaches including ESTIMATE, and others, have been developed to delineate specific signaling networks that arise from discrete cell populations or to estimate differences in cell composition within tumors using bulk sequencing or proteomic data, these methodologies are unable to fully address these challenges [179]. Advances in single-cell omics have had a significant impact on our understanding of tumor characteristics that are not apparent by bulk genomic, proteomic, or metabolomic approaches. These methods have allowed us to identify and characterize unique cell subpopulations, distinguish cell transition states, map molecular markers, identify novel and previously unrecognized biological features, and in combination with other technologies, are beginning to be used to spatially map tumor cell populations, identify circulating tumor cells and provide mechanistic insight into tumorigenic processes including metastasis and therapeutic response. Given spatial limitations, we point our readers to an excellent collection of review articles that discuss these advances in depth [180–191].

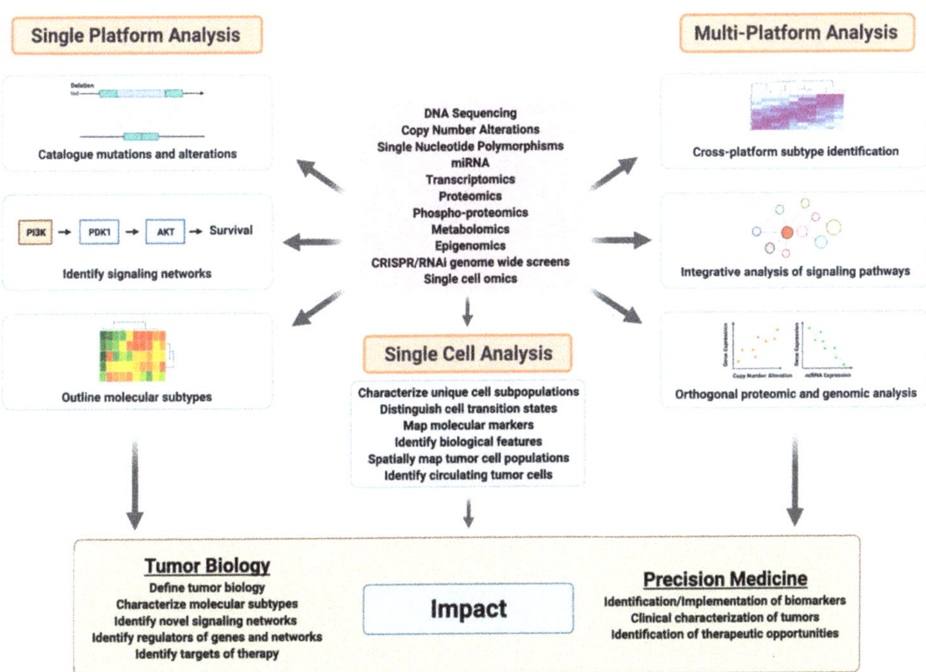

Figure 2. Use of single platform and integrative omics in cancer biology and medicine. The major contributions of single and multi-platform omics studies as well as single cell omics are summarized here. Single platform studies enable cataloging of mutation or alteration patterns, identifying signaling networks of interest and defining certain molecular subtypes. Multiplatform studies can further expand single platform-defined molecular subtypes and identify signaling pathways by identifying mutations in multiple genes representing multiple levels of pathway dysregulation. Single cell analyses allow for analyses of tumor cell subpopulations, identify cell transition states, map molecular markers and cell populations and identify circulating tumor cell populations. Orthogonal analysis of these data provides further context to genomic studies. These approaches contribute to a greater understanding of tumor biology as well as clinical advancements in treating cancer.

A number of recent studies have employed single-cell RNA sequencing (scRNA-seq) analyses to examine tumor immune profiling [192–197]. These approaches have clear implications for both our understanding of the role of the immune system in the tumor microenvironment and for determining or predicting the efficacy of immunotherapy-based treatments. Of note, a recent study by Azizi et al. demonstrated the wide variability in immune cell type composition between breast cancer patient samples in addition to high-lighting the phenotypic expansion of intratumoral immune cells using single-cell RNA and T cell receptor sequencing [198]. Further studies by Savas and colleagues demonstrated the association between tissue-resident memory T cell differentiation signature, developed using single-cell RNA-seq, and prognosis in early stage triple-negative breast [199]. Demonstrating the potential clinical implications and applicability of these approaches, investigators have used these technologies to identity mechanisms of therapeutic resistance [200]. Notably, recent work identified enrichment of immunosuppressive immature myeloid cells (IMC) in anti-Her2 and CDK4/6 inhibitor-resistant HER2-positive breast cancer, while combinatorial treatment with cabozantinib (IMC-targeting tyrosine kinase inhibitor) and immune checkpoint blockade overcame resistance [201]. Moreover, scRNA-seq has been used to develop gene-expression-signature of the myeloid-derived suppressor cells (MDSCs) in addition to identifying CD84 as a surface biomarker for MDSCs in breast cancer [202]. Similarly, Wan et al. reported reprogramming of inert natural killer and T cells to a highly active cytotoxic state following bispecific anti-PD-1orPD-L1 antibody treatment using single-cell RNA-seq analysis of HGSOC organoid co-cultures; this study identified a potential advantage of bispecific antibodies in immune checkpoint blockade therapy in HGSOC [203]. Further analyses have identified inter- and intra-tumor heterogeneity in cancer associated fibroblasts cell states in HGSOC and breast cancer [117,204]. Collectively, immune profiling coupled with imaging and single-cell RNA-seq underscored the importance of the spatial architecture of tumor niches in regulating immune infiltration and activation [205–212].

A number of recent studies have employed single-cell analyses to investigate inter and intra- tumoral heterogeneity [213–218]. Of note, recent work by Chung et al. linked tumor-intrinsic and immune cells diversity with TNBC intratumoral heterogeneity while studies by the Ellisen laboratory identified a connecting between intertumoral heterogeneity and clonality of inferred genomic copy number changes in these tumors. These latter studies suggested that cellular genotype drives gene expression programs, including signatures of treatment resistance and metastasis, in individual tumor cell populations [219,220]. Consistent with this premise, investigators have identified rare plastic pre-adapted cell subpopulations in luminal breast tumors which showed resistance to acute endocrine treatment [221]. Similarly, studies by Izar et al. and Geistlinger et al. used scRNA-seq based analyses to link the transcriptomic-based subtype classification of HGSOC to tumor cell type composition rather than intrinsic difference in gene expression patterns present in tumor epithelial cells further highlighting the importance of considering specific subpopulations of cells and the impact of signaling from the microenvironment on tumor characteristics [117,222]. More complex analyses integrating single-cell RNA-seq coupled with cell lineage tracing has been used to detail tumor cell subpopulations that contribute to various aspects of tumor evolution, including identifying pre-EMT (Epithelial to Mesenchymal Transition) cells that are essential for metastasis initiation [223].

Beyond assessing the transcriptome, single-cell DNA sequencing approaches have been developed and used to identify subpopulations of cells that express unique mutational and CNA patterns in therapeutically actionable genes in a given breast tumor [224]. These findings have clear clinical implications as different subpopulations will be likely be uniquely sensitive or resistant to specific therapeutic regimens and contribute to the evolution of the tumor and therapeutic sensitivity. Consistent with this premise, longitudinal sequencing analyses of tumors have demonstrated the emergence and/or re-emergence of clonal populations following treatment [225–227]. Complementary to these studies, single-cell mass cytometry using CyTOF identified rare tumor subtypes in HGSOC in addition to the dominant subsets and demonstrated that one of the identified rare subtypes was enriched for EMT signaling and associated with increased tumor metastasis [228]. Finally, merging single cell proteomics with other omics analysis has enable investigators to capture tumor-immune interactions in breast tumors [212]. Collectively, single-cell omics underscored intra- and inter-tumoral heterogeneity, identified subpopulation-specific vulnerabilities and emphasized the importance of addressing these vulnerabilities with combinatorial targeted therapeutic options [117,214,220,229–233].

Traditionally genome-wide RNAi and CRPSR/Cas9 screens have identified novel essential genes and pathways [144,234,235]. These studies evolved to include chemo-genetic screens which incorporate loss-of-function screens coupled with drugs or small molecule inhibitors in order to identify drug-gene interactions, cancer genetic vulnerabilities, and potential drug resistance mechanisms [21]. More recently, investigators have begun to incorporate these studies with multi-dimensional genomic analyses of human tumors as an added functional filter to identify clinically relevant cancer vulnerabilities and potential novel therapeutic targets. For instance, Marctotte and colleague integrated a pooled shRNA screen with genomic, transcriptomic and proteomic data from 77 breast cancer cell lines to identify breast cancer subtype-specific vulnerabilities. In this study, PSMB3, PSMA6 and ATP6V1B2 were identified as top ranked "basal-selective" essential genes [234]. Likewise, integrative correlative studies between pathway-specific gene expression signature scores, gene level DNA segment scores and RNAi shRNA abundance led to the identification of 21 amplified, essential and putative driver oncogenes in highly proliferative luminal breast cancers as well as the identification of SOX4 as a driver of PI3K signaling in basal-like breast tumors [99,137]. Similarly, in ovarian cancer, systemic loss-of-function shRNA screen identified 50 essential and amplified genes including CCNE1, PAX8, FRS2, PRKCE, and RPTOR. Of note, PAX8 was found to be amplified in 16% of primary ovarian cancers while shRNA mediated silencing of PAX8 lead to apoptosis in cell lines harboring either PAX8 amplification or overexpression [235]. Likewise, ubiquitin B (UBB) and ubiquitin C (UBC) were identified as a paralog deficiency dependency in ovarian cancers, implying the essentiality of UBC in cell lines with repressed UBB [166]; shRNA mediated silencing of UBC in UBB repressed ovarian cancer xenograft model lead to tumor regression and prolonged survival [160]. More recent studies have evolved to employ machine learning algorithms for predicting functional cancer vulnerabilities while integrating shRNA (DEMETER2) or CRISPR/Cas9 (DepMAp) screens coupled with genomic and proteomic profiling of cancer cell lines [236]. Collectively, genome-wide RNAi and CRPSR/Cas9 loss-of-function screens made a significant contribution in identifying cancer dependencies, and potential novel therapeutic targets [166,237–239].

Unfortunately, spatial limitations prevent an in-depth discussion of the many tools, algorithms, and computational approaches that have been developed for a single platform and integrative analyses. However, biochemical and genetic-based studies as well as large-scale proteogenomic analyses have demonstrated that despite enormous tumor heterogeneity, molecular alterations often converge on a limited number of signaling networks, reflecting pathway activity levels and their role in driving tumor progression [14,102,147,149,240,241]. As a result, a number of tools and approaches have been developed, including the use of gene expression-based signatures, mutational signatures, CNA (copy number alterations) signatures, and protein signatures to quantify pathway activity [7,96,97,242–244]. These approaches include several that incorporate data from multiple technical platforms and use statistical modeling driven by a priori knowledge of signaling pathways and/or protein–protein interaction networks to cluster samples based on similarity networks, detect enriched signaling networks across multi-omics platforms and/or infer pathway activity from the expression or mutation profiles of established pathway components [3,8,23,170,245,246].

As we have outlined, large-scale genomics and proteomics studies including those from the TCGA, METABRIC, and CPTAC projects, as well as many other studies, have enabled the cataloging molecular alterations and signaling pathways in breast and ovarian cancers. While these studies have implications for our understanding of the underlying molecular mechanisms of these diseases, they also highlight the need to personalize therapeutic approaches based on the biology of each patient's disease [6,131,247]. Consistent with this premise, the use of genomic profiling, including DNA sequencing gene panels from Foundation Medicine and others, into clinical trials and practice has identified potential biomarkers, beyond standard immunohistochemistry (IHC)-based assays, to predict response and provided a means to personalize treatment. In addition to DNA sequencing-based assays, multiple molecular biomarkers are currently being used to monitor and track the progression of both ovarian and breast cancer [167,248,249]. For ovarian cancer patients, these biomarkers include single gene markers CEA (Carcinoembryonic Antigen), CA125 (Cancer Antigen 125), and HE4 (Human Epididymis protein 4), as well as multivariate index assays including OVA1, ROMA, and OVERA [250–257]. Similarly, several gene-expression based prognostic tests, including Oncotype DX [258], EndoPredict [259] and MammaPrint [260] as well as the aforementioned Prosigna assays, have been FDA approved to predict risk of recurrence in breast cancer and can be used to help guide clinical decisions. Oncotype DX Recurrence Score is based on the expression level of a panel of 21 genes, which is used to predict the likelihood of the 10-year tumor recurrence and guiding the adjuvant treatment options while weighing the added benefit of adjuvant chemotherapy versus treatment with hormonal therapy alone. Oncotype Recurrence Score stratified patient samples into low-, intermediate- and high-risk groups, predicting high likelihood of added benefit of adjuvant chemotherapy in the high-risk group patient cohort [258,261,262]. MammaPrint on the other hand is based on the gene expression profile of a panel of 70 genes. This test is used in clinics for assessing clinical outcome and predicting recurrence score in early stage breast cancer. Based on recurrence scores, patient samples are stratified into low or high genomic risk. Studies showed that there is an added benefit to adjuvant chemotherapy in the low genomic risk group when compared to patients who did not receive chemotherapy [260,263]. Finally, emerging data supports the role of analyses of circulating tumor DNA in routine clinical care [48,264–268]. The FDA recently approved the FoundationOne Liquid CDx test, which is a circulating cell-free DNA (cfDNA) based-assay as a companion diagnostic for treatment of BRCA mutant (germline or somatic) ovarian cancer patients with the PARP inhibitor rucaparib as well as alpelisib treatment of HR+/HER2-, PIK3CA mutated breast cancer patients [249].

7. Summary

Over the past twenty years, proteogenomic profiling of human tumors has drastically expanded our understanding of breast and ovarian cancer biology. The identification of molecular subtypes; novel oncogenes, tumor suppressor proteins and signaling networks; as well as clinically relevant biomarkers have begun to contribute to the development of novel and more effective treatment strategies. The next challenge will be to effectively translate these efforts into the development of new clinical diagnostic tools, biomarkers, and therapeutic strategies in order to personalize cancer treatment and improve the outcome and quality of life for patients.

Author Contributions: All authors contributed to writing the manuscript. All authors have read and agreed to the published version of the manuscript.

Funding: This work was supported by CA166228 from the National Cancer Institute of the National Institutes of Health, V2016-013 from the V Foundation for Cancer Research, and RSG-19-160-01-TBE from the American Cancer Society to MLG. Additional funding for this project was provided by New Jersey Commission for Cancer Research (DCHS-20PPC-014 to CAK and DHFS-19PPC-019 to GAM), the Cox Foundation for Cancer Research (CAK and GAM) and through a MacMillan Cancer Genetics Summer Undergraduate Research Fellowship from Rutgers University (RNM).

Institutional Review Board Statement: Not applicable.

Informed Consent Statement: Not applicable.

Data Availability Statement: Not applicable.

Acknowledgments: We would like to thank the members of our laboratory and other colleagues for critical review of the manuscript. We apologize to those authors whose work has made essential contributions to this field but that we were unable to discuss due to spatial limitations.

Conflicts of Interest: The authors declare no conflict of interest.

References

1. Siegel, R.L.; Miller, K.D.; Jemal, A. Cancer statistics, 2020. *CA Cancer J. Clin.* **2020**, *70*, 7–30. [CrossRef]
2. Cancer Genome Atlas Network. Comprehensive molecular portraits of human breast tumours. *Nature* **2012**, *490*, 61–70. [CrossRef]
3. Cancer Genome Atlas Research Network. Integrated genomic analyses of ovarian carcinoma. *Nature* **2011**, *474*, 609–615. [CrossRef]
4. Zhang, H.; Liu, T.; Zhang, Z.; Payne, S.H.; Zhang, B.; McDermott, J.E.; Zhou, J.-Y.; Petyuk, V.A.; Chen, L.; Ray, D.; et al. Integrated Proteogenomic Characterization of Human High-Grade Serous Ovarian Cancer. *Cell* **2016**, *166*, 755–765. [CrossRef]
5. Mertins, P.; Mani, D.R.; Ruggles, K.V.; Gillette, M.A.; Clauser, K.R.; Wang, P.; Wang, X.; Qiao, J.W.; Cao, S.; Petralia, F.; et al. Proteogenomics connects somatic mutations to signalling in breast cancer. *Nature* **2016**, *534*, 55–62. [CrossRef]
6. The ICGC/TCGA Pan-Cancer Analysis of Whole Genomes Consortium. Pan-cancer analysis of whole genomes. *Nature* **2020**, *578*, 82–93. [CrossRef]
7. Alexandrov, L.B.; Kim, J.; Haradhvala, N.J.; Huang, M.N.; Tian Ng, A.W.; Wu, Y.; Boot, A.; Covington, K.R.; Gordenin, D.A.; Bergstrom, E.N.; et al. The repertoire of mutational signatures in human cancer. *Nature* **2020**, *578*, 94–101. [CrossRef] [PubMed]
8. Paczkowska, M.; Barenboim, J.; Sintupisut, N.; Fox, N.S.; Zhu, H.; Abd-Rabbo, D.; Mee, M.W.; Boutros, P.C.; Drivers, P.; Functional Interpretation Working Group; et al. Integrative pathway enrichment analysis of multivariate omics data. *Nat. Commun.* **2020**, *11*, 735. [CrossRef] [PubMed]

9. Curtis, C.; Shah, S.P.; Chin, S.F.; Turashvili, G.; Rueda, O.M.; Dunning, M.J.; Speed, D.; Lynch, A.G.; Samarajiwa, S.; Yuan, Y.; et al. The genomic and transcriptomic architecture of 2,000 breast tumours reveals novel subgroups. *Nature* **2012**, *486*, 346–352. [CrossRef] [PubMed]

10. Berger, A.C.; Korkut, A.; Kanchi, R.S.; Hegde, A.M.; Lenoir, W.; Liu, W.; Liu, Y.; Fan, H.; Shen, H.; Ravikumar, V.; et al. A Comprehensive Pan-Cancer Molecular Study of Gynecologic and Breast Cancers. *Cancer Cell* **2018**, *33*, 690–705.e699. [CrossRef]

11. Hoadley, K.A.; Yau, C.; Hinoue, T.; Wolf, D.M.; Lazar, A.J.; Drill, E.; Shen, R.; Taylor, A.M.; Cherniack, A.D.; Thorsson, V.; et al. Cell-of-Origin Patterns Dominate the Molecular Classification of 10,000 Tumors from 33 Types of Cancer. *Cell* **2018**, *173*, 291–304.e296. [CrossRef]

12. Hutter, C.; Zenklusen, J.C. The Cancer Genome Atlas: Creating Lasting Value beyond Its Data. *Cell* **2018**, *173*, 283–285. [CrossRef]

13. Lawrence, M.S.; Stojanov, P.; Polak, P.; Kryukov, G.V.; Cibulskis, K.; Sivachenko, A.; Carter, S.L.; Stewart, C.; Mermel, C.H.; Roberts, S.A.; et al. Mutational heterogeneity in cancer and the search for new cancer-associated genes. *Nature* **2013**, *499*, 214–218. [CrossRef]

14. Vogelstein, B.; Papadopoulos, N.; Velculescu, V.E.; Zhou, S.; Diaz, L.A., Jr.; Kinzler, K.W. Cancer genome landscapes. *Science* **2013**, *339*, 1546–1558. [CrossRef]

15. Venter, J.C.; Adams, M.D.; Myers, E.W.; Li, P.W.; Mural, R.J.; Sutton, G.G.; Smith, H.O.; Yandell, M.; Evans, C.A.; Holt, R.A.; et al. The sequence of the human genome. *Science* **2001**, *291*, 1304–1351. [CrossRef]

16. Tucker, T.; Marra, M.; Friedman, J.M. Massively parallel sequencing: The next big thing in genetic medicine. *Am. J. Hum. Genet.* **2009**, *85*, 142–154. [CrossRef] [PubMed]

17. Bray, F.; Ferlay, J.; Soerjomataram, I.; Siegel, R.L.; Torre, L.A.; Jemal, A. Global cancer statistics 2018: GLOBOCAN estimates of incidence and mortality worldwide for 36 cancers in 185 countries. *CA Cancer J. Clin.* **2018**, *68*, 394–424. [CrossRef]

18. Cardoso, F.; Paluch-Shimon, S.; Senkus, E.; Curigliano, G.; Aapro, M.S.; Andre, F.; Barrios, C.H.; Bergh, J.; Bhattacharyya, G.S.; Biganzoli, L.; et al. 5th ESO-ESMO international consensus guidelines for advanced breast cancer (ABC 5)(dagger). *Ann. Oncol.* **2020**. [CrossRef] [PubMed]

19. Waks, A.G.; Winer, E.P. Breast Cancer Treatment: A Review. *JAMA* **2019**, *321*, 288–300. [CrossRef] [PubMed]

20. Schmid, P.; Adams, S.; Rugo, H.S.; Schneeweiss, A.; Barrios, C.H.; Iwata, H.; Dieras, V.; Hegg, R.; Im, S.A.; Shaw Wright, G.; et al. Atezolizumab and Nab-Paclitaxel in Advanced Triple-Negative Breast Cancer. *N. Engl. J. Med.* **2018**, *379*, 2108–2121. [CrossRef] [PubMed]

21. Burstein, H.J.; Lacchetti, C.; Anderson, H.; Buchholz, T.A.; Davidson, N.E.; Gelmon, K.A.; Giordano, S.H.; Hudis, C.A.; Solky, A.J.; Stearns, V.; et al. Adjuvant Endocrine Therapy for Women With Hormone Receptor-Positive Breast Cancer: ASCO Clinical Practice Guideline Focused Update. *J. Clin. Oncol.* **2019**, *37*, 423–438. [CrossRef] [PubMed]

22. Breast Cancer Treatment (Adult) (PDQ(R)): Health Professional Version. In *PDQ Cancer Information Summaries*; National Cancer Institute: Bethesda, MD, USA, 2002.

23. Ciriello, G.; Gatza, M.L.; Beck, A.H.; Wilkerson, M.D.; Rhie, S.K.; Pastore, A.; Zhang, H.; McLellan, M.; Yau, C.; Kandoth, C.; et al. Comprehensive Molecular Portraits of Invasive Lobular Breast Cancer. *Cell* **2015**, *163*, 506–519. [CrossRef]

24. Keenan, T.; Moy, B.; Mroz, E.A.; Ross, K.; Niemierko, A.; Rocco, J.W.; Isakoff, S.; Ellisen, L.W.; Bardia, A. Comparison of the Genomic Landscape Between Primary Breast Cancer in African American Versus White Women and the Association of Racial Differences With Tumor Recurrence. *J. Clin. Oncol.* **2015**, *33*, 3621–3627. [CrossRef]

25. Huo, D.; Hu, H.; Rhie, S.K.; Gamazon, E.R.; Cherniack, A.D.; Liu, J.; Yoshimatsu, T.F.; Pitt, J.J.; Hoadley, K.A.; Troester, M.; et al. Comparison of Breast Cancer Molecular Features and Survival by African and European Ancestry in The Cancer Genome Atlas. *JAMA Oncol.* **2017**, *3*, 1654–1662. [CrossRef]

26. Ross, J.S.; Gay, L.M.; Wang, K.; Ali, S.M.; Chumsri, S.; Elvin, J.A.; Bose, R.; Vergilio, J.A.; Suh, J.; Yelensky, R.; et al. Nonamplification ERBB2 genomic alterations in 5605 cases of recurrent and metastatic breast cancer: An emerging opportunity for anti-HER2 targeted therapies. *Cancer* **2016**, *122*, 2654–2662. [CrossRef]

27. Chalakur-Ramireddy, N.K.R.; Pakala, S.B. Combined drug therapeutic strategies for the effective treatment of Triple Negative Breast Cancer. *Biosci. Rep.* **2018**, *38*. [CrossRef] [PubMed]

28. Marra, A.; Trapani, D.; Viale, G.; Criscitiello, C.; Curigliano, G. Practical classification of triple-negative breast cancer: Intratumoral heterogeneity, mechanisms of drug resistance, and novel therapies. *NPJ Breast Cancer* **2020**, *6*, 54. [CrossRef]

29. Rozeboom, B.; Dey, N.; De, P. ER+ metastatic breast cancer: Past, present, and a prescription for an apoptosis-targeted future. *Am. J.Cancer Res.* **2019**, *9*, 2821–2831. [PubMed]

30. Haque, M.M.; Desai, K.V. Pathways to Endocrine Therapy Resistance in Breast Cancer. *Front. Endocrinol.* **2019**, *10*, 573. [CrossRef] [PubMed]

31. Razavi, P.; Chang, M.T.; Xu, G.; Bandlamudi, C.; Ross, D.S.; Vasan, N.; Cai, Y.; Bielski, C.M.; Donoghue, M.T.A.; Jonsson, P.; et al. The Genomic Landscape of Endocrine-Resistant Advanced Breast Cancers. *Cancer Cell* **2018**, *34*, 427–438.e426. [CrossRef] [PubMed]

32. Yardley, D.A.; Noguchi, S.; Pritchard, K.I.; Burris, H.A., 3rd; Baselga, J.; Gnant, M.; Hortobagyi, G.N.; Campone, M.; Pistilli, B.; Piccart, M.; et al. Everolimus plus exemestane in postmenopausal patients with HR(+) breast cancer: BOLERO-2 final progression-free survival analysis. *Adv. Ther.* **2013**, *30*, 870–884. [CrossRef]

33. Narayan, P.; Prowell, T.M.; Gao, J.J.; Fernandes, L.L.; Li, E.; Jiang, X.; Qiu, J.; Fan, J.; Song, P.; Yu, J.; et al. FDA Approval Summary: Alpelisib plus fulvestrant for patients with HR-positive, HER2-negative, PIK3CA-mutated, advanced or metastatic breast cancer. *Clin. Cancer Res.* **2020**. [CrossRef]
34. Cardoso, F.; Senkus, E.; Costa, A.; Papadopoulos, E.; Aapro, M.; Andre, F.; Harbeck, N.; Aguilar Lopez, B.; Barrios, C.H.; Bergh, J.; et al. 4th ESO-ESMO International Consensus Guidelines for Advanced Breast Cancer (ABC 4)dagger. *Ann. Oncol.* **2018**, *29*, 1634–1657. [CrossRef]
35. Chopra, N.; Tovey, H.; Pearson, A.; Cutts, R.; Toms, C.; Proszek, P.; Hubank, M.; Dowsett, M.; Dodson, A.; Daley, F.; et al. Homologous recombination DNA repair deficiency and PARP inhibition activity in primary triple negative breast cancer. *Nat. Commun.* **2020**, *11*, 2662. [CrossRef] [PubMed]
36. Lin, P.H.; Chen, M.; Tsai, L.W.; Lo, C.; Yen, T.C.; Huang, T.Y.; Chen, C.K.; Fan, S.C.; Kuo, S.H.; Huang, C.S. Using next-generation sequencing to redefine BRCAness in triple-negative breast cancer. *Cancer Sci.* **2020**, *111*, 1375–1384. [CrossRef]
37. Kobel, M.; Kalloger, S.E.; Boyd, N.; McKinney, S.; Mehl, E.; Palmer, C.; Leung, S.; Bowen, N.J.; Ionescu, D.N.; Rajput, A.; et al. Ovarian carcinoma subtypes are different diseases: Implications for biomarker studies. *PLoS Med.* **2008**, *5*, e232. [CrossRef] [PubMed]
38. Zorn, K.K.; Bonome, T.; Gangi, L.; Chandramouli, G.V.; Awtrey, C.S.; Gardner, G.J.; Barrett, J.C.; Boyd, J.; Birrer, M.J. Gene expression profiles of serous, endometrioid, and clear cell subtypes of ovarian and endometrial cancer. *Clin. Cancer Res.* **2005**, *11*, 6422–6430. [CrossRef]
39. Bashashati, A.; Ha, G.; Tone, A.; Ding, J.; Prentice, L.M.; Roth, A.; Rosner, J.; Shumansky, K.; Kalloger, S.; Senz, J.; et al. Distinct evolutionary trajectories of primary high-grade serous ovarian cancers revealed through spatial mutational profiling. *J. Pathol.* **2013**, *231*, 21–34. [CrossRef] [PubMed]
40. Beckmeyer-Borowko, A.B.; Peterson, C.E.; Brewer, K.C.; Otoo, M.A.; Davis, F.G.; Hoskins, K.F.; Joslin, C.E. The effect of time on racial differences in epithelial ovarian cancer (OVCA) diagnosis stage, overall and by histologic subtypes: A study of the National Cancer Database. *Cancer Causes Control* **2016**, *27*, 1261–1271. [CrossRef] [PubMed]
41. Shih Ie, M.; Kurman, R.J. Ovarian tumorigenesis: A proposed model based on morphological and molecular genetic analysis. *Am. J. Pathol.* **2004**, *164*, 1511–1518. [CrossRef]
42. Jones, S.; Wang, T.L.; Shih Ie, M.; Mao, T.L.; Nakayama, K.; Roden, R.; Glas, R.; Slamon, D.; Diaz, L.A., Jr.; Vogelstein, B.; et al. Frequent mutations of chromatin remodeling gene ARID1A in ovarian clear cell carcinoma. *Science* **2010**, *330*, 228–231. [CrossRef]
43. Wiegand, K.C.; Shah, S.P.; Al-Agha, O.M.; Zhao, Y.; Tse, K.; Zeng, T.; Senz, J.; McConechy, M.K.; Anglesio, M.S.; Kalloger, S.E.; et al. ARID1A mutations in endometriosis-associated ovarian carcinomas. *N. Engl. J. Med.* **2010**, *363*, 1532–1543. [CrossRef]
44. Ahmed, A.A.; Etemadmoghadam, D.; Temple, J.; Lynch, A.G.; Riad, M.; Sharma, R.; Stewart, C.; Fereday, S.; Caldas, C.; Defazio, A.; et al. Driver mutations in TP53 are ubiquitous in high grade serous carcinoma of the ovary. *J. Pathol.* **2010**, *221*, 49–56. [CrossRef] [PubMed]
45. Kurman, R.J.; Shih Ie, M. Molecular pathogenesis and extraovarian origin of epithelial ovarian cancer—Shifting the paradigm. *Hum. Pathol.* **2011**, *42*, 918–931. [CrossRef] [PubMed]
46. Bowtell, D.D.; Bohm, S.; Ahmed, A.A.; Aspuria, P.-J.; Bast, R.C., Jr.; Beral, V.; Berek, J.S.; Birrer, M.J.; Blagden, S.; Bookman, M.A.; et al. Rethinking ovarian cancer II: Reducing mortality from high-grade serous ovarian cancer. *Nat. Rev. Cancer* **2015**, *15*, 668–679. [CrossRef] [PubMed]
47. Vaughan, S.; Coward, J.I.; Bast, R.C.; Berchuck, A.; Berek, J.S.; Brenton, J.D.; Coukos, G.; Crum, C.C.; Drapkin, R.; Etemad-moghadam, D.; et al. Rethinking ovarian cancer: Recommendations for improving outcomes. *Nat. Rev. Cancer* **2011**, *11*, 719–725. [CrossRef] [PubMed]
48. Martignetti, J.A.; Camacho-Vanegas, O.; Priedigkeit, N.; Camacho, C.; Pereira, E.; Lin, L.; Garnar-Wortzel, L.; Miller, D.; Losic, B.; Shah, H.; et al. Personalized ovarian cancer disease surveillance and detection of candidate therapeutic drug target in circulating tumor DNA. *Neoplasia* **2014**, *16*, 97–103. [CrossRef]
49. Matulonis, U.A.; Sood, A.K.; Fallowfield, L.; Howitt, B.E.; Sehouli, J.; Karlan, B.Y. Ovarian cancer. *Nat. Rev. Dis. Primers* **2016**, *2*, 16061. [CrossRef] [PubMed]
50. Bast, R.C., Jr.; Hennessy, B.; Mills, G.B. The biology of ovarian cancer: New opportunities for translation. *Nat. Rev. Cancer* **2009**, *9*, 415–428. [CrossRef]
51. Schwarz, R.F.; Ng, C.K.; Cooke, S.L.; Newman, S.; Temple, J.; Piskorz, A.M.; Gale, D.; Sayal, K.; Murtaza, M.; Baldwin, P.J.; et al. Spatial and temporal heterogeneity in high-grade serous ovarian cancer: A phylogenetic analysis. *PLoS Med.* **2015**, *12*, e1001789. [CrossRef]
52. Pokhriyal, R.; Hariprasad, R.; Kumar, L.; Hariprasad, G. Chemotherapy Resistance in Advanced Ovarian Cancer Patients. *Biomark. Cancer* **2019**, *11*, 1179299X19860815. [CrossRef] [PubMed]
53. Konstantinopoulos, P.A.; Ceccaldi, R.; Shapiro, G.I.; D'Andrea, A.D. Homologous Recombination Deficiency: Exploiting the Fundamental Vulnerability of Ovarian Cancer. *Cancer Discov.* **2015**, *5*, 1137–1154. [CrossRef] [PubMed]
54. Gonzalez-Martin, A.; Pothuri, B.; Vergote, I.; DePont Christensen, R.; Graybill, W.; Mirza, M.R.; McCormick, C.; Lorusso, D.; Hoskins, P.; Freyer, G.; et al. Niraparib in Patients with Newly Diagnosed Advanced Ovarian Cancer. *N. Engl. J. Med.* **2019**, *381*, 2391–2402. [CrossRef] [PubMed]

55. Coleman, R.L.; Fleming, G.F.; Brady, M.F.; Swisher, E.M.; Steffensen, K.D.; Friedlander, M.; Okamoto, A.; Moore, K.N.; Efrat Ben-Baruch, N.; Werner, T.L.; et al. Veliparib with First-Line Chemotherapy and as Maintenance Therapy in Ovarian Cancer. *N. Engl. J. Med.* **2019**, *381*, 2403–2415. [CrossRef]

56. Moore, K.; Colombo, N.; Scambia, G.; Kim, B.G.; Oaknin, A.; Friedlander, M.; Lisyanskaya, A.; Floquet, A.; Leary, A.; Sonke, G.S.; et al. Maintenance Olaparib in Patients with Newly Diagnosed Advanced Ovarian Cancer. *N. Engl. J. Med.* **2018**, *379*, 2495–2505. [CrossRef] [PubMed]

57. Kaufman, B.; Shapira-Frommer, R.; Schmutzler, R.K.; Audeh, M.W.; Friedlander, M.; Balmana, J.; Mitchell, G.; Fried, G.; Stemmer, S.M.; Hubert, A.; et al. Olaparib monotherapy in patients with advanced cancer and a germline BRCA1/2 mutation. *J. Clin. Oncol.* **2015**, *33*, 244–250. [CrossRef]

58. Del Campo, J.M.; Matulonis, U.A.; Malander, S.; Provencher, D.; Mahner, S.; Follana, P.; Waters, J.; Berek, J.S.; Woie, K.; Oza, A.M.; et al. Niraparib Maintenance Therapy in Patients With Recurrent Ovarian Cancer After a Partial Response to the Last Platinum-Based Chemotherapy in the ENGOT-OV16/NOVA Trial. *J. Clin. Oncol.* **2019**, *37*, 2968–2973. [CrossRef]

59. Kristeleit, R.; Shapiro, G.I.; Burris, H.A.; Oza, A.M.; LoRusso, P.; Patel, M.R.; Domchek, S.M.; Balmana, J.; Drew, Y.; Chen, L.M.; et al. A Phase I-II Study of the Oral PARP Inhibitor Rucaparib in Patients with Germline BRCA1/2-Mutated Ovarian Carcinoma or Other Solid Tumors. *Clin. Cancer Res.* **2017**, *23*, 4095–4106. [CrossRef]

60. Swisher, E.M.; Lin, K.K.; Oza, A.M.; Scott, C.L.; Giordano, H.; Sun, J.; Konecny, G.E.; Coleman, R.L.; Tinker, A.V.; O'Malley, D.M.; et al. Rucaparib in relapsed, platinum-sensitive high-grade ovarian carcinoma (ARIEL2 Part 1): An international, multicentre, open-label, phase 2 trial. *Lancet Oncol.* **2017**, *18*, 75–87. [CrossRef]

61. Moore, K.N.; Secord, A.A.; Geller, M.A.; Miller, D.S.; Cloven, N.; Fleming, G.F.; Wahner Hendrickson, A.E.; Azodi, M.; DiSilvestro, P.; Oza, A.M.; et al. Niraparib monotherapy for late-line treatment of ovarian cancer (QUADRA): A multicentre, open-label, single-arm, phase 2 trial. *Lancet Oncol.* **2019**, *20*, 636–648. [CrossRef]

62. Ray-Coquard, I.; Pautier, P.; Pignata, S.; Perol, D.; Gonzalez-Martin, A.; Berger, R.; Fujiwara, K.; Vergote, I.; Colombo, N.; Maenpaa, J.; et al. Olaparib plus Bevacizumab as First-Line Maintenance in Ovarian Cancer. *N. Engl. J. Med.* **2019**, *381*, 2416–2428. [CrossRef]

63. Wang, Q.; Peng, H.; Qi, X.; Wu, M.; Zhao, X. Targeted therapies in gynecological cancers: A comprehensive review of clinical evidence. *Signal Transduct. Target. Ther.* **2020**, *5*, 137. [CrossRef]

64. Peters, M.L.; Garber, J.E.; Tung, N. Managing hereditary breast cancer risk in women with and without ovarian cancer. *Gynecol. Oncol.* **2017**, *146*, 205–214. [CrossRef]

65. Yoshida, R. Hereditary breast and ovarian cancer (HBOC): Review of its molecular characteristics, screening, treatment, and prognosis. *Breast Cancer* **2020**. [CrossRef]

66. King, M.C.; Marks, J.H.; Mandell, J.B.; New York Breast Cancer Study Group. Breast and ovarian cancer risks due to inherited mutations in BRCA1 and BRCA2. *Science* **2003**, *302*, 643–646. [CrossRef] [PubMed]

67. Momozawa, Y.; Iwasaki, Y.; Parsons, M.T.; Kamatani, Y.; Takahashi, A.; Tamura, C.; Katagiri, T.; Yoshida, T.; Nakamura, S.; Sugano, K.; et al. Germline pathogenic variants of 11 breast cancer genes in 7051 Japanese patients and 11,241 controls. *Nat. Commun.* **2018**, *9*, 4083. [CrossRef] [PubMed]

68. Sun, J.; Meng, H.; Yao, L.; Lv, M.; Bai, J.; Zhang, J.; Wang, L.; Ouyang, T.; Li, J.; Wang, T.; et al. Germline Mutations in Cancer Susceptibility Genes in a Large Series of Unselected Breast Cancer Patients. *Clin. Cancer Res.* **2017**, *23*, 6113–6119. [CrossRef] [PubMed]

69. Wen, W.X.; Allen, J.; Lai, K.N.; Mariapun, S.; Hasan, S.N.; Ng, P.S.; Lee, D.S.; Lee, S.Y.; Yoon, S.Y.; Lim, J.; et al. Inherited mutations in BRCA1 and BRCA2 in an unselected multiethnic cohort of Asian patients with breast cancer and healthy controls from Malaysia. *J. Med. Genet.* **2018**, *55*, 97–103. [CrossRef]

70. Tung, N.; Lin, N.U.; Kidd, J.; Allen, B.A.; Singh, N.; Wenstrup, R.J.; Hartman, A.R.; Winer, E.P.; Garber, J.E. Frequency of Germline Mutations in 25 Cancer Susceptibility Genes in a Sequential Series of Patients With Breast Cancer. *J. Clin. Oncol.* **2016**, *34*, 1460–1468. [CrossRef]

71. Song, H.; Cicek, M.S.; Dicks, E.; Harrington, P.; Ramus, S.J.; Cunningham, J.M.; Fridley, B.L.; Tyrer, J.P.; Alsop, J.; Jimenez-Linan, M.; et al. The contribution of deleterious germline mutations in BRCA1, BRCA2 and the mismatch repair genes to ovarian cancer in the population. *Hum. Mol. Genet.* **2014**, *23*, 4703–4709. [CrossRef]

72. Alsop, K.; Fereday, S.; Meldrum, C.; de Fazio, A.; Emmanuel, C.; George, J.; Dobrovic, A.; Birrer, M.J.; Webb, P.M.; Stewart, C.; et al. BRCA mutation frequency and patterns of treatment response in BRCA mutation-positive women with ovarian cancer: A report from the Australian Ovarian Cancer Study Group. *J. Clin. Oncol.* **2012**, *30*, 2654–2663. [CrossRef]

73. Zhang, S.; Royer, R.; Li, S.; McLaughlin, J.R.; Rosen, B.; Risch, H.A.; Fan, I.; Bradley, L.; Shaw, P.A.; Narod, S.A. Frequencies of BRCA1 and BRCA2 mutations among 1,342 unselected patients with invasive ovarian cancer. *Gynecol. Oncol.* **2011**, *121*, 353–357. [CrossRef]

74. Kang, E.; Kim, S.W. The Korean hereditary breast cancer study: Review and future perspectives. *J. Breast Cancer* **2013**, *16*, 245–253. [CrossRef]

75. Kuchenbaecker, K.B.; Hopper, J.L.; Barnes, D.R.; Phillips, K.A.; Mooij, T.M.; Roos-Blom, M.J.; Jervis, S.; van Leeuwen, F.E.; Milne, R.L.; Andrieu, N.; et al. Risks of Breast, Ovarian, and Contralateral Breast Cancer for BRCA1 and BRCA2 Mutation Carriers. *JAMA* **2017**, *317*, 2402–2416. [CrossRef] [PubMed]

76. Chen, S.; Parmigiani, G. Meta-analysis of BRCA1 and BRCA2 penetrance. *J. Clin. Oncol.* **2007**, *25*, 1329–1333. [CrossRef] [PubMed]

77. Antoniou, A.; Pharoah, P.D.; Narod, S.; Risch, H.A.; Eyfjord, J.E.; Hopper, J.L.; Loman, N.; Olsson, H.; Johannsson, O.; Borg, A.; et al. Average risks of breast and ovarian cancer associated with BRCA1 or BRCA2 mutations detected in case Series unselected for family history: A combined analysis of 22 studies. *Am. J. Hum. Genet.* **2003**, *72*, 1117–1130. [CrossRef] [PubMed]

78. Ding, Y.C.; McGuffog, L.; Healey, S.; Friedman, E.; Laitman, Y.; Paluch-Shimon, S.; Kaufman, B.; Swe, B.; Liljegren, A.; Lindblom, A.; et al. A nonsynonymous polymorphism in IRS1 modifies risk of developing breast and ovarian cancers in BRCA1 and ovarian cancer in BRCA2 mutation carriers. *Cancer Epidemiol. Biomark. Prev.* **2012**, *21*, 1362–1370. [CrossRef]

79. Kuchenbaecker, K.B.; Neuhausen, S.L.; Robson, M.; Barrowdale, D.; McGuffog, L.; Mulligan, A.M.; Andrulis, I.L.; Spurdle, A.B.; Schmidt, M.K.; Schmutzler, R.K.; et al. Associations of common breast cancer susceptibility alleles with risk of breast cancer subtypes in BRCA1 and BRCA2 mutation carriers. *Breast Cancer Res. BCR* **2014**, *16*, 3416. [CrossRef]

80. McLaughlin, J.R.; Rosen, B.; Moody, J.; Pal, T.; Fan, I.; Shaw, P.A.; Risch, H.A.; Sellers, T.A.; Sun, P.; Narod, S.A. Long-term ovarian cancer survival associated with mutation in BRCA1 or BRCA2. *J. Natl. Cancer Inst.* **2013**, *105*, 141–148. [CrossRef]

81. Templeton, A.J.; Gonzalez, L.D.; Vera-Badillo, F.E.; Tibau, A.; Goldstein, R.; Seruga, B.; Srikanthan, A.; Pandiella, A.; Amir, E.; Ocana, A. Interaction between Hormonal Receptor Status, Age and Survival in Patients with BRCA1/2 Germline Mutations: A Systematic Review and Meta-Regression. *PLoS ONE* **2016**, *11*, e0154789. [CrossRef]

82. Copson, E.R.; Maishman, T.C.; Tapper, W.J.; Cutress, R.I.; Greville-Heygate, S.; Altman, D.G.; Eccles, B.; Gerty, S.; Durcan, L.T.; Jones, L.; et al. Germline BRCA mutation and outcome in young-onset breast cancer (POSH): A prospective cohort study. *Lancet Oncol.* **2018**, *19*, 169–180. [CrossRef]

83. Pernas, S.; Tolaney, S.M.; Winer, E.P.; Goel, S. CDK4/6 inhibition in breast cancer: Current practice and future directions. *Ther. Adv. Med. Oncol.* **2018**, *10*, 1758835918786451. [CrossRef] [PubMed]

84. Perou, C.M.; Sorlie, T.; Eisen, M.B.; van de Rijn, M.; Jeffrey, S.S.; Rees, C.A.; Pollack, J.R.; Ross, D.T.; Johnsen, H.; Akslen, L.A.; et al. Molecular portraits of human breast tumours. *Nature* **2000**, *406*, 747–752. [CrossRef] [PubMed]

85. Sorlie, T.; Perou, C.M.; Tibshirani, R.; Aas, T.; Geisler, S.; Johnsen, H.; Hastie, T.; Eisen, M.B.; van de Rijn, M.; Jeffrey, S.S.; et al. Gene expression patterns of breast carcinomas distinguish tumor subclasses with clinical implications. *Proc. Natl. Acad. Sci. USA* **2001**, *98*, 10869–10874. [CrossRef]

86. Sorlie, T.; Tibshirani, R.; Parker, J.; Hastie, T.; Marron, J.S.; Nobel, A.; Deng, S.; Johnsen, H.; Pesich, R.; Geisler, S.; et al. Repeated observation of breast tumor subtypes in independent gene expression data sets. *Proc. Natl. Acad. Sci. USA* **2003**, *100*, 8418–8423. [CrossRef]

87. Hu, Z.; Fan, C.; Oh, D.S.; Marron, J.S.; He, X.; Qaqish, B.F.; Livasy, C.; Carey, L.A.; Reynolds, E.; Dressler, L.; et al. The molecular portraits of breast tumors are conserved across microarray platforms. *BMC Genom.* **2006**, *7*, 96. [CrossRef]

88. Parker, J.S.; Mullins, M.; Cheang, M.C.; Leung, S.; Voduc, D.; Vickery, T.; Davies, S.; Fauron, C.; He, X.; Hu, Z.; et al. Supervised risk predictor of breast cancer based on intrinsic subtypes. *J. Clin. Oncol.* **2009**, *27*, 1160–1167. [CrossRef]

89. Wallden, B.; Storhoff, J.; Nielsen, T.; Dowidar, N.; Schaper, C.; Ferree, S.; Liu, S.; Leung, S.; Geiss, G.; Snider, J.; et al. Development and verification of the PAM50-based Prosigna breast cancer gene signature assay. *BMC Med. Genom.* **2015**, *8*, 54. [CrossRef]

90. Cheang, M.C.; Voduc, K.D.; Tu, D.; Jiang, S.; Leung, S.; Chia, S.K.; Shepherd, L.E.; Levine, M.N.; Pritchard, K.I.; Davies, S.; et al. Responsiveness of intrinsic subtypes to adjuvant anthracycline substitution in the NCIC.CTG MA.5 randomized trial. *Clin. Cancer Res.* **2012**, *18*, 2402–2412. [CrossRef]

91. Martin, M.; Prat, A.; Rodriguez-Lescure, A.; Caballero, R.; Ebbert, M.T.; Munarriz, B.; Ruiz-Borrego, M.; Bastien, R.R.; Crespo, C.; Davis, C.; et al. PAM50 proliferation score as a predictor of weekly paclitaxel benefit in breast cancer. *Breast Cancer Res. Treat.* **2013**, *138*, 457–466. [CrossRef]

92. Herschkowitz, J.I.; Simin, K.; Weigman, V.J.; Mikaelian, I.; Usary, J.; Hu, Z.; Rasmussen, K.E.; Jones, L.P.; Assefnia, S.; Chandrasekharan, S.; et al. Identification of conserved gene expression features between murine mammary carcinoma models and human breast tumors. *Genome Biol.* **2007**, *8*, R76. [CrossRef] [PubMed]

93. Prat, A.; Parker, J.S.; Karginova, O.; Fan, C.; Livasy, C.; Herschkowitz, J.I.; He, X.; Perou, C.M. Phenotypic and molecular characterization of the claudin-low intrinsic subtype of breast cancer. *Breast Cancer Res.* **2010**, *12*, R68. [CrossRef]

94. Lehmann, B.D.; Bauer, J.A.; Chen, X.; Sanders, M.E.; Chakravarthy, A.B.; Shyr, Y.; Pietenpol, J.A. Identification of human triple-negative breast cancer subtypes and preclinical models for selection of targeted therapies. *J. Clin. Investig.* **2011**, *121*, 2750–2767. [CrossRef] [PubMed]

95. Lehmann, B.D.; Jovanovic, B.; Chen, X.; Estrada, M.V.; Johnson, K.N.; Shyr, Y.; Moses, H.L.; Sanders, M.E.; Pietenpol, J.A. Refinement of Triple-Negative Breast Cancer Molecular Subtypes: Implications for Neoadjuvant Chemotherapy Selection. *PLoS ONE* **2016**, *11*, e0157368. [CrossRef]

96. Bild, A.H.; Yao, G.; Chang, J.T.; Wang, Q.; Potti, A.; Chasse, D.; Joshi, M.B.; Harpole, D.; Lancaster, J.M.; Berchuck, A.; et al. Oncogenic pathway signatures in human cancers as a guide to targeted therapies. *Nature* **2006**, *439*, 353–357. [CrossRef]

97. Gatza, M.L.; Lucas, J.E.; Barry, W.T.; Kim, J.W.; Wang, Q.; Crawford, M.D.; Datto, M.B.; Kelley, M.; Mathey-Prevot, B.; Potti, A.; et al. A pathway-based classification of human breast cancer. *Proc. Natl. Acad. Sci. USA* **2010**, *107*, 6994–6999. [CrossRef] [PubMed]

98. Gatza, M.L.; Kung, H.N.; Blackwell, K.L.; Dewhirst, M.W.; Marks, J.R.; Chi, J.T. Analysis of tumor environmental response and oncogenic pathway activation identifies distinct basal and luminal features in HER2-related breast tumor subtypes. *Breast Cancer Res. BCR* **2011**, *13*, R62. [CrossRef]

99. Gatza, M.L.; Silva, G.O.; Parker, J.S.; Fan, C.; Perou, C.M. An integrated genomics approach identifies drivers of proliferation in luminal-subtype human breast cancer. *Nat. Genet.* **2014**, *46*, 1051–1059. [CrossRef]

100. Bild, A.H.; Parker, J.S.; Gustafson, A.M.; Acharya, C.R.; Hoadley, K.A.; Anders, C.; Marcom, P.K.; Carey, L.A.; Potti, A.; Nevins, J.R.; et al. An integration of complementary strategies for gene-expression analysis to reveal novel therapeutic opportunities for breast cancer. *Breast Cancer Res. BCR* **2009**, *11*, R55. [CrossRef] [PubMed]

101. Weigman, V.J.; Chao, H.H.; Shabalin, A.A.; He, X.; Parker, J.S.; Nordgard, S.H.; Grushko, T.; Huo, D.; Nwachukwu, C.; Nobel, A.; et al. Basal-like Breast cancer DNA copy number losses identify genes involved in genomic instability, response to therapy, and patient survival. *Breast Cancer Res. Treat.* **2012**, *133*, 865–880. [CrossRef]

102. Ciriello, G.; Miller, M.L.; Aksoy, B.A.; Senbabaoglu, Y.; Schultz, N.; Sander, C. Emerging landscape of oncogenic signatures across human cancers. *Nat. Genet.* **2013**, *45*, 1127–1133. [CrossRef]

103. Valla, M.; Mjones, P.G.; Engstrom, M.J.; Ytterhus, B.; Bordin, D.L.; van Loon, B.; Akslen, L.A.; Vatten, L.J.; Opdahl, S.; Bofin, A.M. Characterization of FGD5 Expression in Primary Breast Cancers and Lymph Node Metastases. *J. Histochem. Cytochem.* **2018**, *66*, 787–799. [CrossRef] [PubMed]

104. Klaestad, E.; Opdahl, S.; Engstrom, M.J.; Ytterhus, B.; Wik, E.; Bofin, A.M.; Valla, M. MRPS23 amplification and gene expression in breast cancer; association with proliferation and the non-basal subtypes. *Breast Cancer Res. Treat.* **2020**, *180*, 73–86. [CrossRef] [PubMed]

105. Valla, M.; Engstrom, M.J.; Ytterhus, B.; Hansen, A.K.; Akslen, L.A.; Vatten, L.J.; Opdahl, S.; Bofin, A.M. FGD5 amplification in breast cancer patients is associated with tumour proliferation and a poorer prognosis. *Breast Cancer Res. Treat.* **2017**, *162*, 243–253. [CrossRef]

106. Desmedt, C.; Zoppoli, G.; Gundem, G.; Pruneri, G.; Larsimont, D.; Fornili, M.; Fumagalli, D.; Brown, D.; Rothe, F.; Vincent, D.; et al. Genomic Characterization of Primary Invasive Lobular Breast Cancer. *J. Clin. Oncol.* **2016**, *34*, 1872–1881. [CrossRef] [PubMed]

107. Guido, L.P.; Gomez-Fernandez, C. Advances in the Molecular Taxonomy of Breast Cancer. *Arch. Med. Res.* **2020**, *51*, 777–783. [CrossRef] [PubMed]

108. Bandura, D.R.; Baranov, V.I.; Ornatsky, O.I.; Antonov, A.; Kinach, R.; Lou, X.; Pavlov, S.; Vorobiev, S.; Dick, J.E.; Tanner, S.D. Mass cytometry: Technique for real time single cell multitarget immunoassay based on inductively coupled plasma time-of-flight mass spectrometry. *Anal. Chem.* **2009**, *81*, 6813–6822. [CrossRef]

109. Ali, H.R.; Jackson, H.W.; Zanotelli, V.R.T.; Danenberg, E.; Fischer, J.R.; Bardwell, H.; Provenzano, E.; Ali, H.R.; Al Sa'd, M.; Alon, S.; et al. Imaging mass cytometry and multiplatform genomics define the phenogenomic landscape of breast cancer. *Nat. Cancer* **2020**, *1*, 163–175. [CrossRef]

110. Giesen, C.; Wang, H.A.; Schapiro, D.; Zivanovic, N.; Jacobs, A.; Hattendorf, B.; Schuffler, P.J.; Grolimund, D.; Buhmann, J.M.; Brandt, S.; et al. Highly multiplexed imaging of tumor tissues with subcellular resolution by mass cytometry. *Nat. Methods* **2014**, *11*, 417–422. [CrossRef]

111. Tothill, R.W.; Tinker, A.V.; George, J.; Brown, R.; Fox, S.B.; Lade, S.; Johnson, D.S.; Trivett, M.K.; Etemadmoghadam, D.; Locandro, B.; et al. Novel molecular subtypes of serous and endometrioid ovarian cancer linked to clinical outcome. *Clin. Cancer Res.* **2008**, *14*, 5198–5208. [CrossRef] [PubMed]

112. Helland, A.; Anglesio, M.S.; George, J.; Cowin, P.A.; Johnstone, C.N.; House, C.M.; Sheppard, K.E.; Etemadmoghadam, D.; Melnyk, N.; Rustgi, A.K.; et al. Deregulation of MYCN, LIN28B and LET7 in a molecular subtype of aggressive high-grade serous ovarian cancers. *PLoS ONE* **2011**, *6*, e18064. [CrossRef] [PubMed]

113. Konecny, G.E.; Wang, C.; Hamidi, H.; Winterhoff, B.; Kalli, K.R.; Dering, J.; Ginther, C.; Chen, H.W.; Dowdy, S.; Cliby, W.; et al. Prognostic and therapeutic relevance of molecular subtypes in high-grade serous ovarian cancer. *J. Natl. Cancer Inst.* **2014**, *106*. [CrossRef] [PubMed]

114. Chen, G.M.; Kannan, L.; Geistlinger, L.; Kofia, V.; Safikhani, Z.; Gendoo, D.M.A.; Parmigiani, G.; Birrer, M.; Haibe-Kains, B.; Waldron, L. Consensus on Molecular Subtypes of High-Grade Serous Ovarian Carcinoma. *Clin. Cancer Res.* **2018**, *24*, 5037–5047. [CrossRef] [PubMed]

115. Verhaak, R.G.; Tamayo, P.; Yang, J.Y.; Hubbard, D.; Zhang, H.; Creighton, C.J.; Fereday, S.; Lawrence, M.; Carter, S.L.; Mermel, C.H.; et al. Prognostically relevant gene signatures of high-grade serous ovarian carcinoma. *J. Clin. Investig.* **2013**, *123*, 517–525. [CrossRef]

116. Karagoz, K.; Mehta, G.A.; Khella, C.A.; Khanna, P.; Gatza, M.L. Integrative proteogenomic analyses of human tumours identifies ADNP as a novel oncogenic mediator of cell cycle progression in high-grade serous ovarian cancer with poor prognosis. *EBioMedicine* **2019**, *50*, 191–202. [CrossRef]

117. Izar, B.; Tirosh, I.; Stover, E.H.; Wakiro, I.; Cuoco, M.S.; Alter, I.; Rodman, C.; Leeson, R.; Su, M.J.; Shah, P.; et al. A single-cell landscape of high-grade serous ovarian cancer. *Nat. Med.* **2020**, *26*, 1271–1279. [CrossRef]

118. Way, G.P.; Rudd, J.; Wang, C.; Hamidi, H.; Fridley, B.L.; Konecny, G.E.; Goode, E.L.; Greene, C.S.; Doherty, J.A. Comprehensive Cross-Population Analysis of High-Grade Serous Ovarian Cancer Supports No More Than Three Subtypes. *G3 (Bethesda)* **2016**, *6*, 4097–4103. [CrossRef]

119. Hu, Y.; Pan, J.; Shah, P.; Ao, M.; Thomas, S.N.; Liu, Y.; Chen, L.; Schnaubelt, M.; Clark, D.J.; Rodriguez, H.; et al. Integrated Proteomic and Glycoproteomic Characterization of Human High-Grade Serous Ovarian Carcinoma. *Cell Rep.* **2020**, *33*, 108276. [CrossRef]

120. Tong, M.; Yu, C.; Zhan, D.; Zhang, M.; Zhen, B.; Zhu, W.; Wang, Y.; Wu, C.; He, F.; Qin, J.; et al. Molecular subtyping of cancer and nomination of kinase candidates for inhibition with phosphoproteomics: Reanalysis of CPTAC ovarian cancer. *EBioMedicine* **2019**, *40*, 305–317. [CrossRef]
121. Yoshihara, K.; Tsunoda, T.; Shigemizu, D.; Fujiwara, H.; Hatae, M.; Fujiwara, H.; Masuzaki, H.; Katabuchi, H.; Kawakami, Y.; Okamoto, A.; et al. High-risk ovarian cancer based on 126-gene expression signature is uniquely characterized by downregulation of antigen presentation pathway. *Clin. Cancer Res.* **2012**, *18*, 1374–1385. [CrossRef]
122. Perets, R.; Drapkin, R. It's Totally Tubular....Riding The New Wave of Ovarian Cancer Research. *Cancer Res.* **2016**, *76*, 10–17. [CrossRef] [PubMed]
123. Karst, A.M.; Drapkin, R. Ovarian cancer pathogenesis: A model in evolution. *J. Oncol.* **2010**, *2010*, 932371. [CrossRef] [PubMed]
124. Jarboe, E.; Folkins, A.; Nucci, M.R.; Kindelberger, D.; Drapkin, R.; Miron, A.; Lee, Y.; Crum, C.P. Serous carcinogenesis in the fallopian tube: A descriptive classification. *Int. J. Gynecol. Pathol.* **2008**, *27*, 1–9. [CrossRef] [PubMed]
125. Engqvist, H.; Parris, T.Z.; Biermann, J.; Ronnerman, E.W.; Larsson, P.; Sundfeldt, K.; Kovacs, A.; Karlsson, P.; Helou, K. Integrative genomics approach identifies molecular features associated with early-stage ovarian carcinoma histotypes. *Sci. Rep.* **2020**, *10*, 7946. [CrossRef]
126. Tsang, T.; Wei, W.; Itamochi, H.; Tambouret, R.; Birrer, M. Integrated genomic analysis of clear cell ovarian cancers identified PRKCI as a potential therapeutic target. *Oncotarget* **2017**, *8*, 96482–96495. [CrossRef]
127. Rehmani, H.; Li, Y.; Li, T.; Padia, R.; Calbay, O.; Jin, L.; Chen, H.; Huang, S. Addiction to protein kinase Ci due to PRKCI gene amplification can be exploited for an aptamer-based targeted therapy in ovarian cancer. *Signal Transduct. Target. Ther.* **2020**, *5*, 140. [CrossRef]
128. Kurimchak, A.M.; Herrera-Montavez, C.; Brown, J.; Johnson, K.J.; Sodi, V.; Srivastava, N.; Kumar, V.; Deihimi, S.; O'Brien, S.; Peri, S.; et al. Functional proteomics interrogation of the kinome identifies MRCKA as a therapeutic target in high-grade serous ovarian carcinoma. *Sci. Signal.* **2020**, *13*. [CrossRef]
129. Coscia, F.; Lengyel, E.; Duraiswamy, J.; Ashcroft, B.; Bassani-Sternberg, M.; Wierer, M.; Johnson, A.; Wroblewski, K.; Montag, A.; Yamada, S.D.; et al. Multi-level Proteomics Identifies CT45 as a Chemosensitivity Mediator and Immunotherapy Target in Ovarian Cancer. *Cell* **2018**, *175*, 159–170.e116. [CrossRef]
130. Hoadley, K.A.; Yau, C.; Wolf, D.M.; Cherniack, A.D.; Tamborero, D.; Ng, S.; Leiserson, M.D.M.; Niu, B.; McLellan, M.D.; Uzunangelov, V.; et al. Multiplatform analysis of 12 cancer types reveals molecular classification within and across tissues of origin. *Cell* **2014**, *158*, 929–944. [CrossRef]
131. Ding, L.; Bailey, M.H.; Porta-Pardo, E.; Thorsson, V.; Colaprico, A.; Bertrand, D.; Gibbs, D.L.; Weerasinghe, A.; Huang, K.L.; Tokheim, C.; et al. Perspective on Oncogenic Processes at the End of the Beginning of Cancer Genomics. *Cell* **2018**, *173*, 305–320.e310. [CrossRef]
132. Wang, Z.C.; Birkbak, N.J.; Culhane, A.C.; Drapkin, R.; Fatima, A.; Tian, R.; Schwede, M.; Alsop, K.; Daniels, K.E.; Piao, H.; et al. Profiles of genomic instability in high-grade serous ovarian cancer predict treatment outcome. *Clin. Cancer Res.* **2012**, *18*, 5806–5815. [CrossRef]
133. Wang, Z.C.; Lin, M.; Wei, L.J.; Li, C.; Miron, A.; Lodeiro, G.; Harris, L.; Ramaswamy, S.; Tanenbaum, D.M.; Meyerson, M.; et al. Loss of heterozygosity and its correlation with expression profiles in subclasses of invasive breast cancers. *Cancer Res.* **2004**, *64*, 64–71. [CrossRef] [PubMed]
134. Gorringe, K.L.; Jacobs, S.; Thompson, E.R.; Sridhar, A.; Qiu, W.; Choong, D.Y.; Campbell, I.G. High-resolution single nucleotide polymorphism array analysis of epithelial ovarian cancer reveals numerous microdeletions and amplifications. *Clin. Cancer Res.* **2007**, *13*, 4731–4739. [CrossRef] [PubMed]
135. Telli, M.L.; Timms, K.M.; Reid, J.; Hennessy, B.; Mills, G.B.; Jensen, K.C.; Szallasi, Z.; Barry, W.T.; Winer, E.P.; Tung, N.M.; et al. Homologous Recombination Deficiency (HRD) Score Predicts Response to Platinum-Containing Neoadjuvant Chemotherapy in Patients with Triple-Negative Breast Cancer. *Clin. Cancer Res.* **2016**, *22*, 3764–3773. [CrossRef] [PubMed]
136. Takaya, H.; Nakai, H.; Takamatsu, S.; Mandai, M.; Matsumura, N. Homologous recombination deficiency status-based classification of high-grade serous ovarian carcinoma. *Sci. Rep.* **2020**, *10*, 2757. [CrossRef] [PubMed]
137. Mehta, G.A.; Parker, J.S.; Silva, G.O.; Hoadley, K.A.; Perou, C.M.; Gatza, M.L. Amplification of SOX4 promotes PI3K/Akt signaling in human breast cancer. *Breast Cancer Res. Treat.* **2017**, *162*, 439–450. [CrossRef] [PubMed]
138. Zhang, Y.; Kwok-Shing Ng, P.; Kucherlapati, M.; Chen, F.; Liu, Y.; Tsang, Y.H.; de Velasco, G.; Jeong, K.J.; Akbani, R.; Hadjipanayis, A.; et al. A Pan-Cancer Proteogenomic Atlas of PI3K/AKT/mTOR Pathway Alterations. *Cancer Cell* **2017**, *31*, 820–832.e823. [CrossRef] [PubMed]
139. Zervantonakis, I.K.; Iavarone, C.; Chen, H.Y.; Selfors, L.M.; Palakurthi, S.; Liu, J.F.; Drapkin, R.; Matulonis, U.; Leverson, J.D.; Sampath, D.; et al. Systems analysis of apoptotic priming in ovarian cancer identifies vulnerabilities and predictors of drug response. *Nat. Commun.* **2017**, *8*, 365. [CrossRef] [PubMed]
140. Hanrahan, A.J.; Schultz, N.; Westfal, M.L.; Sakr, R.A.; Giri, D.D.; Scarperi, S.; Janakiraman, M.; Olvera, N.; Stevens, E.V.; She, Q.B.; et al. Genomic complexity and AKT dependence in serous ovarian cancer. *Cancer Discov.* **2012**, *2*, 56–67. [CrossRef]
141. Cheaib, B.; Auguste, A.; Leary, A. The PI3K/Akt/mTOR pathway in ovarian cancer: Therapeutic opportunities and challenges. *Chin. J. Cancer* **2015**, *34*, 4–16. [CrossRef]
142. Pascual, J.; Turner, N.C. Targeting the PI3-kinase pathway in triple-negative breast cancer. *Ann. Oncol.* **2019**, *30*, 1051–1060. [CrossRef]

143. Ellis, M.J.; Perou, C.M. The genomic landscape of breast cancer as a therapeutic roadmap. *Cancer Discov.* **2013**, *3*, 27–34. [CrossRef]
144. Marcotte, R.; Brown, K.R.; Suarez, F.; Sayad, A.; Karamboulas, K.; Krzyzanowski, P.M.; Sircoulomb, F.; Medrano, M.; Fedyshyn, Y.; Koh, J.L.Y.; et al. Essential gene profiles in breast, pancreatic, and ovarian cancer cells. *Cancer Discov.* **2012**, *2*, 172–189. [CrossRef]
145. Bailey, M.H.; Tokheim, C.; Porta-Pardo, E.; Sengupta, S.; Bertrand, D.; Weerasinghe, A.; Colaprico, A.; Wendl, M.C.; Kim, J.; Reardon, B.; et al. Comprehensive Characterization of Cancer Driver Genes and Mutations. *Cell* **2018**, *173*, 371–385. [CrossRef] [PubMed]
146. Kandoth, C.; Schultz, N.; Cherniack, A.D.; Akbani, R.; Liu, Y.; Shen, H.; Robertson, A.G.; Pashtan, I.; Shen, R.; et al.; Cancer Genome Atlas Research Network. Integrated genomic characterization of endometrial carcinoma. *Nature* **2013**, *497*, 67–73. [CrossRef]
147. Weinstein, J.N.; Collisson, E.A.; Mills, G.B.; Shaw, K.R.; Ozenberger, B.A.; Ellrott, K.; Shmulevich, I.; Sander, C.; Stuart, J.M.; Cancer Genome Atlas Research Network. The Cancer Genome Atlas Pan-Cancer analysis project. *Nat. Genet.* **2013**, *45*, 1113–1120. [CrossRef] [PubMed]
148. Jacobsen, A.; Silber, J.; Harinath, G.; Huse, J.T.; Schultz, N.; Sander, C. Analysis of microRNA-target interactions across diverse cancer types. *Nat. Struct. Mol. Biol.* **2013**, *20*, 1325–1332. [CrossRef] [PubMed]
149. Kandoth, C.; McLellan, M.D.; Vandin, F.; Ye, K.; Niu, B.; Lu, C.; Xie, M.; Zhang, Q.; McMichael, J.F.; Wyczalkowski, M.A.; et al. Mutational landscape and significance across 12 major cancer types. *Nature* **2013**, *502*, 333–339. [CrossRef]
150. Reimand, J.; Wagih, O.; Bader, G.D. The mutational landscape of phosphorylation signaling in cancer. *Sci. Rep.* **2013**, *3*, 2651. [CrossRef]
151. Tomczak, K.; Czerwinska, P.; Wiznerowicz, M. The Cancer Genome Atlas (TCGA): An immeasurable source of knowledge. *Contemp. Oncol.* **2015**, *19*, A68–A77. [CrossRef]
152. Cancer Genome Atlas Research Network. Integrated Genomic Characterization of Pancreatic Ductal Adenocarcinoma. *Cancer Cell* **2017**, *32*, 185–203.e113. [CrossRef] [PubMed]
153. Kreike, B.; van Kouwenhove, M.; Horlings, H.; Weigelt, B.; Peterse, H.; Bartelink, H.; van de Vijver, M.J. Gene expression profiling and histopathological characterization of triple-negative/basal-like breast carcinomas. *Breast Cancer Res. BCR* **2007**, *9*, R65. [CrossRef]
154. Thomas, S.N.; Friedrich, B.; Schnaubelt, M.; Chan, D.W.; Zhang, H.; Aebersold, R. Orthogonal Proteomic Platforms and Their Implications for the Stable Classification of High-Grade Serous Ovarian Cancer Subtypes. *iScience* **2020**, *23*, 101079. [CrossRef] [PubMed]
155. Zhang, Z.; Huang, K.; Gu, C.; Zhao, L.; Wang, N.; Wang, X.; Zhao, D.; Zhang, C.; Lu, Y.; Meng, Y. Molecular Subtyping of Serous Ovarian Cancer Based on Multi-omics Data. *Sci. Rep.* **2016**, *6*, 26001. [CrossRef] [PubMed]
156. Bansal, M.; He, J.; Peyton, M.; Kustagi, M.; Iyer, A.; Comb, M.; White, M.; Minna, J.D.; Califano, A. Elucidating synergistic dependencies in lung adenocarcinoma by proteome-wide signaling-network analysis. *PLoS ONE* **2019**, *14*, e0208646. [CrossRef]
157. Barretina, J.; Caponigro, G.; Stransky, N.; Venkatesan, K.; Margolin, A.A.; Kim, S.; Wilson, C.J.; Lehar, J.; Kryukov, G.V.; Sonkin, D.; et al. The Cancer Cell Line Encyclopedia enables predictive modelling of anticancer drug sensitivity. *Nature* **2012**, *483*, 603–607. [CrossRef]
158. Gordon, V.; Banerji, S. Molecular pathways: PI3K pathway targets in triple-negative breast cancers. *Clin. Cancer Res.* **2013**, *19*, 3738 3744. [CrossRef] [PubMed]
159. Jordan, V.C. Tamoxifen as the first targeted long-term adjuvant therapy for breast cancer. *Endocr. Relat. Cancer* **2014**, *21*, R235–R246. [CrossRef]
160. Kedves, A.T.; Gleim, S.; Liang, X.; Bonal, D.M.; Sigoillot, F.; Harbinski, F.; Sanghavi, S.; Benander, C.; George, E.; Gokhale, P.C.; et al. Recurrent ubiquitin B silencing in gynecological cancers establishes dependence on ubiquitin C. *J. Clin. Investig.* **2017**, *127*, 4554–4568. [CrossRef]
161. Natrajan, R.; Wilkerson, P. From integrative genomics to therapeutic targets. *Cancer Res.* **2013**, *73*, 3483–3488. [CrossRef]
162. Oza, A.M.; Tinker, A.V.; Oaknin, A.; Shapira-Frommer, R.; McNeish, I.A.; Swisher, E.M.; Ray-Coquard, I.; Bell-McGuinn, K.; Coleman, R.L.; O'Malley, D.M.; et al. Antitumor activity and safety of the PARP inhibitor rucaparib in patients with high-grade ovarian carcinoma and a germline or somatic BRCA1 or BRCA2 mutation: Integrated analysis of data from Study 10 and ARIEL2. *Gynecol. Oncol.* **2017**, *147*, 267–275. [CrossRef]
163. Park, J.T.; Chen, X.; Trope, C.G.; Davidson, B.; Shih Ie, M.; Wang, T.L. Notch3 overexpression is related to the recurrence of ovarian cancer and confers resistance to carboplatin. *Am. J. Pathol.* **2010**, *177*, 1087–1094. [CrossRef]
164. Patel, N.; Weekes, D.; Drosopoulos, K.; Gazinska, P.; Noel, E.; Rashid, M.; Mirza, H.; Quist, J.; Braso-Maristany, F.; Mathew, S.; et al. Integrated genomics and functional validation identifies malignant cell specific dependencies in triple negative breast cancer. *Nat. Commun.* **2018**, *9*, 1044. [CrossRef] [PubMed]
165. Shaw, A.T.; Kim, D.W.; Nakagawa, K.; Seto, T.; Crino, L.; Ahn, M.J.; De Pas, T.; Besse, B.; Solomon, B.J.; Blackhall, F.; et al. Crizotinib versus chemotherapy in advanced ALK-positive lung cancer. *N. Engl. J. Med.* **2013**, *368*, 2385–2394. [CrossRef] [PubMed]
166. Tsherniak, A.; Vazquez, F.; Montgomery, P.G.; Weir, B.A.; Kryukov, G.; Cowley, G.S.; Gill, S.; Harrington, W.F.; Pantel, S.; Krill-Burger, J.M.; et al. Defining a Cancer Dependency Map. *Cell* **2017**, *170*, 564–576.e516. [CrossRef] [PubMed]

167. Uzilov, A.V.; Ding, W.; Fink, M.Y.; Antipin, Y.; Brohl, A.S.; Davis, C.; Lau, C.Y.; Pandya, C.; Shah, H.; Kasai, Y.; et al. Development and clinical application of an integrative genomic approach to personalized cancer therapy. *Genome Med.* **2016**, *8*, 62. [CrossRef] [PubMed]

168. Wang, J.; Mouradov, D.; Wang, X.; Jorissen, R.N.; Chambers, M.C.; Zimmerman, L.J.; Vasaikar, S.; Love, C.G.; Li, S.; Lowes, K.; et al. Colorectal Cancer Cell Line Proteomes Are Representative of Primary Tumors and Predict Drug Sensitivity. *Gastroenterology* **2017**, *153*, 1082–1095. [CrossRef] [PubMed]

169. Wang, L.; McLeod, H.L.; Weinshilboum, R.M. Genomics and drug response. *N. Engl. J. Med.* **2011**, *364*, 1144–1153. [CrossRef]

170. Xie, H.; Wang, W.; Sun, F.; Deng, K.; Lu, X.; Liu, H.; Zhao, W.; Zhang, Y.; Zhou, X.; Li, K.; et al. Proteomics analysis to reveal biological pathways and predictive proteins in the survival of high-grade serous ovarian cancer. *Sci. Rep.* **2017**, *7*, 9896. [CrossRef]

171. Yang, J.Y.; Yoshihara, K.; Tanaka, K.; Hatae, M.; Masuzaki, H.; Itamochi, H.; Takano, M.; Ushijima, K.; Tanyi, J.L.; Cancer Genome Atlas Research Network; et al. Predicting time to ovarian carcinoma recurrence using protein markers. *J. Clin. Investig.* **2013**, *123*, 3740–3750. [CrossRef] [PubMed]

172. Yarden, Y.; Pines, G. The ERBB network: At last, cancer therapy meets systems biology. *Nat. Rev. Cancer* **2012**, *12*, 553–563. [CrossRef]

173. Wang, W.; Baladandayuthapani, V.; Morris, J.S.; Broom, B.M.; Manyam, G.; Do, K.A. iBAG: Integrative Bayesian analysis of high-dimensional multiplatform genomics data. *Bioinformatics* **2013**, *29*, 149–159. [CrossRef]

174. Alvarez, M.J.; Shen, Y.; Giorgi, F.M.; Lachmann, A.; Ding, B.B.; Ye, B.H.; Califano, A. Functional characterization of somatic mutations in cancer using network-based inference of protein activity. *Nat. Genet.* **2016**, *48*, 838–847. [CrossRef]

175. Paull, E.O.; Carlin, D.E.; Niepel, M.; Sorger, P.K.; Haussler, D.; Stuart, J.M. Discovering causal pathways linking genomic events to transcriptional states using Tied Diffusion Through Interacting Events (TieDIE). *Bioinformatics* **2013**, *29*, 2757–2764. [CrossRef]

176. Drake, J.M.; Paull, E.O.; Graham, N.A.; Lee, J.K.; Smith, B.A.; Titz, B.; Stoyanova, T.; Faltermeier, C.M.; Uzunangelov, V.; Carlin, D.E.; et al. Phosphoproteome Integration Reveals Patient-Specific Networks in Prostate Cancer. *Cell* **2016**, *166*, 1041–1054. [CrossRef]

177. Lim, B.; Woodward, W.A.; Wang, X.; Reuben, J.M.; Ueno, N.T. Inflammatory breast cancer biology: The tumour microenvironment is key. *Nat. Rev. Cancer* **2018**, *18*, 485–499. [CrossRef] [PubMed]

178. Yang, Y.; Yang, Y.; Yang, J.; Zhao, X.; Wei, X. Tumor Microenvironment in Ovarian Cancer: Function and Therapeutic Strategy. *Front. Cell Dev. Biol.* **2020**, *8*, 758. [CrossRef]

179. Yoshihara, K.; Shahmoradgoli, M.; Martinez, E.; Vegesna, R.; Kim, H.; Torres-Garcia, W.; Trevino, V.; Shen, H.; Laird, P.W.; Levine, D.A.; et al. Inferring tumour purity and stromal and immune cell admixture from expression data. *Nat. Commun.* **2013**, *4*, 2612. [CrossRef] [PubMed]

180. Dai, Z.; Gu, X.Y.; Xiang, S.Y.; Gong, D.D.; Man, C.F.; Fan, Y. Research and application of single-cell sequencing in tumor heterogeneity and drug resistance of circulating tumor cells. *Biomark. Res.* **2020**, *8*, 60. [CrossRef] [PubMed]

181. Carter, B.; Zhao, K. The epigenetic basis of cellular heterogeneity. *Nat. Rev. Genet.* **2020**. [CrossRef] [PubMed]

182. Nam, A.S.; Chaligne, R.; Landau, D.A. Integrating genetic and non-genetic determinants of cancer evolution by single-cell multi-omics. *Nat. Rev. Genet.* **2021**, *22*, 3–18. [CrossRef]

183. Wagner, D.E.; Klein, A.M. Lineage tracing meets single-cell omics: Opportunities and challenges. *Nat. Rev. Genet.* **2020**, *21*, 410–427. [CrossRef]

184. Stuart, T.; Satija, R. Integrative single-cell analysis. *Nat. Rev. Genet.* **2019**, *20*, 257–272. [CrossRef] [PubMed]

185. Kiselev, V.Y.; Andrews, T.S.; Hemberg, M. Challenges in unsupervised clustering of single-cell RNA-seq data. *Nat. Rev. Genet.* **2019**, *20*, 273–282. [CrossRef]

186. Prakadan, S.M.; Shalek, A.K.; Weitz, D.A. Scaling by shrinking: Empowering single-cell 'omics' with microfluidic devices. *Nat. Rev. Genet.* **2017**, *18*, 345–361. [CrossRef] [PubMed]

187. Woodworth, M.B.; Girskis, K.M.; Walsh, C.A. Building a lineage from single cells: Genetic techniques for cell lineage tracking. *Nat. Rev. Genet.* **2017**, *18*, 230–244. [CrossRef] [PubMed]

188. Gawad, C.; Koh, W.; Quake, S.R. Single-cell genome sequencing: Current state of the science. *Nat. Rev. Genet.* **2016**, *17*, 175–188. [CrossRef]

189. Schwartzman, O.; Tanay, A. Single-cell epigenomics: Techniques and emerging applications. *Nat. Rev. Genet.* **2015**, *16*, 716–726. [CrossRef]

190. Stegle, O.; Teichmann, S.A.; Marioni, J.C. Computational and analytical challenges in single-cell transcriptomics. *Nat. Rev. Genet.* **2015**, *16*, 133–145. [CrossRef]

191. Liberali, P.; Snijder, B.; Pelkmans, L. Single-cell and multivariate approaches in genetic perturbation screens. *Nat. Rev. Genet.* **2015**, *16*, 18–32. [CrossRef]

192. Chuah, S.; Chew, V. High-dimensional immune-profiling in cancer: Implications for immunotherapy. *J. Immunother. Cancer* **2020**, *8*, e000363. [CrossRef]

193. Chattopadhyay, P.K.; Gierahn, T.M.; Roederer, M.; Love, J.C. Single-cell technologies for monitoring immune systems. *Nat. Immunol.* **2014**, *15*, 128–135. [CrossRef] [PubMed]

194. Gibellini, L.; De Biasi, S.; Porta, C.; Lo Tartaro, D.; Depenni, R.; Pellacani, G.; Sabbatini, R.; Cossarizza, A. Single-Cell Approaches to Profile the Response to Immune Checkpoint Inhibitors. *Front. Immunol.* **2020**, *11*, 490. [CrossRef] [PubMed]

195. Landhuis, E. Single-cell approaches to immune profiling. *Nature* **2018**, *557*, 595–597. [CrossRef] [PubMed]

196. Jaitin, D.A.; Kenigsberg, E.; Keren-Shaul, H.; Elefant, N.; Paul, F.; Zaretsky, I.; Mildner, A.; Cohen, N.; Jung, S.; Tanay, A.; et al. Massively parallel single-cell RNA-seq for marker-free decomposition of tissues into cell types. *Science* **2014**, *343*, 776–779. [CrossRef] [PubMed]

197. Villani, A.C.; Satija, R.; Reynolds, G.; Sarkizova, S.; Shekhar, K.; Fletcher, J.; Griesbeck, M.; Butler, A.; Zheng, S.; Lazo, S.; et al. Single-cell RNA-seq reveals new types of human blood dendritic cells, monocytes, and progenitors. *Science* **2017**, *356*. [CrossRef]

198. Azizi, E.; Carr, A.J.; Plitas, G.; Cornish, A.E.; Konopacki, C.; Prabhakaran, S.; Nainys, J.; Wu, K.; Kiseliovas, V.; Setty, M.; et al. Single-Cell Map of Diverse Immune Phenotypes in the Breast Tumor Microenvironment. *Cell* **2018**, *174*, 1293–1308.e1236. [CrossRef]

199. Savas, P.; Virassamy, B.; Ye, C.; Salim, A.; Mintoff, C.P.; Caramia, F.; Salgado, R.; Byrne, D.J.; Teo, Z.L.; Dushyanthen, S.; et al. Single-cell profiling of breast cancer T cells reveals a tissue-resident memory subset associated with improved prognosis. *Nat. Med.* **2018**, *24*, 986–993. [CrossRef]

200. Lim, B.; Lin, Y.; Navin, N. Advancing Cancer Research and Medicine with Single-Cell Genomics. *Cancer Cell* **2020**, *37*, 456–470. [CrossRef]

201. Wang, Q.; Guldner, I.H.; Golomb, S.M.; Sun, L.; Harris, J.A.; Lu, X.; Zhang, S. Single-cell profiling guided combinatorial immunotherapy for fast-evolving CDK4/6 inhibitor-resistant HER2-positive breast cancer. *Nat. Commun.* **2019**, *10*, 3817. [CrossRef]

202. Alshetaiwi, H.; Pervolarakis, N.; McIntyre, L.L.; Ma, D.; Nguyen, Q.; Rath, J.A.; Nee, K.; Hernandez, G.; Evans, K.; Torosian, L.; et al. Defining the emergence of myeloid-derived suppressor cells in breast cancer using single-cell transcriptomics. *Sci. Immunol.* **2020**, *5*. [CrossRef]

203. Wan, C.; Keany, M.P.; Dong, H.; Al-Alem, L.F.; Pandya, U.M.; Lazo, S.; Boehnke, K.; Lynch, K.N.; Xu, R.; Zarrella, D.T.; et al. Enhanced efficacy of simultaneous PD-1 and PD-L1 immune checkpoint blockade in high grade serous ovarian cancer. *Cancer Res.* **2020**. [CrossRef]

204. Bartoschek, M.; Oskolkov, N.; Bocci, M.; Lövrot, J.; Larsson, C.; Sommarin, M.; Madsen, C.D.; Lindgren, D.; Pekar, G.; Karlsson, G.; et al. Spatially and functionally distinct subclasses of breast cancer-associated fibroblasts revealed by single cell RNA sequencing. *Nat. Commun.* **2018**, *9*, 5150. [CrossRef]

205. Jimenez-Sanchez, A.; Cybulska, P.; Mager, K.L.; Koplev, S.; Cast, O.; Couturier, D.L.; Memon, D.; Selenica, P.; Nikolovski, I.; Mazaheri, Y.; et al. Unraveling tumor-immune heterogeneity in advanced ovarian cancer uncovers immunogenic effect of chemotherapy. *Nat. Genet.* **2020**, *52*, 582–593. [CrossRef]

206. Zhu, Y.; Zhang, Z.; Jiang, Z.; Liu, Y.; Zhou, J. CD38 Predicts Favorable Prognosis by Enhancing Immune Infiltration and Antitumor Immunity in the Epithelial Ovarian Cancer Microenvironment. *Front. Genet.* **2020**, *11*, 369. [CrossRef]

207. Lyons, Y.A.; Wu, S.Y.; Overwijk, W.W.; Baggerly, K.A.; Sood, A.K. Immune cell profiling in cancer: Molecular approaches to cell-specific identification. *NPJ Precis. Oncol.* **2017**, *1*, 26. [CrossRef] [PubMed]

208. Gonzalez, H.; Hagerling, C.; Werb, Z. Roles of the immune system in cancer: From tumor initiation to metastatic progression. *Genes Dev.* **2018**, *32*, 1267–1284. [CrossRef] [PubMed]

209. Zhu, L.; Narloch, J.L.; Onkar, S.; Joy, M ; Broadwater, G.; Luedke, C.; Hall, A.; Kim, R.; Pogue-Geile, K.; Sammons, S.; et al. Metastatic breast cancers have reduced immune cell recruitment but harbor increased macrophages relative to their matched primary tumors. *J. Immunother. Cancer* **2019**, *7*, 265. [CrossRef] [PubMed]

210. Segovia-Mendoza, M.; Morales-Montor, J. Immune Tumor Microenvironment in Breast Cancer and the Participation of Estrogen and Its Receptors in Cancer Physiopathology. *Front. Immunol.* **2019**, *10*, 348. [CrossRef] [PubMed]

211. Tekpli, X.; Lien, T.; Rossevold, A.H.; Nebdal, D.; Borgen, E.; Ohnstad, H.O.; Kyte, J.A.; Vallon-Christersson, J.; Fongaard, M.; Due, E.U.; et al. An independent poor-prognosis subtype of breast cancer defined by a distinct tumor immune microenvironment. *Nat. Commun.* **2019**, *10*, 5499. [CrossRef] [PubMed]

212. Wagner, J.; Rapsomaniki, M.A.; Chevrier, S.; Anzeneder, T.; Langwieder, C.; Dykgers, A.; Rees, M.; Ramaswamy, A.; Muenst, S.; Soysal, S.D.; et al. A Single-Cell Atlas of the Tumor and Immune Ecosystem of Human Breast Cancer. *Cell* **2019**, *177*, 1330–1345.e1318. [CrossRef] [PubMed]

213. Suva, M.L.; Tirosh, I. Single-Cell RNA Sequencing in Cancer: Lessons Learned and Emerging Challenges. *Mol. Cell* **2019**, *75*, 7–12. [CrossRef] [PubMed]

214. Ding, S.; Chen, X.; Shen, K. Single-cell RNA sequencing in breast cancer: Understanding tumor heterogeneity and paving roads to individualized therapy. *Cancer Commun.* **2020**, *40*, 329–344. [CrossRef]

215. Fan, J.; Slowikowski, K.; Zhang, F. Single-cell transcriptomics in cancer: Computational challenges and opportunities. *Exp. Mol. Med.* **2020**, *52*, 1452–1465. [CrossRef]

216. Hoffman, J.A.; Papas, B.N.; Trotter, K.W.; Archer, T.K. Single-cell RNA sequencing reveals a heterogeneous response to Glucocorticoids in breast cancer cells. *Commun. Biol.* **2020**, *3*, 126. [CrossRef]

217. Fan, J.; Lee, H.O.; Lee, S.; Ryu, D.E.; Lee, S.; Xue, C.; Kim, S.J.; Kim, K.; Barkas, N.; Park, P.J.; et al. Linking transcriptional and genetic tumor heterogeneity through allele analysis of single-cell RNA-seq data. *Genome Res.* **2018**, *28*, 1217–1227. [CrossRef]

218. Ren, X.; Kang, B.; Zhang, Z. Understanding tumor ecosystems by single-cell sequencing: Promises and limitations. *Genome Biol.* **2018**, *19*, 211. [CrossRef]

219. Chung, W.; Eum, H.H.; Lee, H.O.; Lee, K.M.; Lee, H.B.; Kim, K.T.; Ryu, H.S.; Kim, S.; Lee, J.E.; Park, Y.H.; et al. Single-cell RNA-seq enables comprehensive tumour and immune cell profiling in primary breast cancer. *Nat. Commun.* **2017**, *8*, 15081. [CrossRef] [PubMed]

220. Karaayvaz, M.; Cristea, S.; Gillespie, S.M.; Patel, A.P.; Mylvaganam, R.; Luo, C.C.; Specht, M.C.; Bernstein, B.E.; Michor, F.; Ellisen, L.W. Unravelling subclonal heterogeneity and aggressive disease states in TNBC through single-cell RNA-seq. *Nat. Commun.* **2018**, *9*, 3588. [CrossRef] [PubMed]

221. Hong, S.P.; Chan, T.E.; Lombardo, Y.; Corleone, G.; Rotmensz, N.; Bravaccini, S.; Rocca, A.; Pruneri, G.; McEwen, K.R.; Coombes, R.C.; et al. Single-cell transcriptomics reveals multi-step adaptations to endocrine therapy. *Nat. Commun.* **2019**, *10*, 3840. [CrossRef]

222. Geistlinger, L.; Oh, S.; Ramos, M.; Schiffer, L.; LaRue, R.S.; Henzler, C.M.; Munro, S.A.; Daughters, C.; Nelson, A.C.; Winterhoff, B.J.; et al. Multiomic Analysis of Subtype Evolution and Heterogeneity in High-Grade Serous Ovarian Carcinoma. *Cancer Res.* **2020**, *80*, 4335–4345. [CrossRef]

223. Lourenco, A.R.; Ban, Y.; Crowley, M.J.; Lee, S.B.; Ramchandani, D.; Du, W.; Elemento, O.; George, J.T.; Jolly, M.K.; Levine, H.; et al. Differential Contributions of Pre- and Post-EMT Tumor Cells in Breast Cancer Metastasis. *Cancer Res.* **2020**, *80*, 163–169. [CrossRef]

224. Baslan, T.; Kendall, J.; Volyanskyy, K.; McNamara, K.; Cox, H.; D'Italia, S.; Ambrosio, F.; Riggs, M.; Rodgers, L.; Leotta, A.; et al. Novel insights into breast cancer copy number genetic heterogeneity revealed by single-cell genome sequencing. *Elife* **2020**, *9*. [CrossRef]

225. Almendro, V.; Cheng, Y.K.; Randles, A.; Itzkovitz, S.; Marusyk, A.; Ametller, E.; Gonzalez-Farre, X.; Munoz, M.; Russnes, H.G.; Helland, A.; et al. Inference of tumor evolution during chemotherapy by computational modeling and in situ analysis of genetic and phenotypic cellular diversity. *Cell Rep.* **2014**, *6*, 514–527. [CrossRef]

226. Ding, L.; Ley, T.J.; Larson, D.E.; Miller, C.A.; Koboldt, D.C.; Welch, J.S.; Ritchey, J.K.; Young, M.A.; Lamprecht, T.; McLellan, M.D.; et al. Clonal evolution in relapsed acute myeloid leukaemia revealed by whole-genome sequencing. *Nature* **2012**, *481*, 506–510. [CrossRef]

227. Kulkarni, A.; Al-Hraishawi, H.; Simhadri, S.; Hirshfield, K.M.; Chen, S.; Pine, S.; Jeyamohan, C.; Sokol, L.; Ali, S.; Teo, M.L.; et al. BRAF Fusion as a Novel Mechanism of Acquired Resistance to Vemurafenib in BRAF(V600E) Mutant Melanoma. *Clin. Cancer Res.* **2017**, *23*, 5631–5638. [CrossRef]

228. Gonzalez, V.D.; Samusik, N.; Chen, T.J.; Savig, E.S.; Aghaeepour, N.; Quigley, D.A.; Huang, Y.W.; Giangarra, V.; Borowsky, A.D.; Hubbard, N.E.; et al. Commonly Occurring Cell Subsets in High-Grade Serous Ovarian Tumors Identified by Single-Cell Mass Cytometry. *Cell Rep.* **2018**, *22*, 1875–1888. [CrossRef]

229. Kim, C.; Gao, R.; Sei, E.; Brandt, R.; Hartman, J.; Hatschek, T.; Crosetto, N.; Foukakis, T.; Navin, N.E. Chemoresistance Evolution in Triple-Negative Breast Cancer Delineated by Single-Cell Sequencing. *Cell* **2018**, *173*, 879–893.e813. [CrossRef] [PubMed]

230. Mucino-Olmos, E.A.; Vazquez-Jimenez, A.; Avila-Ponce de Leon, U.; Matadamas-Guzman, M.; Maldonado, V.; Lopez-Santaella, T.; Hernandez-Hernandez, A.; Resendis-Antonio, O. Unveiling functional heterogeneity in breast cancer multicellular tumor spheroids through single-cell RNA-seq. *Sci. Rep.* **2020**, *10*, 12728. [CrossRef] [PubMed]

231. Yeo, S.K.; Zhu, X.; Okamoto, T.; Hao, M.; Wang, C.; Lu, P.; Lu, L.J.; Guan, J.L. Single-cell RNA-sequencing reveals distinct patterns of cell state heterogeneity in mouse models of breast cancer. *Elife* **2020**, *9*. [CrossRef] [PubMed]

232. Zhang, A.W.; O'Flanagan, C.; Chavez, E.A.; Lim, J.L.P.; Ceglia, N.; McPherson, A.; Wiens, M.; Walters, P.; Chan, T.; Hewitson, B.; et al. Probabilistic cell-type assignment of single-cell RNA-seq for tumor microenvironment profiling. *Nat. Methods* **2019**, *16*, 1007–1015. [CrossRef]

233. Hu, Z.; Artibani, M.; Alsaadi, A.; Wietek, N.; Morotti, M.; Shi, T.; Zhong, Z.; Santana Gonzalez, L.; El-Sahhar, S.; KaramiNejad-Ranjbar, M.; et al. The Repertoire of Serous Ovarian Cancer Non-genetic Heterogeneity Revealed by Single-Cell Sequencing of Normal Fallopian Tube Epithelial Cells. *Cancer Cell* **2020**, *37*, 226–242.e227. [CrossRef]

234. Marcotte, R.; Sayad, A.; Brown, K.R.; Sanchez-Garcia, F.; Reimand, J.; Haider, M.; Virtanen, C.; Bradner, J.E.; Bader, G.D.; Mills, G.B.; et al. Functional Genomic Landscape of Human Breast Cancer Drivers, Vulnerabilities, and Resistance. *Cell* **2016**, *164*, 293–309. [CrossRef]

235. Cheung, H.W.; Cowley, G.S.; Weir, B.A.; Boehm, J.S.; Rusin, S.; Scott, J.A.; East, A.; Ali, L.D.; Lizotte, P.H.; Wong, T.C.; et al. Systematic investigation of genetic vulnerabilities across cancer cell lines reveals lineage-specific dependencies in ovarian cancer. *Proc. Natl. Acad. Sci. USA* **2011**, *108*, 12372–12377. [CrossRef]

236. Chen, M.M.; Li, J.; Mills, G.B.; Liang, H. Predicting Cancer Cell Line Dependencies from the Protein Expression Data of Reverse-Phase Protein Arrays. *JCO Clin. Cancer Inform.* **2020**, *4*, 357–366. [CrossRef]

237. Lord, C.J.; Quinn, N.; Ryan, C.J. Integrative analysis of large-scale loss-of-function screens identifies robust cancer-associated genetic interactions. *Elife* **2020**, *9*. [CrossRef]

238. Price, C.; Gill, S.; Ho, Z.V.; Davidson, S.M.; Merkel, E.; McFarland, J.M.; Leung, L.; Tang, A.; Kost-Alimova, M.; Tsherniak, A.; et al. Genome-Wide Interrogation of Human Cancers Identifies EGLN1 Dependency in Clear Cell Ovarian Cancers. *Cancer Res.* **2019**, *79*, 2564–2579. [CrossRef] [PubMed]

239. Dempster, J.M.; Pacini, C.; Pantel, S.; Behan, F.M.; Green, T.; Krill-Burger, J.; Beaver, C.M.; Younger, S.T.; Zhivich, V.; Najgebauer, H.; et al. Agreement between two large pan-cancer CRISPR-Cas9 gene dependency data sets. *Nat. Commun.* **2019**, *10*, 5817. [CrossRef]

240. Hofree, M.; Shen, J.P.; Carter, H.; Gross, A.; Ideker, T. Network-based stratification of tumor mutations. *Nat. Methods* **2013**, *10*, 1108–1115. [CrossRef]
241. Tamborero, D.; Gonzalez-Perez, A.; Perez-Llamas, C.; Deu-Pons, J.; Kandoth, C.; Reimand, J.; Lawrence, M.S.; Getz, G.; Bader, G.D.; Ding, L.; et al. Comprehensive identification of mutational cancer driver genes across 12 tumor types. *Sci. Rep.* **2013**, *3*, 2650. [CrossRef]
242. Alexandrov, L.B.; Nik-Zainal, S.; Wedge, D.C.; Aparicio, S.A.; Behjati, S.; Biankin, A.V.; Bignell, G.R.; Bolli, N.; Borg, A.; Borresen-Dale, A.L.; et al. Signatures of mutational processes in human cancer. *Nature* **2013**, *500*, 415–421. [CrossRef]
243. Petljak, M.; Alexandrov, L.B.; Brammeld, J.S.; Price, S.; Wedge, D.C.; Grossmann, S.; Dawson, K.J.; Ju, Y.S.; Iorio, F.; Tubio, J.M.C.; et al. Characterizing Mutational Signatures in Human Cancer Cell Lines Reveals Episodic APOBEC Mutagenesis. *Cell* **2019**, *176*, 1282–1294.e1220. [CrossRef]
244. Ha, M.J.; Banerjee, S.; Akbani, R.; Liang, H.; Mills, G.B.; Do, K.A.; Baladandayuthapani, V. Personalized Integrated Network Modeling of the Cancer Proteome Atlas. *Sci. Rep.* **2018**, *8*, 14924. [CrossRef]
245. Chiu, A.M.; Mitra, M.; Boymoushakian, L.; Coller, H.A. Integrative analysis of the inter-tumoral heterogeneity of triple-negative breast cancer. *Sci. Rep.* **2018**, *8*, 11807. [CrossRef] [PubMed]
246. Vaske, C.J.; Benz, S.C.; Sanborn, J.Z.; Earl, D.; Szeto, C.; Zhu, J.; Haussler, D.; Stuart, J.M. Inference of patient-specific pathway activities from multi-dimensional cancer genomics data using PARADIGM. *Bioinformatics* **2010**, *26*, i237–i245. [CrossRef] [PubMed]
247. Sanchez-Vega, F.; Mina, M.; Armenia, J.; Chatila, W.K.; Luna, A.; La, K.C.; Dimitriadoy, S.; Liu, D.L.; Kantheti, H.S.; Saghafinia, S.; et al. Oncogenic Signaling Pathways in The Cancer Genome Atlas. *Cell* **2018**, *173*, 321–337.e310. [CrossRef]
248. Sicklick, J.K.; Kato, S.; Okamura, R.; Schwaederle, M.; Hahn, M.E.; Williams, C.B.; De, P.; Krie, A.; Piccioni, D.E.; Miller, V.A.; et al. Molecular profiling of cancer patients enables personalized combination therapy: The I-PREDICT study. *Nat. Med.* **2019**, *25*, 744–750. [CrossRef] [PubMed]
249. Woodhouse, R.; Li, M.; Hughes, J.; Delfosse, D.; Skoletsky, J.; Ma, P.; Meng, W.; Dewal, N.; Milbury, C.; Clark, T.; et al. Clinical and analytical validation of FoundationOne Liquid CDx, a novel 324-Gene cfDNA-based comprehensive genomic profiling assay for cancers of solid tumor origin. *PLoS ONE* **2020**, *15*, e0237802. [CrossRef]
250. Whitwell, H.J.; Worthington, J.; Blyuss, O.; Gentry-Maharaj, A.; Ryan, A.; Gunu, R.; Kalsi, J.; Menon, U.; Jacobs, I.; Zaikin, A.; et al. Improved early detection of ovarian cancer using longitudinal multimarker models. *Br. J. Cancer* **2020**, *122*, 847–856. [CrossRef] [PubMed]
251. Russell, M.R.; Graham, C.; D'Amato, A.; Gentry-Maharaj, A.; Ryan, A.; Kalsi, J.K.; Whetton, A.D.; Menon, U.; Jacobs, I.; Graham, R.L.J. Diagnosis of epithelial ovarian cancer using a combined protein biomarker panel. *Br. J. Cancer* **2019**, *121*, 483–489. [CrossRef]
252. Russell, M.R.; Graham, C.; D'Amato, A.; Gentry-Maharaj, A.; Ryan, A.; Kalsi, J.K.; Ainley, C.; Whetton, A.D.; Menon, U.; Jacobs, I.; et al. A combined biomarker panel shows improved sensitivity for the early detection of ovarian cancer allowing the identification of the most aggressive type II tumours. *Br. J. Cancer* **2017**, *117*, 666–674. [CrossRef]
253. Dochez, V.; Caillon, H.; Vaucel, E.; Dimet, J.; Winer, N.; Ducarme, G. Biomarkers and algorithms for diagnosis of ovarian cancer: CA125, HE4, RMI and ROMA, a review. *J. Ovarian Res.* **2019**, *12*, 28. [CrossRef] [PubMed]
254. Fong, P.C.; Boss, D.S.; Yap, T.A.; Tutt, A.; Wu, P.; Mergui-Roelvink, M.; Mortimer, P.; Swaisland, H.; Lau, A.; O'Connor, M.J.; et al. Inhibition of poly(ADP-ribose) polymerase in tumors from BRCA mutation carriers. *N. Engl. J. Med.* **2009**, *361*, 123–134. [CrossRef]
255. Moore, R.G.; Miller, M.C.; Disilvestro, P.; Landrum, L.M.; Gajewski, W.; Ball, J.J.; Skates, S.J. Evaluation of the diagnostic accuracy of the risk of ovarian malignancy algorithm in women with a pelvic mass. *Obstet. Gynecol.* **2011**, *118*, 280–288. [CrossRef] [PubMed]
256. Bristow, R.E.; Smith, A.; Zhang, Z.; Chan, D.W.; Crutcher, G.; Fung, E.T.; Munroe, D.G. Ovarian malignancy risk stratification of the adnexal mass using a multivariate index assay. *Gynecol. Oncol.* **2013**, *128*, 252–259. [CrossRef] [PubMed]
257. Ueland, F.R.; Desimone, C.P.; Seamon, L.G.; Miller, R.A.; Goodrich, S.; Podzielinski, I.; Sokoll, L.; Smith, A.; van Nagell, J.R., Jr.; Zhang, Z. Effectiveness of a multivariate index assay in the preoperative assessment of ovarian tumors. *Obstet. Gynecol.* **2011**, *117*, 1289–1297. [CrossRef]
258. Sparano, J.A.; Gray, R.J.; Makower, D.F.; Pritchard, K.I.; Albain, K.S.; Hayes, D.F.; Geyer, C.E., Jr.; Dees, E.C.; Goetz, M.P.; Olson, J.A., Jr.; et al. Adjuvant Chemotherapy Guided by a 21-Gene Expression Assay in Breast Cancer. *N. Engl. J. Med.* **2018**, *379*, 111–121. [CrossRef]
259. Dubsky, P.; Brase, J.C.; Jakesz, R.; Rudas, M.; Singer, C.F.; Greil, R.; Dietze, O.; Luisser, I.; Klug, E.; Sedivy, R.; et al. The EndoPredict score provides prognostic information on late distant metastases in ER+/HER2- breast cancer patients. *Br. J. Cancer* **2013**, *109*, 2959–2964. [CrossRef]
260. Cardoso, F.; van't Veer, L.J.; Bogaerts, J.; Slaets, L.; Viale, G.; Delaloge, S.; Pierga, J.Y.; Brain, E.; Causeret, S.; DeLorenzi, M.; et al. 70-Gene Signature as an Aid to Treatment Decisions in Early-Stage Breast Cancer. *N. Engl. J. Med.* **2016**, *375*, 717–729. [CrossRef]
261. Losk, K.; Freedman, R.A.; Laws, A.; Kantor, O.; Mittendorf, E.A.; Tan-Wasielewski, Z.; Trippa, L.; Lin, N.U.; Winer, E.P.; King, T.A. Oncotype DX testing in node-positive breast cancer strongly impacts chemotherapy use at a comprehensive cancer center. *Breast Cancer Res. Treat.* **2021**, *185*, 215–227. [CrossRef]
262. Cheng, R.; Kong, X.; Wang, X.; Fang, Y.; Wang, J. Oncotype DX Breast Recurrence Score Distribution and Chemotherapy Benefit Among Women of Different Age Groups With HR-Positive, HER2-Negative, Node-Negative Breast Cancer in the SEER Database. *Front. Oncol.* **2020**, *10*, 1583. [CrossRef]

263. Cardoso, F.; Veer, L.v.t.; Poncet, C.; Cardozo, J.L.; Delaloge, S.; Pierga, J.-Y.; Vuylsteke, P.; Brain, E.; Viale, G.; Kuemmel, S.; et al. MINDACT: Long-term results of the large prospective trial testing the 70-gene signature MammaPrint as guidance for adjuvant chemotherapy in breast cancer patients. *J. Clin. Oncol.* **2020**, *38*, 506. [CrossRef]
264. Pereira, E.; Camacho-Vanegas, O.; Anand, S.; Sebra, R.; Catalina Camacho, S.; Garnar-Wortzel, L.; Nair, N.; Moshier, E.; Wooten, M.; Uzilov, A.; et al. Personalized Circulating Tumor DNA Biomarkers Dynamically Predict Treatment Response and Survival In Gynecologic Cancers. *PLoS ONE* **2015**, *10*, e0145754. [CrossRef] [PubMed]
265. Lee, J.H.; Long, G.V.; Menzies, A.M.; Lo, S.; Guminski, A.; Whitbourne, K.; Peranec, M.; Scolyer, R.; Kefford, R.F.; Rizos, H.; et al. Association Between Circulating Tumor DNA and Pseudoprogression in Patients With Metastatic Melanoma Treated with Anti-Programmed Cell Death 1 Antibodies. *JAMA Oncol.* **2018**, *4*, 717–721. [CrossRef] [PubMed]
266. Ogasawara, A.; Hihara, T.; Shintani, D.; Yabuno, A.; Ikeda, Y.; Tai, K.; Fujiwara, K.; Watanabe, K.; Hasegawa, K. Evaluation of Circulating Tumor DNA in Patients with Ovarian Cancer Harboring Somatic PIK3CA or KRAS Mutations. *Cancer Res. Treat.* **2020**, *52*, 1219–1228. [CrossRef] [PubMed]
267. Noguchi, T.; Sakai, K.; Iwahashi, N.; Matsuda, K.; Matsukawa, H.; Yahata, T.; Toujima, S.; Nishio, K.; Ino, K. Changes in the gene mutation profiles of circulating tumor DNA detected using CAPP-Seq in neoadjuvant chemotherapy-treated advanced ovarian cancer. *Oncol. Lett.* **2020**, *19*, 2713–2720. [CrossRef]
268. Liggett, T.E.; Melnikov, A.; Yi, Q.; Replogle, C.; Hu, W.; Rotmensch, J.; Kamat, A.; Sood, A.K.; Levenson, V. Distinctive DNA methylation patterns of cell-free plasma DNA in women with malignant ovarian tumors. *Gynecol. Oncol.* **2011**, *120*, 113–120. [CrossRef] [PubMed]

Journal of
*Personalized
Medicine*

Article

Identification of Somatic Structural Variants in Solid Tumors by Optical Genome Mapping

David Y. Goldrich [1,†], Brandon LaBarge [1,†], Scott Chartrand [2], Lijun Zhang [2], Henry B. Sadowski [3], Yang Zhang [3], Khoa Pham [3], Hannah Way [3], Chi-Yu Jill Lai [3], Andy Wing Chun Pang [3], Benjamin Clifford [3], Alex R. Hastie [3], Mark Oldakowski [3], David Goldenberg [1] and James R. Broach [2,*]

[1] Department of Otolaryngology—Head and Neck Surgery, Pennsylvania State University College of Medicine, Hershey, PA 17033, USA; dgoldrich@pennstatehealth.psu.edu (D.Y.G.); blabarge@pennstatehealth.psu.edu (B.L.); dgoldenberg@pennstatehealth.psu.edu (D.G.)

[2] Department of Biochemistry and Molecular Biology, Pennsylvania State University College of Medicine, Hershey, PA 17033, USA; schartrand@pennstatehealth.psu.edu (S.C.); lzhang6@pennstatehealth.psu.edu (L.Z.)

[3] Bionano Genomics, San Diego, CA 92121, USA; hsadowski@bionanogenomics.com (H.B.S.); yzhang@bionanogenomics.com (Y.Z.); kpham@bionanogenomics.com (K.P.); hway@bionanogenomics.com (H.W.); jlai@bionanogenomics.com (C.-Y.J.L.); apang@bionanogenomics.com (A.W.C.P.); bclifford@bionanogenomics.com (B.C.); ahastie@bionanogenomics.com (A.R.H.); moldakowski@bionanogenomics.com (M.O.)

* Correspondence: jbroach@pennstatehealth.psu.edu

† These authors contributed equally to this work.

Citation: Goldrich, D.Y.; LaBarge, B.; Chartrand, S.; Zhang, L.; Sadowski, H.B.; Zhang, Y.; Pham, K.; Way, H.; Lai, C.-Y.J.; Pang, A.W.C.; et al. Identification of Somatic Structural Variants in Solid Tumors by Optical Genome Mapping. *J. Pers. Med.* **2021**, *11*, 142. https://doi.org/10.3390/jpm11020142

Academic Editor: Silvia Fantozzi

Received: 22 January 2021
Accepted: 15 February 2021
Published: 18 February 2021

Publisher's Note: MDPI stays neutral with regard to jurisdictional claims in published maps and institutional affiliations.

Abstract: Genomic structural variants comprise a significant fraction of somatic mutations driving cancer onset and progression. However, such variants are not readily revealed by standard next-generation sequencing. Optical genome mapping (OGM) surpasses short-read sequencing in detecting large (>500 bp) and complex structural variants (SVs) but requires isolation of ultra-high-molecular-weight DNA from the tissue of interest. We have successfully applied a protocol involving a paramagnetic nanobind disc to a wide range of solid tumors. Using as little as 6.5 mg of input tumor tissue, we show successful extraction of high-molecular-weight genomic DNA that provides a high genomic map rate and effective coverage by optical mapping. We demonstrate the system's utility in identifying somatic SVs affecting functional and cancer-related genes for each sample. Duplicate/triplicate analysis of select samples shows intra-sample reliability but also intra-sample heterogeneity. We also demonstrate that simply filtering SVs based on a GRCh38 human control database provides high positive and negative predictive values for true somatic variants. Our results indicate that the solid tissue DNA extraction protocol, OGM and SV analysis can be applied to a wide variety of solid tumors to capture SVs across the entire genome with functional importance in cancer prognosis and treatment.

Keywords: optical genome mapping; solid tumors; cancer genomics

1. Introduction

One of the hallmarks of cancer is genomic instability, which often affects genes controlling cell division and genome integrity. The resulting alterations include single-nucleotide variant (SNV) point mutations as well as structural variants (SVs), in which larger DNA segments undergo chromosomal perturbations such as deletions, insertions, duplications, inversions, and translocations. For instance, recurrent translocations, such as the Philadelphia chromosome, can activate oncogenes but at the same time reveal avenues for implementing or developing effective targeted drug therapies [1–4]. Likewise, SV identification plays an increasingly important role in cancer diagnosis and prognosis [5,6], and SVs have been shown to play a crucial role in intra-tumoral genetic heterogeneity [7]. Therefore, SV identification and analysis are important to understanding oncogenesis and tumor behavior.

Short-read sequencing can readily detect many SNVs, but is less successful in detecting SVs, by either alignment-based or assembly-based methods [8]. Since alignment-based approaches rely on mapping reads to unique positions, repetitive and low-complexity genomic regions can lead to misalignment and false-positive SV calls. Additionally, homologous alleles may be incorrectly combined, leading to haploid assembly only representing a single allele or chimeric assemblies mixing alleles. Whole-genome and cytogenetic approaches such as whole-genome sequencing (WGS), karyotyping, fluorescent in situ hybridization (FISH) and CNV microarrays also contain significant limitations. Karyotyping provides a comprehensive view of the entire genome but carries limited resolution of ~5 Mb and in most cases requires culturing cells before preparing chromosomes. FISH has a higher resolution but requires prior knowledge as to which loci to test and has limited throughput. CNV microarrays offer a resolution down to multiple Kb but are insensitive to balanced chromosomal aberrations such as translocations and inversions, are unable to detect low-frequency allelic changes, and cannot distinguish tandem duplications from insertions in trans. Finally, WGS has difficulty with de novo genome assembly and resolving duplications and repeated sequences [8–10]. Therefore, alternative methods are required to preserve long-range genomic structural information.

Optical genome mapping (OGM) has emerged as a viable option for analyzing large genomes for SVs. OGM preserves long-range information by imaging entire intact molecules of DNA in their native state and, as a result, has contributed to constructing reference genome assemblies, including those for maize, mouse, goat, and humans [11–28]. OGM can detect large (>500 bp) and complex SVs, such as chromothrypsis, that are difficult to detect using traditional short-read sequencing alone. OGM preparation and analysis workflow has been successfully applied to liquid-phase tumor and cell culture SV analyses. For instance, investigators have analyzed primary leukemic cells with OGM to identify previously unrecognized SVs implicated in oncogenesis and patients' survival and have combined OGM with chromosome conformation capture to demonstrate enhancer highjacking resulting from SVs [5,29,30]. Similarly, investigators used OGM to visualize complex gene fusions and novel somatic SVs in liposarcoma, melanoma and other well-studied cancer cell lines [31,32].

Despite its success in visualizing SVs in liquid tumors and cell lines, OGM has not yet seen widespread application in solid tissue tumors, due primarily to the difficulty of obtaining high-quality, high-molecular-weight DNA from solid tumor samples. Nonetheless, previous work has shown the feasibility of high-quality high-molecular-weight DNA isolation and analysis using earlier workflow iterations [33], and recent feasibility studies have shown the importance of OGM application to solid tumor analysis [7,34,35]. Peng et al. demonstrated large SVs not detected by WGS implicated in metastatic lung squamous cell carcinoma [7], and Jaratlerdiri et al. and Crumbaker et al. similarly found SVs impacting oncogenic and tumor-suppressing genes not identified by NGS or WGS alone in prostate cancer [34,35]. However, these previous methods for extracting gDNA from solid tissue were either prohibitively expensive or yielded low quantities of DNA [36]. We demonstrate here the successful implementation of a workflow to generate ultra-high-molecular-weight gDNA and subsequent SV analysis for 20 solid tumor samples comprising a wide variety of solid tissue organ systems.

2. Materials and Methods

2.1. Tumor Samples

Solid tissue was collected following surgical resection for 10 tumors: four squamous cell carcinomas of the tongue, three anaplastic carcinomas of the thyroid, one liver hepatocellular carcinoma, one lung pleomorphic carcinoma, and one bladder tumor. Patients consented under protocols approved by the Penn State Health Institution Review Board and tissue was flash frozen and stored at −80 °C in the Penn State Institute for Personalized Medicine (IPM). Ten additional fresh frozen solid tumor samples were acquired from BioIVT for the following tumor types: lung adenosquamous carcinoma, liver hepatocellular carcinoma, bladder papillary urothelial carcinoma, kidney renal cell carcinoma, breast ductal carcinoma in situ, prostate invasive adenocarcinoma, brain anaplastic astrocytoma, ovarian serous carcinoma, colon adenocarcinoma, and papillary thyroid carcinoma. For some of the samples, two or three separate sections of the tumor were excised and processed independently to provide duplicate or triplicate biological replicates.

2.2. Bionano Optical Genome Mapping

Ultra-High-Molecular-Weight gDNA Isolation from Solid Tissue. The following protocol is diagrammed in Figure 1 and described in greater detail in a support document from Bionano Genomics (https://bionanogenomics.com/support-page/sp-tissue-and-tumor-dna-isolation-kit/). Briefly, tissue sections with a target mass of 10 mg were sliced from a frozen parent piece on a sterilized aluminum block over dry ice. The tissues were minced briefly and placed into a 15 mL conical tube on ice containing homogenization buffer (HB) for subsequent blending with a Tissueruptor II (Qiagen). Following tissue disruption, samples were washed in additional HB, poured through a 40 µm filter, and centrifuged to pellets, from which the supernatants were decanted.

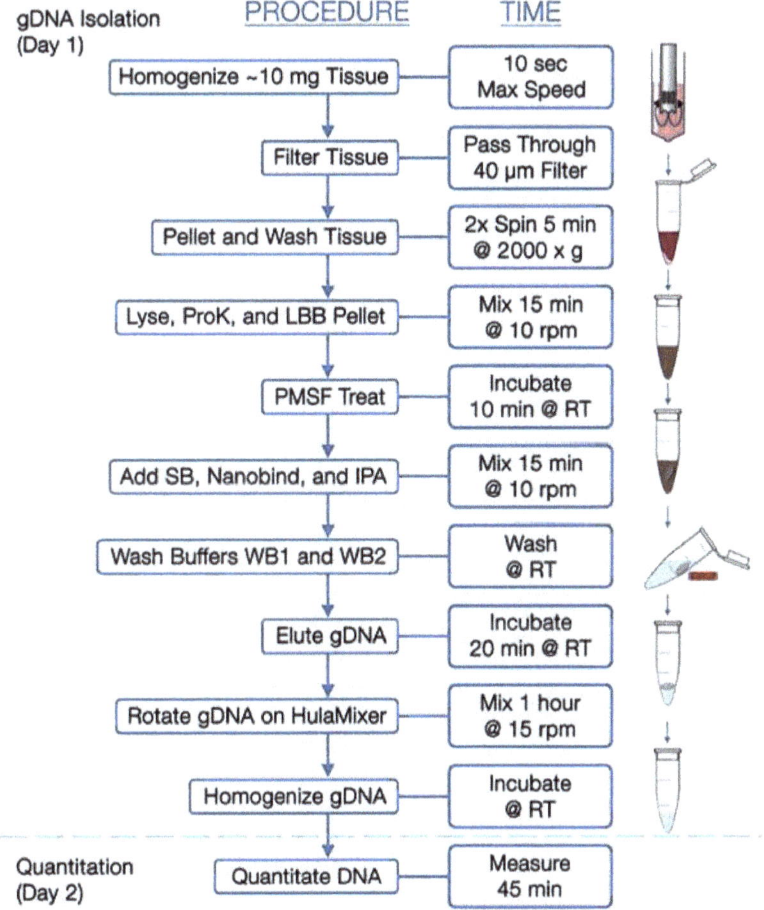

Figure 1. Workflow for isolation of high-molecular-weight DNA from solid tumors.

Pellets were resuspended in Wash Buffer A (Bionano, San Diego, CA, USA) and transferred to microcentrifuge tubes for additional washing. Supernatants were then decanted, and pellets resuspended in residual volume. Proteinase K (Bionano Genomics, San Diego, CA, USA) was added to samples, followed by Lysis and Binding Buffer (LBB, Bionano Genomics, San Diego, CA, USA) and mixed to produce a lysate containing high-molecular-weight DNA. Phenylmethylsulfonyl Fluoride Solution (PMSF, Millipore Sigma) was added to inactivate Proteinase K, followed by Salting Buffer (SB, Bionano Genomics, San Diego, CA, USA).

A single paramagnetic Nanobind Disc (Bionano Genomics, San Diego, CA, USA) was added to the lysate with 100% isopropanol, to facilitate binding and washing of gDNA strands. With gDNA captured on the disc, the supernatants were carefully removed and discs were washed with rounds of ethanol-based wash buffer. Discs were then transferred to clean tubes, where gDNA was eluted in buffer and homogenized at room temperature.

Ultra-High-Molecular-Weight gDNA Isolation from Blood. Previously frozen EDTA-stabilized blood aliquots were thawed, inverted to mix, and measured for white blood cell counts (HemoCue, Brea, CA USA, WBC). Blood volumes corresponding to 1.5×10^6 cells were transferred to a microcentrifuge tubes, then spun to obtain cell pellets. After removing supernatants, pellets were resuspended in 40 µL Stabilizing Buffer and 50 µL Proteinase K (Bionano Genomics, San Diego, CA, USA). Lysis and Binding Buffer (LBB, Bionano Genomics, San Diego, CA, USA) was then added and mixed to produce a lysate, after which isolation of DNA was performed essentially as described above for tumor tissue.

Direct Label and Staining (DLS). For both tumor- and blood-derived samples, gDNA was labeled in Direct Label and Stain reactions, in which fluorescent labels are enzymatically conjugated to a six-base pair recognition sequence followed by DNA counterstaining. Briefly, 750 ng gDNA was diluted and mixed with a labeling master mix containing DLE-1 Enzyme and DL-Green (Bionano Genomics, San Diego, CA, USA). Reactions were shielded from light and incubated at 37 °C for 2 h. A Proteinase K solution then inactivated the enzyme, and successive membrane adsorption steps were used for cleanup. A portion of each sample was then carried forward into a staining master mix addition, slowly homogenized, and incubated overnight at room temperature.

The DNA concentration of each labeled sample was confirmed within 4–12 ng/µL by High-Sensitivity dsDNA Qubit Assay and then loaded onto a Bionano Saphyr® Chip (Bionano Genomics, San Diego, CA, USA, Part#20366) and run on the Bionano Saphyr® instrument, targeting approximately 300× human genome coverage.

2.3. Bionano Access and Solve Pipeline

Genome analysis was performed using Rare Variant Analysis in Bionano Access 1.6 and Bionano Solve 3.6, which captures somatic SVs occurring at low allelic fractions. Briefly, molecules of a given sample dataset were first aligned against the public Genome Reference Consortium GRCh38 human assembly. SVs were identified based on discrepant alignment between sample molecules and GRCh38, with no assumptions about ploidy. Consensus genome maps (*.cmaps) were then assembled from clustered sets of at least three molecules that identify the same variant. Finally, the genome maps were realigned to GRCh38, with SV data confirmed by consensus forming final SV calls. SVs were then annotated with known canonical gene set present in GRCh38, as well as estimated population frequency for each structural variant detected by comparing to a custom control database ($n = 297$) from Bionano Genomics.

2.4. Data Comparison

Whole-genome imaging data were compared to the human reference genome GRCh38 (hg38) to retain only those SVs not present in the reference genome. SVs were further filtered to eliminate any variant observed in any of the Bionano control samples or, if available, patient-matched blood. Bionano Access-created csv files containing filtered SVs were analyzed to compare SV content across samples. For tissue samples with associated blood samples, control database filtration efficacy was compared to blood-filtering efficacy at identification of somatic mutations. For duplicate/triplicate samples, filtered SVs were compared to determine intra-sample reliability. For identification of cancer-related genes, the set of genes affected by SVs in each of the samples was compared to the list of genes causally implicated in cancer available in the Cosmic Cancer Gene Census database (v92) [37] (https://cancer.sanger.ac.uk/census).

3. Results

Patient Clinical Characteristics. Clinical data for the patients from whom tumor samples were acquired are shown in Table 1. A total of 60% (12/20) patients were male, with a mean age of 73.5 years at sample acquisition. A total of 45% (9/20) patients identified as Caucasian, 40% (8/20) as Asian, and 5% (1/20) as Hispanic, with 10% (2/20) not identifying. The majority of IPM-sourced tumor samples were obtained from Caucasian patients (7/10), while the majority of the BioIVT-sourced tumor samples were obtained from patients of Asian ethnicity (8/10). In terms of overall risk factors, 55% (11/20) of patients were self-described current or former tobacco users and 45% (9/20) endorsed some history of alcohol use.

Table 1. Patient demographics and tumor characteristics.

Study ID	Cancer Type *	Age †	M/F	Ethnicity	Smoking History	Alcohol History	Pathologic TNM ‡	Cancer Stage
7528	Tongue (SCC)	25	M	Caucasian	None	Rare	T3N2bM0	IVa
7052	Tongue (SCC)	35	M	Caucasian	None	None	T2N3M0	IVb
7622	Tongue (SCC)	60	F	Caucasian	50 pack years	1–2 drinks/week	T3N0M0	III
7403	Tongue (SCC)	65	M	Caucasian	45 pack years	Rare	T2N3bM0	IVb
7518	Thyroid (AP)	70	F	Caucasian	20 pack years	2 drinks per day	T4bN1bM1	IVc
7708	Thyroid (AP)	65	M	Caucasian	None	None	4aN1bM1	IVc
3717	Thyroid (AP)	80	M	Hispanic	25 pack years	Rare	T4aN1aM1	IV
14369	Lung (pleomorphic carcinoma)	60	M	N/A	60 pack years	None	T2bN1M0	IIa
10974	Liver (metastic adenocarcinoma of colon)	65	F	N/A	Former	None	T3N2aM1	IVB
3096	Bladder (urothelial carcinoma)	55	M	Caucasian	60 pack years	None	T2N0M0	II
73432	Lung (adeno-squamous carcinoma)	35	M	Asian	Former (5 pack years)	Former (1 per day, 10 years)	T2aN1M0	IIA
94894	Liver (hepato-cellular carcinoma)	70	M	Asian	7 pack years	1 per day, 35 years	T1NxM0	I
101558	Bladder (papillary urothelial carcinoma)	65	M	Asian	Former (5 pack years)	1 per day, 20 years	T2NxM0	II
69033	Kidney (renal cell carcinoma)	60	F	Asian	None	None	T2bNxM0	II

Table 1. *Cont.*

Study ID	Cancer Type *	Age [†]	M/F	Ethnicity	Smoking History	Alcohol History	Pathologic TNM [‡]	Cancer Stage
79379	Breast (ductal carcinoma in situ)	50	F	Asian	None	None	T3N0M0	IIB
102095	Prostate (invasive adeno-carcinoma)	60	M	Caucasian	40 pack years	None	T3bN1M0	IV
80384	Brain (anaplastic astrocytoma)	40	F	Caucasian	None	None	NA	NA
81347	Ovarian (serous carcinoma)	75	F	Asian	None	None	T1aN0M0	IA
119664	Colon Cancer (adenocarci-noma)	80	M	Asian	2 pack years	1 per day, 40 years	TXNXMX	UNK
128019	Thyroid (papillary)	35	F	Asian	None	None	T3bNxM0	I

* SCC: squamous cell carcinoma; AP: anaplastic.[†] ~Age ($\geq$Age-3 and $\leq$Age+3) [‡]. Pathologic Staging: Tumor, Node Metastasis (TNM) staging is the internationally accepted system set forth by the American Joint Committee on Cancer (AJCC) used to determine cancer disease stage and guide prognosis and treatment (https://www.cancerstaging.org) [38].

The tumor samples consisted of a variety of stages (Table 1). A total of 75% (3/4) of tongue cancer samples and 100% (3/3) anaplastic thyroid cancers were stage IV cancers, while 100% (2/2) lung and (2/2) bladder cancers were stage II. Limited tumor data were available for the commercially available BioIVT-sourced tumor samples.

DNA Quality Metrics: All 20 solid tumors yielded high-molecular-weight gDNA (Table 2). The average concentration across all samples following gDNA isolation was 120 ng/μL by Broad Range dsDNA Qubit Assay. All eluted gDNA were well above the minimal concentration required for DLS labeling (35 ng/μL) and the average final DNA yields for each tumor ranged from 1.2 to 16.4 μg/10 mg input tissue. Analysis on a Saphyr instrument following DLS labeling revealed that samples achieved an average label density of 14.4/100 Kbp, average filtered N50 (>20 Kbp) DNA size of 242 Kbp, average filtered N50 (>150 Kbp) DNA size of 315 Kbp, map rate of 82.62%, effective reference coverage of 320× and average effective DNA throughput ($\geq$150 Kbp) of 50 Gbp/scan. Rare Variant Pipeline Analysis of the samples yielded an average of 82.4% of molecules aligning to the reference genome. These values are all well above the acceptable range for obtaining high-quality data and none of the samples failed any of these quality control metrics.

Identification of somatic structural variants. Rare Variant analysis of the samples revealed a large numbers of variants in each sample, only a fraction of which were likely somatic. The unfiltered analysis yielded an average of 1633 total SVs per sample (range 1241–2000), which include both somatic and germline polymorphic variants (Figure 2, upper panel). These consisted predominantly of insertions and deletions, with an average of 712 insertions and 604 deletions, a fewer number of inversion (an average of 153) and duplications (an average of 123), and relatively few translocations (an average of 41). Eliminating those SVs found in Bionano's control database of known polymorphic SVs reduced the number of putative somatic structural mutations by 91% to an average of 124 total SVs per sample (Figure 2, lower panel). Most of the variants eliminated were insertions and deletions, of which on average 97% and 94%, respectively, were removed. On the other hand, less than 0.2% of the translocations were flagged as polymorphic, consistent with the fact that almost no translocations persist in the population as polymorphisms.

Table 2. Single-molecule quality report metrics.

Tissue	No. of Duplicates	Input (mg)	DNA (ng/μL)	DNA Yield (μg/mg)	N50 Kbp (>20 Kbp)	N50 Kbp (>150 Kbp)	Labels/100 kbp	Map Rate (%)	Gbp/Scan	Effective Coverage
7528 (tongue)	1	17.5	37	0.12	211	317	12.3	58.8	53	237×
7052 (tongue)	1	17.1	81	0.28	179	287	15.2	82.4	37	345×
7622 (tongue)	1	18.7	160	0.51	315	361	13.4	75.8	64	317×
7403 (tongue)	1	18	79	0.26	148	272	14.8	72.8	33	304×
7518 (thyroid)	1	8.6	28	0.20	143	265	14.4	76.6	26	312×
7708 (thyroid)	1	10.6	85	0.47	269	356	13.1	61.2	35	253×
3717 (thyroid)	1	13.2	49	0.22	250	320	14.5	88.2	58	371×
14369 (lung)	1	11.4	87	0.45	268	323	14.0	89.6	36	372×
10974 (liver)	1	6.5	82	0.74	235	289	15.2	87.9	49	360×
3096 (bladder)	1	9.4	59	0.37	265	319	13.8	78.3	39	325×
73432 (lung)	3	9.6	128	0.86	248	304	15.0	90.4	51	339×
94894 (liver)	2	9.0	196	1.41	265	306	14.9	89.3	84	325×
101558 (bladder)	3	9.7	245	1.64	313	357	15.2	91.8	66	338×
69033 (kidney)	3	10	96	0.63	201	269	14.6	83.5	41	296×
79379 (breast)	3	13.3	183	1.04	317	395	14.2	84.1	77	288×
102095 (prostate)	3	10.3	113	0.72	273	361	14.8	85.1	62	295×
80384 (brain)	2	10.5	168	1.06	228	292	14.6	90.2	42	306×
81347 (ovary)	2	10.5	168	1.05	228	292	14.6	90.2	42	330×
119664 (colon)	2	11.3	231	1.33	263	330	14.9	88.6	42	274×
128019 (thyroid)	4	10	126	0.77	213	294	14.5	87.6	64	294×
Average	1.9	11.8	120.	0.71	241.	315.	14.4	82.6	50.0	314×

Average values are presented for samples with multiple replicates.

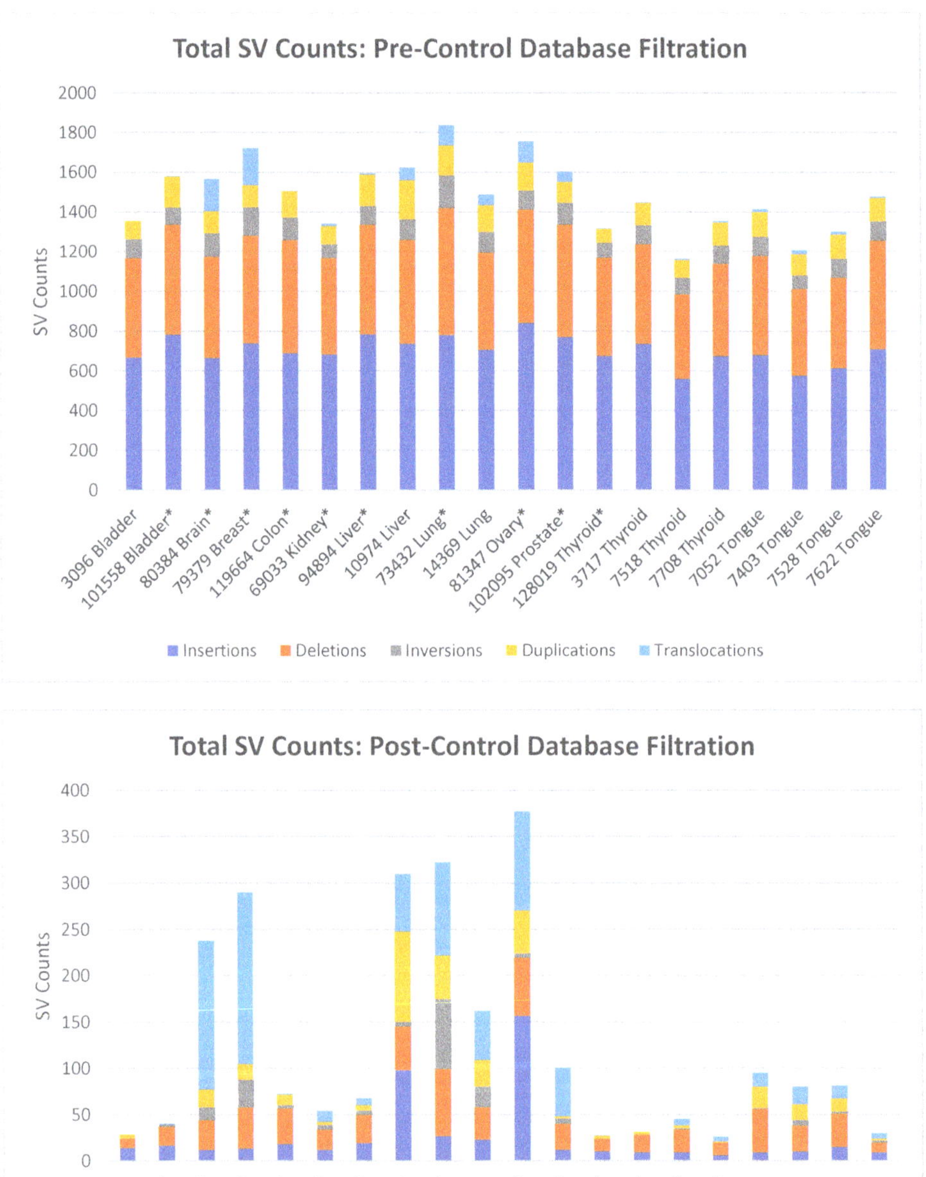

Figure 2. Total and somatic structural variants present in tumor samples. Upper panel: SV counts as determined using the Bionano Rare Variant pipeline, before control database filtration. SV counts are averages for duplicate and triplicate samples. Lower panel: SV counts after filtering total SVs to remove known polymorphic SV found in Bionano's GRCh38 control database. SV counts are averages for duplicate and triplicate samples, which are indicated by (*).

To determine the efficacy of identifying somatic SVs by filtering against Bionano's database of known polymorphisms, we used as a gold standard the blood samples from four patients from whom we had obtained tongue tumors. That is, we determined the true somatic mutations in each of these four tumors by eliminating those SVs identified in each of the tumors that were also present in the corresponding blood sample. We could then compare those true somatic variants to the list of somatic variants predicted by filtering against the database of polymorphisms. For these four tongue tumor samples, we identified an average of 1474 total SVs per sample. Filtering these SVs using the Rare Variant Analysis pipeline for SVs not found in the Bionano control database yielded an average of 72 total SVs per sample, consisting of 11 insertions (range 9–15), 31 deletions (range 11–47), 3 inversions (range 1–6), 14 duplications (range 2–23), and 14 translocations (6–19) (Figure 3, right upper panel). Filtering against the variants found in the corresponding blood samples returned an average of 58 total SVs per sample, consisting of 10 insertions (range 9–10), 20 deletions (range 7–35), 2 inversions (range 0–4), 13 duplications (range 4–24), and 14 translocations (range 6–19) (Figure 3, left upper panel). Comparing the residual SV sets obtained by filtering against Bionano's control database to the sets of true somatic SVs for each sample demonstrated that the control database filtration exhibited strong statistical accuracy (Figure 3, lower panel). Across the four separate samples, the control database exhibited an average sensitivity of 92% (83–96%) and specificity of 98% (range 97–99%). That is, filtering with the control database retained most of the true somatic mutations while eliminating almost all of the polymorphic SVs. Similarly, the average negative predictive value of the filter was 99.6%, demonstrating that an SV identified as germline was indeed a germline variant, while the positive predictive value of 74% (range 60–81%) indicates that a majority, but not all, the variants identified as somatic are in fact somatic. In other words, the results obtained by filtering SVs against Bionano's control database retained almost all the true somatic mutations. However, several of the SVs identified as somatic were actually germline. Those SVs inaccurately identified as somatic were rare germline variants, predominantly insertions or deletions, essentially private to the patient's genome. As above, we noted that the filtering process did not affect all SV types equally: while most deletions and insertions were flagged as polymorphic and eliminated from the list of somatic mutations, very few duplications and essentially no translocations were identified as polymorphic. This is consistent with observation that few translocations or duplications are stable through meiosis. *Duplicate Sample Analysis.* We compared SV calls from separate isolates of the same sample to assess consistency and reproducibility of the method, albeit without knowing the extent of tumor heterogeneity of the individual samples. Six samples underwent triplicate analysis, and four samples underwent duplicate analysis (Table 3). After identifying SVs using the Rare Variant Analysis pipeline and filtering them against the Bionano control database of known polymorphisms, we recovered an average of 116 somatic SVs shared among the separate isolates of the same tumor. These comprised an average of 23 insertions, 29 deletions, 10 inversions, 11 duplications and 43 translocations (Table 3). As noted above, the number of SVs identified in a tumor varied widely across the different tumors examined, with lung, breast, brain and ovarian tumors showing a high level of somatic SVs while the others containing a relative low number of SVs. Moreover, the percentage of SVs shared among different isolates of the same tumor also varied among the different tumor types. However, the percentage of shared SVs and the total number of SVs were uncorrelated. Assuming that the higher values for shared SVs reflect the reproducibility of the method, then we might postulate that the lower shared values represent both the reproducibility and the tumor heterogeneity. That is, we would suggest that the reproducibility of the method across multiple biological replicates is 85–95%, corresponding to the values obtained from those samples with the least variability. Thus, we would suggest that the residual variability in those samples with lower reproducibility (50–75%) reflects heterogeneity of SVs in the tumors. This would suggest that these brain, liver, lung and prostate tumors had a relatively high level of tumor heterogeneity.

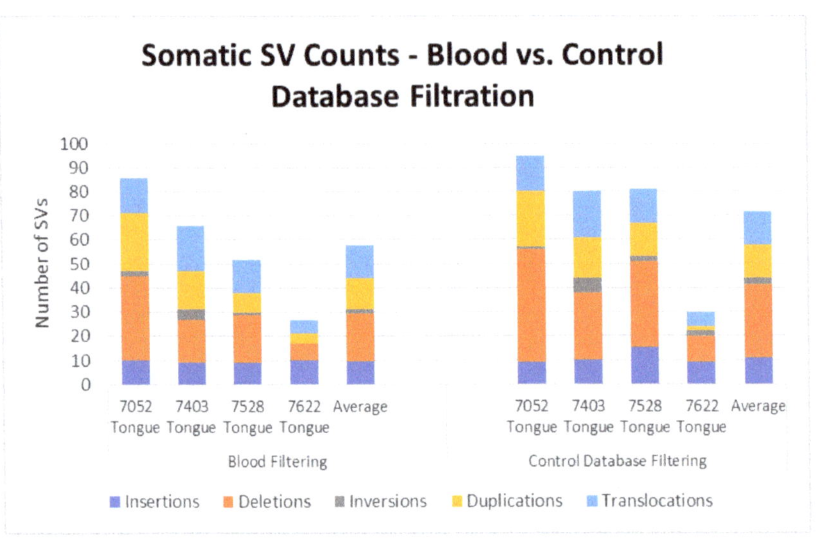

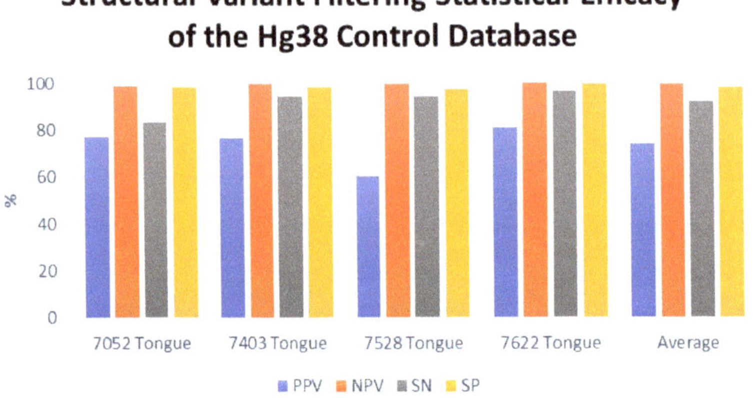

Figure 3. Efficacy of the somatic variant identification using a control database of known polymorphisms. Upper Panel: Number and distribution of somatic structural variant in four tongue tumors as determined by filtering against SVs in the patient's genome from peripheral blood (left) or against Bionano's control database of known polymorphisms. Lower Panel: Values for sensitivity (SN), specificity (SP) and positive (PPV) and negative predictive values (NPV) for identification of somatic structural variants obtained by filtering total identified SVs to remove those present in a control database of know human polymorphisms. Data obtained by filtering against the control database were compared to those obtained by filtering total SVs to remove those present in the genomes obtained from peripheral blood from the each of the patients from whom the tumors were removed.

Table 3. Duplicate Sample Analysis. Shown are the number of somatic structural variants shared among the multiple isolates of the same sample and the percentage of those relative the total number of somatic variants found in all the isolates of the same sample.

	Total	%	Insertion	%	Deletion	%	Inversion	%	Duplication	%	Translocation	%
Brain *	134	70	5	63	21	78	9	69	13	72	86	69
Colon *	63	93	15	83	36	97	2	100	9	90	1	100
Liver *	45	70	14	74	21	81	1	17	4	67	5	71
Ovary ‡	338	86	136	82	59	87	4	80	40	85	99	91
Bladder ‡	30	88	11	79	18	100	1	100	0	100	0	0
Breast ‡	221	92	9	82	33	85	23	88	14	88	142	95
Kidney ‡	19	76	6	67	11	85	2	100	0	0	0	100
Lung ‡	221	66	18	69	53	75	59	73	26	50	65	63
Prostate ‡	69	48	8	47	22	61	3	38	1	25	35	44
Thyroid ‡	19	86	7	88	10	91	0	100	2	100	0	0
Average (all)	116	78	23	73	28	84	10	76	11	68	43	63
Duplicate Average	145	80	43	75	34	86	4	66	17	78	48	83
Triplicate Average	97	76	10	72	25	83	15	83	7	60	40	50

* = duplicate sample, ‡ = triplicate sample. % = % of SV calls shared among duplicate/triplicate samples.

The number and types of somatic variants in a tumor varied substantially across the collection of samples (Figure 4). Several tumor samples, including those from colon, bladder, kidney and all four from thyroid, contained relatively few somatic SVs whereas others, including those from prostate, ovaries, lung and brain, carried a large number of somatic SVs. Since these samples for the most part serve as single representatives of each tumor type, we cannot extrapolate to the tumor types as a whole the contribution of SVs to cancer onset and development for each class of tumor. However, it is noteworthy that the SNV mutational burden in thyroid cancers is among the lowest among all tumor types and that measure of genome instability is mirrored in the low number of somatic SVs in all four of the samples examined [39]. Similarly, the SNV mutational burden in lung cancers is among the highest across all tumor types and both of the lung tumors examined here also carry a high level of somatic SV. Finally, the extent of somatic SVs observed in our collection of tumors does not correlate with either cancer stage nor with obvious lifestyle characteristics (Table 1). For instance, neither smoking nor drinking history has a stronger influence on SV mutation burden than does site of origin of the tumor. However, further data examining the correlation of lifestyle characteristics and tumor stages with SV mutational burden are warranted to assess the impact of these behaviors on SV formation and persistence.

Identification of Cancer Gene Mutations. While, as noted above, we cannot generalize regarding the role of structural variants in onset and progression of different tumor types, our results indicate that we can extract from the structural variant list clinically relevant data on individual tumors that might inform prognosis or treatment options. We examined the somatic structural variants in each tumor sample for those that affected genes previously associated with cancer. In particular, we annotated those genes altered by a structural variant, either by disruption, duplication, deletion or fusion, and intersected that list with the set of cancer-related genes in the Cosmic database (v92) [37]. The resultant list by tumor type is provided in Table 4 and subdivided into oncogenes, tumor suppressor genes and gene fusions. We included only those oncogenes that were potentially activated by duplication or gene fusion and only those tumor suppressor genes that were potentially inactivated by deletion, insertion or fusion. As evident, every tumor sample carried at least one such cancer gene mutation and most contained multiple hits. Several of these genes offer the opportunity for targeted therapies, focused either directly on the oncogene, as would be the case for CDK6 and ERBB2, or at the pathway downstream of the affected gene, as would be the case for BRAF and CDKN2A. Other affected genes, such as MSH2, RAD51B, RAD21 and RAD18, suggest the potential of therapy based on possible ensuing genome instability, such as immunotherapy or PARP inhibitors. Many of these variants would not be readily identified by targeted gene panels generally used for clinical assessment of tumor genomes. Moreover, in many cases, the cancer genes altered by SVs were not previously associated with the cancer type in which we observed it. For instance, we observed a fusion of CDK6 in one of the tongue tumors while it has previously been associated predominantly only with ALL. Similarly, LRP1B is often inactivated in CLL or ovarian cancer, while we find it inactivated by deletion in one of the lung tumors. Thus, the identification of somatic structural variants by OGM could provide useful clinical insights not readily available through standard next-generation sequencing or targeted panels.

Figure 4. Global view of structural variants in solid tumor samples. Diagrams of somatic structural variants in all the solid tumor genomes, filtered to remove known polymorphisms, showing translocations and inversions in the center, copy number on the inner ring and insertions (green), deletions (orange) inversions (light blue) and duplications (violet) on the next to most outer ring. Chromosomes are ordered sequentially in a clockwise orientation in the outer ring on which are indicated cytological banding patterns and the centromere (red bar).

Diagrams of somatic structural variants in all the solid tumor genomes, filtered to remove known polymorphisms, showing translocations and inversions in the center, copy number on the inner ring and insertions (green), deletions (orange) inversions (light blue) and duplications (violet) on the next to most outer ring. Chromosomes are ordered sequentially in a clockwise orientation in the outer ring on which are indicated cytological banding patterns and the centromere (red bar).

In addition to identifying individual cancer-related genes in tumor types, our results provide a panoramic view of the entire tumor genome and reveal large-scale genomic features not readily available from standard sequencing techniques. As evident in the results in Figure 4, our data provide a rapid snapshot of the extent of genomic instability in each of the tumors. Such images present an integrated picture of the aneuploidies, translocations, inversions, deletions and insertions, which offers a readily digestible impression of the extent of genetic instability underlying a tumor. Moreover, several large-scale features are evident in these data. For instance, chromothripsis is a massive cluster of chromosomal rearrangements localized to a restricted region of a chromosome, which often results from a single catastrophic event [40]. Figure 5 details a chromothripsis event on a portion of chromosome 5 in one of the lung tumor samples. In fact, such events are readily evident in four of the Circos plots in Figure 4, consistent with previous estimates of 2–3% prevalence across all cancers, albeit with different frequency in different cancers [41]. The detection and mapping of such a feature are difficult to achieve by short-read sequencing [41] but can indicate poor prognosis and the corresponding need for aggressive therapy.

Table 4. Structural variants affecting cancer relevant genes.

Sample	Oncogene	Tumor Suppressor	Gene Fusion
Prostate	ERBB2 (Dup)	PTEN (Del)	PTEN-LINC01374
	GATA2 (T)	NF1 (Del)	DHX30-GATA2
	NUP98 (T)		CASC15-NUP98
			PRKAR1A-FRMPD4
			ERG-TMPRSS2
			FREM1-MYH9
Ovarian		NUMA1 (T)	NBEA-ZFHX3
		NF1 (I)	HMGN2P46-BLOC1S6
		SMARCA4 (I)	LPP-PIEZO1
Kidney	PRKAR1A (T)	CDKN2A (Del)	PRKAR1A-FRMPD4
	ERBB2 (Dup)	ZFHX3 (Del)	
Colon		FHIT (Del)	
Breast	ERBB4 (Dup)	USP8 (T)	USP8-PRPSAP
	ERBB2 (Dup)	PRKAR1A (T)	PRKAR1A-FRMPD4
		RAD51B (Del)	LINC01476-BRIP1
		CDKN2A (Del)	SYK-CFAP77
Brain	SETBP1 (T)	LARP4B (T)	CCDC158-LARP4B
		CSMD3 (T)	CA13-CSMD3
		LRP1B (Del)	DPYD-SETBP1
		RAD21 (Del)	CNBD1-AC083836.1
Bladder		DDX10 (Del)	
Tongue	BRAF (T)	CDKN2A (Del)	EPHB1-BRAF
	CDK6 (T)	PTPRD (Del)	CDK6–AC091551.1
	CCND2 (T)	RAD51B (T)	PCLO-RAD51B
	CCND1 (Dup)	LRP1B (Del)	
		CDKN1B (Del)	

Table 4. *Cont.*

Sample	Oncogene	Tumor Suppressor	Gene Fusion
Thyroid		YWHAE (T)	ABR-YWHAE
		PTPRD (Del)	CDK12-CSF3
			RAD18-SRGAP3
			SHROOM3-AFF1
Liver	VTI1A (T)	RMI2 (T)	VTI1A-NHLRC2
	MAP3K13 (T)	NCOR (T)	C3orf70-MAP3K13
	MACC1 (T)	CBLC (T)	AC005062.1-MACC1
	NSD3 (T)	MSH2 (T)	NSD3-AC087623.2
			RASGEF1B-VTI1A
			RMI2-TOX3
			NCOR1-LRRC75A
			MSH2-CYP3A43
Lung	CTNND2 (Del,T)	PTPRD (Del)	CTNND2-TRIO
	IKBKB (T)	RAD51B (T)	DUSP10-CTNND2
		FUS (T)	IKBKB-FAM91A1
		LRP1B (T)	FUS-CNOT1
			PDE6D-RAD51B
			PRKCH-HIF1A
			GAS7-LYRM9
			EHBP1-LRP1B

T, translocation; Dup, duplication; I, insertion; Del, deletion.

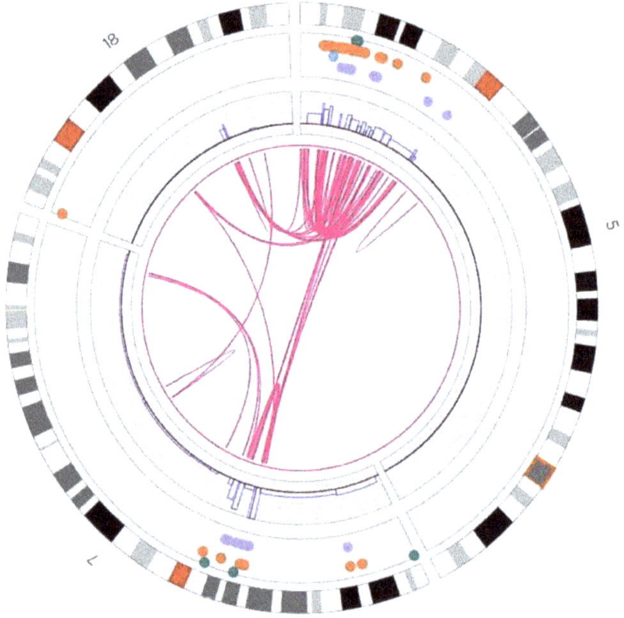

Figure 5. Chromothrypsis of chromosome 5p in a lung tumor. Shown is a truncated Circos plot of the lung tumor, focused on the region of chromosome 5, highlighting the chromothrypsis event that occurred on its p arm. The organization of the Circos plot is as indicated in the legend to Figure 4.

4. Discussion

In this report, we described the application of optical genome mapping to solid tumors, which we suggest can significantly augment the genomic analysis of such tumors obtained by next-generation sequencing. Genomic analysis of tumors has stimulated major advances in cancer diagnosis, prognosis and treatment, shifting the focus from morphological and histochemical characterization to consideration of the landscape of driver mutations in the tumor [42–44]. Somatic driver events in a tumor—point mutations and structural variants (SVs) including insertions, deletions, inversions, translocations and copy number changes—are currently identified in solid tumors by some combinations of RNA sequencing and genome sequencing of either targeted gene panels, whole exomes or whole genomes. As noted in this report, OGM can provide a pervasive view of the structural variants in a tumor and the cancer-related genes on which they impinge, thus identifying affected genes agnostically, without prior bias imposed by gene panels.

Some prior studies have begun to demonstrate the utility of Bionano DNA isolation protocols in solid tissue tumor analysis. These include studies of lung squamous cell carcinoma and metastatic prostate carcinoma [7,34,35]. This current report demonstrates the utility of the DNA isolation protocol and SV analysis in a wide variety of solid tissue types, and expands the feasibility of such analysis for previously unused human tissue types. The high DNA yield, high effective coverage, map rate and other molecular quality metrics shown across tumor types confirm how our extraction and analysis workflow can be effectively applied to many solid tissue tumors.

This current DNA isolation protocol carries a number of advantages. Tissue handling can be performed at room temperature. The current protocol showed successful DNA isolation in solid tissue samples of <20 mg, and even as low as 6 mg. The low tissue input requirement carries important applications for rare cancer samples, human tissue biopsy testing and other low-quantity specimen acquisition. Additionally, utilizing the novel paramagnetic Nanobind disks rather than prior agarose gel plugs greatly decreases time needed to complete DNA isolation to only 5 h. The ability to isolate DNA from up to eight simultaneous samples using the current protocol greatly amplifies throughput and reduces tissue-to-data processing time, increasing both laboratory convenience as well as expanding potential for clinical utility where rapid data turnaround is paramount. Furthermore, the strong inter-sample SV correspondence shown by most tissue types in duplicate/triplicate sample analysis demonstrates the reproducibility of this technique; intra-sample heterogeneity of select samples may be attributed to non-tumor normal tissue within some tissue fragments, or attributed to specific cancer subtype, and merits further investigation. Although the isolation protocol described here affords many advantages, there are some limitations to this protocol. While high-quality DNA isolation and OGM SV analysis was obtained for a wide variety of tumor types that were tested, it may not be generalizable to every additional untested solid tumor type. Future directions include continuing to validate this protocol in additional tissue types, and assessing additional tumor samples to assess broader trends in the role of specific OGM-identified SVs in individual cancer subtypes.

In clinical evaluation of liquid tumors such as leukemia, genomic analysis is augmented by karyotyping, which gives a panoramic, albeit low resolution, view of the entire genome. Despite the low resolution, the genome wide view of the structural changes afforded by karyotyping reveals diagnostic features of the tumor that have strong prognostic value. Given the consistent correlation of clinical outcomes with specific mutation classes, the World Health Organization (WHO), National Comprehensive Cancer Network (NCCN) and European Leukemia Net (ELN) agencies developed recommendations for diagnosis and management of acute myeloid leukemia in adults based on the spectrum of somatic point mutations and SVs generally revealed by karyotyping [45]. SVs, particularly translocations and inversions, are major considerations in this diagnosis. Since karyotyping is a very challenging technique to apply to solid tumors, the clinician does not have access to a comparable global view of a solid tumor's genome and the role of SVs in prognosis has likely been underappreciated. Applying OGM broadly to cancer types and correlating SVs revealed by that analysis with clinical outcomes could provide new genomic markers for prognosis and treatment selection.

5. Conclusions

We demonstrate the utility of a DNA isolation protocol for high-molecular-weight DNA extraction and OGM SV analysis of a wide variety of solid human tumor types on the Bionano Saphyr system, including breast, colon, liver, brain, bladder, kidney, lung, ovary, prostate and thyroid cancer tissue. The system can be used to accurately detect genetic mutation hallmarks in cancer tissue samples, including rearrangements such as translocations, gene fusions and copy number alterations. Somatic SVs can be determined by comparison filtering with the Bionano control sample database, or against a matched pair sample. Importantly, Bionano SV pipelines can detect SVs with complex breakpoint structures that are difficult to detect with other technologies. Our results indicate that the solid tissue DNA extraction protocol can be applied to a wide variety of solid tumors, and that the Saphyr system can capture, in a streamlined workflow, a broad spectrum of SVs. These SVs have functional importance and provide great utility in cancer prognosis and treatment.

Author Contributions: Conceptualization, D.G., A.R.H., and J.R.B.; methodology, H.B.S., H.W., K.P. M.O., and Y.Z.; software, L.Z.; validation, C.-Y.J.L.; formal analysis, A.W.C.P., B.C., C.-Y.J.L., D.Y.G., and L.Z.; investigation, D.Y.G., B.L., and S.C.; writing—initial draft, D.G. and A.R.H.; writing—editing, D.G., A.W.C.P., and A.R.H.; writing—final draft, J.R.B.; visualization, D.Y.G., A.R.H., and B.C.; supervision, A.R.H., D.G., J.R.B., H.B.S., and M.O.; project administration, A.R.H., J.R.B.; funding acquisition, D.G. All authors have read and agreed to the published version of the manuscript.

Funding: Funding for this study was provided in part by a grant from the George Laverty Foundation (DG).

Institutional Review Board Statement: Patients consented under protocol 40532 approved by the Penn State Health Institution Review Board.

Informed Consent Statement: Informed consent was obtained from all subjects involved in the study.

Data Availability Statement: Primary Bionano Saphyr data are available on request (jbroach@pennstatehealth.psu.edu).

Conflicts of Interest: H.B.S., Y.Z., K.P., H.W., C.Y.J.L., A.W.C.P., B.C., M.O., and A.R.H. are employees of Bionano Genomics. The authors have no other conflicts of interest.

References

1. Futreal, P.A.; Coin, L.; Marshall, M.; Down, T.; Hubbard, T.; Wooster, R.; Rahman, N.; Stratton, M.R. A census of human cancer genes. *Nat. Rev. Cancer* **2004**, *4*, 177–183. [CrossRef] [PubMed]
2. Kantarjian, H.; Sawyers, C.; Hochhaus, A.; Guilhot, F.; Schiffer, C.; Gambacorti-Passerini, C.; Niederwieser, D.; Resta, D.; Capdeville, R.; Zoellner, U.; et al. Hematologic and Cytogenetic Responses to Imatinib Mesylate in Chronic Myelogenous Leukemia. *N. Engl. J. Med.* **2002**, *346*, 645–652. [CrossRef]
3. Kwak, E.L.; Bang, Y.-J.; Camidge, D.R.; Shaw, A.T.; Solomon, B.; Maki, R.G.; Ou, S.H.; Dezube, B.J.; Janne, P.A.; Costa, D.B.; et al. Anaplastic Lymphoma Kinase Inhibition in Non-Small-Cell Lung Cancer. *N. Engl. J. Med.* **2010**, *363*, 1693–1703. [CrossRef]
4. Soda, M.; Choi, Y.L.; Enomoto, M.; Takada, S.; Yamashita, Y.; Ishikawa, S.; Fujiwara, S.-I.; Watanabe, H.; Kurashina, K.; Hatanaka, H.; et al. Identification of the transforming EML4–ALK fusion gene in non-small-cell lung cancer. *Nature* **2007**, *448*, 561–566. [CrossRef]
5. Xu, J.; Song, F.; Schleicher, E.; Pool, C.; Bann, D.; Hennessy, M.; Sheldon, K.; Batchelder, E.; Annageldiyev, C.; Sharma, A.; et al. An Integrated Framework for Genome Analysis Reveals Numerous Previously Unrecognizable Structural Variants in Leukemia Patients' Samples. *bioRxiv* **2019**, 563270. [CrossRef]
6. Zhu, Y.; Brown, H.N.; Zhang, Y.; Stevens, R.G.; Zheng, T. Period3 structural variation: A circadian biomarker associated with breast cancer in young women. *Cancer Epidemiol. Biomark. Prev.* **2005**, *14*, 268–270.
7. Peng, Y.; Yuan, C.; Tao, X.; Zhao, Y.; Yao, X.; Zhuge, L.; Huang, J.; Zheng, Q.; Zhang, Y.; Hong, H.; et al. Integrated analysis of optical mapping and whole-genome sequencing reveals intratumoral genetic heterogeneity in metastatic lung squamous cell carci-noma. *Transl. Lung Cancer Res.* **2020**, *9*, 670–681. [CrossRef] [PubMed]
8. Alkan, C.; Sajjadian, S.; Eichler, E.E. Limitations of next-generation genome sequence assembly. *Nat. Methods* **2010**, *8*, 61–65. [CrossRef] [PubMed]

9. Liu, S.; Song, L.; Cram, D.S.; Xiong, L.; Wang, K.; Wu, R.; Liu, J.; Deng, K.; Jia, B.; Zhong, M.; et al. Traditional karyotyping vs copy number variation sequencing for detection of chromosomal abnormalities associated with spontaneous miscarriage. *Ultrasound Obstet. Gynecol.* **2015**, *46*, 472–477. [CrossRef]

10. Trask, B.J. Human cytogenetics: 46 chromosomes, 46 years and counting. *Nat. Rev. Genet.* **2002**, *3*, 769–778. [CrossRef]

11. Bickhart, D.M.; Rosen, B.D.; Koren, S.; Sayre, B.L.; Hastie, A.R.; Chan, S.; Lee, J.; Lam, E.T.; Liachko, I.; Sullivan, S.T.; et al. Single-molecule sequencing and chromatin conformation capture enable de novo reference assembly of the domestic goat genome. *Nat. Genet.* **2017**, *49*, 643–650. [CrossRef]

12. Cao, H.; Hastie, A.R.; Cao, D.; Lam, E.T.; Sun, Y.; Huang, H.; Liu, X.; Lin, L.; Andrews, W.; Chan, S.; et al. Rapid detection of structural variation in a human genome using nanochannel-based genome mapping technology. *GigaScience* **2014**, *3*, 34. [CrossRef] [PubMed]

13. Cao, H.; Wu, H.; Luo, R.; Huang, S.; Sun, Y.; Tong, X.; Xie, Y.; Liu, B.; Yang, H.; Zheng, H.; et al. De novo assembly of a haplo-type-resolved human genome. *Nat. Biotechnol.* **2015**, *33*, 617–622. [CrossRef] [PubMed]

14. Chaisson, M.J.P.; Sanders, A.D.; Zhao, X.; Malhotra, A.; Porubsky, D.; Rausch, T.; Gardner, E.J.; Rodriguez, O.L.; Guo, L.; Collins, R.L.; et al. Multi-platform discovery of haplotype-resolved structural variation in human genomes. *Nat. Commun.* **2019**, *10*, 1784. [CrossRef]

15. Cockburn, I.A.; Mackinnon, M.J.; O'Donnell, A.; Allen, S.J.; Moulds, J.M.; Baisor, M.; Bockarie, M.; Reeder, J.C.; Rowe, J.A. A human com-plement receptor 1 polymorphism that reduces Plasmodium falciparum rosetting confers protection against severe malaria. *Proc. Natl. Acad. Sci. USA* **2004**, *101*, 272–277. [CrossRef]

16. Conrad, D.F.; The Wellcome Trust Case Control Consortium; Pinto, D.; Redon, R.; Feuk, L.; Gokcumen, O.; Zhang, Y.; Aerts, J.; Andrews, T.D.; Barnes, C.; et al. Origins and functional impact of copy number variation in the human genome. *Nature* **2010**, *464*, 704–712. [CrossRef] [PubMed]

17. Cooper, G.M.; Zerr, T.; Kidd, J.M.; Eichler, E.E.; Nickerson, D.A. Systematic assessment of copy number variant detection via ge-nome-wide SNP genotyping. *Nat. Genet.* **2008**, *40*, 1199–1203. [CrossRef] [PubMed]

18. De Cid, R.; Riveira-Munoz, E.; Zeeuwen, P.L.; Robarge, J.; Liao, W.; Dannhauser, E.N.; Giardina, E.; Stuart, P.E.; Nair, R.; Helms, C.; et al. Deletion of the late cornified envelope LCE3B and LCE3C genes as a susceptibility factor for psoriasis. *Nat. Genet.* **2009**, *41*, 211–215. [CrossRef]

19. English, A.C.; Salerno, W.J.; Hampton, A.O.; Gonzaga-Jauregui, C.; Ambreth, S.; Ritter, D.I.; Beck, C.R.; Davis, C.F.; Dahdouli, M.; Mahmoud, D.; et al. Assessing structural variation in a personal genome—towards a human reference diploid genome. *BMC Genom.* **2015**, *16*, 286. [CrossRef] [PubMed]

20. Hastie, A.R.; Dong, L.; Smith, A.; Finklestein, J.; Lam, E.T.; Huo, N.; Cao, H.; Kwok, P.Y.; Deal, K.R.; Dvorak, J.; et al. Rapid genome mapping in nanochannel arrays for highly complete and accurate de novo sequence assembly of the complex Aegilops tauschii genome. *PLoS ONE* **2013**, *8*, e55864. [CrossRef]

21. Jiao, Y.; Peluso, P.; Shi, J.; Liang, T.; Stitzer, M.C.; Wang, B.; Campbell, M.S.; Stein, J.C.; Wei, X.; Chin, C.-S.; et al. Improved maize reference genome with single-molecule technologies. *Nature* **2017**, *546*, 524–527. [CrossRef]

22. Mak, A.C.; Lai, Y.Y.; Lam, E.T.; Kwok, T.P.; Leung, A.K.; Poon, A.; Mostovoy, Y.; Hastie, A.R.; Stedman, W.; Anantharaman, T.; et al. Genome-Wide Structural Variation Detection by Genome Mapping on Nanochannel Arrays. *Genetics* **2016**, *202*, 351–362. [CrossRef] [PubMed]

23. Pendleton, M.; Sebra, R.; Pang, A.W.C.; Ummat, A.; Franzen, O.; Rausch, T.; Stütz, A.M.; Stedman, W.; Anantharaman, T.; Hastie, A.; et al. Assembly and diploid architecture of an individual human genome via single-molecule technologies. *Nat. Methods* **2015**, *12*, 780–786. [CrossRef]

24. Sarsani, V.K.; Raghupathy, N.; Fiddes, I.T.; Armstrong, J.; Thibaud-Nissen, F.; Zinder, O.; Bolisetty, M.; Howe, K.; Hinerfeld, D.; Ruan, X.; et al. The Genome of C57BL/6J "Eve", the Mother of the Laboratory Mouse Genome Reference Strain. *G3* **2019**, *9*, 1795–1805. [CrossRef] [PubMed]

25. Seo, J.-S.; Rhie, A.; Kim, J.; Lee, S.; Sohn, M.-H.; Kim, C.-U.; Hastie, A.; Cao, A.H.H.; Yun, J.-Y.; Kim, J.; et al. De novo assembly and phasing of a Korean human genome. *Nature* **2016**, *538*, 243–247. [CrossRef]

26. Shi, L.; Guo, Y.; Dong, C.; Huddleston, J.; Yang, H.; Han, X.; Fu, A.; Li, Q.; Li, N.; Gong, S.; et al. Long-read sequencing and de novo assembly of a Chinese genome. *Nat. Commun.* **2016**, *7*, 12065. [CrossRef]

27. Usher, C.L.; Handsaker, R.E.; Esko, T.; Tuke, M.A.; Weedon, M.N.; Hastie, A.R.; Cao, H.; Moon, J.E.; Kashin, S.; Fuchsberger, C.; et al. Structural forms of the human amylase locus and their relationships to SNPs, haplotypes and obesity. *Nat. Genet.* **2015**, *47*, 921–925. [CrossRef]

28. Zook, J.M.; Hansen, N.F.; Olson, N.D.; Chapman, L.; Mullikin, J.C.; Xiao, C.; Sherry, S.; Koren, S.; Phillippy, A.M.; Boutros, P.C.; et al. A robust benchmark for detection of germline large deletions and insertions. *Nat. Biotechnol.* **2020**, *38*, 1347–1355. [CrossRef] [PubMed]

29. Neveling, K.; Mantere, T.; Vermeulen, S.; Oorsprong, M.; van Beek, R.; Kater-Baats, E.; Pauper, M.; van der Zande, G.; Smeets, D.; Weghuis, D.O.; et al. Next generation cytogenetics: Comprehensive assessment of 48 leukemia genomes by genome imaging. *bioRxiv* **2020**, preprint. [CrossRef]

30. Dixon, J.R.; Xu, J.; Dileep, V.; Zhan, Y.; Song, F.; Le, V.T.; Yardimci, G.G.; Chakraborty, A.; Bann, D.V.; Wang, Y.; et al. Integrative detection and analysis of structural variation in cancer genomes. *Nat. Genet.* **2018**, *50*, 1388–1398. [CrossRef] [PubMed]

31. Chan, S.; Lam, E.; Saghbini, M.; Bocklandt, S.; Hastie, A.; Cao, H.; Holmlin, E.; Borodkin, M. Structural Variation Detection and Analysis Using Bionano Optical Mapping. *Methods Mol. Biol.* **2018**, *1833*, 193–203. [CrossRef] [PubMed]
32. Pang, A.; Lee, J.; Anantharaman, T.; Lam, E.; Hastie, A.; Borodkin, M. Comprehensive Detection of Germline and Somatic Structural Mutation in Cancer Genomes by Bionano Genomics Optical Mapping. *J. Biomol. Technol.* **2019**, *30*, S9.
33. Zhang, Y.; Broach, J. Abstract 5125: A novel method for isolating high-quality UHMW DNA from 10 mg of freshly frozen or liquid-preserved animal and human tissue including solid tumors. *Cancer Res.* **2019**, *79*, 5125.
34. Crumbaker, M.; Chan, E.K.F.; Gong, T.; Corcoran, N.; Jaratlerdsiri, W.; Lyons, R.J.; Haynes, A.M.; Kulidjian, A.A.; Kalsbeek, A.M.F.; Petersen, D.C.; et al. The Impact of Whole Genome Data on Therapeutic Decision-Making in Metastatic Prostate Cancer: A Retrospective Analysis. *Cancers* **2020**, *12*, 1178. [CrossRef]
35. Jaratlerdsiri, W.; Chan, E.K.; Petersen, D.C.; Yang, C.; Croucher, P.I.; Bornman, M.R.; Sheth, P.; Hayes, V.M. Next generation mapping reveals novel large genomic rearrangements in prostate cancer. *Oncotarget* **2017**, *8*, 23588–23602. [CrossRef] [PubMed]
36. Whitlock, R.; Hipperson, H.; Mannarelli, M.; Burke, T. A high-throughput protocol for extracting high-purity genomic DNA from plants and animals. *Mol. Ecol. Resour.* **2008**, *8*, 736–741. [CrossRef]
37. Sondka, Z.; Bamford, S.; Cole, C.G.; Ward, S.A.; Dunham, I.; Forbes, S.A. The COSMIC Cancer Gene Census: Describing genetic dysfunction across all human cancers. *Nat. Rev. Cancer* **2018**, *18*, 696–705. [CrossRef]
38. Amin, M.B.; Greene, F.L.; Edge, S.B.; Compton, C.C.; Gershenwald, J.E.; Brookland, R.K.; Meyer, L.; Gress, D.M.; Byrd, D.R.; Winchester, D.P. The Eighth Edition AJCC Cancer Staging Manual: Continuing to build a bridge from a population-based to a more "per-sonalized" approach to cancer staging. *CA Cancer J. Clin.* **2017**, *67*, 93–99. [CrossRef]
39. Lawrence, M.S.; Stojanov, P.; Polak, P.; Kryukov, G.V.; Cibulskis, K.; Sivachenko, A.; Carter, S.L.; Stewart, C.; Mermel, C.H.; Roberts, S.A.; et al. Mutational heterogeneity in cancer and the search for new cancer-associated genes. *Nature* **2013**, *499*, 214–218. [CrossRef] [PubMed]
40. Stephens, P.J.; Greenman, C.D.; Fu, B.; Yang, F.; Bignell, G.R.; Mudie, L.J.; Pleasance, E.D.; Lau, K.W.; Beare, D.; Stebbings, L.A.; et al. Massive Genomic Rearrangement Acquired in a Single Catastrophic Event during Cancer Development. *Cell* **2011**, *144*, 27–40. [CrossRef]
41. Cortés-Ciriano, I.; Lee, J.J.K.; Xi, R.; Jain, D.; Jung, Y.L.; Yang, L.; Gordenin, D.; Klimczak, L.J.; Zhang, C.Z.; Pellman, D.S.; et al. Comprehensive analysis of chromothripsis in 2,658 human cancers using whole-genome sequencing. *Nat. Genet.* **2020**, *52*, 331–341. [CrossRef] [PubMed]
42. Berger, M.F.; Mardis, E.R. The emerging clinical relevance of genomics in cancer medicine. *Nat. Rev. Clin. Oncol.* **2018**, *15*, 353–365. [CrossRef] [PubMed]
43. Vogelstein, B.; Papadopoulos, N.; Velculescu, V.E.; Zhou, S.; Diaz, L.A., Jr.; Kinzler, K.W. Cancer genome landscapes. *Science* **2013**, *339*, 1546–1558. [CrossRef]
44. Zack, T.I.; Schumacher, S.E.; Carter, S.L.; Cherniack, A.D.; Saksena, G.; Tabak, B.; Lawrence, M.S.; Zhang, C.-Z.; Wala, J.; Mermel, C.H.; et al. Pan-cancer patterns of somatic copy number alteration. *Nat. Genet.* **2013**, *45*, 1134–1140. [CrossRef] [PubMed]
45. Döhner, H.; Estey, E.; Grimwade, D.; Amadori, S.; Appelbaum, F.R.; Büchner, T.; Dombret, H.; Ebert, B.L.; Fenaux, P.; Larson, R.A.; et al. Diagnosis and management of AML in adults: 2017 ELN recommendations from an international expert panel. *Blood* **2017**, *129*, 424–447. [CrossRef] [PubMed]

Journal of *Personalized Medicine*

Article

Gene Regulatory Network of ETS Domain Transcription Factors in Different Stages of Glioma

Yigit Koray Babal [1], Basak Kandemir [2] and Isil Aksan Kurnaz [1,*]

[1] Institute of Biotechnology, Gebze Technical University, 41400 Gebze, Kocaeli, Turkey; ykbabal@gtu.edu.tr
[2] Department of Molecular Biology and Genetics, Baskent University, 06790 Ankara, Turkey; bkandemir@baskent.edu.tr
* Correspondence: ikurnaz@gtu.edu.tr

Abstract: The ETS domain family of transcription factors is involved in a number of biological processes, and is commonly misregulated in various forms of cancer. Using microarray datasets from patients with different grades of glioma, we have analyzed the expression profiles of various *ETS* genes, and have identified *ETV1, ELK3, ETV4, ELF4,* and *ETV6* as novel biomarkers for the identification of different glioma grades. We have further analyzed the gene regulatory networks of ETS transcription factors and compared them to previous microarray studies, where Elk-1-VP16 or PEA3-VP16 were overexpressed in neuroblastoma cell lines, and we identify unique and common regulatory networks for these ETS proteins.

Keywords: Ets; Elk-1; PEA3; Ets-1; glioma; biomarker

Citation: Babal, Y.K.; Kandemir, B.; Kurnaz, I.A. Gene Regulatory Network of ETS Domain Transcription Factors in Different Stages of Glioma. *J. Pers. Med.* **2021**, *11*, 138. https://doi.org/10.3390/jpm11020138

Academic Editor: Greg Gibson

Received: 21 January 2021
Accepted: 13 February 2021
Published: 17 February 2021

Publisher's Note: MDPI stays neutral with regard to jurisdictional claims in published maps and institutional affiliations.

1. Introduction

ETS proteins are present in metazoan lineages [1] and play a role in diverse biological processes. Intriguingly, ETS proteins also exhibit extensive overlaps in their tissue expression profiles, with many members of this superfamily having ubiquitous expression [2]. Not surprisingly, members of this family also tend to exhibit overlapping and sometimes redundant DNA binding, as analyzed by genome-wide occupancy and other assays [3].

The ETS (E26 transformation specific) domain transcription factor superfamily includes 27 members in humans in 11 subfamilies [2,4]. Taking its name from the founding member of this superfamily, namely oncogenic v-ets [5], ETS domain transcription factors are typically defined by their DNA binding domain, called the ETS domain, which binds the consensus motif 5′-GGA(A/T)-3′, called the ets motif of the ETS binding site (EBS) [6]. Their DNA binding property, as well as transactivation function, is regulated by the MAPK signaling pathway [7,8].

In addition to their roles in normal growth and development, ETS proteins are commonly involved in cancer formation and progression through the regulation of cell proliferation, adhesion, migration, or vascularization, as well as regulation of epithelial–stromal interactions and epithelial–mesenchymal transition [9]. The expressions of several ETS family members, such as PEA3, ETS-1, and ETS-2, are upregulated in tumors, playing a role in different aspects of tumorigenesis, including tumor initiation, epithelial-mesenchymal transition, metastasis, and angiogenesis [4,10]. In some cases, ETS members are amplified and/or rearranged, such as c-ETS1 in acute myelomonocytic leukemia, or undergo chromosomal relocations that result in fusions, like in the case of chromosomal translocation of 5′ TMPRSS2 to the ETS genes, resulting in TMPRSS:ERG fusion proteins in nearly half of prostate cancers, or chromosomal translocation that yields EWS-FLI1 fusion in Ewing sarcoma. The transcriptional potency of ETS proteins is also often increased in various cancers as a result of changes in protein–protein interactions, post-translational modifications, and/or protein stabilization [4].

Ets-1 was found to be overexpressed in breast cancer, which was reported to be associated with poor prognosis [10]. Ets-1 is not only a critical regulator of invasion [10], but is also involved in regulating cancer energy metabolism in ovarian and breast cancer cell lines [11]. It has also been shown to play a role in telomere maintenance through the regulation of hTERT expression [12]. T417 phosphorylation of Elk-1, a member of the ternary complex factor (TCF) subfamily, was found to be associated with the differentiation grade of colonic adenocarcinomas [13]. Additionally, in high clinical stage prostate cancer, ELK-1, not TCF members ELK3 or ELK4, was found to be associated with disease recurrence [14]. In fact, ELK1 expression was reported to be higher than ELK3 in many cancer cell lines, including brain, skin, and myeloid tumors and sarcomas [15]. PKCα expression, parallel to cell migration and tumorigenicity, of hepatocellular carcinoma was increased by MZF/Elk-1 transcription factor complex [16]. The PEA3 subfamily of ETS domain transcription factors was also involved in a number of cancers, such as in lung tumors with MET amplification, and PEA3 subfamily members were found to play a role in migration and invasion [17]. In colorectal carcinoma, PEA3 was shown to promote invasiveness and metastatic potential [18]. In ovarian cancer, the loss of repressors of the PEA3 subfamily was shown to cause overaccumulation of ETV4 and ETV5 [19].

Malignant gliomas are the most common and lethal primary tumors of the brain. Grading of diffuse gliomas is based largely on the mitotic activity and vascular proliferation states, and molecular markers are also used as diagnostic entities. Still, further molecular information will be important in a more detailed description and categorization of central nervous system (CNS) tumors, in particular for reliable and reproducible classification of grade II and grade III diffuse gliomas [20,21]. Glioblastoma multiforme (GBM), WHO grade IV, is the most aggressive and lethal among all gliomas. High-grade gliomas are composed of a highly proliferative tumor core, with highly invasive cells surrounding them [22].

ETV2, an early regulator of vascular development, was found to be overexpressed in high-grade gliomas, and was reported to play a critical role in endothelial transdifferentiation of CD133+ GBM stem cells, which is thought to render them resistant to anti-angiogenic therapy [23]. Another ETS-related gene, ERG, was found to be a novel and highly specific marker for endothelial cells within CNS tumors, a feature that can be used in studying the vascularization of gliomas [24]. A transposon-based study of gliomagenesis identified friend leukemia integration 1 transcription factor (Fli1), among other genes, to be expressed in gliomas, although Fli1 expression is limited to a subset of glioma cells [25], and ETS protein PU.1, known for its critical role in hematopoietic development, was also reported to be highly expressed in glioma patients, indicating its role in the progression of glioma [26]. In addition to their role in tumor vascularization, ETS proteins can also regulate other aspects of tumorigenesis. In a network analysis based on complexity, as measured by betweenness, Etv5 was identified as a regulator in low-grade optic gliomas in Nf1 mutant mice, and experiments validated the increased expression of both Etv5 and its target genes in optic gliomas [27].

Gliomas can be broadly classified as diffuse and non-diffuse (circumscribed) gliomas. Diffuse gliomas, namely oligodendrogliomas and astrocytomas, exhibit similarities to glial precursors, and are identified and categorized based on the WHO classification of CNS tumors [28]. Due to their rather heterogeneous nature, the reproducibility in diagnosis of low-grade (WHO grade II/III) diffuse gliomas can be a challenge. Several molecular markers, such as isocitrate dehydrogenase (IDH) mutations or telomerase reverse transcriptase (TERT) promoter mutations, which create ETS binding sites [12], are used to assist in differential diagnosis [20]. Previously, ETS gene status in clinical prostate tumor samples has been determined, and ERG+ and ETV1/4/5+ cases were found to be associated with worse prognosis, indicating that ETS status may act as a prognostic biomarker and be used in combination with other existing molecular determinants [29]. In this study, we have analyzed microarray data from patients with different grades of glioma for relative expression of ETS genes, and identified different ETS genes that are upregulated at different glioma grades. We show that, while ETV1 is expressed at high levels in grade 2 glioma, its expression gradually decreases with glioma stage, and on the other hand, ELK3 and ETV4 expressions are increased with progression of the glioma stage. Furthermore, both ELF4 and ETV6 expressions are downregulated at grade 2 glioma, but upregulated at increasing levels in grades 3 and 4, indicating that these genes can also be utilized as additional molecular determinants to distinguish glioma grades. We further compare these data to microarray results from Elk-1-VP16 or PEA3-VP16 overexpression SH-SY5Y cells in order to narrow down transcriptional regulons, and identify common and unique transcriptional regulatory networks for these ETS proteins.

2. Materials and Methods

2.1. Data Collection

Microarray datasets related to glioma were searched from the Gene Expression Omnibus data repository [30], and GSE4290 datasets, including expression data of brain tissue from glioma patients, were selected in this study [31]. The dataset contains the brain tissue of three glioma grades (grade 2–4) from glioma patients and epilepsy patients as a non-tumor control by obtaining brain tissue from surgery patients from Henry Ford Hospital. Patient classification and tissue preparation for microarray were described in [30]. A preprocessed expression matrix was imported into an R programming interface by using R package GEOquery from the Bioconductor project [32]. The methodological flow chart of the study is shown in Figure 1.

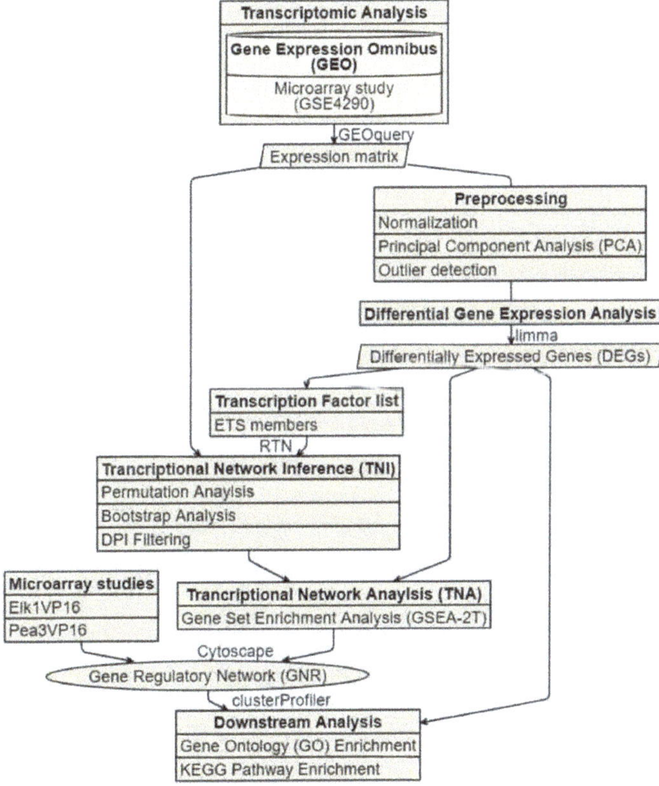

Figure 1. Methodology of transcriptomic profiling and gene regulatory network inference algorithm.

2.2. Data Processing and Differentially Gene Expression Analysis

The expression matrix of the study was Log2 normalized, used in principal component analysis (PCA) to investigate the relationship between samples. Outlier samples were determined by using the PCA plot and eliminated from the expression matrix before the differential gene expression analysis. Differential gene expression analysis was performed with the R package limma from Bioconductor with the contrast of each glioma grade versus non-tumor samples [33]. Differentially expressed genes (DEGs) were determined with a Benjamini & Hochberg corrected p-value < 0.05 significance level and absolute Log2 fold change > 0.6. DEGs were visualized with a volcano plot by using R package Enhanced-Volcano from Bioconductor [34]. Additionally, the intersection of DEGs was performed between limma results, Elk-1-VP16, and the PEA3-VP16 overexpression microarray, which were previously published from our laboratory for use in further analysis.

2.3. Gene Regulatory Network Construction

Transcriptional gene regulatory network (GNR) mediated by the ETS transcription factor family was constructed by using the R package RTN from Bioconductor [35–37]. The expression matrix and ETS transcription factor list from differentially expressed genes were used as an input for the transcriptional network inference (TNI) algorithm with a Benjamini & Hochberg corrected p-value < 0.05 significance level and 1000 permutation number parameters. Regulon activity scores were calculated from TNI by using a two-tailed gene set enrichment analysis (GSEA) algorithm built-in RTN package and visualized as a heatmap. To construct an edge-weighted gene regulatory network mediated with ETS members (Regulons), transcriptional regulatory networks from TNI and differentially expressed genes were integrated by a transcriptional regulatory analysis (TNA) algorithm with two-tailed GSEA. A gene regulatory network of ETS members was constructed by using significant network interaction from TNA (p-value < 0.05). Additionally, the constructed network was filtered with the intersection of DEGs and previous microarray studies (Elk-1-VP16 and PEA3-VP16 overexpression studies). The final GNR was visualized by using Cytoscape software [38].

2.4. Functional Enrichment Analysis

Gene ontology (GO) and KEGG pathway enrichment analysis were performed by using R package clusterProfiler from Bioconductor to analyze the functional profile of gene clusters from differentially expressed genes (up- and downregulated genes of individual glioma grade) and transcription factor regulon clusters (positively and negatively regulated targets for each ETS regulon) [39]. The enriched GO term and KEGG pathways were determined by using a Benjamini & Hochberg corrected p-value < 0.05 significance level and context manner term filtration.

3. Results

3.1. Identification of Differentially Expressed Genes in Gliomas

Before analyzing the expression of genes within the ETS superfamily, we have obtained and analyzed microarray datasets corresponding to different stages of glioma from patients (grades 2–4), using epilepsy patients as the non-tumor control [31].

Analysis of the transcriptome profiles of these glioma samples yielded a set of differentially expressed genes (DEGs), and the performance of the differential expression levels in discriminating the tumor cells from non-tumor cells was verified through principal component analysis (Figure 2). According to the PCA analysis, it was found that, while non-tumor and glioma samples were readily separated, discrimination of glioma stages was less pronounced; grade 2 and grade 4 gliomas were relatively separate, however, grade 3 tumor samples were not clearly separated from the other tumor groups (Figure 2A). Therefore, any outlier data were eliminated for differential gene expression analysis. According to differential gene expression analysis, 8402 of 19,225 genes were found to be significantly changed (with adjusted p-value < 0.05 and absolute log2 fold change > 0.6 thresholds) in the tumor samples (grades 2–4) (Supplementary Table S1). Up- and downregulated genes in all three grades of gliomas were obtained by differential gene expression analysis, and the volcano plot was used to visualize the DEGs, which shows that the distribution of differentially expressed genes was compatible (Figure 2B–D).

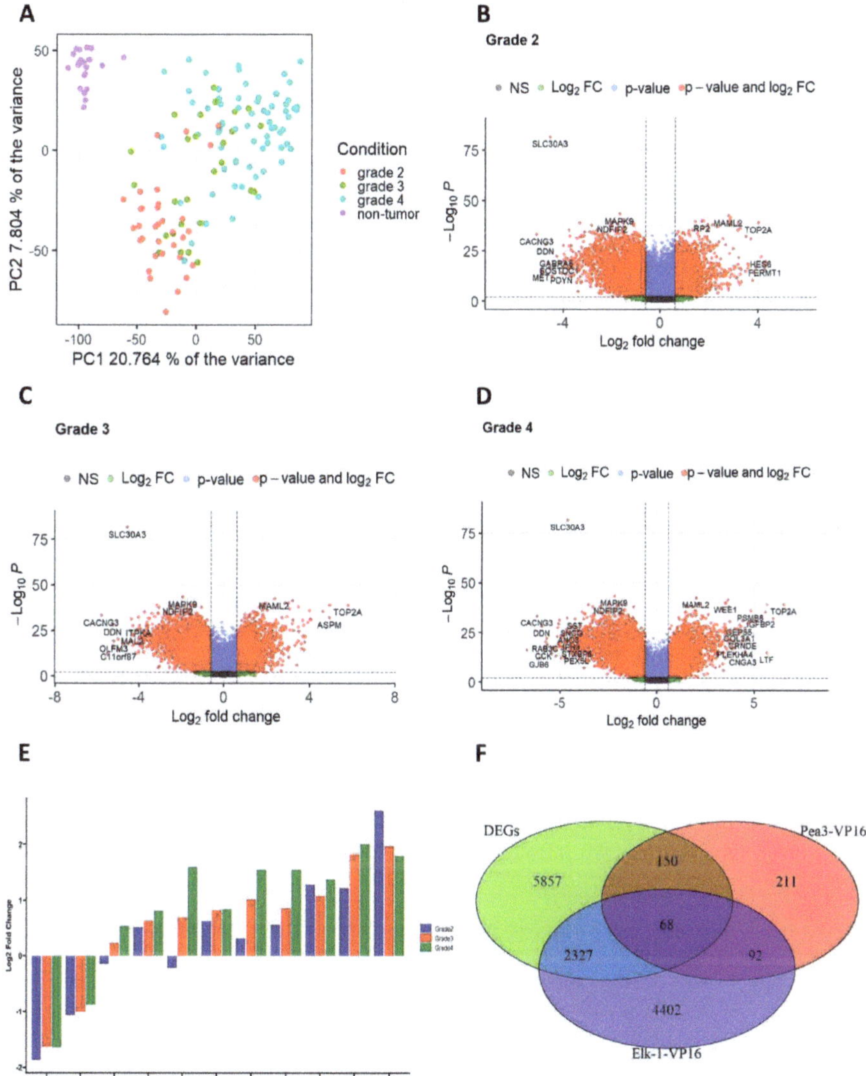

Figure 2. (**A**). Differential gene expression analysis results. Principal component analysis (PCA) plot of the normalized expression matrix. Each point represents individual samples. (**B–D**). Corresponding differentially expressed genes (DEGs) were obtained from comparisons of non-tumor vs. individual glioma grades by using the limma package with a 0.6 absolute log2 fold change and adjusted *p*-value < 0.01 with FDR cutoff, which is indicated with red points. (**E**). Relative fold change of significantly changed ETS members for glioma grade compared with non-tumor samples by differential gene expression analysis. (**F**). Intersection of DEGs between limma results, Elk1VP16 and PEA3VP16, which are our previously published micro arrays, were represented as venn diagrams.

We have next asked whether ETS genes were among the DEGs, and, if so, how their expressions were affected in different glioma stages compared to the non-tumor control (Figure 2E). To that end, we have focused our studies to ETS subfamilies that have been previously reported to be involved in gliomas, namely the ETS subfamily [40–42], TCF subfamily [43–45], ELF subfamily [46,47], PEA3 subfamily [27,48], and TEL subfamily [49,50]. The results showed that the expressions of ETS2 and ELK-1 were downregulated at all grades, while the expressions of ELK4, ELF1, ELK3, ETS1, ETV4, and ETV1 were upregulated at all grades; intriguingly, ETV6 and ELF4 expressions were downregulated at grade 2, but upregulated at grades 3 and 4. The expression of ELK3, ETV4, and to some extent ELK4 was found to increase gradually with glioma grade, while ETV1 expression was highest in grade 2 glioma, and progressively decreased with glioma stage (Figure 2E). The only member of the SPI subfamily of ETS transcription factors represented was PU.1, which was previously shown to be involved in glioma progression, and its levels were indeed found to be increased with glioma grades (Figure 2E; [26]).

Previous microarray studies where either constitutively active Elk-1-VP16 or PEA3-VP16 was overexpressed in SH-SY5Y neuroblastoma cells have identified a number of transcriptional targets for these ETS proteins. In order to narrow down our search and focus on gene regulatory networks of ETS proteins, we have identified overlapping genes by comparing DEGs from glioma grades 2–4 with the microarray results from Elk-1-VP16 and PEA3-VP16 overexpressing cells; 2637 genes were found to be at the intersections of these two experiments, with 63 genes commonly regulated in both glioma tumor samples (DEGs), as well as cell line studies (Elk-1-VP16 and PEA3-VP16) (Figure 2F).

To clarify the functional profiles of the identified DEGs in glioma grades, enrichment analysis was performed, and significant GO and KEGG annotations were obtained (Figure 3). For the GO enrichment analysis of biological processes, initially, up- and down-regulated genes of the different glioma grades were analyzed. While the upregulated gene clusters of grades were observed with cell cycle related phenotype in all glioma grades, the downregulated gene clusters of grades showed neuronal phenotype, such as synaptic transmission, as would be expected (Figure 3A, Supplementary Table S2). In the KEGG enrichment analysis, while the upregulated genes of glioma grades were enriched in p53, TGF-β, and Notch signaling pathways, prominent downregulated gene clusters fell into synaptic function related pathways, such as glutamatergic, GABAergic, serotonergic, and dopaminergic pathways (Figure 3B, Supplementary Table S3). For instance, the TGF-β signaling pathway was found to be altered through glioma progression, as observed by an increase in the level of BMP molecules, including BMP3 and BMP4, as well as their targets, such as Smad1/5/8 and Id. Additionally, the expression of cMyc and p15 associated with cell cycle were significantly increased. On the other hand, Notch signaling pathway genes, such as Delta, Notch, and Fringe, were observed to be upregulated in glioma. The MAPK signaling pathway was found to be enriched in downregulated clusters, with the expression of genes, such as Ras, MEK1, ERK, JNK, and Elk-1, being downregulated. All of these functional enrichment analysis results confirmed that, in all of the glioma grades, cells had downregulated pathways directly related with neuronal function, but upregulated signaling pathways related to cell proliferation and survival. It is interesting to note that the grade 4 glioma samples did not exhibit significant TGF-β pathway upregulation, but PI3K pathway upregulation instead (Figure 3B).

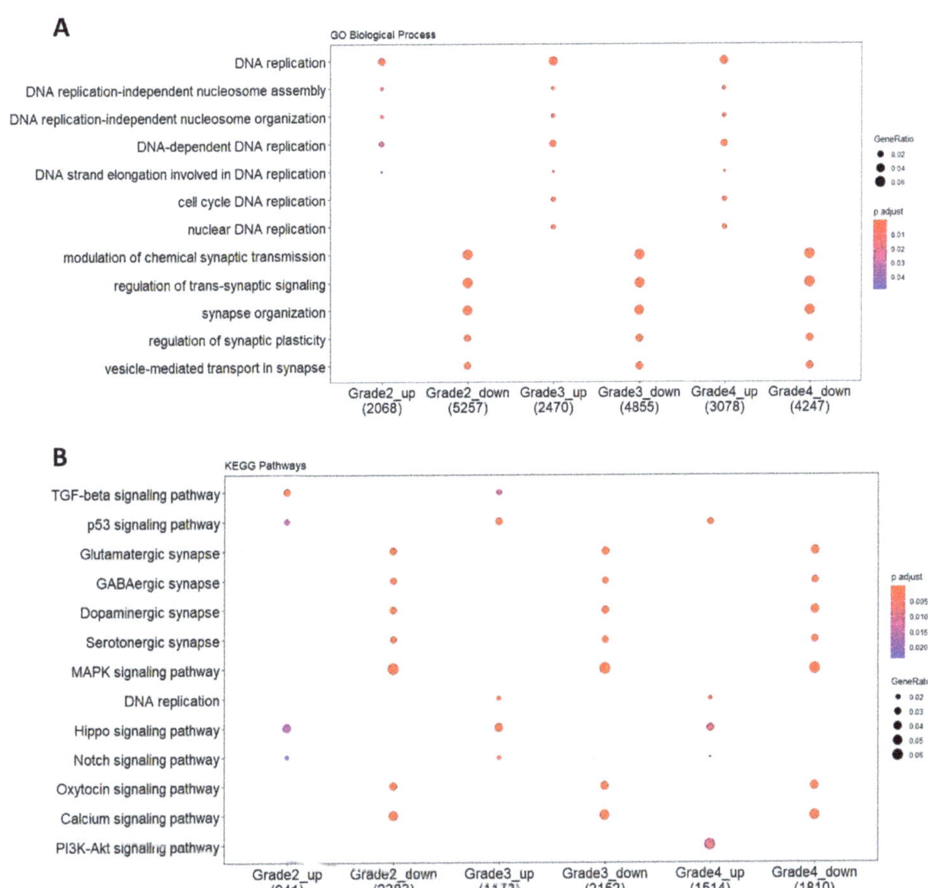

Figure 3. GO and KEGG enrichment analysis of differentially expressed genes for glioma grades. Each glioma grade was clustered as up- and downregulated gene clusters. (**A**). GO and (**B**). KEGG analysis were performed by clusterProfiler with an adjusted *p*-value < 0.05. The gene ratio indicates the number of genes enriched with a corresponding GO or KEGG term among the total gene number introduced into the enrichme.nt analysis.

3.2. Transcriptional Gene Regulatory Network Construction

In order to investigate ETS transcriptional regulation networks specific for each glioma grade, we have integrated DEGs obtained from the analysis of glioma grades with the normalized expression matrix, where significantly changed ETS members are referred as regulons. The initial network obtained using the transcriptional network inference (TNI) algorithm contained 10 ETS member regulons, 11,762 target genes, and 23,181 total interactions (Figure 4A). We have focused on the expression changes of ETS members in different grades of glioma, and regulon activity scores were calculated from the initial network. The analysis of regulon activity scores showed that, while ELK-1 and ETS2 showed high regulon activity in the non-tumor condition (magenta, Figure 4A), ETV1 showed a high activity score on mainly grade 2 glioma (green, Figure 4A). The other ETS proteins showed high regulon activity in the grade 4 glioma cluster, however, regulon activity of grade 3 glioma was dispersed between grade 2 and grade 4 (Figure 4A). After including DEGs into the initial network using the transcriptional network analysis (TNA) algorithm, a focused network of glioma grades was constructed. According to TNA algorithm, nine significant ETS regulons were found to be enriched with different numbers of target genes. ETV1, which expressed the least significance among the significant ETS regulons, was marked in blue (Figure 4B).

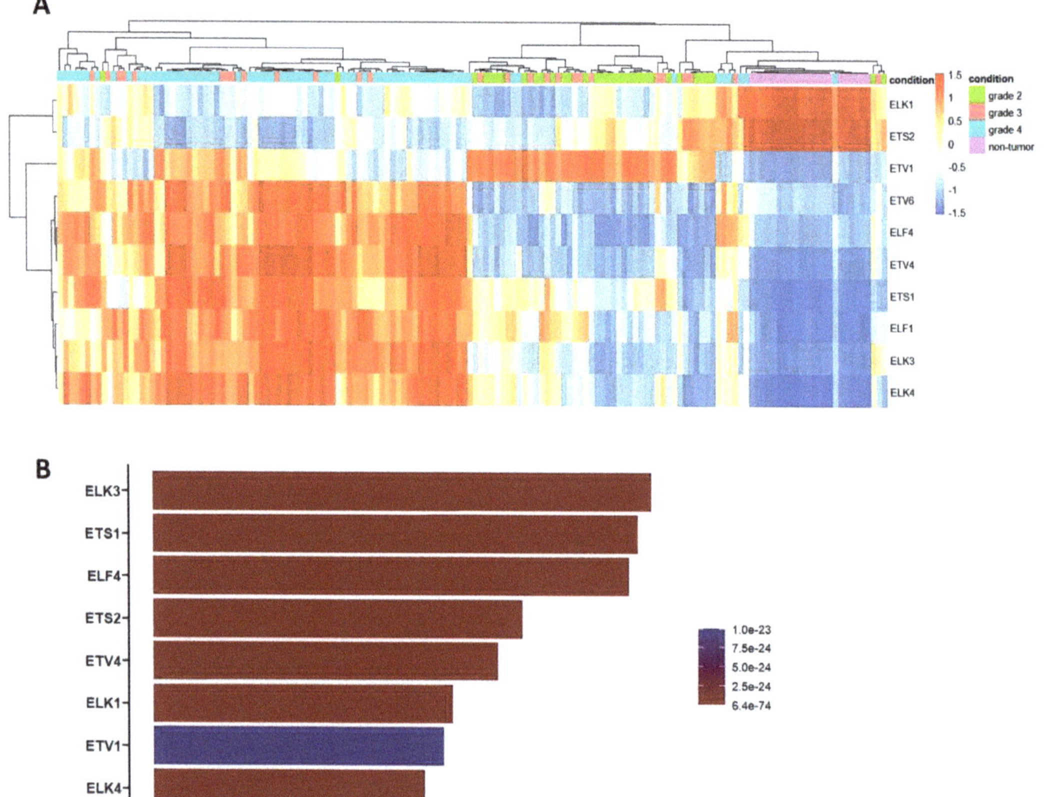

Figure 4. Regulon activity and size of the ETS-mediated gene regulatory network. (**A**). The correlation distance heatmap of regulon activity for non-tumor and glioma grades. (**B**). Regulon size of the individual transcription factor in the gene regulatory network resulting from the transcriptional network analysis (TNA).

To determine the direction of regulation between each regulon and its targets, two-tailed GSEA was performed, and positively and negatively regulated co-expression patterns in target gene distribution were constructed for individual ETS regulons (Figure 5). In this analysis, genes were ranked for their fold changes in the x-axis, and enrichment scores were given in the y-axis; the peak of each plot is the enrichment score for the gene indicated (dotted lines), while the colored bar shows the positively and negatively correlated genes. These results suggest that enriched ETS regulons have both unique and common gene targets in gliomas, as indicated by a clear separation of negatively and positively correlated targets in regulons such as ETS2 and ELK1 (unique), and overlapping negative and positive regulons, such as those of ELF4 and ELK3 (common).

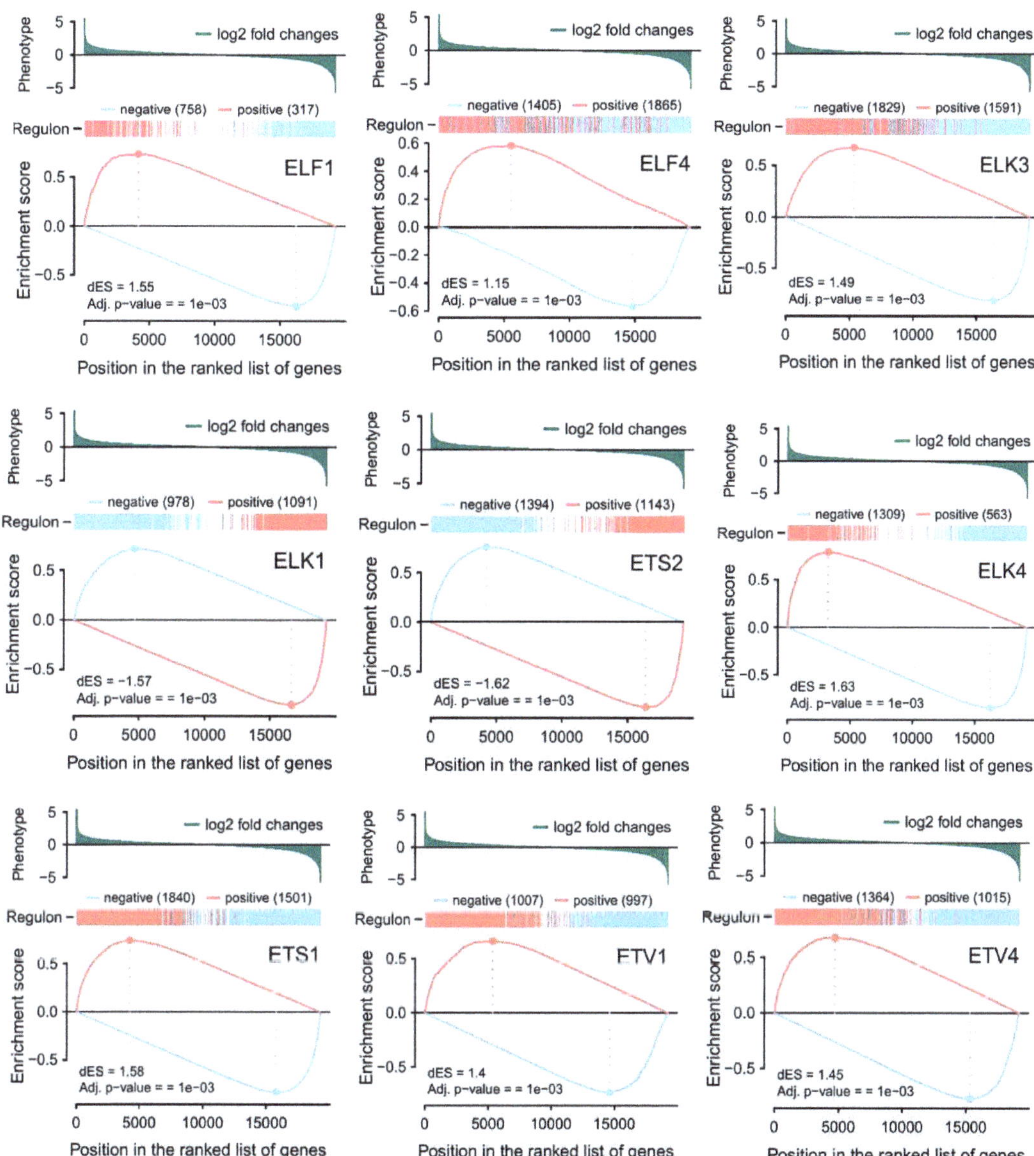

Figure 5. Two-tailed GSEA analysis associated positively and negatively regulated targets of individual regulons. Target genes are ranked by gene expression analysis, and scored by enrichment analysis that indicates the edge weight of the gene regulatory network.

The network from the TNA algorithm was filtered by DEGs from Elk-1-VP16 and PEA3-VP16 overexpression microarray results to create a much more unique regulatory network of ETS members. As a result of filtering, a final regulatory network was constructed with 3366 target genes and 6610 interactions with ETS regulons, and the gene regulatory network was visualized with Cytoscape (Figure 6). This network representation shows fold changes of DEGs, as well as their interaction with the ETS regulons, showing the common targets to be clustered in the middle (Figure 6).

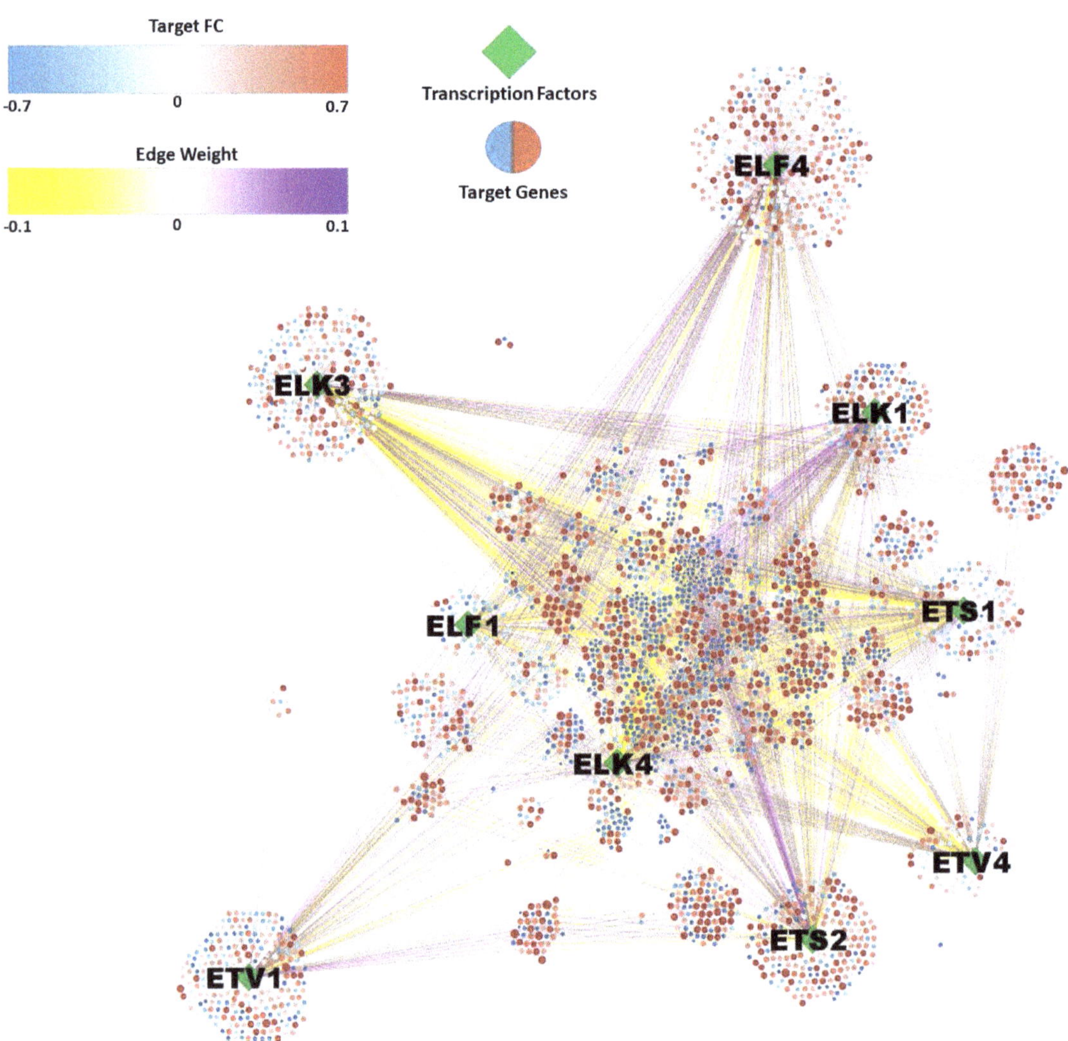

Figure 6. Gene regulatory network of glioma grades under the regulation of ETS transcription factors. Diamond nodes represent ETS members as a regulator, and circle nodes correspond to the target genes. The color of the circle indicates the mean fold change of glioma grades compared to non-tumor samples, resulting from differential gene expression analysis. Edge colors show the enrichment score of each target gene with corresponding regulators resulted from GSEA analysis.

Focusing on the functional investigation of the gene regulatory network, GO and KEGG enrichment analysis was performed with positively and negatively regulated targets of ETS regulons on the regulatory network (positive regulation indicates similar coexpression patterns, i.e., when ETS protein is downregulated, its targets are also downregulated, and vice versa). It was observed that positive and negative cluster targets of ETS regulons were enriched in biological processes, such as cell–cell adhesion, synapse formation, and protein localization, some of which are common across members, while some are unique for one or few family member(s) (Figure 7A,B, Supplementary Tables S4 and S5). ELF1 and ELF4 regulons appear to belong to similar biological processes; ELK1 and ETS2 also were found to have targets within the same biological pathways (Figure 7A). The ETV1 regulon appears to have a distinct set of positively regulated targets, while ELK3, ELK4,

ETS1, and, to some extent, ETV4 appear to regulate similar biological processes (Figure 7A). This classification is not conserved for negatively regulated targets, however; here, ELF1, ELK3, ELK4, ETS1, and ETV4 appear to regulate similar biological processes, while the ELF4 regulon and ETS2 regulon each are comprised of distinct targets (Figure 7B). It is important to note that, in positively regulated targets of the ETV1 regulon, nucleosome and chromatin disassembly related processes were prominent, while no significant negatively regulated targets were identified for the ETV1 regulon. The ELK3 regulon included positively regulated targets in ECM organization, protein maturation, and processing pathways, and negatively regulated targets in synaptic vesicle signaling, synaptic transmission, and synaptic plasticity pathways.

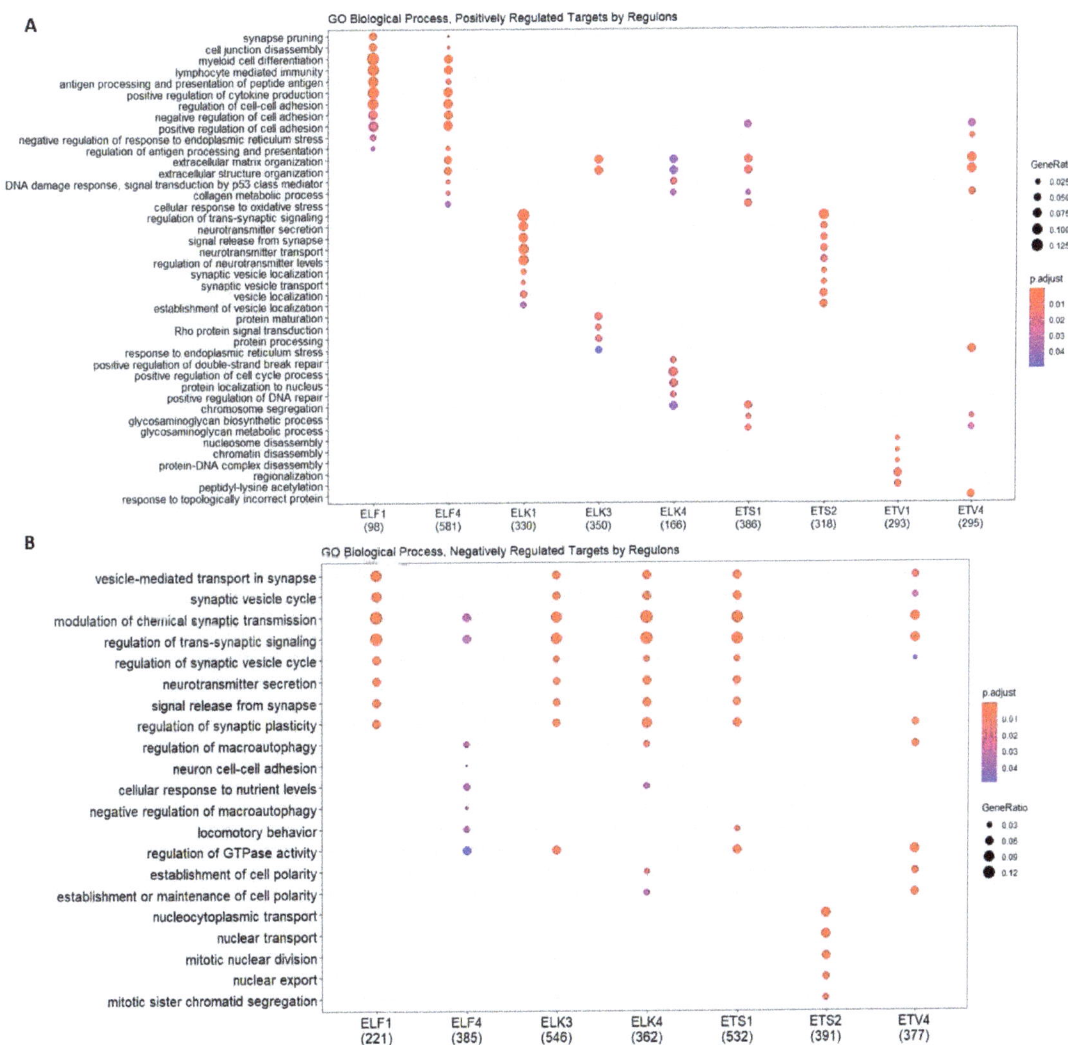

Figure 7. GO enrichment analysis of individual regulons and their (**A**). positively and (**B**). negatively regulated targets. GO analysis was performed by clusterProfiler with an adjusted *p*-value < 0.05. The gene ratio indicates the number of genes enriched with corresponding GO terms among the total gene number introduced into the enrichment analysis.

Similar comparative analysis using KEGG pathway enrichment showed that, unlike the GO biological processes analyzed above, distinct signaling pathways were regulated by each ETS regulon, while a positively regulated cluster of the ETV4 regulon was enriched for the MAPK and PI3K-Akt signaling pathways, and a negatively regulated cluster of this protein was enriched for endocytosis and the synaptic vesicle cycle (Figure 8A,B, Supplementary Tables S6 and S7); the positively regulated cluster of ELK1 was enriched for cholinergic and dopaminergic synapses, as well as calcium signaling, while its negatively regulated cluster was enriched Hippo signaling, signaling of pluripotent stem cells, and cell cycle (Figure 8A,B). Interestingly, endocytosis and the synaptic vesicle cycle were common signaling pathways in almost all ETS regulons, except for ELK1 (Figure 8B).

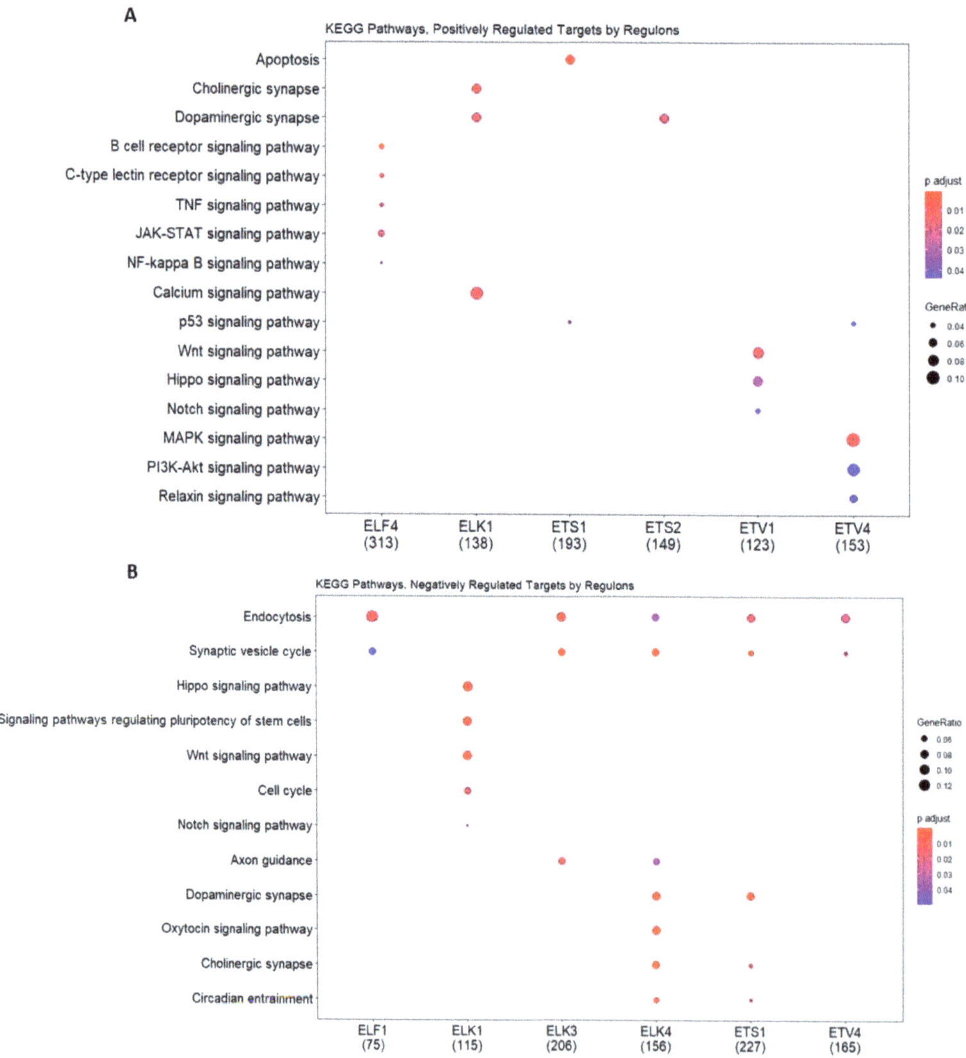

Figure 8. KEGG enrichment analysis of individual regulons and their (**A**). positively and (**B**). negatively regulated targets. KEGG analysis was performed by clusterProfiler with an adjusted *p*-value < 0.05. The gene ratio indicates the number of genes enriched with corresponding KEGG pathways among the total gene number introduced into the enrichment analysis.

4. Discussion

The five stages of gliomagenesis are the initial growth stage, oncogene-dependent senescence stage, growth stage, replicative senescence stage, and, finally, the immortalization stage [28]. Disease stage classification and identification of stage-dependent or grade-dependent biomarkers is important in the accurate diagnosis of gliomas.

Graph complexity analysis in low-grade glioma has shown Etv5 and its network expression to be critical features of the neoplastic state [27]. Unfortunately, ETV5 of the PEA3 subfamily does not appear to be significantly altered in tumor vs. non-tumor samples in the microarray datasets used in this study (data not shown). However, we have identified another PEA3 subfamily member, ETV1, to be expressed at high levels in low-grade glioma and decrease in expression in higher grades (Figure 2). ELK-1 protein is known to be a critical partner for the androgen receptor (AR) in prostate cancer, and its expression was found to be associated with a higher clinical stage and prognostic marker of disease recurrence in prostate cancer [14]. No such distinction was apparent in our study on glioma grades 2–4. However, we have identified ELF4 and ETV6 to be downregulated in grade 2 gliomas, and upregulated in increasing amounts in grades 3 and 4 (Figure 2). It should be noted, however, that the ETS expression profile is also different in epilepsy; ELF1, ELK1, ELK4, ETS1, ETS2, and ETV1 are expressed at higher levels than ELF4, ELK3, and ETV4, and there is also variability in expression among different types of epilepsy (Supplementary Figure S1). However, since the type of epilepsy used in the datasets analyzed in this study were not known, it was not possible to normalize for ETS gene expression (Supplementary Figure S1) [51].

ETS proteins focused on in this study (namely, class I subfamilies ETC, TCF, and PEA3 and class II subfamilies ELF and TEL) exhibit little tissue specificity, and, in fact, many family members are ubiquitously expressed [2,15]. It is therefore not surprising that gene regulatory network analysis of ETS transcription factors exhibits extensive overlap of targets, confirming that functional redundancy exists at least to a certain extent (Figure 6). However when positively and negatively regulated targets of ETS regulons were analyzed, negatively regulated targets were found to extensively overlap (mostly related to synaptic vesicle trafficking and synaptic transmission, Figure 7B), while positively regulated targets appeared to be selective for groups of ETS family members, with ELF1 and ELF4 comprising one class (targets in synapse pruning, immune function, and cell to cell adhesion related biological processes), ELK1 and ETS2 forming another class (targets in synapse function and synaptic vesicle trafficking related processes), and ELK3, ELK4, ETS1, and ETV4 forming a third class (targets in extracellular matrix related processes, as well as protein processing and localization, ER stress, and cell cycle related processes); ETV1 appeared to be a class by itself (targets in nucleosome and chromatin disassembly) (Figure 7A). These regulon classes were not directly related to ETS subfamily assignments.

Although more physiologically relevant non-tumor controls are required to fine-tune the results in the future, this is a proof of concept study that shows that expression levels of ETS genes can be used as diagnostic markers for glioma grade identification, in addition to already existing molecular markers. In addition, the gene regulatory network analysis for ETS regulons can be used to identify target gene clusters in positively and negatively regulated pathways and processes, which can help in understanding the molecular mechanisms of transcriptional redundancy among family members. We propose that such network analysis can also be extended to differentiate stages of tumorigenesis in other types of tumors, as well as to developmental stages of various tissues.

Supplementary Materials: The following are available online at https://www.mdpi.com/2075-4426/11/2/138/s1, Figure S1: Expression level of ETS members on the neocortex and hippocampus in epilepsy patients and the healthy control from GSE134697; Table S1: GO biological process enrichment analysis result table of glioma grades from DEGs; Table S2: KEGG pathway enrichment analysis result table of glioma grades from DEGs; Table S3: GO biological process enrichment analysis result table of positively regulated targets of ETS regulons; Table S4: GO biological process enrichment analysis result table of negatively regulated targets of ETS regulons; Table S5: KEGG pathway enrichment analysis result table of positively regulated targets of ETS regulons; Table S6: KEGG pathway enrichment analysis result table of negatively regulated targets of ETS regulons; Table S7: Table formation of the gene regulatory network in Figure 6.

Author Contributions: Conceptualization, B.K. and I.A.K.; Data curation, Y.K.B.; Formal analysis, Y.K.B. and B.K.; Investigation, Y.K.B. and I.A.K.; Methodology, B.K.; Project administration, I.A.K.; Supervision, B.K. and I.A.K.; Writing—original draft, Y.K.B., B.K., and I.A.K. All authors have read and agreed to the published version of the manuscript.

Funding: This research received no external funding.

Institutional Review Board Statement: Not applicable.

Informed Consent Statement: Not applicable.

Data Availability Statement: The data is available in Supplementary Files and upon request.

Acknowledgments: We would like to thank members of the AxanLab for helpful discussions. IAK is a GYA member.

Conflicts of Interest: The authors declare no conflict of interest.

References

1. Wang, Z.; Zhang, Q. Genome-wide identification and evolutionary analysis of the animal specific ETS transcription factor family. *Evol. Bioinf.* **2009**, *5*, 119–131. [CrossRef] [PubMed]
2. Hollenhorst, P.; McIntosh, L.P.; Graves, B.J. Genomic and biochemical insights into the specificity of ETS transcription factors. *Annu. Rev. Biochem.* **2011**, *80*, 437–471. [CrossRef] [PubMed]
3. Boros, J.; O'Donnell, A.; Donaldson, I.J.; Kasza, A.; Zeef, L.; Sharrocks, A.D. Overlapping promoter targeting by Elk-1 and other divergent ETS-domain transcription factor family members. *Nucleic Acids Res.* **2009**, *37*, 7368–7380. [CrossRef] [PubMed]
4. Kar, A.; Gutierrez-Hartmann, A. Molecular mechanisms of ETS transcription factor mediated tumorigenesis. *Crit. Rev. Biochem. Mol. Biol* **2013**, *48*, 522–543. [CrossRef]
5. Yuan, C.C.; Dunn, K.J.; Papas, T.S.; Blair, D.G. Properties of a murine retroviral recombinant of avian acute leukemia virus E26: A murine fibroblast assay for v-ets function. *J. Virol.* **1989**, *63*, 205–215. [CrossRef]
6. Shore, P.; Whitmarsh, A.J.; Bhaskaran, R.; Davis, R.J.; Waltho, J.P.; Sharrocks, A.D. Determinants of DNA-binding specificity of ETS-domain transcription factors. *Mol. Cell. Biol.* **1996**, *16*, 3338–3349. [CrossRef]
7. Yang, S.-H.; Whitmarsh, A.J.; Davis, R.J.; Sharrocks, A.D. Differential targeting of MAP kinases to the ETS-domain transcription factor Elk-1. *EMBO J.* **1998**, *17*, 1740–1749. [CrossRef]
8. Yang, S.-H.; Shore, P.; Willingham, N.; Lakey, J.H.; Sharrocks, A.D. The mechanism of phosphorylation-inducible activation of the ETS domain transcription factor Elk-1. *EMBO J.* **1999**, *18*, 5666–5674. [CrossRef]
9. Hsu, T.; Trojanowska, M.; Watson, D.K. Ets proteins in biological control and cancer. *J Cell. Biochem.* **2004**, *91*, 896–903. [CrossRef]
10. Furlan, A.; Vercamer, C.; Heliot, L.; Wernert, N.; Desbiens, X.; Pourtier, A. Ets-1 drives breast cancer cell angiogenic potential and interactions between breast cancer and endothelial cells. *Int. J. Oncol.* **2019**, *54*, 29–40. [CrossRef]
11. Verschoor, M.L.; Verschoor, C.P.; Singh, G. Ets-1 global gene expression profile reveals associations with metabolism and oxidative stress in ovarian and breast cancers. *Cancer Metab.* **2013**, *1*, 17. [CrossRef]
12. Fry, E.A.; Inoue, K. Aberrant expression of ETS1 and ETS2 proteins in cancer. *Cancer Rep. Rev.* **2018**, *2*. [CrossRef] [PubMed]
13. Morris, J.F.; Sul, J.-Y.; Kim, M.-S.; Klein-Szanto, A.J.; Schochet, T.; Rustgi, A.; Eberwine, J.H. Elk-1 phosphorylated at threonine-417 is present in diverse cancers and correlates with differentiation grade of colonic adenocarcinoma. *Hum. Pathol.* **2013**, *44*, 766–776. [CrossRef] [PubMed]
14. Pardy, L.; Rosati, R.; Soave, C.; Huang, Y.; Kim, S.; Ratnam, M. The ternary complex factor protein ELK-1 is an independent prognosticator of disease recurrence in prostate cancer. *Prostate* **2019**, *80*, 198–208. [CrossRef]
15. Ahmad, A.; Hayat, A. Expression of oncogenes ELK1 and ELK3 in cancer. *Annal. Colorec. Cancer Res.* **2019**, *1*, 1–6.
16. Yue, C.-H.; Huang, C.-Y.; Tsai, J.-H.; Hsu, C.-W.; Hsieh, Y.-H.; Liu, J.-Y. MZF-1/Elk-1 complex binds to protein kinase Calpha promoter and is involved in hepatocellular carcinoma. *PLoS ONE* **2015**, *10*, e0127420. [CrossRef]
17. Kherrouche, Z.; Monte, D.; Werkmeister, E.; Stoven, L.; de Launoit, Y.; Cortot, A.B.; Tulasne, D.; Chotteau-Lelievre, A. PEA3 transcription factors are downstream effectors of Met signaling involved in migration and invasiveness of Met-addicted tumor cells. *Mol. Oncol.* **2015**, *9*, 1852–1867. [CrossRef]
18. Mesci, A.; Taeb, S.; Huang, X.; Jairath, R.; Sivaloganathan, D.; Liu, S.K. PEA3 expression promotes the invasive and metastatic potential of colorectal carcinoma. *World J. Gastroenterol.* **2014**, *20*, 17376–17387. [CrossRef]
19. Qi, T.; Qu, Q.; Li, G.; Wang, J.; Zhu, H.; Yang, Z.; Sun, Y.; Lu, Q.; Qu, J. Function and regulation of the PEA3 subfamily of ETS transcription factors in cancer. *Am. J. Cancer Res.* **2020**, *10*, 3083–3105.
20. Olar, A.; Sulman, E.P. Molecular markers in low grade glioma—toward tumor reclassification. *Semin. Radiat. Oncol.* **2015**, *25*, 155–163. [CrossRef]

21. Wood, M.D.; Halfpenny, A.M.; Moore, S.R. Applications of molecular neurooncology- a review of diffuse glioma integrated diagnosis and emerging molecular entities. *Diag. Pathol.* **2019**, *14*, 29. [CrossRef]

22. Mehta, S.; Lo Cascio, C. Developmentally regulated signaling pathways in glioma invasion. *Cell. Mol. Life Sci.* **2018**, *75*, 385–402. [CrossRef]

23. Zhao, C.; Gomez, G.A.; Zhao, Y.; Yang, Y.; Cao, D.; Lu, J.; Yang, H.; Lin, S. ETV2 mediates endothelial transdifferentiation of glioblastoma. *Signal Transduc. Tar. Ther.* **2018**, *3*, 4. [CrossRef]

24. Haber, M.A.; Iranmahboob, A.; Thomas, C.; Liu, M.; Najjar, A.; Zagzag, D. ERG is a novel and reliable marker for endothelial cells in central nervous system tumors. *Clin. Neuropathol.* **2015**, *34*, 117–127. [CrossRef]

25. Vyazunova, I.; Maklakova, V.I.; Berman, S.; De, I.; Steffen, M.D.; Hong, W.; Lincoln, H.; Morrissy, A.S.; Taylor, M.D.; Akagi, K.; et al. Sleeping beauty mouse models identify candidate genes involved in gliomagenesis. *PLoS ONE* **2014**, *9*, e113489. [CrossRef]

26. Xu, Y.; Gu, S.; Bi, Y.; Qi, X.; Yan, Y.; Lou, M. Transcription factor PU.1 is involved in the progression of glioma. *Oncol. Lett.* **2018**, *15*, 3753–3759. [CrossRef]

27. Pan, Y.; Duron, C.; Bush, E.C.; Ma, Y.; Sims, P.A.; Gutmann, D.H.; Radunskaya, A.; Hardin, J. Graph complexity analysis identifies an ETV5 tumor-specific network in human and murine low-grade glioma. *PLoS ONE* **2018**, *13*, e0190001. [CrossRef]

28. Barthel, F.P.; Wesseling, P.; Verhaak, R.G.W. Reconstructing the molecular life history of gliomas. Acta Neuropathol. **2018**, *135*, 649–670. murine low-grade glioma. *PLoS ONE* **2018**, *13*, e0190001. [CrossRef]

29. Torres, A.; Alshalalfa, M.; Tomlins, S.A.; Erho, N.; Gibb, E.A.; Chelliserry, J.; Lim, L.; Lam, L.L.C.; Faraj, S.F.; Bezerra, S.M.; et al. Comprehensive determination of prostate tumor ETS gene status in clinical samples using the CLIA decipher assay. *J. Mol. Diag.* **2017**, *19*, 475–484. [CrossRef] [PubMed]

30. Barrett, T.; Wilhite, S.E.; Ledoux, P.; Evangelista, C.; Kim, I.F.; Tomashevsky, M.; Marshall, K.A.; Phillippy, K.H.; Sherman, P.M.; Holko, M.; et al. NCBI GEO: Archive for functional genomics data sets—Update. *Nucleic Acids Res.* **2013**, *41*, D991–D995. [CrossRef] [PubMed]

31. Sun, L.; Hui, A.-M.; Su, Q.; Vortmeyer, A.; Kotliarov, Y.; Pastorino, S.; Passaniti, A.; Menon, J.; Walling, J.; Bailey, R.; et al. Neuronal and glioma-derived stem cell factor induces angiogenesis within the brain. *Cancer Cell* **2006**, *9*, 287–300. [CrossRef] [PubMed]

32. Davis, S.; Meltzer, P.S. GEOquery: A bridge between the Gene Expression Omnibus (GEO) and BioConductor. *Bioinformatics* **2007**, *23*, 1846–1847. [CrossRef]

33. Ritchie, M.E.; Phipson, B.; Wu, D.; Hu, Y.; Law, C.W.; Shi, W.; Smyth, G.K. Limma powers differential expression analyses for RNA-sequencing and microarray studies. *Nucleic Acids Res.* **2015**, *43*, e47. [CrossRef]

34. Blighe, K.; Rana, S.; Lewis, M. EnhancedVolcano: Publication-Ready Volcano Plots with Enhanced Colouring and Labeling. 2020. Available online: https://github.com/kevinblighe/EnhancedVolcano (accessed on 14 December 2020).

35. Campbell, T.M.; Castro, M.A.A.; de Santiago, I.; Fletcher, M.N.C.; Halim, S.; Prathalingam, R.; Ponder, B.A.J.; Meyer, K.B. FGFR2 risk SNPs confer breast cancer risk by augmenting oestrogen responsiveness. *Carcinogenesis* **2016**, *37*, 741–750. [CrossRef]

36. Castro, M.A.A.; de Santiago, I.; Campbell, T.M.; Vaughn, C.; Hickey, T.E.; Ross, E.; Tilley, W.D.; Markowetz, F.; Ponder, B.A.J.; Meyer, K.B. Regulators of genetic risk of breast cancer identified by integrative network analysis. *Nat. Genet.* **2016**, *48*, 12–21. [CrossRef]

37. Fletcher, M.N.C.; Castro, M.A.A.; Wang, X.; de Santiago, I.; O'Reilly, M.; Chin, S.-F.; Rueda, O.M.; Caldas, C.; Ponder, B.A.J.; Markowetz, F.; et al. Master regulators of FGFR2 signalling and breast cancer risk. *Nat. Commun.* **2013**, *4*, 2464. [CrossRef]

38. Shannon, P.; Markiel, A.; Ozier, O.; Baliga, N.S.; Wang, J.T.; Ramage, D.; Amin, N.; Schwikowski, B.; Ideker, T. Cytoscape. A software environment for integrated models of biomolecular interaction networks. *Genome Res.* **2003**, *13*, 2498–2504. [CrossRef] [PubMed]

39. Yu, G.; Wang, L.; Han, Y.; He, Q. clusterProfiler: An R package for comparing biological themes among gene clusters. *OMICS J. Integr. Biol.* **2012**, *16*, 284–287. [CrossRef]

40. Xu, H.; Zhao, G.; Zhang, Y.; Jiang, H.; Wang, W.; Zhao, D.; Yu, H.; Qi, L. Long non-coding RNA PAXIP1-AS1 facilitates cell invasion and angiogenesis of glioma by recruiting transcription factor ETS1 to upregulate KIF14 expression. *J. Exp. Clin. Cancer Res.* **2019**, *38*, 486. [CrossRef]

41. Cam, M.; Charan, M.; Welker, A.M.; Dravid, P.; Studebaker, A.W.; Leonard, J.R.; Pierson, C.R.; Nakano, I.; Beattie, C.E.; Hwang, E.I.; et al. DeltaNp73/ETS2 complex drives glioblastoma pathogenesis-targeting downstream mediators by rebastinib prolongs survival in preclinical models of glioblastoma. *Neuro-oncology* **2020**, *22*, 345–356. [CrossRef] [PubMed]

42. Giese, A. Glioma invasion-pattern of dissemination by mechanisms of invasion and surgical intervention, pattern of gene expression and its regulatory control by tumor suppressor p53 and proto-oncogene ETS-1. *Acta Neurochir. Suppl.* **2003**, *88*, 153–162. [PubMed]

43. Mut, M.; Lule, S.; Demir, O.; Kurnaz, I.A.; Vural, I. Both mitogen-activated protein kinase (MAPK)/extracellular signal regulated kinases (ERK) $\frac{1}{2}$ and phosphatidylinositide-3-OH kinase (PI3K)/Akt pathways regulate activation of E-twenty-six (ETS)-like transcription factor-1 (Elk-1) in U138 glioblastoma cells. *Int. J. Biochem. Cell. Biol.* **2012**, *44*, 302–310. [CrossRef] [PubMed]

44. Wang, Z.; Yuan, H.; Sun, C.; Xu, L.; Chen, Y.; Zhu, Q.; Zhao, H.; Huang, Q.; Dong, J.; Lan, Q. GATA2 promotes glioma progression through EGFR/ERK/Elk-1 pathway. *Med. Oncol.* **2015**, *32*, 87. [CrossRef]

45. Savasan Sogut, M.; Venugopal, C.; Kandemir, B.; Dag, U.; Mahendram, S.; Singh, S.; Gulfidan, G.; Arga, K.Y.; Yilmaz, B.; Aksan Kurnaz, I. ETS-domainn transcription factor Elk-1 in brain tumors and CD133+ brain tumor initiating cells. *J. Per. Med.* **2021**, *11*, 125. [CrossRef]

46. Cheng, M.; Zeng, Y.; Zhang, T.; Xu, M.; Li, Z.; Wu, Y. Transcription factor ELF1 activates MEIS1 transcription factor and then regulates the GFI1/FBW7 axis to promote the development of glioma. *Mol. Ther. Nucleic Acids* **2020**, *23*, 418–430. [CrossRef] [PubMed]
47. Wang, M.; Yang, C.; Liu, X.; Zheng, J.; Xue, Y.; Ruan, X.; Shen, S.; Wang, D.; Li, Z.; Cai, H.; et al. An upstream open reading frame regulates vasculogenic mimicry of glioma via ZNRD1-AS1/miR-499a-5p/ELF1/EMI1 pathway. *Cell. Mol. Med.* **2020**, *24*, 6120–6136. [CrossRef]
48. Padul, V.; Epari, S.; Moiyadi, A.; Shetty, P.; Shirsat, N.V. ETV/PEA3 family transcription factor-encoding genes are overexpressed in CIC-mutant oligodendrogliomas. *Genes Chrom. Cancer* **2015**, *54*, 725–733. [CrossRef] [PubMed]
49. Gatalica, Z.; Xiu, J.; Swensen, J.; Vranic, S. Molecular characterization of cancers with MTRK gene fusions. *Mod. Pathol.* **2019**, *32*, 147–153. [CrossRef]
50. Busse, T.M.; Roth, J.J.; Wilmoth, D.; Wainright, L.; Tooke, L.; Biegel, J.A. Copy number alterations determined by single nucleotide polymorphism array testing in the clinical laboratory are indicative of gene fusions in pediatric cancer patients. *Genes Chrom. Cancer* **2017**, *56*, 730–749. [CrossRef]
51. Kjær, C.; Barzaghi, G.; Bak, L.K.; Goetze, J.P.; Yde, C.W.; Woldbye, D.; Pinborg, L.H.; Jensen, L.J. Transcriptome analysis in patients with temporal lobe epilepsy. *Brain* **2019**, *142*, e55. [CrossRef]

Journal of
Personalized
Medicine

Article

ETS-Domain Transcription Factor Elk-1 Regulates Stemness Genes in Brain Tumors and CD133+ BrainTumor-Initiating Cells

Melis Savasan Sogut [1,2,3], Chitra Venugopal [4], Basak Kandemir [1,2,3,†], Ugur Dag [3,‡], Sujeivan Mahendram [4], Sheila Singh [4], Gizem Gulfidan [5], Kazim Yalcin Arga [5], Bayram Yilmaz [6,*] and Isil Aksan Kurnaz [1,2,*]

[1]   Institute of Biotechnology, Gebze Technical University, 41400 Kocaeli, Turkey; melis.savasan.sogut@gmail.com (M.S.S.); bkandemir@baskent.edu.tr (B.K.)
[2]   Molecular Neurobiology Laboratory (AxanLab), Department of Molecular Biology and Genetics, Gebze Technical University, 41400 Kocaeli, Turkey
[3]   Biotechnology Graduate Program, Graduate School of Sciences, Yeditepe University, 26 Agustos Yerlesimi, Kayisdagi, 34755 Istanbul, Turkey; dagu@janelia.hhmi.org
[4]   Stem Cell and Cancer Research Institute, McMaster University, Hamilton, ON L8S 4K1, Canada; venugop@mcmaster.ca (C.V.); smahendram@gmail.com (S.M.); ssingh@mcmaster.ca (S.S.)
[5]   Department of Bioengineering, Marmara University, 34722 Istanbul, Turkey; gizemgulfidn@gmail.com (G.G.); kazim.arga@marmara.edu.tr (K.Y.A.)
[6]   Department of Physiology, Faculty of Medicine, Yeditepe University, 26 Agustos Yerlesimi, Kayisdagi, 34755 Istanbul, Turkey
*   Correspondence: byilmaz@yeditepe.edu.tr (B.Y.); ikurnaz@gtu.edu.tr (I.A.K.)
†   Present address: Department of Molecular Biology and Genetics, Baskent University, 06790 Ankara, Turkey.
‡   Present address: Department of Brain and Cognitive Sciences, Massachusetts Institute of Technology, Cambridge, MA 02139-4307, USA.

check for
updates

Citation: Sogut, M.S.; Venugopal, C.; Kandemir, B.; Dag, U.; Mahendram, S.; Singh, S.; Gulfidan, G.; Arga, K.Y.; Yilmaz, B.; Kurnaz, I.A. ETS-Domain Transcription Factor Elk-1 Regulates Stemness Genes in Brain Tumors and CD133+ BrainTumor-Initiating Cells. *J. Pers. Med.* **2021**, *11*, 125. https://doi.org/10.3390/jpm11020125

Academic Editor: Chiara Villa

Received: 20 January 2021
Accepted: 9 February 2021
Published: 14 February 2021

Publisher's Note: MDPI stays neutral with regard to jurisdictional claims in published maps and institutional affiliations.

Abstract: Elk-1, a member of the ternary complex factors (TCFs) within the ETS (E26 transformation-specific) domain superfamily, is a transcription factor implicated in neuroprotection, neurodegeneration, and brain tumor proliferation. Except for known targets, *c-fos* and *egr-1*, few targets of Elk-1 have been identified. Interestingly, *SMN*, *SOD1*, and *PSEN1* promoters were shown to be regulated by Elk-1. On the other hand, Elk-1 was shown to regulate the CD133 gene, which is highly expressed in brain-tumor-initiating cells (BTICs) and used as a marker for separating this cancer stem cell population. In this study, we have carried out microarray analysis in SH-SY5Y cells overexpressing Elk-1-VP16, which has revealed a large number of genes significantly regulated by Elk-1 that function in nervous system development, embryonic development, pluripotency, apoptosis, survival, and proliferation. Among these, we have shown that genes related to pluripotency, such as *Sox2*, *Nanog*, and *Oct4*, were indeed regulated by Elk-1, and in the context of brain tumors, we further showed that Elk-1 overexpression in CD133+ BTIC population results in the upregulation of these genes. When Elk-1 expression is silenced, the expression of these stemness genes is decreased. We propose that Elk-1 is a transcription factor upstream of these genes, regulating the self-renewal of CD133+ BTICs.

Keywords: ETS; Elk-1; stem cell; microarray; brain-tumor-initiating cell (BTIC)

1. Introduction

The ternary complex factor (TCF) Elk-1 of the ETS domain superfamily is a ubiquitous transcription factor, yet it interacts with neuronal microtubules and motor proteins, is found mainly in neuronal axons and dendrites, and is phosphorylated at Serine 383 residue in fear conditioning or synaptic plasticity paradigms [1–6]. Phosphorylation of Elk-1 by MAPKs, in particular Serine 383 and Serine 389 within the activation domain, was shown to induce its binding to DNA [7,8].

Elk-1 transcription factor has been widely studied with respect to its mitogen-induced activation through phosphorylation by mitogen-activated protein kinases (MAPKs) and regulation of the *c-fos* promoter in complex with serum response factor (SRF) [9]. However,

Elk-1 and other ternary complex factor (TCF) members have a rather large number of targets, some of which have a high degree of redundancy [10]. A thousand new promoters were identified for Elk-1 binding using a ChIP-chip assay, with two distinct binding modes: SRF-dependent and SRF-independent; furthermore, it was shown that there was a redundancy of promoter occupancy by other ETS proteins in a subset of promoters [10]. Elk-1 was also shown to regulate survival in neuronal cell models by regulating the Survival of Motor Neuron (SMN) promoter as a novel target [11]. CD133, a widely-accepted Cancer Stem Cell (CSC) marker [12–14], was also shown to be regulated through *ets* motifs as well as hypoxia-inducible elements, through the interaction of HIF-1α and Elk-1 on the promoter [15].

Elk-1 was recently found to have both activating and repressive role in human embryonic stem cells (hESCs), particularly through SRF interaction, and found to be upregulated in mesoderm differentiation [16].

In this study, we have first aimed to identify novel targets of Elk-1 using SH-SY5Y neuroblastoma cell line in a transcriptomics approach. We have identified novel pathways and genes that were up- or downregulated upon Elk-1-VP16 overexpression, and when promoters of a subset of these genes were analyzed, several *ets* motifs were identified. Among these, genes related to pluripotency or early neuronal development were particularly interesting, hence we have further analyzed and verified the regulation of a selected set of genes by Elk-1 using qPCR and investigated the regulation of *SOX2*, *NANOG*, and *POU5F1* promoters by Elk-1 and its binding to predicted *ets* motifs in neuroblastoma and glioblastoma (GBM) cell lines. Considering Elk-1 was previously shown to regulate CD133 expression [15], we have also studied Elk-1 expression levels in CD133− and CD133+ cell lines as well as primary brain tumors, indicating Elk-1 was indeed overexpressed in CD133+ cells, and when Elk-1 expression was silenced by RNAi, *SOX2*, and *NANOG* expression were reduced in both CD133+ primary GBMs, as well as CD133+ cell lines in a cell context-dependent manner.

2. Materials and Methods

2.1. Cell Culture and BTIC Isolation from Cell Lines and Primary Tumors

SK-N-BE (2) (ATCC CRL-2271) and SH-SY5Y (ATCC CRL-2266) human neuroblastoma cell lines as well as U-87 MG (ATCC® HTB-14), A172 (ATCC CRL-1620), and T98G (ATCC CRL-1690) human GBM cell lines were used. U87-MG, A172, and T98G cell lines were provided by Assist. Prof. Tugba Bagci Onder from Koc University. For all the stated cell lines for monolayer culture, DMEM high-glucose (4.5 g/L) medium (Gibco, #41966029, Waltham, MA, USA) was used as a basal medium and supplemented with one percent penicillin-streptomycin solution (Gibco, #15140122, Rockville, MA, USA) and 10 percent fetal bovine serum (FBS) (Life Technologies, #10500064, Carlsbad, CA, USA). Cells were grown in 37 °C and 5 percent CO2 incubator.

To form tumorsphere cultures from monolayer cells and support brain-tumor-initiating cells after conducting CD133+ isolation, initial proliferation media (IPM), N2 media, and coated culture plates were used. Plates were prepared by coating with poly-HEMA (poly (2-hydroxyethyl methacrylate) solution. To prepare poly-HEMA solution, 38 mL absolute ethanol was mixed with two mL double distilled water. Following the addition of 1.2 g of poly-HEMA (Sigma Aldrich, #P3932, Taufkirchen, Germany) powder into the mixture, it was placed in a shaker at 37 °C with a vigorous shake for four-five hours until no powder could be seen with the naked eye. This poly-HEMA solution was filtered through a 0.22-micron filter and kept at 4 °C up to six months. Initial proliferation medium (IPM) is necessary for culturing tumorspheres and isolated brain-tumor-initiating cells up to three passages. IPM is made up of neurobasal medium (Gibco, #21103049, Waltham, MA, USA), 1X B27 (Gibco, #17504044, Waltham, MA, USA), 1X GlutaMAX (Gibco, #35050061, Waltham, MA, USA), one percent penicillin-streptomycin solution (Gibco, #15140122, Waltham, MA, USA), 20 ng/mL FGF-2 (Gibco, #13256029, Waltham, MA, USA), and 20 ng/mL EGF (Gibco, #SRP3027, Waltham, MA, USA). N2 medium is necessary for culturing spheroids and isolated brain-tumor-initiating cells over three passages. N2 medium is made up of neurobasal medium (Gibco, #21103049), 1X N2 (Gibco, #17502048, Waltham, MA, USA), 1X GlutaMAX (Gibco, 35050061, Waltham, MA, USA), one percent penicillin-streptomycin solution (Gibco, #15140122, Waltham, MA, USA), 20 ng/mL FGF-2 (Gibco, #13256029, Waltham, MA, USA), and 20 ng/mL EGF (Waltham, MA, USA, #SRP3027, Waltham, MA, USA).

For brain-tumor-initiating cells' (BTICs) isolation from cell lines, SK-N-BE (2) neuroblastoma cells were grown as monolayer cells up to 80 percent confluency, and on the day of isolation, the media was removed, cells were washed with five mL PBS/flask, and three mL of StemPro Accutase/flask was added onto the cells. The suspension was centrifuged at $300\times g$ for five minutes. The cells were resuspended with MACS buffer [two percent bovine serum albumin (BSA), two mM EDTA, and phosphate-buffered saline (PBS) pH 7.2]. To prevent the clogging of the columns at the ongoing isolation procedure, cells were passed through the first 70-micron cell strainer several times until they could pass freely through it. Then, they were passed through a 30-micron filter several times, cell aggregates were removed, and the single-cell suspension was prepared. Cells could be counted at this stage of the procedure. The viable cell number was determined by staining the cells with 0.4 percent Trypan Blue Solution (Gibco, #15250061, Waltham, MA, USA). To continue, cells were centrifuged at $300\times g$ for 10 min, and the supernatant was removed. Cells were resuspended in 60 µL MACS buffer/10^7 cells and 20 µL FcR Blocking Agent/10^7 cells, and 20 µL CD133 Microbeads/10^7 cells were added (CD133 MicroBead Kit—Tumor Tissue, human, Miltenyl Biotec, #130-100-857, Gladbach, Germany). The cells were incubated at 4 °C for 30 min at a constant, slow rotation (12 rpm). Following incubation, two mL buffer/10^7 cells were added to wash the cells, and then, they were centrifuged at $300\times g$ for 10 min again. The supernatant was aspirated, and the pellet was resuspended in 500 µL MACS buffer/10^7 cells and continued with magnetic separation part.

MACS MS column (Miltenyl Biotec, #130-042-201, Germany) was placed on the MACS Mini Separation stand and was equilibrated with 500 µL MACS buffer. The cells prepared in the previous step were loaded onto that column, and with gravity effect, the suspended cells flow through the column for positive selection. That is, the cells labeled for CD133 (CD133+) were kept in the column, while marker-free cells (CD133−) would not bind to the column and were collected in a tube. The column was washed three times with 500 µL of the buffer to wash column-retaining CD133+ cells, the flowing liquid was collected again, and the resulting cells were combined to assemble CD133− cells. For elution of CD133+ cells, the column was separated from the magnetic stand and allowed to stand in the non-magnetic field for about two minutes and flushed out with one mL MACS buffer with the supplied plunger. Cells were counted with Trypan Blue, centrifuged for five minutes at $150\times g$ and resuspended in complete IPM and cultured for 7–10 days in a humidified incubator at 37 °C and five percent CO_2, replacing the medium with freshly prepared IPM every three–four days until their size reached 200 microns, or they started dying from the center. When they reached the limitations, they were passaged. For passaging the cells, the suspension cells were collected from the dishes to a falcon and centrifuged at $300\times g$ for 10 min. Following the centrifugation, the medium was aspirated, and one mL StemPro Accutase Cell Dissociation Reagent (Gibco, #A1110501, Waltham, MA, USA) was added onto the cells, and cells were incubated at 37 °C for five minutes. Cells were triturated about 40 times until the spheroids become single-cell suspension. Onto this single-cell suspension, five mL of PBS with antibiotics was added. Cells were counted at this stage if necessary or to continue cells were centrifuged at $300\times g$ for five minutes. The cells were resuspended in complete IPM at the proper volume.

2.2. Dissociation and Culture of Primary GBM Tissue

Human GBM samples were obtained from consenting patients, as approved by the Hamilton Health Sciences/McMaster Health Sciences Research Ethics Board. Brain tumor samples were dissociated as previously described [17] and cultured as neurospheres in Neurocult complete (NCC) media, a chemically defined serum-free neural stem cell medium (STEMCELL Technologies, Vancouver, BC, Canada), supplemented with human recombinant epidermal growth factor (20 ng/mL: STEMCELL Technologies, Vancouver, Canada), basic fibroblast growth factor (20 ng/mL; STEMCELL Technologies, Canada), heparin (2 µg/mL 0.2% Heparin Sodium Salt in PBS; STEMCELL Technologies, Vancouver, Canada), antibiotic-antimycotic (10 mg/mL; Wisent Bioproducts, Saint Bruno, QC, Canada) in ultra-low attachment plates (Corning, New York, NY, USA). Primary GBM cells (BT 428, BT 458 and BT 624) were cultured in NSC complete media and flow-sorted for CD133+ and CD133− populations as described previously [18,19]. Transfections were carried out by Lipofectamine 2000 as per the manufacturer's instructions.

2.3. Transient Transfection of Cells

For transfection of adherent cells, single-cell suspensions of adherent cell cultures were prepared and seeded at 0.3–0.6×10^6 cells/cm^2 density in complete DMEM medium, and they were incubated in 37 °C, five percent CO_2 incubator, so that they would be 85–90 percent confluent at the time of transfection. On day one, for the formation of the carrier liposome complex, the desired plasmid and PEI were mixed at the determined ratio for each cell line in serum-free DMEM and incubated at room temperature for 20 min. At the end of the period, a complete DMEM medium with 10 percent FBS was added to the mixture at half the volume of the mix. Two hours later, complete DMEM medium containing 10 percent FBS was added to the wells/dishes and the cells were incubated for 48 h in 37 °C, five percent CO_2 incubator for the transgene expression. Cells were transfected with empty pCDNA3 or pCMV plasmids, pCMV-Elk-1 and pRSV-Elk-1-VP16 (courtesy of Prof. A.D. Sharrocks) using the PEI reagent (CellnTech), in 3 replicas per sample. psiSTRIKE hMGFP-scrRNA (from here on referred to as scrRNA) and psiSTRIKE hMGFP-siElk-1 (from here on referred to as siElk-1) has been described elsewhere [11].

For transfection of BTICs, the suspension cells were collected from the dishes to a falcon and centrifuged at $300\times g$ for 10 min. Following the centrifugation, the medium was aspirated, and one mL StemPro Accutase Cell Dissociation Reagent (Gibco, #A1110501) was added onto the cells, and cells were incubated 37 °C for five minutes. Cells were triturated for about 40 times until the spheroids become single-cell suspension. Onto this single-cell suspension, five mL of PBS with antibiotics was added and centrifuged at $300\times g$ for five minutes. The cells were resuspended in complete IPM at the proper volume. Cell density and Lipofectamine 2000 (Thermo Fischer Scientific, Waltham, MA, USA) ratio were determined. Cells were seeded at 0.3–0.6×10^6 cells/cm^2 density in complete IPM without antibiotics on the day of transfection. Following the cell seeding, Lipofectamine 2000 and the nucleic acids were diluted in neurobasal medium without antibiotics, incubated at room temperature for five minutes, then the diluted Lipofectamine 2000 was gently combined with the dilute nucleic acids, and the mixture was incubated at room temperature for 20 min to form liposome. Then, the mixture was added directly onto wells containing cells, and the cells were incubated for 24–72 h in 37 °C, 5% CO_2 incubator for the transgene expression

2.4. Soft Agar Assay

For softy agar assay, 100 cells in 100 µL IPM and an equal volume of 2.8% low-melting-point (LMP) agarose solution were mixed to generate 1.4% agarose-cell solution per well in a 96-well plate, and the mixture was incubated at 37 °C, 5% CO_2 incubator for 14 days. At the end of 14 days, colonies were counted under a $10\times$ magnification or sterco microscope. For staining, crystal violet was dissolved in PBS with two percent ethanol at a final concentration of 0.04 percent, filtered with 0.45 µm filter, and dishes were stained with 50 µL of this solution for one hour at room temperature. The plates were checked every ten minutes to prevent the staining of the background. Then, the staining solution was removed carefully, and the wells were washed with water three times for 30 min. At the last wash, water was kept in the wells overnight to remove the background. The assay was performed in quadruplicate; colonies ≥ 20 µm were counted and analyzed using MS Excel software; results were reported as mean $\pm$ standard deviation.

2.5. Limiting Dilution Analysis (LDA)

Limiting dilution analysis (LDA) has been extensively used to find out differences within multiple groups for a particular trait. In our case, LDA was used for determining the cancer cell initiating frequency of CD133+ and CD133− SKNBE (2) cells; in other words, to evaluate the self-renewing capacity of BTICs. For LDA, following the BTIC isolation procedure, cells were counted so that 10,000 cells/50 μL complete IPM would be present in the first tube. Through serial dilution by factor two up to 1 cell/50 μL, cells were seeded on poly-HEMA coated 96-well plates. For each condition/cell number, samples were seeded in quintuplet. Twenty-five microliters of culture media were added to each well every three–four days, and cells were examined for the presence/absence of spheres and quantified on day 10.

2.6. RNA Isolation, cDNA Synthesis, Reverse Transcription Polymerase Chain Reaction (RT-PCR), and Real-Time PCR

PureLink RNA Mini Kit (Life Technologies Ambion, #12183-018) and PicoPure RNA Isolation Kit (Arcturus, #KIT0202) were used for RNA isolation throughout the experiments. In summary, adherent cells grown in cell culture plates (usually $1.5x10^6$ cells/10 cm culture dish) were washed with cold PBS; then, the resuspended cells were centrifuged at $300\times g$ for 5 min at 4 °C. An amount of 0.3–0.6 mL of lysis solution with beta-mercaptoethanol was added onto the cells depending on the number of cells, and they were mechanically burst and homogenized by triturating through an insulin syringe 15 times. The cells were centrifuged at $2000\times g$ for five minutes at 4 °C, followed by the addition of 70 percent ethanol equal to the volume of the present cell lysate. The lysates were transferred to the filter cartridges and were centrifuged at $12,000\times g$ for 30 s. This step was repeated until the whole sample was finished, and the washing process was started. For washing, 700 μL of wash buffer I was added and centrifuged at $12,000\times g$ for 15 s. Following the first washing step, 500 μL of wash buffer II was added and repeated twice after centrifugation at $12,000\times g$ for 30 s. Tubes were centrifuged for two minutes to dry the membrane. In the elution stage, the cartridges were transferred to new Eppendorf tubes, and depending on the starting number of cells, 20–35 μL of nuclease-free water was put onto the membrane surface and incubated for three minutes at room temperature. Total RNA isolation was completed with centrifugation at $12,000\times g$ for one minute. The concentrations of RNA samples obtained were determined with NanoDrop Spectrophotometer (Thermo Fisher Scientific, Paisley, UK), and the samples were stored at −80 °C in the presence of RNase inhibitors or used for further experiments.

Following total RNA isolation, cDNA synthesis was performed using modified MMLV-derived reversible transcriptase using the iScript cDNA Synthesis Kit (BioRad, #1708891, Hercules, CA, USA). For this purpose, a maximum of one μg total RNA sample was diluted to a maximum volume of 15 μL. The RNA sample was denatured at 70 °C for five minutes and centrifuged briefly. After the addition of the 5X reaction buffer and iScript reversible transcriptase, the mix was ready for the cycling. Prepared cDNA samples were diluted with nuclease-free water to the desired concentration immediately before use in qPCR and/or stored at −20 °C for a maximum of one month.

PrimerQuest (Integrated DNA Technologies, IDT, Coralville, IA, USA), a free online software, was used for the qPCR primer design. The mRNA sequences of the target genes were obtained from the NCBI Gene (https://www.ncbi.nlm.nih.gov/gene/, accessed on 20 January 2021) database, the exon regions of the respective genes were determined, and the primers were designed to be at the exon–exon boundary (if possible). Potential primer pairs were evaluated for GC content, melting temperatures (Tm), and the hairpin formation and appropriate primers were determined. The NCBI BLAST database (https://blast.ncbi.nlm.nih.gov/, accessed on 20 January 2021) was used to check the specificity of the designed primers. The designed primers are listed in Table 1.

Table 1. qPCR primers used.

Gene	Site	Sequence (5′-3′)
GAPDH	Frw	CAT CTT CCA GGA GCG AGA TCC
	Rev	AAA TGA GCC CCA GCC TTC TCC
ACTB	Frw	ACG AAA CTA CCT TCA ACT CC
	Rev	GAT CTT GAT CTT CAT TGT GCT GG
ELK1	Frw	GCT TCC TAC GCA TAC ATT GAC C
	Rev	ACT GGA TGG AAA CTG GAA GG
SOX2	Frw	GGG AAA TGG GAG GGG TGC AAA AGA GG
	Rev	TTG CGT GAG TGT GGA TGG GAT TGG TG
POU5F1	Frw	AAG GAT GTG GTC CGA GTG TGG
	Rev	CCT GAG AAA GGA GAC CCA GCA G
NANOG	Frw	TTC AGA GAC AGA AAT ACC TCA GCC
	Rev	CCT TCT GCG TCA CAC CAT TGC
WNT3A	Frw	GACAAAGCTACCAGGGAGTC
	Rev	CTGCTGCAGCCACAGAT
IRAK3	Frw	ACATACTAGAGTTGGCTGCATATT
	Rev	TGTCACCTACACACTGCAATC
MEF2B	Frw	CAACCGCCTCTTCCAGTATG
	Rev	TCAGCGTCTCGAGGATGT
TCF7L1	Frw	TGAGCGTGAAATCACCAGTC
	Rev	TGGCCCTCATCTCCTTCATA
RHO	Frw	CATGATGAACAAGCAGTTCCG
	Rev	AGAGTCCTAGGCAGGTCTTAG
HES7	Frw	CGGGATCGAGCTGAGAATAG
	Rev	GTTCCGGAGGTTCTGGTC
NOTO	Frw	GCTGGAAGAGTTGGAGAAAGT
	Rev	ACTCTCACCTGGTTCTCTGTA
SIX3	Frw	CAGCAAGAAACGCGAACTG
	Rev	GTGCTGGAGCCTGTTCTT
CREB3	Frw	ACCTGCATCTTGGTCCTACTA
	Rev	GGACAACACTCCATGCTCAG
CREM	Frw	ATCCCAGCATGATGGAAGTATAA
	Rev	ATTGCTGCTACCTGAGCTAAA
LIFR	Frw	GCTCTGGACAAGTTAAATCCATAC
	Rev	CCCTTTGAAGGACTGGCT
FRZB	Frw	AAGTTAAGCGCTGGGATATGA
	Rev	GGGATTTAGTTGCGTGCTTG
GLUT3	Frw	AGCTCTCTGGGATCAATGCTGTGT
	Rev	ATGGTGGCATAGATGGGCTCTTGA
RXRB	Frw	GATGTGAAGCCACCAGTCTTAG
	Rev	GTAGTGTTTGCCTGAGCTTCT

Table 1. *Cont.*

Gene	Site	Sequence (5′-3′)
NODAL	Frw	TACATCCAGAGTCTGCTGAAAC
	Rev	CTAGGAGCACTCTGCCATTATC
PAX6	Frw	GTGAATGGGCGGAGTTATGA
	Rev	ATGAGTCCTGTTGAAGTGGTG
GSK3B	Frw	CCGAGGAGAACCCAATGTTT
	Rev	GCCAGCAGACCATACATCTATAC
FGF11	Frw	CAAAGGCATCGTCACCAAAC
	Rev	GATCAGGTTGAAGTGGGTGAA
FRIT1	Frw	GTGCAGGAAACCGAGTAGAA
	Rev	GCGCCTTTAGAGTGAGTGAA
GLI4	Frw	CTCGGAAGGTCCCAGGT
	Rev	CCCGGTGATGAGAGACTGA
BRACHYURY	Frw	GTAAACTCCACCAGTCCTACTTT
	Rev	TCTGTCCTTAACAGCTCAACTC
NOTCH4	Frw	GAGGATATCGATGAGTGCAGAAG
	Rev	TTCAAAGCCTGGGAGACAC
ZIC1	Frw	GAGCGACAAGCCCTATCTTT
	Rev	GGATTCGTGGACCTTCATGT
ARC	Frw	TCAGCTCATGACTCACCCA
	Rev	CTTGAGACCTGTTGTCACTCTC
ALS2	Frw	GGACTCAAAGAAGAGAAGCTCAA
	Rev	TGGCAATCTCTCTGGTGTTATG
SOX10	Frw	CTTCATGGTGTGGGCTCA
	Rev	CGTTCAGCAGCCTCCAG
SMAD6	Frw	CCTACCGTGTGCTGCAA
	Rev	GGAATCGGACAGATCCAGTG
NGFR	Frw	CATAGCCTTCAAGAGGTGGAAC
	Rev	CACTGTCGCTGTGGAGTTT
MAPK6	Frw	AGGAGCTTCTCAGCGTAATTC
	Rev	CCAGGAAATCCAGTGCTTCT
NGFR	Frw	CATAGCCTTCAAGAGGTGGAAC
	Rev	CACTGTCGCTGTGGAGTTT
CD133	Frw	GCGTCTTCCTCATGGTTGGAG
	Rev	CTTGCTCGTGTAAGGTTCACAG

All the qPCR experiments were performed using SSOAdvanced Universal SYBR Green Supermix (Biorad, #1725274, Hercules, CA, USA) and Applied Bioscience StepOne Plus Real-Time PCR System (Thermo Fisher Scientific, Waltham, MA, USA); essentially, 1–10 ng cDNA was used as template; primers were used at 300 nM each, and the reaction was carried out at 60 °C for 40 cycles. The differences between the expression of target genes were normalized by the expressions of *β-actin* and *gapdh* genes. Each setup was prepared in triplicate and analyzed by the $\Delta\Delta$CT method as described previously [20]. The fold changes in target gene expressions were calculated based on the mean of the reference gene expression and logarithm-transformed. All qPCR experiments were repeated at least 3 times, unless otherwise noted. The mean and standard deviation values were calculated for each group, and the differences in the gene expression levels were determined considering the control group. For statistical analysis, depending on the context, one-way ANOVA with Tukey post hoc test or Student's t-test depending on the context with Prism 5 GraphPad software was used. p-value under 0.05 was considered statistically significant.

Total RNA was extracted using a Norgen Total RNA isolation kit and quantified using the NanoDrop Spectrophotometer ND-1000. Complementary DNA was synthesized from 0.5–1 µg RNA by using qScript cDNA Super Mix (Quanta Biosciences, Beverly, MA, USA) and a C1000 Thermo Cycler (Bio-Rad, Hercules, CA, USA) with the following cycle parameters: 4 min at 25 °C, 30 min at 42 °C, 5 min at 85 °C, hold at 4 °C. qRT-PCR was performed by using Perfecta SybrGreen (Quanta Biosciences, Waltham, MA, USA) and an Opticon Chroma4 instrument (Bio-Rad, Hercules, CA, USA). Gene expression was quantified by using Opticon software, and expression levels were normalized to 28srRNA expression. For statistical analysis, multiple Student's t-tests with Prism 5 GraphPad software were used. p-value under 0.05 was considered statistically significant.

2.7. Microarray and Data Analysis

For microarray analysis, SH-SY5Y cells were transfected with Elk-1-VP16 expression plasmid or empty pCDNA3 plasmid as described above, and 48 h after transfection, RNA samples were isolated using Ambion Tri-pure RNA isolation kit, checked for quality, converted to cDNA, and confirmed for Elk-1 expression as described above. Thereafter, RNA was converted to cDNA using the Superscript Double-Stranded cDNA Synthesis (Invitrogen, Carlsbad, CA, USA) Kit and labeled with NimbleGen One Color DNA Labeling (NimbleGen, Roche, Madison, WI, USA). The labeled cDNA was hybridized to NimbleGen Human Gene Expression Array 12x135K (NimbleGen, Roche, Wisconsin, USA), which covers 45.033 genes with 3 probes per gene, containing 12 arrays per slide. After hybridization, slides were scanned using Genepix 4000B scanner and analyzed with NimbleScan 2.5 software using three arrays from the pCDNA3-transfected cell as reference samples. The expression datasets were normalized using the Robust Multi-Array Average expression measure [21], and differentially expressed genes (DEGs) and their fold-changes were identified from the normalized expression values using two-tailed Student's t-test assuming equal variances and Benjamini-Hochberg's method as the multiple testing option to control the false discovery rate. An adjusted p-value threshold of 0.15 was used to determine the statistical significance of differential expression. The dataset is accessible from EBI ArrayExpress, with the accession number of E-MTAB-9938.

Gene IDs were converted to official gene symbol, and gene set enrichment analyses of DEGs were performed through ConsensusPathDb (r.32) [22] using KEGG [23], Reactome [24], and Biocarta [25] as the data source for molecular pathways, and Gene Ontology Biological Process annotations [26] as the data source for biological processes. Whole-genome annotation for the human genome was used as the background reference set. p-values were determined through a modified Fisher exact test and adjusted via Benjamini-Hochberg's method. A threshold of adjusted p-value < 0.05 was used to determine the statistical significance of the enrichment results. Besides, to characterize the molecular functions of each gene product, and their association with diseases, we manually searched GeneCards Human Gene Database [27].

2.8. Promoter Clonings and Site-Directed Mutagenesis

To identify the putative Elk-1 transcription factor binding sites in selected stemness gene promoters (*SOX2, NANOG, POU5F1*), the Cold Spring Harbor Laboratory—Transcriptional Regulatory Element Database (TRED), Swiss Institute of Bioinformatics—The Eukaryotic Promoter Database (EPD), and Alggen-Promo algorithmic analysis program were used. The promoter sequences that correspond to the genes of interest were retrieved from either the Transcriptional Regulatory Element Database (TRED) (http:

//rulai.cshl.edu/cgi-bin/TRED/tred.cgi?process=home, accessed on 20 January 2021), or the Eukaryotic Promoter Database (EPD) (http://epd.vital-it.ch/, accessed on 20 January 2021). The obtained promoter sequences were analyzed with Promo 3.0 (http://alggen.lsi.upc.es/cgi-bin/promo_v3/promo/promoinit.cgi?dirDB=TF_8.3, accessed on 20 January 2021). The promoter binding regions for transcription factors can be analyzed by the Promo 3.0 tool, and the results are displayed as "dissimilarity rate". The dissimilarity matrix expresses the similarity pair to pair between Elk-1 DNA binding sequence and the putative sequences at analyzed genes. From this point of view, the smaller dissimilarity rates are the indicators of a higher possibility for the interaction between Elk-1 and the promoter of interest. The binding ability of Elk-1 to the predicted sites on the promoters could be confirmed by luciferase and chromatin immunoprecipitation assays, thereby verifying the microarray results (Table 2).

Table 2. Number of ets motifs predicted on selected promoters and their dissimilarity score (DS) range; DS of 0% means perfect match to consensus; TRED, Transcriptional Regulatory Element Database; EPD, Eukaryotic Promoter Database. *

	Number of Predicted *ets* Binding Motifs with Different Dissimilarity Scores (DS) in Promo 3.0		
	DS: 0–1 Percent	**DS: 1–5 Percent**	**DS: 5–10 Percent**
SRF	-	2	1
MCL1	2	-	3
LIF	-	2	1
SOX2 (TRED)	-	1	1
NANOG (TRED)	-	1	-
POU5F1 (TRED)	-	1	2
SOX2 (EPD)	1	2	3
NANOG (EPD)	-	1	1
POU5F1 (EPD)	1	2	1

* For dissimilarity scores of individual ets motifs, see Supplementary Table S3 for URL of databases, please refer to text.

Cloning primers for human *SOX2*, *NANOG*, and *POU5F1* promoters were designed and analyzed with NetPrimer (http://www.premierbiosoft.com/netprimer/, accessed on 20 January 2021) and PrimerBlast (http://www.ncbi.nlm.nih.gov/tools/primer-blast/, accessed on 20 January 2021) softwares. The designed cloning primers are listed in Table 3. Gradient PCR with five different annealing temperatures was performed to detect the optimum annealing temperature of the primers. The PCR reactions were prepared with i-Taq DNA Polymerase (Intron, #25024, Seoul, Korea) kit using the genomic DNA isolated from the SH-SY5Y cell line as a template. After optimization, the preparation of the insert was carried out using Pfu DNA Polymerase (Thermo Scientific, #EP0571, Waltham, MA, USA) suitable annealing temperatures as indicated in text for 30 cycles. Following amplification, PCR products were purified by PureLink PCR Purification Kit (Invitrogen, # K3100-01) and cloned into pGL3luciferase reporter plasmid.

Intentional deletion mutations were made on cloned promoter sequences with site-directed mutagenesis (SDM). The promoter sequences were analyzed with Promo 3.0, as stated previously. Potential Elk-1 binding sites on stemness promoters were chosen according to the dissimilarity rate Promo 3.0. Accordingly, *ets1* motif on *NANOG* promoter, *ets1* and *ets2* motifs on *SOX2* promoter, and *ets1*, *ets2*, and *ets3* motifs on *POU5F1* promoter were deleted in corresponding pGL3 luciferase reporter constructs. SDM primers were designed using the NEB Base Changer (http://nebasechanger.neb.com/, accessed on 20 January 2021) website, and Q5® Site-Directed Mutagenesis Kit Protocol (NEB, #E0554) was followed for the mutations. The primer pairs designed flanking the region to be deleted and eventually forming deletion mutants from the cloned promoter sequences are given in Table 4. The mutagenesis was carried out according to the manufacturer's instructions.

Table 3. Cloning primers for chosen Homo sapiens stemness gene promoters.

Promoter	Forward Primer (5′-3′)	RE Site	Reverse Primer (5′-3′)	RE Site	Product (bp)
POU5F1	AGACggtaccAGGGCTG TTGGCTTTGGACA	*Kpn*I	CTGTagatctAGCCATTTAA GAATTCCAGAGTAGG	*Bgl*II	993
SOX2	CTGTggtaccGGGGAGTG ATTATGGGAAGAA	*Kpn*I	CTGTagatctCACTAGACTG TCTTCATTCAACCGTAGC	*Bgl*II	993
NANOG	CTGTggtaccTTTCTGCC TAAACTAGCCA	*Kpn*I	CTGTagatctAGGTGAAGA TTCTTTACAGTCG	*Bgl*II	988

Table 4. Primers used for site-directed mutagenesis of cloned promoters.

Primer	Site	Oligo	Length	Tm (°C)	Ta (°C)
*NANOG-ets*Δ	Fwd	TACTAACATGAGTGTGGATC	20	59	58
	Rev	AGGAGGAAAAAATTTAAGAGG	21	57	
*POU5F1- ets*Δ1	Fwd	CCTTTCCCCCTGTCTCTG	18	64	65
	Rev	CAGGGAAAGGGACCGAGG	18	68	
*POU5F1- ets*Δ2	Fwd	GAATTGGGAACACAAAGG	18	57	57
	Rev	TGAATGAAGAACTTAATCCC	20	56	
*POU5F1- ets*Δ3	Fwd	GTGAAGTTCAATGATGCTCTTG	22	61	62
	Rev	AACCAGTTGCCCCAAACT	18	64	
*SOX2- ets*Δ1	Fwd	TTGAAATCACCCTCCCCC	18	64	65
	Rev	ATCCCACGGCACTGTATG	18	65	
*SOX2- ets*Δ2	Fwd	GTGTCTTTCCCCAGCCCC	18	69	68
	Rev	GGCGCTCAAAAGTGCAGG	18	67	

2.9. Luciferase Reporter Assay

For each cell line, the necessary optimization experiments were performed, and cell numbers and DNA: PEI ratios were determined for co-transfections. For 24-well cell culture plates for luciferase analysis, for SK-N-BE (2), T98G, and A172 cells 80×10^4 cells/well, and for SH-SY5Y and U87-MG cells, 60×10^4 cells/well were seeded with triplicates for each transfection group. The following day, *SOX2*-luc, *NANOG*-luc, *POU5F1*-luc, or one of the deletion mutant of these plasmids, one of the Elk-1 series plasmids (pCDNA3.1, Elk1-VP16, Elk1-EN, siElk1, or scrRNA), Renilla-luc plasmid (pRL-TK (Promega, #E2241, Madison, WI, USA)) and the proper ratio of PEI mixture was prepared. After transfection, the cells were incubated for 42 h in a normoxic medium and subjected to one percent hypoxia for the last six hours for the normoxia–hypoxia experiments. At the end of hypoxia treatment, luciferase analysis was performed with Thermo Luminoskan Ascent device by using Dual-Glo Luciferase kit (Promega, Wisconsin, USA) with some modifications. For luciferase analysis of monolayer cell lines, 48 h of incubation was necessary before performing luciferase analysis.

On the day of the luciferase assay, the medium on the cells was aspirated, and the wells were washed with PBS. Cells were lysed with 100 µL of 5X Passive Lysis Buffer (PLB) (Promega, #E1941, Wisconsin, USA) diluted to 1X. Seventy-five microliters of the cells were transferred to luminometer compatible white-bottomed 96-well plates. To measure the Firefly luciferase activity, 75 µL of Dual-Glo® Luciferase Reagent was added onto the lysed cells. For at least 15 min, the plates were incubated at room temperature, and the luminescence for Firefly luciferase activity was measured. To measure the Renilla luciferase activity, 75 µL of Dual-Glo® Stop&Glo Luciferase Reagent was added to the wells. They were incubated at room temperature for the equal time that was done for Firefly luciferase, and the luminescence for Renilla luciferase activity was measured. Firefly/Renilla ratios were calculated, normalizations were done, and the results were graphed as relative luciferase activity. For statistical analysis, one-way ANOVA with Tukey post hoc test or Student's t-test depending on the context with Prism 5 GraphPad software was used. *p*-value under 0.05 was considered significant.

2.10. Chromatin Immunoprecipitation (ChIP) Assay

In this assay, proteins and interacting DNA are crosslinked with formaldehyde; the chromatin is sheared with either sonication mechanically or micrococcal nuclease enzymatically. The nucleoprotein complex is enriched by immunoprecipitation, and through the reversal of the crosslinking, DNA and the interacting protein are separated. In the end, the interacting DNA fragment is purified and quantified with ChIP-qPCR. To determine the promoter fragment to be amplified in ChIP PCR, Promo3.0 analysis used for predicting *ets* motifs and their dissimilarity scores was used (Supplementary Table S3). The amplicon size was arranged between 75–150 bp; CpG islands were checked for the potential binding sites of Elk-1, and the position of Elk-1 to those sequences was considered for primer design. The UCSC in silico PCR tool was used to verify the amplicon (https://genome.ucsc.edu/cgi-bin/hgPcr, accessed on 20 January 2021); primers used for ChIP PCR are listed in Table 5.

Table 5. The list of primers used in chromatin immunoprecipitation (ChIP) assay.

Name	Primer	Sequence	PCR Product Size
ChIP_MCL1	Frw	GCCGCCCTAAAAACCGTGATA	99
	Rev	CGCCTGGCTGAGAAAACTG	
ChIP_SRF	Frw	TGACAGCAACGAGTTCGGTA	130
	Rev	CCCCCATATAAAGAGATACAATGTT	
ChIP_SOX2_ETS1	Frw	TGGGAGGGAGTTTGTGACT	97
	Rev	AAAGTGCAGGCGATGGG	
ChIP_SOX2_ETS2	Frw	GTGGGATGCCAGGAAGTT	102
	Rev	GTCGTGCGGCTTTCAAATG	
ChIP_SOX2_ETS3	Frw	AGACAGTCTAGTGGGAGATGTG	138
	Rev	CGGACCATAAGGCAGACTCTA	
ChIP_SOX2_ETS4	Frw	CTTATGGTCCGAGCAGGATTT	103
	Rev	TCCCGACTAGAAGTTAGGAGAC	
ChIP_SOX2_ETS5	Frw	CGCACCTTAGCTGCTTCC	143
	Rev	GTCACACCACACGCCTTT	
ChIP_NANOG_ETS1	Frw	CTGGAGGTCCTATTTCTCTAACATC	155
	Rev	ATGCTTCAAAGCAAGGCAAG	
ChIP_NANOG_ETS2	Frw	GCAGAGGAGAATGAGTCAAAGA	131
	Rev	CCCAAACCCAACATTCAAGAAA	
ChIP_NANOG_ETS3	Frw	CTTAGTCCAGCCTGTTCCAAA	136
	Rev	AGTGAAAGACCAAAGGGAAGG	
ChIP_POU5F1_ETS1	Frw	CTTCACTGCACTGTACTCCTC	101
	Rev	CACCTCAGTTTGAATGCATGG	
ChIP_POU5F1_ETS2	Frw	GGAGTTTGTGCCAGGGTT	105
	Rev	CCCTCCAACCAGTTGCC	
ChIP_POU5F1_ETS3	Frw	GTTGGAGGGAAGGTGAAGTT	93
	Rev	TACTGTGTCCCAAGCTTCTTTAT	

Table 5. *Cont.*

Name	Primer	Sequence	PCR Product Size
ChIP_MCL1	Frw	GCCGCCCTAAAAACCGTGATA	99
	Rev	CGCCTGGCTGAGAAAACTG	
ChIP_SRF	Frw	TGACAGCAACGAGTTCGGTA	130
	Rev	CCCCCATATAAAGAGATACAATGTT	
ChIP_SOX2_ETS1	Frw	TGGGAGGGAGTTTGTGACT	97
	Rev	AAAGTGCAGGCGATGGG	
ChIP_SOX2_ETS2	Frw	GTGGGATGCCAGGAAGTT	102
	Rev	GTCGTGCGGCTTTCAAATG	
ChIP_SOX2_ETS3	Frw	AGACAGTCTAGTGGGAGATGTG	138
	Rev	CGGACCATAAGGCAGACTCTA	
ChIP_SOX2_ETS4	Frw	CTTATGGTCCGAGCAGGATTT	103
	Rev	TCCCGACTAGAAGTTAGGAGAC	
ChIP_SOX2_ETS5	Frw	CGCACCTTAGCTGCTTCC	143
	Rev	GTCACACCACACGCCTTT	
ChIP_NANOG_ETS1	Frw	CTGGAGGTCCTATTTCTCTAACATC	155
	Rev	ATGCTTCAAAGCAAGGCAAG	
ChIP_NANOG_ETS2	Frw	GCAGAGGAGAATGAGTCAAAGA	131
	Rev	CCCAAACCCAACATTCAAGAAA	
ChIP_NANOG_ETS3	Frw	CTTAGTCCAGCCTGTTCCAAA	136
	Rev	AGTGAAAGACCAAAGGGAAGG	
ChIP_POU5F1_ETS1	Frw	CTTCACTGCACTGTACTCCTC	101
	Rev	CACCTCAGTTTGAATGCATGG	
ChIP_POU5F1_ETS2	Frw	GGAGTTTGTGCCAGGGTT	105
	Rev	CCCTCCAACCAGTTGCC	
ChIP_POU5F1_ETS3	Frw	GTTGGAGGGAAGGTGAAGTT	93
	Rev	TACTGTGTCCCAAGCTTCTTTAT	

Essentially, cells were seeded in three separate 150 mm cell culture dishes of 2×10^6 cells/dish per experimental group on day zero. On day 1, cells were transfected with either an empty pCDNA3.1 plasmid or an expression plasmid for Elk1-VP16 plasmids and incubated 48 h at 37 °C, 5% CO_2. Cells were then treated with 1% formaldehyde at room temperature for 20 min; glycine was then added to the dishes to a final concentration of 0.125 M and incubated for 5 min at room temperature. The dishes were washed three times with cold PBS on ice and then centrifuged at $400 \times g$ for five minutes at 4 °C with $1 \times$ protease inhibitor cocktail (PIC) (Roche, 4693159001). The supernatant was aspirated, and lysis buffer was added onto the cells with a volume of at least 10 times the pellet obtained. The suspension was incubated on ice for 10 min and passed through an insulin needle 20 times. One volume of the sample was mixed with an equal volume of 0.4 percent Trypan Blue Dye, and the cell nuclei were checked under the microscope. The volume of the sonication buffer to be used to dissolve the pellet was adjusted to $2\text{--}3 \times 10^6$ nuclei/mL and sonicated in the Biorupter UCD-200 Sonicator (Diagenode, Denville, NJ, USA). Following the sonication, cell lysates were centrifuged at $22,000 \times g$ for 20 min at 4 °C to remove insoluble materials. The supernatant was then diluted five-fold with dilution buffer and pre-cleared for 4 h with slow rotation with protein A/G mixture beads. After incubation, the samples were precipitated at $150 \times g$ for 5 min at 4 °C, and 10% of the total supernatant was removed as total input control and kept in −20 °C. The rest of the supernatant was divided into two fractions of the negative control (IgG-mock) and immunoprecipitation (IP) per group.

Sixty microliters of ANTI-FLAG® M2 Affinity Gel (Sigma Aldrich, #A2220, Taufkirchen, Germany) resin per group were washed and equilibrated with five volumes of dilution buffer and centrifuged three times at $400 \times g$ for one minute each at 4 °C. The negative control and

IP fractions separated from the dilution in the previous step were mixed with Protein G-Plus agarose beads and anti-Flag M2 resin, respectively. The tubes were incubated at 4 °C overnight with slow rotation. The following day, the mix was centrifuged at 4 °C and $600 \times g$ for five minutes, and the pellet was collected. The beads were washed with one mL of low salt, high salt, LiCl, and TE buffers at 4 °C with rotation, respectively. Following each of the washing steps, the beads were centrifuged at 4 °C and $600 \times g$ for five minutes.

At the elution step, the inputs that were collected and frozen a day before were thawed and added as the third fraction of each group. After the last wash, 250 µL fresh elution buffer, pre-heated at 65 °C, was added onto the beads, and they were incubated on a shaker for 15 min. The tubes were vortexed with five-minute intervals and then centrifuged at 4 °C and $18,000 \times g$ for five minutes. The supernatant was collected for each fraction of each group, and the elution step was repeated with another 250 µL elution buffer. After elution of the crosslinked DNA–protein complex, 10 µL of RnaseA (10 mg/mL) (Intron, #BR003) and 25 µL of 5 M NaCl was added onto the elutes and incubated for at least five hours or overnight at 65 °C. The following day, 10 µL of 0.5 M EDTA, 20 µL 1 M Tris–HCl (pH 6.5), and two µL Proteinase K (20 mg/mL) (Invitrogen, #25530049, Carlsbad, CA, USA) mix were added and incubated again at 65 °C for two more hours. Using MEGAquick-spin™ Plus Total Fragment DNA Purification Kit (Intron Bio, #17290, Sungnam, Korea), the DNA was cleaned up. The resulting fractions were used for qPCR analysis.

SSOAdvanced Universal SYBR Green Supermix (Bio-Rad, #1725274, Hercules, CA, USA) and Applied Biosciences StepOne Plus Real-Time System were used for qPCR analysis with DNA isolated from ChIP. Ten microliters of PCR reaction were prepared by mixing 2X SSO Advanced Universal SYBR Green Supermix, 300 nM forward and reverse primers each, and 1 µL template. In the analysis phase, qPCR signals obtained from the ChIP samples were normalized by the signals obtained from the input, and the mock samples and the results are presented as fold change. For statistical analysis, one-way ANOVA with Tukey post hoc test or Student's *t*-test depending on the context with Prism 5 GraphPad software was used. *p* value under 0.05 was considered significant.

3. Results

3.1. Microarray Analyses Reveal Novel Targets in Elk-1-VP16 Overexpressing SH-SY5Y Cells

Elk-1 is a ubiquitous transcription factor, yet it has been implicated in different biological processes in the nervous system. In order to identify novel target genes of Elk-1 with respect to survival in neurons, we have overexpressed Elk-1-VP16 constitutively active fusion protein in SH-SY5Y neuroblastoma cells. The comparative analysis of the transcriptome profiles indicated 11,018 differentially expressed genes (DEGs), of which 4212 were downregulated and 6806 were upregulated, when SH-SY5Y neuroblastoma cells were transfected with Elk1-VP16. The gene set enrichment analysis (GSEA) of these genes up- or downregulated by exogenous Elk-1-VP16 presented overrepresentation of quite a high number of biological processes such as anatomical structure development, cell proliferation, single-organism developmental process, developmental growth, and organ and tissue development, including forebrain and midbrain development (Supplement Tables S1 and S2). When a subset of these genes was analyzed further, stemness genes such as *POU5F1*, *SOX2*, and *NANOG*, as well as growth factors and receptors or transcription factors including FGFR1, WNT16, WNT 3, PDGFA, PAX6, PAX7, HIF3A, NOTO, among many others were found to be upregulated, whereas genes such as EGLN2, FEV, JUNB, and GLI4 were found to be downregulated upon overexpression of Elk-1-VP16 (Figure 1A,B).

Prediction of putative Elk-1 binding sites (i.e., ets motifs) on the promoters of these genes was assessed via Alggen PROMO 3.0 online software [28]. Among the genes of interest for which human promoter sequences were available, the analysis was performed for human ELK-1 (TRANSFAC database accession no. T00250) binding, thereby limiting the number of promoters investigated, and out of these, promoters with at least one motif are listed (Supplement Table S3). Among the selected subset of genes, *SOX2* promoter was found to contain one ets motif with a dissimilarity score of 2.16, *NANOG* was found to contain one ets motif with a dissimilarity score of 2.3, and *POU5F1* contained one ets motif with a dissimilarity score of 3.12, among other potential ets binding sites, indicating a high probability of binding (Supplement Table S3). Other promoters of the microarray-determined set of putative Elk-1 target genes, whose promoters contained low dissimilarity score ets motifs, included transcription factors such as *RXRB, TCF7L1, MEF2B, PAX6, SOX10, CREB3, SMAD6, CREM,* and *HES7* and signal transduction pathway elements such as *RHO, IRAK3, WNT3A, LIFR, FRZB, NGFR, MAPK6, NOTCH4, FGF11,* and *NODAL,* among many others (Supplement Table S3).

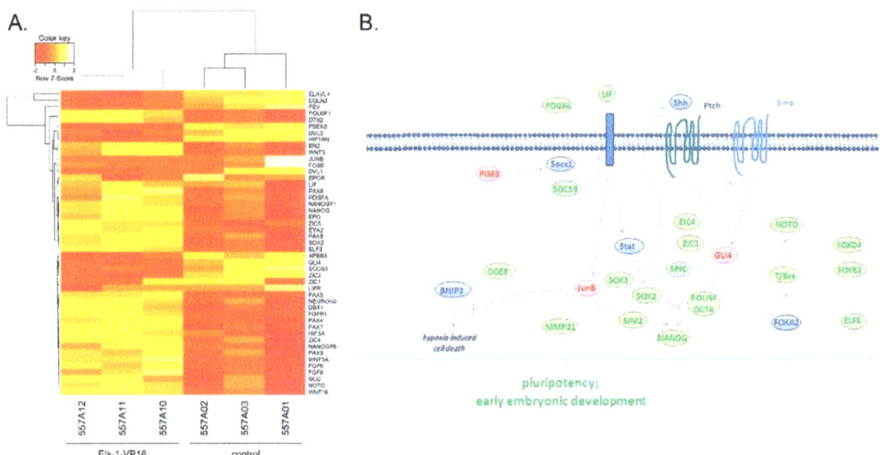

Figure 1. (**A**) Heatmap of a subset of genes regulated by Elk-1-VP16 showing increased (green) or decreased (red) expression in Elk1-VP16 overexpressing SH-SY5Y neuroblastoma cells; 557A10, 557A11, 557A12 correspond to SH-SY5Y cells transfected with Elk-1-VP16 expression plasmid, 557A01,557A02,557A03 control SH-SY5Y cells transfected with empty plasmid; color key shows up- and downregulation levels. (**B**) Schematic representation of the relation between selected genes in pluripotency and early embryonic development pathways that were found to be regulated by Elk-1-VP16 in microarray analysis.

3.2. Regulation of Nervous System Development Related Genes by Elk-1

To validate regulation of selected candidate genes identified through microarray experiments by Elk-1 transcription factor, we have either overexpressed Elk-1-VP16 constitutively active fusion protein or knocked down endogenous Elk-1 expression in SH-SY5Y and SK-N-BE (2) neuroblastoma cell lines and A172 and T98G GBM cell lines (Figure 2).

qPCR results in SH-SY5Y cells were parallel to those obtained from the microarray analysis, especially in the genes related to pluripotency such as SOX2, NANOG, POU5F1, RXRB, GLUT3, TCF7L1, NODAL, and CREB3 (Figure 2A,B). SOX2 was upregulated in SH-SY5Y overexpressing Elk-1-VP16 protein, similar to microarray, but not in other cell types, while it was repressed when Elk-1 was knocked down (siElk-1) in all cell types (Figure 2). Similarly, NANOG and POU5F1 was upregulated in SH-SY5Y cell overexpressing Elk-1-VP16, but downregulated in cells transfected with siElk-1 plasmid (Figure 2A,B), whereas both genes were repressed in SK-N-BE (2) cells overexpressing Elk-1-VP16 and upregulated in siElk-1 knockdown (Figure 2C,D; Table 6), indicating a cell context-dependent regulation. TCF7L1 and NODAL expression increased in Elk-1-VP16 overexpressing SH-SY5Y and SK-N-BE (2) but decreased in siElk-1 silencing; BRACHYURY (T) expression was upregulated in Elk-1-VP16 overexpressing but decreased in siElk-1 silenced SK-N-BE (2) cells (Figure 2; Table 6). GLUT3 expression was upregulated in all cell types overexpressing Elk-1-VP16, and decreased in all cells with siElk-1 silencing, paralleling the microarray results (Figure 2, Table 6). The expression of ARC and CREB3 increased in A172 and T98G cells overexpressing Elk-1-VP16 but decreased in siElk-1 knockdown cells (Figure 2E–H; Table 6). GLI4 and ALS genes increased in A172 cells overexpressing Elk-1-VP16 and decreased with siElk-1 silencing (Figure 2E,F).

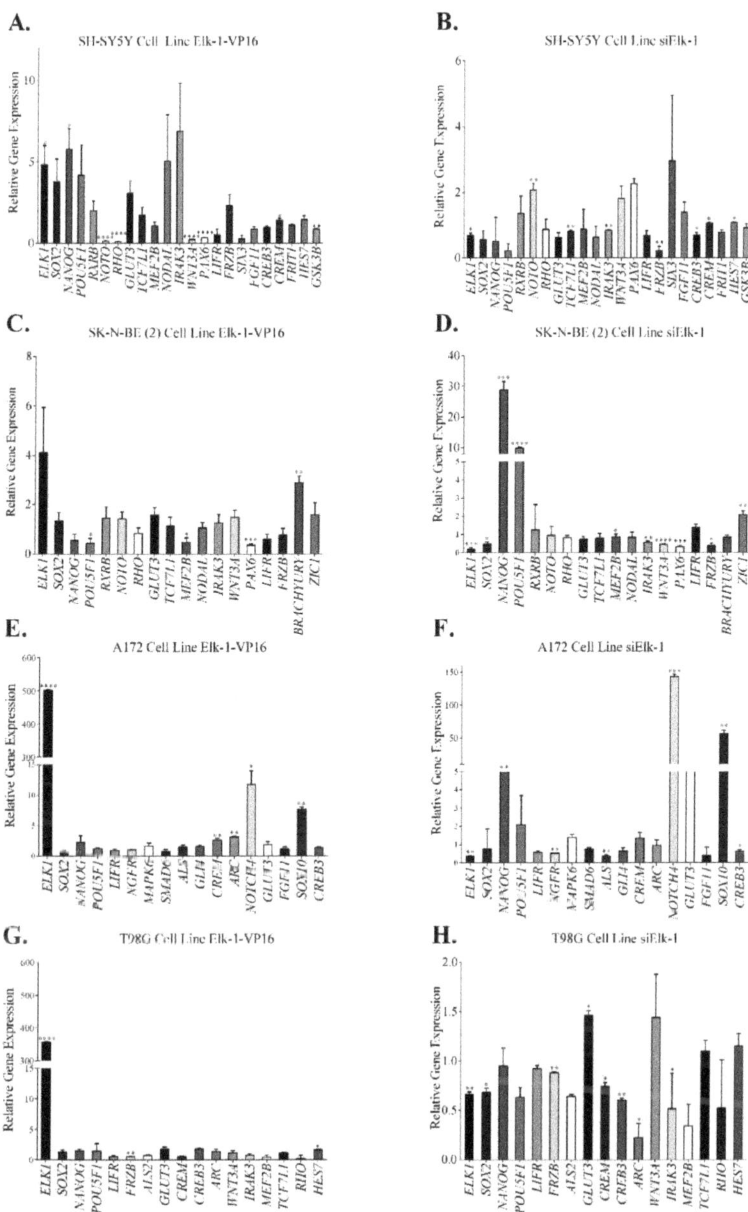

Figure 2. qPCR expression profiles of selected genes in different cell lines upon overexpression of Elk-1-VP16 (**A**,**C**,**E**,**G**) or knockdown of endogenous Elk-1 (**B**,**D**,**F**,**H**). Expression profiles after (**A**). over-expression with Elk1-VP16 and (**B**). after knock-down with siElk1 in SH-SY5Y neuroblastoma cell line; expression profiles after (**C**). over-expression with Elk1-VP16 and (**D**). after knock-down with siElk1 in SK-N-BE(2) neuroblastoma cell line; expression profiles after (**E**). over-expression with Elk1-VP16 and (**F**). after knock-down with siElk1 in A172 GBM cell line; expression profiles after (**G**). over-expression with Elk1-VP16 and (**H**). after knock-down with siElk1 in T98G GBM cell line. Unpaired t-test; **** $p < 0.0001$, *** $p < 0.001$, ** $p < 0.01$, * $p < 0.05$.

Table 6. Summary of qPCR and microarray comparisons of selected potential Elk-1 target genes after Elk1-VP16 overexpression or siElk-1 silencing in neuroblastoma and glioblastoma cell lines.

	Elk-1-VP16 Overexpression					siElk-1 Silencing			
Gene ID	*SH-SY5Y*	*SK-N-BE (2)*	*T98G*	*A172*	*Microarray Data*	*SH-SY5Y*	*SK-N-BE (2)*	*T98G*	*A172*
ALS2	N/A*	N/A	−0.5	0.60	−6.06	N/A	N/A	−0.64	−1.44
ARC	N/A	N/A	0.44	1.67	−6.87	N/A	N/A	−2.17	−0.06
BRACHYURY	N/A	0.32	N/A	N/A	1.54	N/A	−0.25	N/A	N/A
CREB3	−0.01	N/A	0.83	0.46	−1.92	0.09	N/A	−0.74	−0.67
CREM	N/A	N/A	−0.91	1.40	N/A	−0.51	N/A	−0.43	0.44
ELK-1	1.50	2.04	8.48	8.98	13.11	−0.54	−2.36	−0.60	−1.49
FGF11	−0.23	N/A	N/A	0.28	−2.40	0.48	N/A	N/A	−1.31
FRIT1	0.17	N/A	N/A	N/A	1.82	−0.34	N/A	N/A	N/A
FRZB	1.21	−0.72	−0.86	N/A	1.91	−2.20	−1.34	−0.18	N/A
GLI4	N/A	N/A	N/A	0.695	−3.81	N/A	N/A	N/A	0.61
GLUT3	1.62	0.07	0.83	0.92	2.43	−0.67	−0.46	0.54	2.64
GSK3B	−0.22	N/A	N/A	N/A	−1.61	−0.12	N/A	N/A	N/A
HES7	N/A	N/A	0.68	N/A	−1.77	0.12	N/A	0.20	N/A
IRAK3	2.78	0.51	−0.46	N/A	1.70	−0.23	−0.81	−0.97	N/A
LIFR	−0.91	−1.52	−0.84	−0.08	−2.01	−0.53	0.48	−0.12	−0.82
MAPK6	N/A	N/A	N/A	0.76	1.64	N/A	N/A	N/A	N/A
MEF2B	0.09	−0.41	−1.19	N/A	−2.74	−0.19	−0.33	−1.57	N/A
NANOG	1.96	−0.91	0.58	1.18	2.54	−0.99	4.85	−0.07	3.12
NODAL	2.33	0.66	N/A	N/A	1.64	−0.66	−0.26	N/A	N/A
NOTCH4	N/A	N/A	N/A	3.56	3.39	N/A	N/A	N/A	7.16
NOTO	−2.94	0.18	−0.34	N/A	2.15	1.05	−0.08	−0.76	N/A
PAX6	−1.54	−0.28	N/A	N/A	2.61	1.18	−1.62	N/A	N/A
POU5F1	1.62	−1.20	0.56	0.32	3.68	−2.19	3.31	−0.67	1.06
RHO	−3.56	−1.07	−2.26	N/A	2.27	−0.20	−0.30	−0.93	N/A
RXRB	1.01	0.66	0.83	0.46	5.95	0.45	0.31	N/A	N/A
SIX3	−1.94	N/A	0.61	N/A	−5.72	1.56	N/A	0.41	N/A
SMAD6	N/A	N/A	N/A	−0.29	−3.78	N/A	N/A	N/A	−0.42
SOX2	1.54	0.41	0.37	−0.59	2.75	−0.84	−1.08	−0.56	−0.36
SOX10	N/A	N/A	N/A	2.94	2.41	N/A	N/A	N/A	5.83
TCF7L1	0.79	0.56	0.15	N/A	2.34	−0.29	−0.25	0.14	N/A
WNT3A	−2.11	0.53	0.11	N/A	2.25	0.86	−1.11	0.52	N/A
ZIC1	N/A	1.53	N/A	N/A	2.26	N/A	1.07	N/A	N/A

* N/A: the expression level is not available.

The promoters of a subset of genes have been selected for chromatin immunoprecipitation to address whether predicted binding sites were indeed binding to Elk-1 (Figure 3). To that end, we have transfected SK-N-BE (2) neuroblastoma and T98G GBM cell lines with Elk-1-Flag expression vector and pulled down exogenous Elk-1 using Flag-agarose beads. The known targets *SRF* ($p = 0.0451$) and *MCL1* ($p = 0.0102$) showed significant binding in SK-N-BE (2) cells, but the binding was not statistically significant in T98G cells (Figure 3A). Among the novel promoters identified in this study, *GLUT3* promoter showed Elk-1 binding in both cell types, albeit not to the same extent, while *KLF4* ($p = 0.0496$) only showed significant binding in T98G cells (Figure 3B). LIF1, however, did not show significant Elk-1 binding in either cell type.

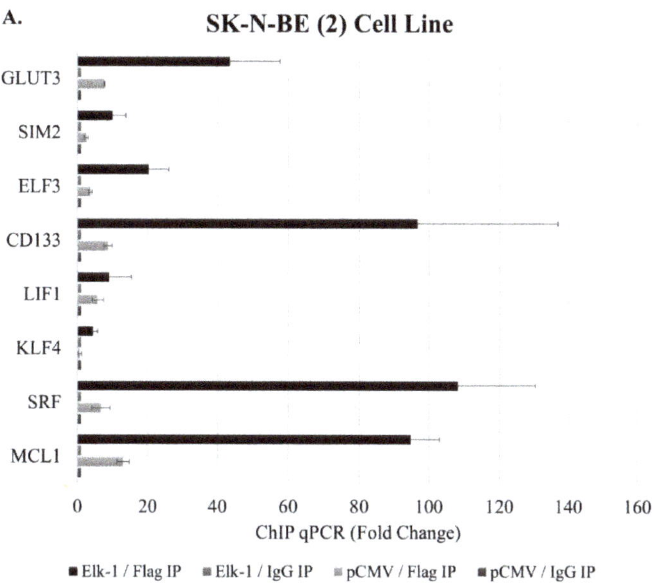

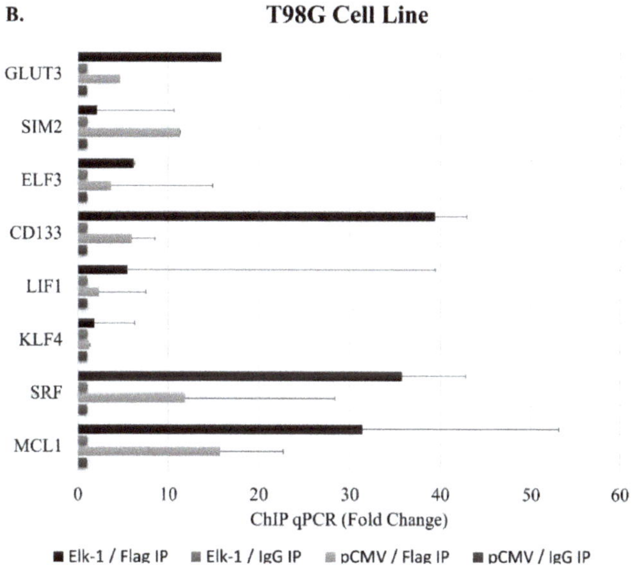

Figure 3. Chromatin immunoprecipitation assay for the identification of Elk-1 binding sites on the target gene promoters in pCMV-transfected (pCMV) vs. Elk-1 over-expressing cells (Elk-1) in (**A**). SK-N-BE (2) cells and (**B**). T98G cells. Lysates were immunoprecipitated with either Flag antibody (Flag IP) for exogenous Elk-1 or IgG (IgG IP) as control. qPCR results were analyzed as explained in Materials and Methods and reported as average fold change.

3.3. Regulation of SOX2, NANOG, and POU5F1 by Elk-1 in CD133+ Cells

Since Elk-1 was previously shown to be important in human embryonic stem cells (hESCs) maintenance of self-renewal capacity through co-occupation of promoters with *ERK2* [29], and to regulate the promoter of CD133, a cell surface protein commonly used as a cancer stem cell marker [15], we addressed whether the cell context-dependent regulation was due to heterogenous nature of some cell lines used in terms of their tumorsphere forming abilities. SK-N-BE (2) neuroblastoma cells were shown to form CD133+ tumorspheres, unlike SH-SY5Y cells, hence we have first sorted CD133− and CD133+ SK-N-BE (2) cells and showed that expression of *CD133, ELK-1, SOX2, NANOG,* and *POU5F1* were all significantly more in CD133+ cells than in CD133− cells (Figure 4A). Intriguingly, ELK-1 levels increased in different passages (p1 and p2) of CD133+ sorted cells, while *CD133* levels declined with each passage; *NANOG* and *POU5F1* levels also increased slightly in p2 cells, albeit not significantly (Figure 4B). Both passages (p1 and p2) of CD133+ SK-N-BE (2) cells were shown to be Nestin+ (data not shown). To address whether this coexpression of ELK-1 with stemness genes studied is through direct regulation, we have silenced endogenous Elk-1 expression in CD133+ SK-N-BE (2) cells, and observed that *NANOG* and *SOX2* but not *POU5F1* were downregulated significantly upon silencing (Figure 4C). It must be noted, however, that overexpression of Elk-1-VP16 in CD133− cells did not yield upregulation of *NANOG, SOX2* or *POU5F1* in SK-N-BE (2) cells (data not shown).

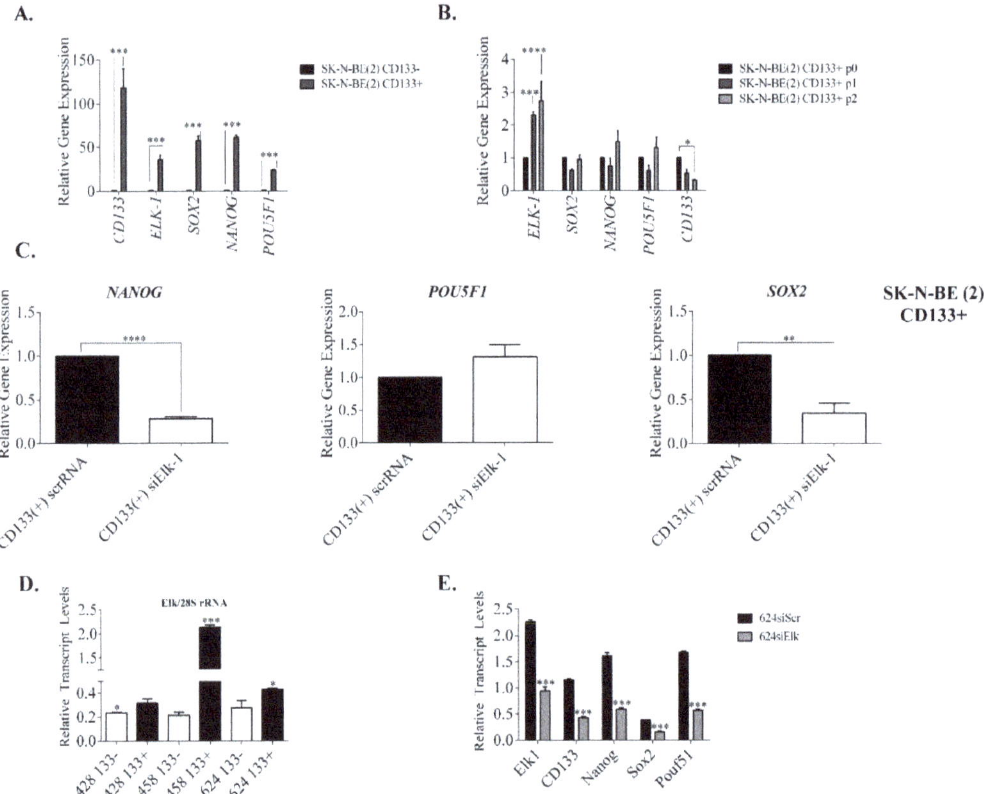

Figure 4. qPCR expression profiles of stemness genes in CD133− vs. CD133+ SK-N-BE(2) cells and in primary brain tumors. (**A**). Stemness gene expression analysis of SKNBE(2) passage 0, passage 1, and passage 2 cells (*** $p < 0.001$, two-way ANOVA w/Dunnett multiple comparison test); (**B**). Stemness gene expression analysis of SKNBE(2) CD133+ BTICs vs. CD133− spheroids (unpaired t-test, * $p < 0.05$, *** $p < 0.001$, **** $p < 0.0001$); (**C**). *left, NANOG, middle POU5F1* and *right, SOX2* gene expressions in CD133+ cells upon silencing of Elk-1 expression (unpaired t-test; **** $p < 0.0001$, ** $p = 0.0051$); (**D**). endogenous Elk-1 expression levels in CD133− and CD133+ primary brain tumor samples (sample no 428, 458 and 624); relative gene expression is reported as normalized to 28S rRNA level (unpaired t-tests; * $p < 0.05$, *** $p < 0.0001$); (**E**). primary brain tumor cells from sample no 624 were transfected with either scrRNA or siElk-1 plasmids and analyzed for expression level of endogenous *ELK-1, CD133, NANOG, SOX2,* and *POU5F1* normalized to 28S rRNA level (unpaired t-tests; *** $p < 0.0001$).

To investigate whether similar regulation could be observed in primary GBM, primary brain tumor samples from three different patients (patient no. 428, 458, 624) were analyzed for ELK-1 expression in CD133− vs. CD133+ cells. Although there was variability between samples, in all three GBMs, CD133+ cells expressed significantly more ELK-1 than CD133− cells (Figure 4D). This was parallel to our analysis of GBM cell lines, where tumorspheres of A172, T98G, and U87 GBM cells expressed significantly more Elk-1 protein than the monolayer cultures did, whereas ELK-1 expression level did not alter significantly in SH-SY5Y tumorsphere vs. monolayer cultures (data not shown). Furthermore, when endogenous ELK-1 was silenced in the primary tumor culture of patient 624 (middle level of ELK-1 expression), *CD133, NANOG, SOX2,* and *POU5F1* levels were all downregulated as compared to scramble RNA control (Figure 4E).

3.4. Effect of Elk-1 Expression on Colony Formation of SK-N-BE (2) Cells on Soft Agar

The ability of transformed cells to grow in anchorage-free conditions is one of the hallmarks of cancer formation, and soft agar colony assay is a commonly used tool to assay for this feature [30]. It was shown in endometrial tumors, for instance, that CD133+ cells exhibited higher colony formation than CD133− cells in soft agar assay [31]. We have, therefore, addressed whether the same scenario was true for CD133+/CD133− SK-N-BE (2) cells, and whether overexpression of Elk-1-VP16 or silencing of endogenous Elk-1 would affect the number of colonies. To that end, we have sorted SK-N-BE (2) cells into CD133+ BTICs and CD133− cells, and both CD133+ and CD133− spheroids were grown in IPM culture conditions for three days in limiting dilution assay (LDA), and the frequency of spheroid formation was found to be almost tenfold more in CD133+ BTIC cells, indicating that sorting of cells was successful.

Next, the effects of Elk-1 overexpression or silencing were studied; to that end, we have transfected CD133− cells with Elk-1-VP16 expression vector, while CD133+ cells were transfected with siElk-1 silencing vector as described in Materials and Methods, and colony formation frequencies were determined in soft agar assay. In untransfected SK-N-BE (2) cells, CD133− cells and unsorted cells showed a similar number of colonies (24 ± 11 vs. 25 ± 11, respectively), whereas CD133+ BTICs had almost 50% more colonies formed (33 ± 9 colonies). CD133− cells transfected with either pCDNA3-Elk-1-VP16 (37 ± 10 colonies) or pCMV6-Flag-Elk-1-VP16 (50 ± 15 colonies) showed higher colony number than their counterparts transfected with empty vectors, pCDNA3.1 (32 ± 3 colonies) or pCMV-Flag (24 ± 10 colonies). On the other hand, CD133+ cells where endogenous Elk-1 was silenced by RNAi exhibited a decreased colony number (18 ± 5) compared to scrambled RNA control (26 + 9) (Supplement Table S4).

3.5. Regulation of NANOG, POU5F1, and SOX2 Promoters by Elk-1

To assess whether the regulation of these genes by Elk-1 was direct or indirect, the promoters for *NANOG, POU5F1,* and *SOX2* were cloned to luciferase reporter vectors and tested for Elk-1 regulation in different cell lines.

Initially, SK-N-BE (2) (Figure 5A) and SH-SY5Y (Figure 5B) neuroblastoma cells and U87-MG (Figure 5C), A172 (Figure 5D), and T98G (Figure 5E) GBM cells were either transfected with constitutively active Elk-1-VP16 and/or dominant-negative Elk-1-EN fusion protein expression vectors for overexpression (i), or with siElk-1 or scrRNA vectors for silencing (ii) experiments to study the regulation of *SOX2* promoter by Elk-1 protein (Figure 5). Although there appear to be cell type-specific variations, *SOX2* promoter appeared to be upregulated upon Elk-1-VP16 overexpression in all cell types (Figure 5Ai–Ei), whereas only SH-SY5Y and U87 cells exhibited downregulation of SOX2-dependent luciferase activity upon the silencing of endogenous Elk-1, indicating that other proteins are involved in the regulation of this promoter (Figure 5Bii,Cii).

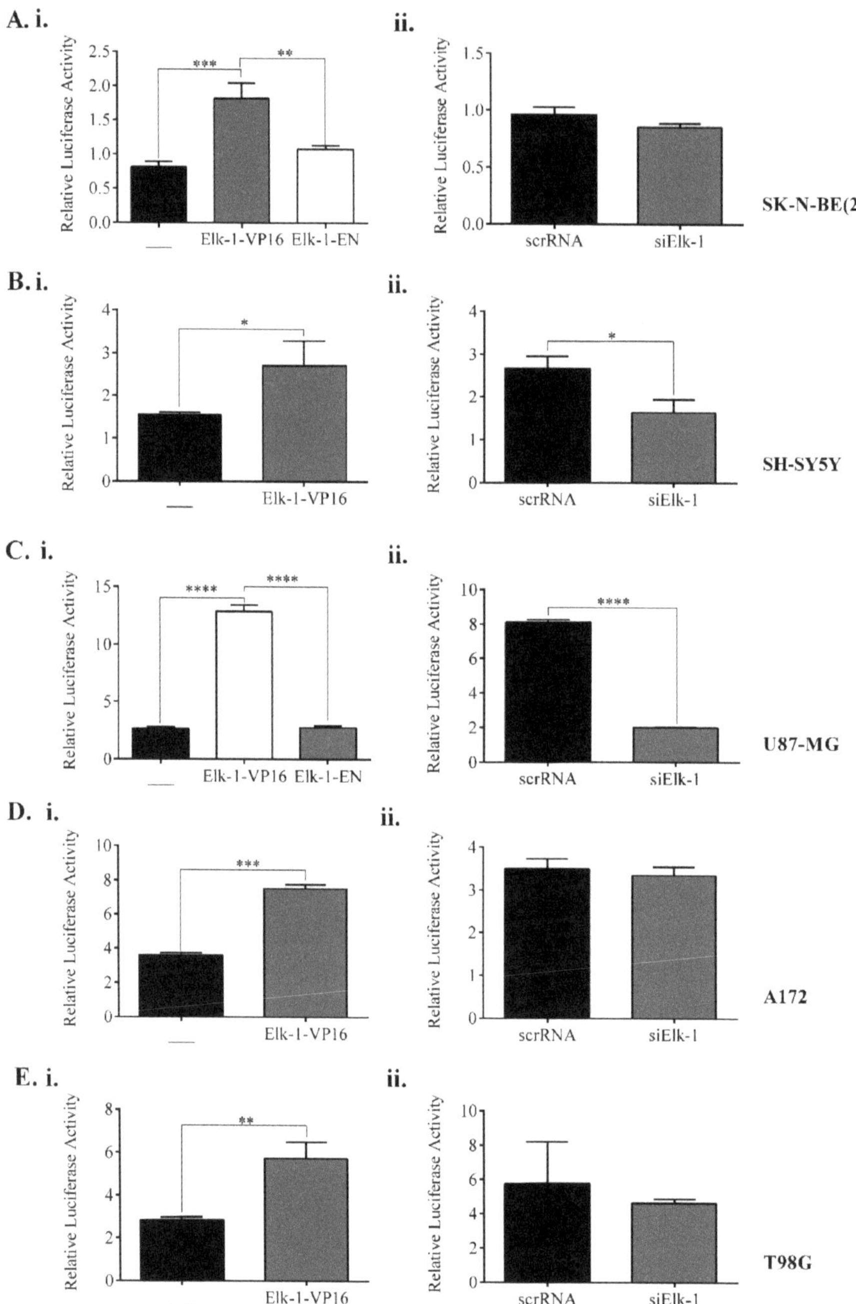

Figure 5. *SOX2* promoter activity analysis with respect to (**i**) Elk-1 variants over-expression and (**ii**) endogenous Elk-1 silencing in (**A**) SK-NBE (2), (**B**) SH-SY5Y, (**C**) U87-MG, (**D**) A172, and (**E**) T98G cell lines. Luminometric measurements were normalized to *Renilla*-luc activity. ANOVA, Tukey's multiple comparative tests, ** $p < 0.01$, *** $p < 0.001$, **** $p < 0.0001$ for Ai, Ci; unpaired two-tailed t-test, * $p < 0.5$, ** $p < 0.01$, **** $p < 0.0001$ was done for Bi, Bii, Cii, Di, Ei.

We next studied *NANOG* promoter; while *SOX2* promoter was found to have 1 consensus Elk-1 binding motif with dissimilarity score (DS) of less than 1%, and 5 ets motifs with DS 1–10%, *NANOG* promoter was found to contain three consensus ets motifs (Figure 6A), two of which had DS of 1–10% (Table 2). We have constructed a wildtype *NANOG* promoter reporter vector (*NANOG*-Luc), and one where the higher similarity consensus ets motif (ets1) was deleted (*NANOGΔ*-Luc), and studied the regulation of this promoter by Elk-1 in different cell lines (Figure 6).

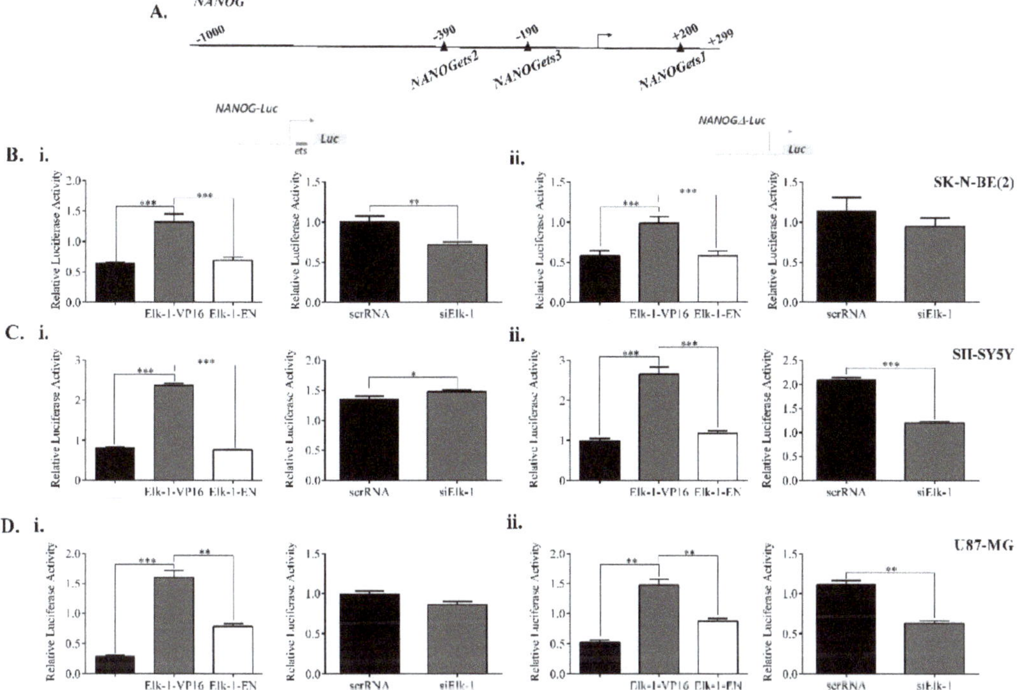

Figure 6. Regulation of *NANOG* promoter by Elk-1. (**A**) Schematic diagram of predicted *ets* motifs *ets1-3* on *NANOG* promoter. *Ets1* was predicted to be a stronger binding motif for Elk-1 and was deleted to generate *NANOGΔ*-Luc reporter plasmid. (**B**) Luciferase assay for (**i**) wildtype *NANOG*-Luc and (**ii**) *NANOGΔ*-Luc reporters in SK-N-BE (2) cells after transfection of expression plasmids with Elk1-VP16, Elk1-EN, or empty control plasmid pCDNA3.1 (**left graphs**) or co-transfection of silencing plasmids for scrRNA control or siElk-1 (**right graphs**). Luminometric measurements were normalized to Renilla-luc activity. ANOVA, Tukey's multiple comparative tests, *** $p < 0.001$ for (**i**) and (**ii**) left graphs; unpaired two-tailed t-test, ** $p < 0.01$ was done for (**i**) and (**ii**) right graphs. (**C**) Luciferase assay for (**i**) wildtype *NANOG*-Luc and (**ii**) *NANOGΔ*-Luc reporters in SH-SY5Y cells after transfection of expression plasmids with Elk1-VP16, Elk1-EN or empty control plasmid pCDNA3.1 (**left graphs**) or co-transfection of silencing plasmids for scrRNA control or siElk-1 (**right graphs**). Luminometric measurements were normalized to *Renilla*-luc activity. ANOVA, Tukey's multiple comparative tests, *** $p < 0.001$ for (**i**) and (**ii**) left graphs; unpaired two-tailed t-test; * $p < 0.5$, *** $p < 0.001$ for (**i**) and (**ii**) right graphs. (**D**) Luciferase assay for (**i**) wildtype *NANOG*-Luc and (**ii**) *NANOGΔ*-Luc reporters in U87-MG cells after transfection of expression plasmids with Elk1-VP16, Elk1-EN, or empty control plasmid pCDNA3.1 (**left graphs**) or co-transfection of silencing plasmids for scrRNA control or siElk-1 (**right graphs**). Luminometric measurements were normalized to *Renilla*-luc activity. ANOVA, Tukey's multiple comparative tests, ** $p < 0.01$, *** $p < 0.001$ for (**i**) and (**ii**) left graphs; unpaired t-test; ** $p < 0.01$ for (**i**) and (**ii**) right graphs.

Elk-1-VP16 overexpression in SK-N-BE (2) cells resulted in upregulation from wildtype *NANOG* promoter, but the upregulation was slightly less in *NANOGΔ*-Luc reporter; Elk-1-EN repressed both promoter activities to control levels (Figure 6Bi vs. Figure 6Bii). Parallel to this, when Elk-1 was silenced using siElk-1, *NANOG*-Luc reporter activity was decreased (Figure 6Bi), whereas there was no significant change in NANOGΔ-Luc activity in SK-N-BE (2) cells (Figure 6Bii). On the other hand, there was no significant difference between *NANOG* vs. *NANOGΔ* promoter activity by Elk-1-VP16 overexpression in SH-SY5Y or U87 cells, while Elk-1-EN repressed both wildtype and mutant promoter activities (Figure 6C,D). There was a slight albeit significant increase in *NANOG* promoter activity in siElk-1 SH-SY5Y cells, whereas *NANOGΔ* promoter activity was decreased upon siElk-1 silencing (Figure 6Ci vs. Figure 6Cii); in U87 silencing, endogenous Elk-1 did not significantly alter wildtype *NANOG*-Luc activity but resulted in a decrease in *NANOGΔ*-Luc (Figure 6Di vs. Figure 6Dii). There was no significant change in either Elk-1-VP16 overexpression or siElk-1 silencing in A172 and T98G cells (**data not shown**).

In *POU5F1* promoter, of the four predicted ets motifs, three of them were predicted to have DS score of 1-10% DS (Table 2; Figure 7A). Wildtype *POU5F1* promoter was cloned, and deletion constructs for these motifs (ets1-ets3) were generated for luciferase reporter assays as described in Materials and Methods. When wildtype *POU5F1* promoter activity was compared to deletion constructs in SK-N-BE (2) cells transfected with Elk-1-VP16 expression plasmid, *POU5F1Δets2*-Luc deletion construct exhibited less upregulation (around 2.4 units) than wildtype, *POU5F1Δets1*, and *POU5F1Δets3* promoters (around 3 units), while Elk-1-EN overexpression resulted in similar level of activation to control in all cases (Figure 7B). On the other hand, siElk-1 silencing did not result in a significant change in wildtype *POU5F1* promoter activity or the *POU5F1Δets3* deletion mutant, whereas it resulted in a decrease in luciferase activity in both *POU5F1Δets1* and *POU5F1Δets2* constructs in SK-N-BE (2) cells (Figure 7B).

Elk-1-VP16 overexpression upregulated wildtype *POU5F1*-Luc reporter activity, while Elk-1-EN repressed it in SH-SY5Y cells; there was no significant change in this profile in either of the three ets deletion constructs, indicating the regulation might be through a different motif or could be indirect (Figure 7C). Interestingly, siElk-1 silencing upregulated wildtype *POU5F1*-Luc and *POU5F1Δets1*-Luc reporter activity, while decreasing *POU5F1Δets2*-Luc and *POU5F1Δets3*-Luc reporter activity (Figure 7C). In U87-MG GBM cells, however, wildtype *POU5F1*-Luc and *POU5F1Δets2*-Luc reporters were upregulated to similar levels in Elk-1-VP16 overexpression (1.2 units in Figure 7Di and 1.5 units in Figure 7Diii), while *POU5F1Δets1*-Luc was upregulated more (2.4 units, Figure 7Dii), and upregulation was significantly less in *POU5F1Δets3*-Luc reporter (Figure 7Div). Elk-1-EN overexpression did not significantly alter promoter activity (Figure 7D), and while siElk-1 silencing did not cause any change in wildtype promoter, it resulted in a downregulation in all deletion constructs to a different extent (Figure 7D). Wildtype *POU5F1*-Luc promoter was upregulated by Elk-1-VP16 overexpression in both A172 and T98G cells, although siElk-1 silencing did not significantly change with respect to scrambled RNA control.

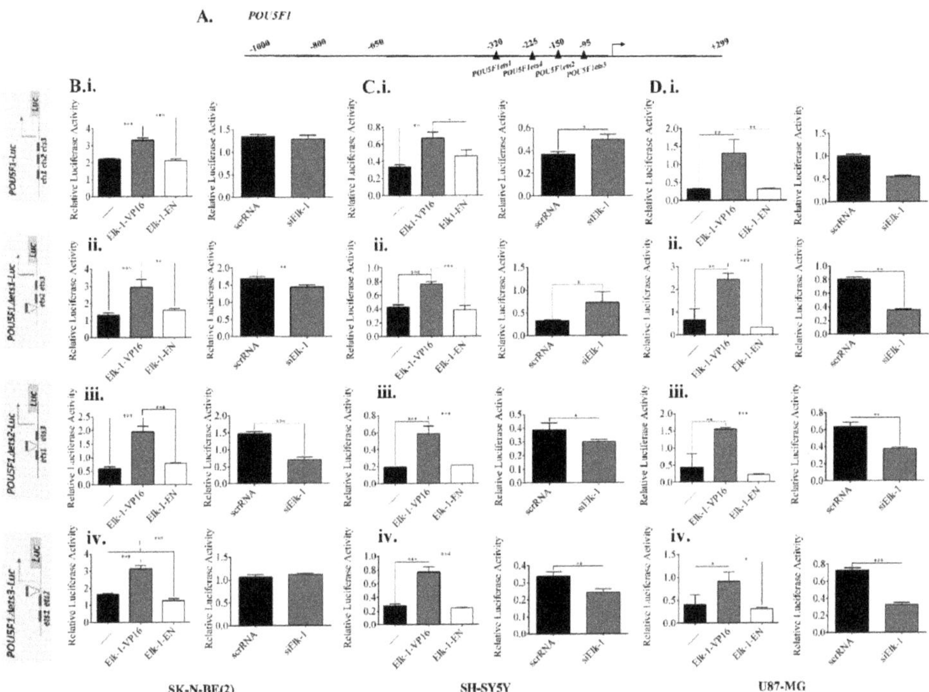

Figure 7. Regulation of *POU5F1* promoter by Elk-1. (**A**) Schematic diagram of predicted *ets* motifs *ets1-4* on *POU5F1* promoter. Motifs *ets1-3* were predicted to be stronger binding motifs for Elk-1 and were individually deleted to generate *POU5F1Δets1*-Luc, *POU5F1Δets2*-Luc, and *POU5F1Δets3*-Luc reporter plasmids. (**B**) Luciferase assay for (**i**) wildtype *POU5F1*-Luc and its deletion mutant reporters (**ii**) *POU5F1Δets1*-Luc, (**iii**) *POU5F1Δets2*-Luc, and (**iv**) *POU5F1Δets3*-Luc in SK-N BE (2) cells after transfection of expression plasmids with Elk1-VP16, Elk1-EN, or empty control plasmid pCDNA3.1 (**left graphs**) or co-transfection of silencing plasmids for scrRNA control or siElk-1 (**right graphs**). Luminometric measurements were normalized to Renilla-luc activity. ANOVA, Tukey's multiple comparative tests, ** $p < 0.01$, *** $p < 0.001$ for left graphs (**i–iv**); unpaired two-tailed t-test, ** $p < 0.01$ and *** $p < 0.001$ for right graphs (**i–iv**). **C.** Luciferase assay for (**i**) wildtype *POU5F1*-Luc and its deletion mutant reporters (**ii**) *POU5F1Δets1*-Luc, (**iii**) *POU5F1Δets2*-Luc, and (**iv**) *POU5F1 ets3*-Luc in SH-SY5Y cells after transfection of expression plasmids with Elk1-VP16, Elk1-EN, or empty control plasmid pCDNA3.1 (**left graphs**) or co-transfection of silencing plasmids for scrRNA control or siElk-1 (**right graphs**). Luminometric measurements were normalized to Renilla-luc activity. ANOVA, Tukey's multiple comparative tests, * $p < 0.5$, ** $p < 0.01$, *** $p < 0.001$ for left graphs (**i–iv**); unpaired two-tailed t-test, * $p < 0.5$, ** $p < 0.01$ for right graphs (**i–iv**). (**D**) Luciferase assay for (**i**) wildtype *POU5F1*-Luc and its deletion mutant reporters (**ii**) *POU5F1Δets1*-Luc, (**iii**) *POU5F1Δets2*-Luc, and (**iv**) *POU5F1Δets3*-Luc in U87-MG cells after transfection of expression plasmids with Elk1-VP16, Elk1-EN, or empty control plasmid pCDNA3.1 (**left graphs**) or co-transfection of silencing plasmids for scrRNA control or siElk-1 (**right graphs**). Luminometric measurements were normalized to Renilla-luc activity. ANOVA, Tukey's multiple comparative tests, * $p < 0.5$, ** $p < 0.01$, *** $p < 0.001$ for left graphs (**i–iv**); unpaired two-tailed t-test, ** $p < 0.01$, *** $p < 0.001$ for right graphs (**i–iv**).

3.6. Binding of Elk-1 to Predicted ets Motifs on SOX2, NANOG, and POU5F1 Promoters

Elk-1-VP16 overexpression was found to upregulate expression of *SOX2*, *NANOG*, and *POU5F1* expression in qPCR analysis, and wildtype promoter luciferase reporters were found to be upregulated by Elk-1-VP16 in a cell context-dependent manner, yet deletion of predicted ets motifs did not significantly change reporter activities, indicating that either there are other ets motifs in distal promoters that are not cloned in this study, or that the regulation is not through direct Elk-1 binding to these predicted ets motifs. To address this second point, we have carried out chromatin immunoprecipitation (ChIP) experiments in SK-N-BE (2) neuroblastoma and T98G GBM cell lines (Figure 8).

The cells were transfected with pCMV-Flag-Elk-1 (empty pCMV was used as control), and immunoprecipitation was carried out using Flag agarose beads (Flag IP); IgG beads were used as control (IgG IP). Elk-1 binding motifs on *SRF* and *MCL*-1 promoters were used as a positive control for Elk-1 binding. All three of the predicted ets motifs on the *NANOG* promoter exhibited Elk-1 binding in SK-N-BE (2) cells (Figure 8A) but not on T98G cells (Figure 8B). Similarly, all four predicted ets motifs on *POU5F1* promoter showed Elk-1 binding, albeit to different extents, in SK-N-BE (2) cells (Figure 8A) but not on T98G cells (Figure 8B). Likewise, all five predicted ets motifs showed Elk-1 binding in SK-N-BE (2) cells (Figure 8A), whereas only the ets3 motif showed significant binding to Elk-1 in T98G (Figure 8B). This indicates that, while Elk-1 is capable of binding to these predicted motifs, this binding is affected by cell-dependent circumstances, which may be a transcriptional partner or posttranslational modification status of the Elk-1 protein in that particular cell type.

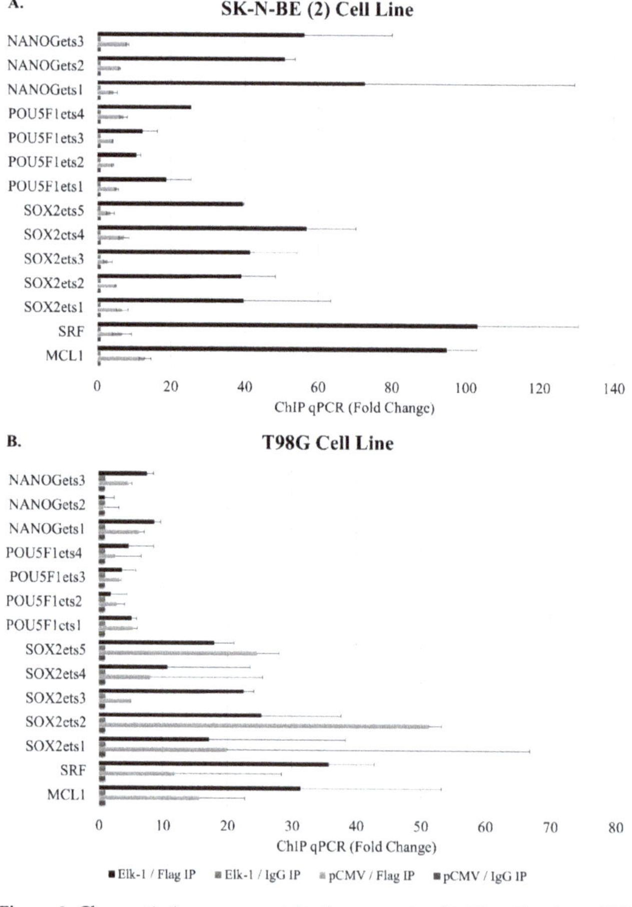

Figure 8. Chromatin immunoprecipitation assay for the identification of Elk-1 binding sites on the target gene promoters in pCMV-transfected (pCMV) vs. Elk-1 over-expressing cells (Elk-1) in (**A**). SK-N-BE (2) cells and (**B**). T98G cells. Lysates were immunoprecipitated with either Flag antibody (Flag IP) for exogenous Elk-1 or IgG (IgG IP) as control. qPCR results were analyzed as explained in Materials and Methods and reported as average fold change. All predicted *ets* motif sequences on *NANOG*, *SOX2*, and *POU5F1* promoters were screened for Elk-1 binding; *SRF* and *MCL1* promoter sequences were used as positive control.

4. Discussion

ETS transcription factors are involved in a number of biological processes in different tissues, and it was shown that in embryonic development, expression of several ETS proteins including Elf3 and SpiC increased after fertilization until the blastocyst stage, and silencing of ETS expression affected Oct3/4 gene expression [32]. It was shown in human pluripotent stem cells (hPSCs) with different X chromosome inactivation states (Xa, active, Xi, inactive) that Elk-1 overexpression mimicked XaXa in terms of decreased pluripotency, the differences being diminished in low oxygen [33].

One study has shown Elk-1 to be essential for human embryonic stem cells, and that it co-occupies promoters of genes in cell proliferation pathways with *ERK2*, and in the absence of *ERK2*, the promoters were repressed by Polycomb proteins [29]. In fact, Elk-1 was further found to be upregulated, while Nanog, Oct4, and Sox2 were found to be repressed during mesoderm differentiation of hESCs, and it was shown to bind to and activate promoters such as *EGR-1* while repressing a subset of promoters such as *FOSL1* [16]. Intriguingly, mice deficient for Elk-1 were viable albeit with mild neuronal impairment, indicating other Ets proteins may act redundantly and compensate for its embryonic functions [34]. During neuronal differentiation of mES cells, Sox2 chromatin interaction profiles were altered, and promoters of neuronal differentially expressed gene clusters were enriched in Elk-1, among other transcription factors [35]. Similarly, during reprogramming of fibroblasts into neural stem cells (NSCs) using pharmacological molecules, Elk-1 was found to be one of the transcription factors to regulate reprogramming, particularly through binding Sox2 promoter [36].

Another Ets protein, Pea3/ETV4, was shown to regulate Nanog and Oct4 expression in pluripotent NCCIT embryonic carcinoma cells [37,38]. Interestingly, members of the Pea3 subfamily of ETS proteins, ETV4 and ETV5, were found to be expressed in undifferentiated ES cells, and suppression of Oct3/4 was found to result in downregulation of their expression, and ETV4 and ETV5 were found to be important for proliferation of undifferentiated ES cells through regulation of stem cell-related genes such as Tcf15, Gbx2, and Zic3 [39]. A transcriptional partner of Elk-1, namely serum response factor (SRF), was shown to repress the reprogramming induced by ERK pathway inhibition, and to negatively regulate pluripotency [40], which may be independent of Elk-1 interaction.

CD133 is a cell surface protein that has been used alone [12] or in combination with CD15 [41] to isolate and culture brain-tumor-initiating cells from a variety of tumors. ERK/MAPK pathway was shown to be required for CD133 expression [42], and HIF-1α was shown to bind to the CD133 promoter through Elk-1 [15], which is supported in our study by overexpression of Elk-1 in CD133+ BTIC subpopulation.

In a genome-wide study in the human embryonic stem cell (hESC) population, ELK1 was found to be essential for hESCs, and some of the promoters bound by ELK1 were determined to be important in the maintenance of embryonic identity, spinal cord development, and neuron fate development [29]. Furthermore, induced neural stem cells were found to contain relatively high levels of phosphorylated Elk-1, along with Gli2, and both were shown to bind to *Sox2* promoter upon neural reprogramming [36], and distinct GABPA/Elk-1 motifs were found in Sox2 promoter, identified as a neuronal cluster gene involved in differentiation of embryonic stem cells to neuronal precursors [35]. It is intriguing whether tumorigenesis reactivates this mechanism in a cell context-dependent manner.

5. Conclusions

We propose that not only does ELK1 present a novel target for tumor therapy directed at eliminating BTIC population, but also can be used as a molecular diagnostic molecule to identify potential for tumor recurrence. It should be noted, however, that posttranslational modifications such as phosphorylation and SUMOylation regulate ELK1 protein, which can differ among gliomas and must be studied in more detail.

Supplementary Materials: The following are available online at https://www.mdpi.com/xxx/s1, supppl1: Biological processes among enriched pathways that were upregulated in Elk-1-VP16-transfected SH-SY5Y cells, Table S2: Biological processes among enriched pathways that were downregulated in Elk-1-VP16-transfected SH-SY5Y cells, Table S3: Transcription factor binding site analysis for promoters of genes identified in microarray analysis, Table S4: Soft agar assay colony formation assay.

Author Contributions: Conceptualization, I.A.K.; data curation, U.D.; formal analysis, M.S.S., C.V., B.K., U.D., S.M., G.G., and I.A.K.; funding acquisition, I.A.K.; investigation, M.S.S., C.V., B.K., U.D., and S.S.; methodology, M.S.S., B.K., S.M., S.S., G.G., and K.Y.A.; project administration, B.Y. and I.A.K.; resources, S.S. and I.A.K.; software, G.G.; supervision, K.Y.A., B.Y., and I.A.K.; writing—original draft, M.S.S., C.V., B.K., and S.S.; writing—review and editing, M.S.S., C.V., S.S., K.Y.A., and B.Y. All authors have read and agreed to the published version of the manuscript.

Funding: This research was funded by TUBITAK project no 211T167 as part of the COST TD901 Action, and TUBITAK project no 115Z804 as part of COST TN1301 Action.

Institutional Review Board Statement: Human GBM samples were obtained from consenting patients, as approved by the Hamilton Health Sciences/McMaster Health Sciences Research Ethics Board (REB# 07-366; approved 2007, last updated September 2020).

Informed Consent Statement: Not applicable.

Data Availability Statement: Transcriptomics raw data have been uploaded to and available at EBI ArrayExpress, with accession number of E-MTAB-9938. The rest of the data are available upon request.

Acknowledgments: We would like to thank members of the Kurnaz and Singh laboratories for helpful discussions and Eray Sahin for initial data analysis of the Elk-1-VP16 microarray with demo IPA software. IAK is a GYA member and GG is a YOK 100/2000 student.

Conflicts of Interest: The authors declare no conflict of interest.

References

1. Davis, S.P.; Vanhoutte, C.; Pages, J. Caboche and S. Laroche. The MAPK/ERK cascade targets both Elk-1 and cAMP response element-binding protein to control long-term potentiation-dependent gene expression in the dentate gyrus in vivo. *J. Neurosci.* **2000**, *20*, 4563–4572. [CrossRef]
2. Demir, O.; Ari, O.; Kurnaz, I.A. Elk-1 interacts with dynein upon serum stimulation but independent of serine 383 phosphorylation. *Cell. Mol. Neurobiol.* **2012**, *32*, 185–189. [CrossRef]
3. Demir, O.; Korulu, S.; Yildiz, A.; Karabay, A.; Kurnaz, I.A. Elk-1 interacts with neuronal microtubules and relocalizes to the nucleus upon phosphorylation. *Mol. Cell. Neurosci.* **2009**, *40*, 111–119. [CrossRef]
4. Kaminska, B.; Kaczmarek, L.; Zangenehpour, S.; Chaudhuri, A. Rapid phosphorylation of Elk-1 transcription factor and activation of MAP kinase signal transduction pathways in response to visual stimulation. *Mol. Cell. Neurosci.* **1999**, *13*, 405–414. [CrossRef]
5. Sananbenesi, F.; Fischer, A.; Schrick, C.; Spiess, J.; Radulovic, J. Phosphorylation of hippocampal Erk-1/2, Elk-1, and p90-Rsk-1 during contextual fear conditioning: Interactions between Erk-1/2 and Elk-1. *Mol. Cell. Neurosci.* **2002**, *21*, 463–476. [CrossRef]
6. Sgambato, V.; Vanhoutte, P.; Pages, C.; Rogard, M.; Hipskind, R.; Besson, M.J.; Caboche, J. In vivo expression and regulation of Elk-1, a target of the extracellular-regulated kinase signaling pathway, in the adult rat brain. *J. Neurosci.* **1998**, *18*, 214–226. [CrossRef]
7. Yang, S.H.; Shore, P.; Willingham, N.; Lakey, J.H.; Sharrocks, A.D. The mechanism of phosphorylation-inducible activation of the ETS-domain transcription factor Elk-1. *EMBO J.* **1999**, *18*, 5666–5674. [CrossRef] [PubMed]
8. Yang, S.H.; Whitmarsh, A.J.; Davis, R.J.; Sharrocks, A.D. Differential targeting of MAP kinases to the ETS-domain transcription factor Elk-1. *EMBO J.* **1998**, *17*, 1740–1749. [CrossRef]
9. Sharrocks, A.D. The ETS-domain transcription factor family. *Nat. Rev. Mol. Cell Biol.* **2001**, *2*, 827–837. [CrossRef] [PubMed]
10. Boros, J.; Donaldson, I.J.; O'donnell, A.; Odrowaz, Z.A.; Zeef, L.; Lupien, M.; Meyer, C.A.; Liu, X.S.; Brown, M.; Sharrocks, A.D. Elucidation of the ELK1 target gene network reveals a role in the coordinate regulation of core components of the gene regulation machinery. *Genome Res.* **2009**, *19*, 1963–1973. [CrossRef]
11. Demir, O.; Aysit, N.; Onder, Z.; Turkel, N.; Ozturk, G.; Sharrocks, A.D.; Kurnaz, I.A. ETS-domain transcription factor Elk-1 mediates neuronal survival: SMN as a potential target. *Biochim. Biophys. Acta* **2011**, *1812*, 652–662. [CrossRef]
12. Lenkiewicz, M.; Li, N.; Singh, S.K. Culture and isolation of brain tumor initiating cells. *Curr. Protoc. Stem Cell Biol.* **2009**, *11*, 3.3.1–3.3.10. [CrossRef]

13. Mahller, Y.Y.; Williams, J.P.; Baird, W.H.; Mitton, B.; Grossheim, J.; Saeki, Y.; Cancelas, J.A.; Ratner, N.; Cripe, T.P. Neuroblastoma cell lines contain pluripotent tumor initiating cells that are susceptible to a targeted oncolytic virus. *PLoS ONE* **2009**, *4*, e4235. [CrossRef]

14. Singh, S.K.; Hawkins, C.; Clarke, I.D.; Squire, J.A.; Bayani, J.; Hide, T.; Henkelman, R.M.; Cusimano, M.D.; Dirks, P.B. Identification of human brain tumour initiating cells. *Nature* **2004**, *432*, 396–401. [CrossRef]

15. Ohnishi, S.; Maehara, O.; Nakagawa, K.; Kameya, A.; Otaki, K.; Fujita, H.; Higashi, R.; Takagi, K.; Asaka, M.; Sakamoto, N.; et al. hypoxia-inducible factors activate CD133 promoter through ETS family transcription factors. *PLoS ONE* **2013**, *8*, e66255. [CrossRef]

16. Prise, I.; Sharrocks, A.D. ELK1 has a dual activating and repressive role in human embryonic stem cells. *Wellcome Open Res.* **2019**, *4*, 41. [CrossRef]

17. Vora, P.; Seyfrid, M.; Venugopal, C.; Qazi, M.A.; Salim, S.; Isserlin, R.; Subapanditha, M.; O'farrell, E.; Mahendram, S.; Singh, M.; et al. Bmi1 regulates human glioblastoma stem cells through activation of differential gene networks in CD133+ brain tumor initiating cells. *J. Neurooncol.* **2019**, *143*, 417–428. [CrossRef]

18. Venugopal, C.; Hallett, R.; Vora, P.; Manoranjan, B.; Mahendram, S.; Qazi, M.A.; Mcfarlane, N.; Subapanditha, M.; Nolte, S.M.; Singh, M.; et al. Pyrvinium Targets CD133 in Human Glioblastoma Brain Tumor-Initiating Cells. *Clin. Cancer Res.* **2015**, *21*, 5324–5337. [CrossRef] [PubMed]

19. Venugopal, C.; Mcfarlane, N.M.; Nolte, S.; Manoranjan, B.; Singh, S.K. Processing of primary brain tumor tissue for stem cell assays and flow sorting. *J. Vis. Exp.* **2012**. [CrossRef]

20. Kandemir, B.; Dag, U.; Bakir Gungor, B.; Durasi, I.M.; Erdogan, B.; Sahin, E.; Sezerman, U.; Aksan Kurnaz, I. In silico analyses and global transcriptional profiling reveal novel putative targets for Pea3 transcription factor related to its function in neurons. *PLoS ONE* **2017**, *12*, e0170585. [CrossRef]

21. Bolstad, B.M.; Irizarry, R.A.; Astrand, M.; Speed, T.P. A comparison of normalization methods for high density oligonucleotide array data based on variance and bias. *Bioinformatics* **2003**, *19*, 185–193. [CrossRef]

22. Kamburov, A.; Stelzl, U.; Lehrach, H.; Herwig, R. The ConsensusPathDB interaction database: 2013 update. *Nucleic Acids Res.* **2013**, *41*, D793–D800. [CrossRef]

23. Kanehisa, M.; Sato, Y.; Kawashima, M.; Furumichi, M.; Tanabe, M. KEGG as a reference resource for gene and protein annotation. *Nucleic Acids Res.* **2016**, *44*, D457–D462. [CrossRef]

24. Fabregat, A.; Sidiropoulos, K.; Garapati, P.; Gillespie, M.; Hausmann, K.; Haw, R.; Jassal, B.; Jupe, S.; Korninger, F.; Mckay, S.; et al. The Reactome pathway Knowledgebase. *Nucleic Acids Res.* **2016**, *44*, D481–D487. [CrossRef]

25. Nishimura, A. BioCarta. *Biotech. Softw. Internet Rep.* **2001**, *2*, 117–120. [CrossRef]

26. Ashburner, M.; Ball, C.A.; Blake, J.A.; Botstein, D.; Butler, H.; Cherry, J.M.; Davis, A.P.; Dolinski, K.; Dwight, S.S.; Eppig, J.T.; et al. Gene ontology: Tool for the unification of biology. The Gene Ontology Consortium. *Nat. Genet.* **2000**, *25*, 25–29. [CrossRef]

27. Fishilevich, S.; Zimmerman, S.; Kohn, A.; Iny Stein, T.; Olender, T.; Kolker, E.; Safran, M.; Lancet, D. Genic insights from integrated human proteomics in GeneCards. *Database* **2016**, *2016*, baw030. [CrossRef] [PubMed]

28. Messeguer, X.; Escudero, R.; Farre, D.; Nunez, O.; Martinez, J.; Alba, M.M. PROMO: Detection of known transcription regulatory elements using species-tailored searches. *Bioinformatics* **2002**, *18*, 333–334. [CrossRef] [PubMed]

29. Goke, J.; Chan, Y.S.; Yan, J.; Vingron, M.; Ng, H.H. Genome-wide kinase-chromatin interactions reveal the regulatory network of ERK signaling in human embryonic stem cells. *Mol. Cell* **2013**, *50*, 844–855. [CrossRef]

30. Borowicz, S.; Van Scoyk, M.; Avasarala, S.; Karuppusamy Rathinam, M.K.; Tauler, J.; Bikkavilli, R.K.; Winn, R.A. The soft agar colony formation assay. *J. Vis. Exp.* **2014**, e51998. [CrossRef]

31. Ding, D.C.; Liu, H.W.; Chang, Y.H.; Chu, T.Y. Expression of CD133 in endometrial cancer cells and its implications. *J. Cancer* **2017**, *8*, 2142–2153. [CrossRef]

32. Kageyama, S.; Liu, H.; Nagata, M.; Aoki, F. The role of ETS transcription factors in transcription and development of mouse preimplantation embryos. *Biochem. Biophys. Res. Commun.* **2006**, *344*, 675–679. [CrossRef]

33. Bruck, T.; Yanuka, O.; Benvenisty, N. Human pluripotent stem cells with distinct X inactivation status show molecular and cellular differences controlled by the X-Linked ELK-1 gene. *Cell Rep.* **2013**, *4*, 262–270. [CrossRef]

34. Cesari, F.; Brecht, S.; Vintersten, K.; Vuong, L.G.; Hofmann, M.; Klingel, K.; Schnorr, J.J.; Arsenian, S.; Schild, H.; Herdegen, T.; et al. Mice deficient for the ets transcription factor elk-1 show normal immune responses and mildly impaired neuronal gene activation. *Mol. Cell. Biol.* **2004**, *24*, 294–305. [CrossRef]

35. Bunina, D.; Abazova, N.; Diaz, N.; Noh, K.M.; Krijgsveld, J.; Zaugg, J.B. Genomic Rewiring of SOX2 Chromatin Interaction Network during Differentiation of ESCs to Postmitotic Neurons. *Cell Syst.* **2020**, *10*, 480–494.e8. [CrossRef]

36. Zhang, L.; Gong, Y.; Zhao, X.; Zhou, H. Comparative study between hypoxia and hypoxia mimetic agents on osteogenesis of bone marrow mesenchymal stem cells in mouse. *Zhongguo Xiu Fu Chong Jian Wai Ke Za Zhi* **2016**, *30*, 903–908.

37. Park, S.W.; Do, H.J.; Choi, W.; Song, H.; Chung, H.J.; Kim, J.H. NANOG gene expression is regulated by the ETS transcription factor ETV4 in human embryonic carcinoma NCCIT cells. *Biochem. Biophys. Res. Commun.* **2017**, *487*, 532–538. [CrossRef]

38. Park, S.W.; Do, H.J.; Ha, W.T.; Han, M.H.; Park, K.H.; Song, H.; Kim, N.H.; Kim, J.H. Transcriptional activation of OCT4 by the ETS transcription factor PEA3 in NCCIT human embryonic carcinoma cells. *FEBS Lett.* **2014**, *588*, 3129–3136. [CrossRef]

39. Akagi, T.; Kuure, S.; Uranishi, K.; Koide, H.; Costantini, F.; Yokota, T. ETS-related transcription factors ETV4 and ETV5 are involved in proliferation and induction of differentiation-associated genes in embryonic stem(ES) cells. *J. Biol. Chem.* **2015**, *290*, 22460–22473. [CrossRef]
40. Huh, S.; Song, H.R.; Jeong, G.R.; Jang, H.; Seo, N.H.; Lee, J.H.; Yi, J.Y.; Lee, B.; Choi, H.W.; Do, J.T.; et al. Suppression of the ERK-SRF axis facilitates somatic cell reprogramming. *Exp. Mol. Med.* **2018**, *50*, e448. [CrossRef]
41. Vora, P.; Venugopal, C.; Mcfarlane, N.; Singh, S.K. Culture and Isolation of Brain Tumor Initiating Cells. *Curr. Protoc. Stem Cell Biol.* **2015**, *34*, 3.3.1–3.3.13. [CrossRef]
42. Tabu, K.; Kimura, T.; Sasai, K.; Wang, L.; Bizen, N.; Nishihara, H.; Taga, T.; Tanaka, S. Analysis of an alternative human CD133 promoter reveals the implication of Ras/ERK pathway in tumor stem-like hallmarks. *Mol. Cancer* **2010**, *9*, 39. [CrossRef]

Journal of
Personalized
Medicine

MDPI

Review

Current State of "Omics" Biomarkers in Pancreatic Cancer

Beste Turanli [1,†], Esra Yildirim [1,†], Gizem Gulfidan [1], Kazim Yalcin Arga [1,2,*] and Raghu Sinha [3,*]

[1]   Department of Bioengineering, Marmara University, 34722 Istanbul, Turkey; turanlibeste@gmail.com (B.T.);
      essyildirim@gmail.com (E.Y.); gizemgulfidn@gmail.com (G.G.)
[2]   Turkish Institute of Public Health and Chronic Diseases, 34718 Istanbul, Turkey
[3]   Department of Biochemistry and Molecular Biology, Penn State College of Medicine, Hershey, PA 17033, USA
*   Correspondence: kazim.arga@marmara.edu.tr (K.Y.A.); rsinha@pennstatehealth.psu.edu (R.S.)
†   These authors contributed equally to this work.

Abstract: Pancreatic cancer is one of the most fatal malignancies and the seventh leading cause of cancer-related deaths related to late diagnosis, poor survival rates, and high incidence of metastasis. Unfortunately, pancreatic cancer is predicted to become the third leading cause of cancer deaths in the future. Therefore, diagnosis at the early stages of pancreatic cancer for initial diagnosis or postoperative recurrence is a great challenge, as well as predicting prognosis precisely in the context of biomarker discovery. From the personalized medicine perspective, the lack of molecular biomarkers for patient selection confines tailored therapy options, including selecting drugs and their doses or even diet. Currently, there is no standardized pancreatic cancer screening strategy using molecular biomarkers, but CA19-9 is the most well known marker for the detection of pancreatic cancer. In contrast, recent innovations in high-throughput techniques have enabled the discovery of specific biomarkers of cancers using genomics, transcriptomics, proteomics, metabolomics, glycomics, and metagenomics. Panels combining CA19-9 with other novel biomarkers from different "omics" levels might represent an ideal strategy for the early detection of pancreatic cancer. The systems biology approach may shed a light on biomarker identification of pancreatic cancer by integrating multi-omics approaches. In this review, we provide background information on the current state of pancreatic cancer biomarkers from multi-omics stages. Furthermore, we conclude this review on how multi-omics data may reveal new biomarkers to be used for personalized medicine in the future.

Keywords: pancreatic cancer; systems biology; omics; biomarker; genomics; transcriptomics; proteomics; metabolomics; glycomics; metagenomics; personalized medicine

 check for
updates

Citation: Turanli, B.; Yildirim, E.;
Gulfidan, G.; Arga, K.Y.; Sinha, R.
Current State of "Omics" Biomarkers
in Pancreatic Cancer. *J. Pers. Med.*
2021, *11*, 127. https://doi.org/
10.3390/jpm11020127

Academic Editor: Raghu Sinha
Received: 22 January 2021
Accepted: 11 February 2021
Published: 14 February 2021

Publisher's Note: MDPI stays neutral
with regard to jurisdictional claims in
published maps and institutional affiliations.

1. Introduction

Pancreatic cancer is one of the most fatal malignancies and the seventh leading cause of cancer-related deaths considering both sexes worldwide according to the latest global cancer statistics reported in 2018 [1]. Pancreatic cancer has a difficult diagnosis at an early stage and a 5 year survival rate of 10% at the time of diagnosis in the United States, where the poor survival rates have hardly changed for almost 40 years since most patients reporting to the hospital have either unresectable or metastatic disease. Only 10.8% of these patients are at a locally advanced stage at the time of diagnosis [2,3]. Unfortunately, pancreatic cancer is projected to become the third leading cause of cancer deaths in the future [1].

It is a great challenge to intervene at the early stages of pancreatic cancer that is in initial diagnosis or postoperative recurrence because of the difficulties in early diagnosis and inadequacy in precise prognostic biomarkers, and this challenge may result in undesirable overdiagnosis and/or overtreatment, causing the high mortality rate [4–7].

Pancreatic cancer can be divided into two large groups; (a) endocrine pancreatic tumors, including gastrinoma, glucagonoma, and insulinoma, and (b) exocrine (non-endocrine) pancreatic tumors, including adenoma, ductal adenocarcinoma, acinar cell carcinoma, cystadenocarcinoma, adenosquamous carcinoma, signet ring cell carcinoma, hepatoid carcinoma, colloid carcinoma, undifferentiated carcinoma, pancreatoblastoma, and pancreatic mucinous cystic neoplasm [8,9]. Most of the pancreatic cancers are exocrine types—namely, ductal adenocarcinoma, which comprises 80–90% of all pancreatic cancers; whereas endocrine (neuroendocrine) pancreatic tumors are rare with 1–2% of all pancreatic cancers [7].

Moreover, pancreatic neoplasms can be categorized by their gross appearance as solid, cystic, or intraductal. The solid pancreatic tumors contain pancreatic ductal adenocarcinoma (PDAC), neuroendocrine (islet cell) neoplasms, acinar cell carcinomas, and pancreatoblastoma. The cystic types of pancreatic tumors tend to be less aggressive and include mucinous cystic neoplasms, serous cystadenoma, intraductal papillary mucinous neoplasms, and solid-pseudopapillary neoplasms [10]. Pancreatoblastoma is mostly observed in childhood, and it has a poor prognosis if an adult is diagnosed with it. Mucinous cystic neoplasms consist of a range from benign to malignant [7].

The World Health Organization (WHO) classifies the morphological variants of PDAC differently from the conventional pancreatic adenocarcinoma classification. These variants have different histological features besides molecular signatures and prognosis. According to WHO, the different subtypes of PDAC are adenosquamous carcinoma, colloid/mucinous carcinoma, undifferentiated/anaplastic carcinoma, signet ring cell carcinoma, medullary carcinoma, and hepatoid carcinoma [11].

Like most cancer types, pancreatic cancer has also several known risk factors, such as cigarette smoking, diabetes, obesity, lack of physical activity, and chronic pancreatitis [12,13]. Currently, computed tomography (CT), magnetic resonance imaging (MRI), endoscopic ultrasound (EUS), positron emission tomography (PET), and other imaging methods are used in the diagnosis and prognosis of pancreatic cancer [12–14].

Unsurprisingly, early detection of PDAC by effective screening approaches is crucial to improve a better prognosis of the disease. The absence of clinical symptoms in the early stage of pancreatic cancer could lead to a delay in confirmed diagnosis even though tumor biomarkers and imaging techniques are being developed. Therefore, using circulating biomarkers for primary screening and its combination with imaging and histopathologic results might be the future strategy for diagnosing PDAC. Candidate circulating biomarkers in PDAC are not limited to circulating tumor cells (CTC) but also consist of metabolites, cell-free DNA and non-coding RNA, exosomes, autoantibodies, and inflammatory or growth factors, which are recently summarized [15]. The presence of CTCs in the blood usually correlates with the systemic spread of the tumor, and the characteristics of these CTCs could be used as potential biomarkers. Moreover, the challenging tasks of CTC isolation and detection are being overcome [16,17], and the emerging area of profiling CTCs has been recognized in prognosis of pancreatic cancer [18].

Sample source is very critical in the identification of biomarkers for the detection and diagnosis of early-stage pancreatic cancer [19]. The pancreas is located in the back of the abdomen and is surrounded by the stomach, small intestine, liver, and spleen, so it becomes a big challenge in getting a biopsy. The most common way to get pancreatic tumor samples is by fine-needle aspiration (FNA). However, a core needle biopsy using a larger needle than an FNA can provide a larger sample, often useful for molecular profiling. These biopsies can be taken with an EUS. Other biopsy types, like brush biopsy or forceps biopsy, can be done during an endoscopic cholangiopancreatography (ERCP). However, body fluids such as blood, cyst fluid, pancreatic juice, bile, as well as urine are characteristically enriched with biomarkers that can be a potential source of diagnostic, predictive, and/or prognostic biomarkers in PDAC. As a source of pancreatic cancer biomarker, saliva has also been used. In omics biomarker studies, blood is a frequently preferred sample source due to its easy accessibility, noninvasiveness, and cost-effectiveness [20]. As an alternative rich source for the discovery of biomarkers, pancreatic juice has recently been identified. Pancreatic juice contains pancreatic cancer-specific markers such as DNA, RNA, proteins, and cancer cells, but the collection procedure for this sample source is invasive [19]. Although urine contains limited protein, DNA, and RNA, it can be considered as an ideal source sample for proteomic and genomic biomarkers [21]. Furthermore, accurate staging is very important for providing appropriate treatment. The majority of the time, surgical excision is used for treatment, and traditional chemoradiotherapy has very restricted effectiveness, despite the development of novel therapy options [7]. In this review, we present a systems-level outlook of PDAC biomarkers from different "omics" levels (Figure 1) as well as a comprehensive overview of methodology and sampling used in biomarker studies for PDAC (Table 1).

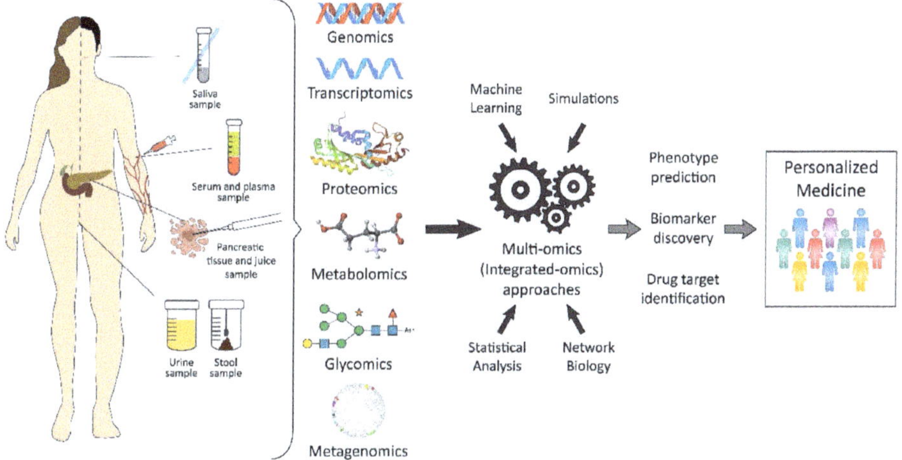

Figure 1. A conceptual review of pancreatic cancer biomarkers from a variety of "omics" levels.

2. Recent Insights from Different Omics Levels

Despite the substantial advancement in pancreatic cancer research, there has not been any remarkable reduction in the mortality-to-incidence ratio. This is mainly a result of the limited early diagnostic characteristic symptoms and reliable biomarkers, besides the unresponsiveness to the treatments due to the tumor heterogeneity, plasticity, and the aggressive metastasis that presents in more than 50% of the diagnosed patients [22].

Systems biology studies of pancreatic cancer rely on the integration of omics data from different biological levels. With the frequently arising challenges regarding cancer diagnosis and treatment—mainly due to its complex pathogenic landscape and cellular heterogeneity—the holistic view provided by the systems biology approach allowed for having a global understanding of the mechanisms of the disease and gaining more insight toward diagnostic or prognostic biomarkers and drug target discovery [23,24].

Likewise, systems biology also augments current diagnosis and therapy options. Aggressiveness and chemoresistance of PDAC are caused by the desmoplastic reactions induced by immune cells, stromal cells, neural cells, and the extracellular matrix surrounding and forming the bulk of the tumor mass. Therefore, single-cell sequencing may shed a better insight into cellular differences. Moreover, altered metabolism is caused by limited delivery of the needed oxygen and nutrients in such a hypoxic and acidic microenvironment; a direct impact on the drug delivery mechanisms is common [25,26].

3. Genomic Signatures

Next-generation sequencing (NGS) provides support for the early diagnosis and screening of PDAC as well as many other diseases. Genomics techniques may assist in the early diagnosis of pancreatic cancer in patients with specific alleles that predispose them to cancer development. Different potential biomarkers discovered by genomics methods can be categorized as chromosomal aberrations, driver changes, single nucleotide polymorphisms (SNPs), or copy-number alterations.

Previous studies pointed out the most prominent genetic features of PDAC, such as oncogenic activation of K-RAS, which is a standard feature in more than 90% of the patients, and with the early onset mutation of that gene, it is considered a critical driver of PDAC initiation and progression [27]. Along with the oncogenic activation, inactivating mutations of the tumor suppressor gene CDKN2A/2B are also observed in more than 80% of the early-stage lesions, while later stages of PDAC exhibit inactivating mutations and deletions of tumor suppressor genes most prominently including TP53 and SMAD4 [28].

Metabolic reprogramming is considered a prominent hallmark of PDAC. Therefore, tackling this aggressive cancer might be possible through establishing a clear understanding regarding its metabolism in addition to genomics [29]. Recent studies have shown the crucial role of both glucose and glutamine metabolism in the progression of PDAC tumors that are regulated by the K-RAS oncogene to maintain tumor growth [30–32]. Inducible oncogenic K-RAS mouse model of PDAC showed—in addition to being a key driver of PDAC initiation—that it plays a central role in rewiring the tumor glucose metabolism by stimulating the glucose uptake and driving glycolysis intermediates toward nonoxidative pentose phosphate pathways [31]. It was also reported that the PDAC cells maintain the tumor growth by relying on the distinct pathway of glutamine metabolism and that this reprogramming is mediated by K-RAS [30].

Therefore, not only genomics biomarkers but also network reconstructions [33], including different omics levels, become an essential tool for exploring the disease under the systems biology perspective. Network models and computational platforms for integrating and analyzing these data, as well as investigating more thoroughly into these networks by simulations, are prominent efforts.

4. Coding and Noncoding RNA Signatures of Pancreatic Cancer

Initial transcriptome studies were performed for analysis of the mRNA profiles, which focused on protein-coding genes in PDAC. Thereafter, researchers compared gene expression levels between tumors and normal pancreas tissues and determined the genes with altered expression profiles in the disease state; this assisted in discovering potential diagnostic or prognostic biomarkers [34]. Over the years, microarray and RNAseq technology have been utilized not only to obtain coding but also non-coding RNA signatures. Although transcriptomic studies of non-coding RNAs are mainly focused on microRNAs (miRNA) and long non-coding RNAs (lncRNAs), other non-coding RNA types such as piwi interacting RNA (piRNAs), circular (circRNAs), small nucleolar RNA (snoRNA), and small nuclear RNA (snRNA) [35] are also promising biomarker candidates as they are quantitatively assessed, providing opportunities for noninvasive and early diagnosis of PDAC [20].

miRNAs involve in the expression of posttranscriptional regulatory mechanisms [36] and act as oncogenes or inhibit tumor suppressors in PDAC. Overexpression of the oncogene miRNAs (oncomir) increases in tumor progression, while tumor suppressors inhibit cell proliferation and induce apoptosis [37] by inactivating TP53, P16, and SMAD4 in PDAC [38]. miRNAs have the advantage of being stable in serum, hence these show remarkable potential as diagnostic biomarkers or a prognostic tool for noninvasive detection and convenient screening [39]. Therefore, the use of miRNA expression profiling has gained importance for the early detection of cancer [40,41].

Dysregulation of miRNAs in PDAC has been investigated not only in pancreatic tumors but also in blood samples, pancreatic juice, stool, urine, and saliva [39,42]. In several studies, the expression levels of miR-21, miR-155, and miR-196 have been reported to be upregulated in PDAC [43–46]. The higher concentration of miR-155 and miR-210 in the sera of pancreatic cancer patients as compared to normal healthy individuals has been proposed as a potential diagnostic marker in the early stages of pancreatic cancer [47,48]. Moreover, miR-155 and miR-21 were also found to have increased expression in pancreatic juices, while expressions are linked with histological progression characteristics [49]. In addition, the evaluation of more than 700 miRNAs in a study using blood samples compared between pancreatic cancer patients and healthy individuals emphasized miR-1290 as a promising biomarker [50]. Likewise, multiple studies have proposed not only miR-21, miR-155, miR-196, and miR-1290 but also miR-200, miR-18a, miR-210, miR-192, miR-22, miR-642b, miR-885-5p, and miR-375 as candidate biomarkers for PDAC patients [47,51–55]. Another comparison between cancer patients and healthy individuals clearly showed a distinct miRNA expression profile that included upregulation of miR-21, miR-23a, miR-31, miR-100, miR-143, miR-155, miR-2214, and downregulation of miR-148a, miR-375, and miR-217 [43].

The combination of various biomarkers such as CA19-9 with miR-16 and miR-196a provoked distinct improvement to distinguish between PDAC patients and healthy controls [56]. Similarly, the miR-27a-3p expression profile coupled with CA19-9 differentiated PDAC patients and healthy controls with a sensitivity and specificity of more than 80% [57,58]. Among diagnostic features of miRNAs, poor survival in PDAC patients was determined regarding overexpression of miR-221/222 and miR-744 levels in tumor tissue and plasma, respectively, as well as low-expression levels of miR-218 and miR-494 in tumor tissue [59–62].

In addition to microRNAs, other non-coding RNAs—such as long non-coding RNAs (lncRNAs), small nuclear RNAs (snRNAs), or circular RNAs (circRNAs)—have also been identified that might have potential as diagnostic or prognostic markers for PDAC. Long non-coding RNAs (lncRNAs) consist of more than 200 nucleotides, and some of them are circulating in body fluids which makes them promising markers for disease detection [63]. Although the biological functions of lncRNAs are not fully understood, the expression of lncRNAs (HOTAIR, MALAT-1, GAS5, MEG3, HULC, BC008363, and HSATII) showed significant alterations in pancreatic cancer cell lines. Besides, HOTAIR and PVT1 had higher concentrations in saliva in PDAC patients than saliva taken from healthy individuals. Therefore, these lncRNAs in saliva offer a potential noninvasive detection method for PDAC [35]. To date, U2snRNA, which is overexpressed in PDAC, has been the only reported snRNA biomarker in PDAC patients [64].

Circular RNAs (circRNAs), as another type of non-coding RNAs, have drawn increased attention through their regulatory roles in cancer. Generally, these are generated from precursor mRNA (pre-mRNA) by canonical splicing and head-to-tail back splicing, which makes them circular. Moreover, their structure without a polyA tail makes circRNAs favorably insensitive to ribonuclease and more desirable as clinically useful biomarkers. These function as miRNA sponges and overwhelm the ability of the miRNA to bind its mRNA targets [65]. Therefore, the associations of miRNAs and circRNAs with their potential regulatory role were also investigated in PDAC. For instance, hsa_circ_0005785 is potentially able to bind miR181a and miR181b as "oncomiRs" in pancreatic cancer, while miR-181a plays a critical role in regulating cancer growth and migration [66]. In another study, two upregulated circRNAs (hsa_circ_0001946, hsa_circ_0005397) and five downregulated circRNAs (hsa_circ_0006913, hsa_circ_0000257, hsa_circ_0005785, hsa_circ_0041150, and hsa_circ_0008719) were proposed as biomarkers after microarray analysis. They also validated the expression pattern of the above seven proposed circRNAs via qRT-PCR in PDAC tissues and adjacent normal tissues [67]. More recently, circRNAs expression in PDAC was explored by comparing PDAC tissues versus normal tissues by using microarray again. As a result, 256 differentially expressed circRNAs and 20 differentially expressed miRNAs were proposed to be associated with PDAC development [68].

Seimiya and coworkers [69] applied circular RNA-specific RNA sequencing and determined more than 40,000 previously unknown circRNAs that were altered in PDAC. Their research resulted in a novel circRNA, named circPDAC RNA, with no peptide production but the aberrant expression in PDAC tissues as well as patient serum. Another recent study involving a 208-case cohort of patients with PDAC identified a novel circRNA, named circBFAR or hsa_circ_0009065. The expression of circBFAR correlated positively with the tumor-node-metastasis stage and was related to the poor prognosis of patients with PDAC. Likewise, circBFAR knockdown dramatically inhibited the proliferation and motility of PDAC cells in vitro and their tumor-promoting and metastatic properties in the in vivo models [70]. A recent systematic review designating the roles of circRNAs in pancreatic and biliary tract cancers gathered detailed information and provided an understanding of the role of circRNAs in pancreatic cancer [71].

In recent studies, single-cell transcriptomics has paved the way to elucidate molecular biomarkers for early diagnosis of PDAC. Peng et. al. [72] found that a subset of ductal cells with unique proliferative features were associated with an inactivation state in tumor-infiltrating T cells, providing novel markers for the prediction of an antitumor immune response. EGLN3, MMP9, and PLAU have been reported as participating in PDAC carcinogenesis regarding dysregulated gene expression in malignant ductal cells [72]. In another single-cell RNA-sequencing study, sampling was from the mouse pancreas during the progression from preinvasive stages to tumor formation. While metaplastic cells were found to express two transcription factors, ONECUT2 and FOXQ1, the altered expression profiles of MARCKSL1, MMP7, and IGFBP7 were also observed, which could be accomplished as candidate markers for early detection of PDAC [72].

Consequently, findings provided by transcriptomic analysis of PDAC have been a valuable resource not only for deciphering the intra-tumoral heterogeneity and disease mechanism but also suggesting potential biomarkers for diagnosis, targeted therapy, or immunotherapy.

5. Proteomic Signatures of Pancreatic Cancer

Proteomics is a powerful approach that encompasses an extensive range involving the systematic analysis of protein structure, function, expression, protein–protein interactions, and posttranslational modifications [73]. Over many years, proteomics has been a key player for researchers to pinpoint biomarkers, which can be used as a tool for a faster disease diagnosis, prognosis, and enhanced treatment [74,75]. In terms of making contributions to clinical disease prediction, protein-based biomarkers are promising. The analysis and verification of unique protein biomarkers have been achieved by using highly sensitive and reliable mass spectrometry-based proteomics. Moreover, this technique is crucial in terms of querying protein modifications [20]. Numerous clinical specimens of pancreatic cancer such as pancreatic juice, pancreatic tumor tissue, pancreatic cyst fluid, urine, and plasma/serum have become targets for the proteomics field to dig into mechanisms of disease, improve novel biomarkers, and enhance drug development [76–78]. Identifying proteins or peptides detected in body fluids in cases of cancer might be useful for the early diagnosis of PDAC [78].

Sample type is a critical concern for the study of biomarkers. Since blood serum or plasma is convenient for periodic collections and includes a reproducible quantification, it is presumably the most preferred option. Although blood samples are easily accessible and noninvasive, the fundamental disadvantage of blood collection for the discovery of novel biomarkers is that not every protein carrying diagnostic potential is secreted into the bloodstream [79]. Investigation of the human pancreatic proteome has been done in patients with premalignant neoplasia, PDAC, and benign pancreatic disease. Although one of the most potent samples from the pancreas is the pancreatic juice, involving a high amount of proteins that might display the disease status, its collection is onerous since this procedure requires an endoscopy and cannulation of the pancreatic duct [80–86]. Collecting and conserving the intact tumor tissue and adjacent normal tissue is challenging due to the presence of digestive enzymes secreted by the pancreas. Nonetheless, pancreatic tissue is considered an excellent specimen for investigation of the pathological mechanisms underlying PDAC as well as for determining drug targets in virtue of its proximity to the lesion and its greater ingredient of tumor-related proteins [87]. Pancreatic cysts, which possess peculiarly stagnated fluids, are extensively seen as the most hopeful origin for the discovery of potential biomarkers since these tend to turn into pancreatic cancer [88]. In terms of urine, this is an effortlessly approachable biological specimen for biomarker detection, and its proteins are generated from both glomerular filtration and kidney [89]. Due to their accessibility and noninvasiveness, various urinary protein biomarkers have been examined to improve clinical assays for the diagnosis of several cancer types. As yet, merely a restricted amount of proteomics studies have been carried out to investigate the urinary proteome [90].

A retrospective study using a comprehensive proteomic analysis of pancreatic juice and pancreatic cell line samples from PDAC patients demonstrated that regenerating Family Member 1 Beta (REG1B) and syncollin (SYCN) could represent potential PDAC biomarkers [84,91]. Sogawa et al. [92] carried out a comparative proteomics analysis using a tandem mass tag (TMT) labeling and demonstrated that C4b-binding protein α-chain (C4BPA) is a novel serum biomarker for the early diagnosis of PDAC as well as for discrimination between PDAC and other gastroenterological cancers. Based on the results of a combinatorial proteomics strategy, Yoneyama et al. [93] indicated that insulin-like growth factor-binding proteins, IGFBP2 and IGFBP3, are compensatory biomarkers that can allow more accuracy through the combination with CA19-9 for the early detection of PDAC. In an MS-based proteomic study, Guo et al. [94] have demonstrated that dysbindin as a potential biomarker improved the accuracy of diagnosis in distinguishing PDAC from other pancreatic diseases. In a recent study, Cohen et al. [95] observed that the combination of testing circulating tumor DNA (ctDNA) with protein biomarkers (CA19-9, CEA, hepatocyte growth factor (HGF), and osteopontin) shows better performance than the CA19-9 test alone to distinguish PDAC from healthy controls. The improved accuracy of the biomarker panel—which is composed of a gold standard biomarker CA19-9, tissue factor pathway inhibitor (TFPI), and an isoform of tenascin C (TNC- FNIII-B)—in the differentiation of early-stage PDAC from different diseases was also demonstrated in a clinical cohort study [96]. In addition, Capello et al. [97] reported that the combination of TIMP1, LRG1, and CA19-9 performed better diagnostic accuracy than CA19-9 alone in differentiating early-stage PDAC from benign PDAC. Kim et al. [98] identified another biomarker panel that has high plasma THBS-2 and CA19-9 concentrations, which showed a remarkable differentiation ability between PDAC and healthy patients with 87% sensitivity and 98% specificity. The clinical significance of serum survivin was also reported in PDAC patients [99].

The pancreatic ductal fluid has been proposed as a good biological fluid for identifying prognostic biomarkers [100]. Focusing on the content of the ductal fluid, high concentrations of mucins and S100A8 or S100A9 were associated with the low survival rate in PDAC [100]. Ger et al. [101] recently investigated the proteome of 37 samples from pancreatic cancer and healthy subjects and identified that FLT3 and PCBP3 are promising prognostic biomarkers of pancreatic cancer.

Targeted proteomics is a rapidly evolving technological tool that conceptually represents an important advancement in alleviating the bottleneck in the preclinical biomarker assessment processes. In a targeted proteomics pilot study [102], five pancreatic cancer biomarker candidates—including 14-3-3 protein sigma, gelsolin, lumican, transglutaminase 2, and tissue inhibitor of metalloproteinase 1—were investigated in 60 plasma samples using a simple and robust selected reaction monitoring (SRM) multiplexed assay. Their results showed that gelsolin, lumican, and tissue inhibitor of metalloproteinase 1 have better area under curve (AUC) values than CA19-9 to discriminate pancreatic cancer from healthy controls and chronic pancreatitis controls. Yoneyama and colleagues [103] developed a quantification method specific for α-fibrinogen hydroxylated at proline residues 530 and 565 by SRM/multiple reaction monitoring (SRM/MRM). To validate these modifications as pancreatic cancer biomarkers, they quantified these posttranscriptional modifications in plasma samples from 70 pancreatic cancer patients and 27 healthy controls. They demonstrated that the plasma concentration of proline-hydroxylated α fibrinogen is significantly greater in pancreatic cancer patients.

In light of the rapidly developing accuracy and efficiency of proteomic approaches, our knowledge of the underlying molecular mechanism of pancreatic cancer has greatly increased [104,105]. However, there are still various limitations and analytic challenges that have resulted from the dynamic nature of the proteome of tissues and cells and the variation in the forms and functions of proteins due to several modifications [106]. Although several standardizations and improvements are required, proteomics is certainly a promising approach for the early diagnosis, prognosis, and discovery of targets for the treatment of pancreatic cancer.

6. Metabolomic Signature of Pancreatic Cancer

Metabolomics or metabolite profiling is a novel promising approach for the identification of robust biomarkers for diagnosis, prognosis, and assessment of treatment in pancreatic cancer [107–111]. Although there is currently no clinically validated metabolic biomarker that can help to provide early diagnosis of pancreatic cancer, the number of studies focusing on metabolic profiling and phenotyping of pancreatic cancer is increasing drastically [111–114]. As compared to other omics technologies, metabolic phenotyping is a sensitive indicator due to rapid and more precise results for new biomarker discovery [115]. The largest case-control study to discover a blood-derived metabolic biomarker signature that enables one to distinguish PDAC from chronic pancreatitis (ChP) was conducted by Mayerle et al. [114]. They investigated metabolomic profiles of plasma and serum samples from 914 subjects (patients with PDAC, ChP, liver cirrhosis, healthy, and non-pancreatic disease control), and a tumor biomarker signature (nine metabolites and additionally CA 19-9) was identified for differential diagnosis between PDAC and ChP with an AUC of 0.96. In a retrospective study investigating tissue metabolomics from 25 pancreatic cancer patients who had to undergo tumor resection surgery and gemcitabine-based adjuvant therapy, high lactic acid levels were observed in patients with poor clinical outcomes after gemcitabine therapy. Moreover, the combined evaluation of hENT1 with lactic acid showed superior performance in differentiating patients according to their overall survival [116]. In another study, Battini et al. [117] investigated tissue samples from 106 patients after PDAC resection to find metabolic biomarkers associated with long-term survival using metabolomic analysis methods. While the network analysis results revealed that higher levels of glucose, ascorbate, and taurine associated with long term survivors, decreased levels of choline, ethanolamine, glycerophosphocholine, phenylalanine, tyrosine, aspartate, threonine, succinate, glycerol, lactate, glycine, glutamate, glutamine, and creatine were estimated in long-term survivors. Due to the association of higher ethanolamine levels with worse survival, the metabolite with the highest accuracy in distinguishing between long-term and short-term survivors was ethanolamine.

An animal study was conducted to obtain metabolite profiling of pancreatic intraepithelial neoplasia (PanIN) and PDAC tissue samples from rats. They observed that the levels of kynurenate and methionine decreased in PDAC but increased in PanIN, demonstrating the potential of these metabolites to be biomarkers to differentiate PDAC from PanIN [116,118]. Laconti et al. identified that circulatory metabolite signatures can be used to differentiate animals with early-stage lesions with a diagnostic accuracy of 81.5% and 73.2% respectively [110].

Since the metabolic changes are quite important to detect and treat cancer regardless of the disease stage [119], genome-scale metabolic models (GEMs) might be a very helpful source to create and/or test the hypothesis for the elucidation of physiological mechanisms or novel biomarkers [120,121] so that GEMs can be used as a tool in both "top-down" and "bottom-up" methods in the context of biomarker discovery. GEMs have been employed for studying cancer metabolism utilizing either generic/personalized or tumor/cell-specific methods, which may translate into clinically relevant applications. They can also be used to identify drug targets leading to inhibition of cancer-related phenotypes or drug resistance in cancer therapy. Furthermore, the fortification of GEMs can be obtained via the integration of omics data like genomic, transcriptomic, and proteomic data, as well as the incorporation of regulatory molecules to the metabolism [122]. GEMs also provide valuable insight into the interaction between cancer cells and supporting cells in their niches as paving the way for whole-cell modeling [123,124].

In addition to all these, there are still some challenges in metabolomic studies. Whether significant changes in the metabolite level are due to the occurrence of the targeted disease, the use of non-confirmed metabolites with small sample size and the variability of patients' parameters would affect the accuracy and reliability of the results [125]. Therefore, further standardization and improvement of currently available metabolomics techniques is a prospective requirement for the designation of highly accurate biomarkers that will provide significant clinical benefits and may help to obtain new target signatures for accurate diagnosis, imaging, and possible therapeutic options [126,127].

7. Glycomic Signatures of Pancreatic Cancer

Cancer studies are performed mostly based on alterations in genome, transcriptome, proteome, and metabolome levels, with a relatively small number of studies in alterations in glycan compositions and/or structures and glycoproteins [128]. However, the glycan studies have been increasing day by day to identify potential glycan alterations and glycoprotein biomarkers for cancer owing to the developments in glycans profiling [129]. In cancer cells, alterations in carbohydrate structures of secreted proteins are functionally significant and may offer promising targets to develop potential diagnostic and therapeutic strategies [130–132].

Since pancreatic cancer does not indicate any noticeable symptom during the early stages, it is a very difficult cancer type to diagnose [131]. It is an important challenge to detect new diagnostic biomarkers for pancreatic cancer. The glycoproteome occurring after co-translational or posttranslational modifications (PTM) and its role in the mechanism of pathogenesis have not been explained completely in pancreatic cancer. Besides, the available information about glycoproteome in normal pancreas and pancreatic cancer is very limited [133,134].

Glycosylation—the covalent attachment of a glycan to protein, lipid, carbohydrate, or other organic molecules—is the most common and complex PTM of proteins and significantly affects the function of proteins. Glycosylation of proteins plays an important role in various biological functions, including immune response and cellular regulation. Abnormal glycosylation is accepted as a molecular characteristic of transformation into malignant tumors for many epithelial cancers, including PDAC. Therefore, targeting aberrant glycosylation associated with cancer would be a useful approach to improve accurate diagnosis and possibly therapeutic strategies [129,133].

Several studies were published about glycan alterations and glycoproteome in pancreatic cancer. Pan et al. investigated protein N-glycosylation in pancreatic tumor tissue compared to the normal pancreas and chronic pancreatitis tissue through a quantitative glycoproteomics approach using HPLC and MS. This study presented a set of glycoproteins having aberrant N-glycosylation levels in pancreatic cancer, including mucin-5AC (MUC5AC), carcinoembryonic antigen-related cell adhesion molecule 5 (CEACAM5), insulin-like growth factor binding protein (IGFBP3), and galectin-3-binding protein (LGALS3BP) [133]. MUC5AC and CEACAM5 have been shown to play a role in tumor progression and metastasis in pancreatic cancer [133,135,136]. On the other hand, LGALS3BP was significantly hyperglycosylated in tumor tissue. Additionally, increased N-glycosylation on many cancer-associated aberrant glycoproteins was reported on pancreatic cancer-associated pathways such as TGF-β, TNF, NF-kappa-B, and TFEB-related lysosomal changes [133].

Yue et al. studied sera from pancreatic cancer patients to determine certain glycan alterations and their possible usage in the diagnosis of pancreatic cancer. To that end, they characterized glycan and protein levels of specific mucins and carcinoembryonic antigen-related proteins of these patients through the antibody-lectin sandwich array method previously developed. They found that MUC16 protein was frequently increased (65% of the patients) in the cancer patients, whereas MUC1 (30%) and MUC5AC (35%) proteins were less frequently elevated. In addition to this, MUC1 and MUC5AC proteins indicated highly extensive and diverse glycan alterations, while MUC16 protein did not. The most frequent glycan elevations that affected these proteins involved the Thomsen–Friedenreich antigen, fucose, and Lewis antigens. Additionally, they reported an unanticipated enhancement in the exposure of alpha-linked mannose on MUC1 and MUC5AC. Moreover, the CA19-9 on MUC1 had the most important increase (87%) in cancer patients with 4% of the control subjects [130].

In another study, N-glycosylation at Asn88 in serum human pancreatic ribonuclease 1 (RNase1) was substantially elevated in pancreas cancer patients compared with normal human subjects [131]. Similarly, increased fucosylation levels of serum α-1-acid glycoprotein (AGP) glycoforms were reported in pancreatic cancer compared to healthy controls and pancreatitis patients via numerous analytical methods consisting of MS, capillary zone electrophoresis (CZE), and enzyme-linked lectin assays (ELLA) [134].

As an alternative therapy option having fewer adverse effects than others, regional intra-arterial chemotherapy (RIAC) is preferred for advanced pancreatic cancer. Qian and colleagues [137] took advantage of the presence of Glypican-1 (GPC1) in extracellular vesicles (EVs) to determine if the change in GPC1+ cells in EVs could be a predictor of the consequences of RIAC for advanced pancreatic cancer patients. They concluded that patients with advanced pancreatic cancer who displayed a decrease in GPC1+ EVs experienced enhanced overall survival rates with the aid of RIAC therapy.

Another cell-surface glycoprotein, CD44 is a known prognostic biomarker and therapeutic target in pancreatic cancer [138]. The overexpression of CD44 was shown to be associated with aggressive malignant attitudes, cell migration, and distance metastasis, therefore with poor overall survival in patients with pancreatic cancer [138]. On the other hand, the reduction in CA19-9 levels envisaged a good prognosis after neoadjuvant therapy with a low incidence of recurrence after surgery [139].

All of these studies provide an insight into the potential biomarker candidates for effective diagnosis, prognosis, and treatment in pancreas cancer using measurements in glycan alterations on precise glycoproteins.

8. Metagenomic Biomarkers of Pancreatic Cancer

In recent studies, the interaction between microbiomes and the initiation and progression of pancreatic cancer has become recognized, raising the possibility of identifying novel diagnostic and prognostic factors for PDAC [140]. The existence of intratumoral microbiota is considered to have a potential etiologic impact on pancreatic carcinogenesis, including inflammation, immunosuppression, and stimulation of cellular carcinogenic pathways [141–143].

It is becoming clear that there is a correlation between oral microbiota and PDAC, and the abnormalities of oral microbiota have been proposed to appear before the development of cancer [144]. Available literature data provide knowledge on the oral bacteria that might play a pathogenic role in the progression of PDAC, and these are *Porphyromonas gingivalis*, *Fusobacterium*, *Neisseria elongata*, and *Streptococcus mitis* [145]. In this context, a large metagenomic study comparing PDAC patients and healthy controls revealed that *P. gingivalis* was associated with an approximately 60% greater risk of PDAC [146]. Mitsuhashi et al. [147] indicated that the existence of approximately 10% *Fusobacterium* in pancreatic cancer tissue is independently associated with poor prognosis of PDAC but not with its clinical and molecular features. It is also thought that *Fusobacterium* species may be a candidate prognostic biomarker for pancreatic cancer and should be considered for further oral microbiota studies. On the other hand, some studies have revealed that *Fusobacteria* are associated with reduced risk of PDAC, revealing that the role of *Fusobacteria* on PDAC could be controversial [144,146,148].

Fecal microbial transplantation (FMT) possesses an enormous amount of microbiota compared to usually preferred probiotic supplements and might provide a significant movement in reducing the immunosuppression and in increasing the response rate to treatment in cancer patients having a probable low survival [149]. In a recent cohort study,

Riquelme and colleagues [150] made a metagenomic analysis from 68 tumor samples of tumor microbiome composition of PDAC patients with short-term survival (STS) and long-term survival (LTS) phenotypes using 16S rRNA gene sequencing. They reported that the tumor microbiome diversity of long-term survivors was higher than that of short-term survivors, potentially representing a strong interrelation between the gut microbiome and patients' survival rate. Besides, animal studies by human-into-mice FMT experiments from STS, LTS, and healthy donors conspicuously confirmed that the transference of the long-term survivors' gut microbiome can modulate the intratumoral microbiome. According to a study encompassing a comparative analysis of fecal microbiota from PDAC patients and control donors in murine models, a certain type of bacteria—namely, Proteobacteria, Actinobacteria, Fusobacteria, and Verrucomicrobia—are found in higher amounts in the gut of PDAC patients. Specifically, the gut microbiota of PDAC patients contains greater amounts of Proteobacteria (45%), Bacteroidetes (31%), and Firmicutes (22%). This study remarkably highlights that the intratumoral microbiome associated with pancreatic cancer has relatively distinct proportions in comparison to the microbiome of normal pancreatic tissue [143].

In an animal study, Mendez et al. [151] demonstrated a substantial correlation between microbial dysbiosis and the release of tumor-inducing metabolites in the early-stage, while showing significantly elevated serum polyamine concentrations in PDAC patients; this may be postulated as a predictive biomarker for early detection of pancreatic cancer. It is among the current assumptions that bacteria in the pancreatic microbiome may contribute to the resistance of gemcitabine, which is widely used in the treatment of PDAC. Based on this assumption, 76% of the tested pancreatic tissue was found to be positive for bacteria, particularly Gammaproteobacteria [152].

Several studies also suggest that the composition of oral [146,148,153], fecal [154], and pancreatic microbiome [143,155] may be used for early diagnosis of PDAC. With the accumulation and advanced evaluations of data on the pancreas, gut, and oral microbiota, it might be possible to develop microbiome screening methods that can be considered as a promising tool in the prediction of PDAC risk and treatment of disease progression.

9. Biomarkers Leading to Improved Personalized Medicine

On the way to personalized medicine, there are promising and on-going efforts for the integration of multi-omic data. As an aim of precision medicine, the first attempt is to stratify patients according to their disease subtypes, biomarkers, clinical features, or demography. Later, in addition to the stratification process, more features such as environment, medication history, behaviors, and habits are utilized to create smaller groups. In theory, this stratification technique should avoid failures in clinical trials since the suitable diagnosis and targeted treatments are applied to small patient populations or directly to individuals. Instead of "one-size-fits-all" treatment approaches, the best therapy options or medications for each individual or a small group can be achieved through disease stratification and then personalization by the integration of multi-omics networks. In addition, personalized medicine treatment necessitates the co-development of diagnostic tools (preferably within noninvasive methods) to characterize the ideal therapy for patients. There is an urgent need for multi-omic data integration not only for pancreatic cancer but also for many other diseases from the personalized medicine perspective in the future (Figure 1).

According to the present clinical data, using only chemotherapeutic approaches in the treatment of pancreatic cancer will likely be insufficient in terms of the increase in survival time and response rate in the near future. Therefore, there is an urgent need for precision medicine, which aims at tailoring the best treatment option for individual patients based on their genomic information, together with molecular, environmental, and lifestyle factors, to identify the suitable biomarkers and targeted therapies for cancer patients. Personalized medicine stratifies the patients by considering the individual differences among cancer patients, unlike conventional therapy. As in other types of cancer, studies on precision medicine in pancreatic cancer have increased in recent years [156,157].

There are several precision medicine programs and clinical studies run by various initiatives from different countries to offer the best personalized treatment options for pancreatic cancer patients according to their molecular tumor profiling [156]. These programs have demonstrated that a small patient cohort had better progression-free survival after switching their therapies from standard-of-care treatment to molecular-targeted therapy [158]. Further, molecular profiling of tumors from patients with all stages of pancreatic cancer was performed using NGS to develop response rates and therapeutic biomarkers [159]. Besides, different clinical studies were performed to discover biomarkers for prognosis or treatment response [160], focusing on alterations in genome and epigenome in tumor tissue [161]. The Comprehensive Molecular Characterization of Advanced Pancreatic Ductal Adenocarcinoma for Better Treatment Selection (COMPASS) trial was the prospective translational study that investigated the feasibility of comprehensive real-time genomic analysis of advanced PDAC, integrating genomic and transcriptomic subtypes and chemotherapy response [162].

The alterations in the genome, epigenome, proteome, and metabolome cause the changes in the phenotype in pancreatic cancer, and thus studies carried out on these alterations could help with the stratification of pancreatic cancer. The identification of new biomarkers for subtyping, diagnosing cancer, and predicting therapy response is an ongoing process in preclinical studies. However, the difficulties in the translation of promising preclinical findings into clinical practice make the application of precision medicine approaches in clinics a great challenge. These difficulties arise from the evaluation of basic science findings in the clinical settings and the selection of the best effective scientific data for clinical trials [156]. Moreover, it is very important and vital to building collaborations among basic scientists, clinicians, and bioinformaticians to overcome these challenges.

For patients with pancreatic cancer, CA19-9 is the only routinely used serum biomarker in prognosis and early diagnosis of recurrence after therapy [156]. Although the increase in CA19-9 level indicates advanced pancreatic cancer and poor prognosis [139], this elevation can be only observed in 65% of the patients with resectable pancreatic cancer, in addition to patients with other diseases such as pancreatitis or cirrhosis [163]. Besides, 10% of patients with pancreatic cancer cannot synthesize CA19-9 even if they are in the advanced stage, since they are negative for Lewis antigen a or b. Moreover, it is not a screening biomarker for pancreatic cancer to be used alone [156].

Numerous gene alterations that play important roles in tumorigenesis can provide the development of novel treatments that target specific genes for pancreatic cancer patients. Personalized medicine can certainly improve the management of patients and outcomes of novel treatments with the administration of the right therapy using the right dose at the right time to the right patient when applied to pancreatic cancer patients. The generation of well-designed clinical trials allowing the construction of molecular profiling of tumors of patients will further guide the development of novel and effective strategies for the overall survival of patients in this highly lethal cancer [157,160].

10. Conclusions

There are big initiatives, various research programs, and databases in which researchers are able to collect different omics datasets of pancreatic cancer. However, many biomarker studies have been challenged by low case numbers, non-specificity of molecular markers and their low reproducibility, and the absence of preclinical or clinical as well as feasibility studies.

The well-known example of pancreatic cancer biomarkers is CA19-9, but as a single biomarker it cannot offer a potential to be used in the clinic. Recent studies on non-coding RNAs such as miRNAs, circRNAs, and lncRNAs hold great promise not only as biomarkers but also for understanding the regulatory network components in pancreatic cancer. Targeted or shotgun proteomic approaches also provide an opportunity for more sensitive or novel biomarker identification. Metagenomics is another emerging technique that measures altered microorganism abundance and may act as a potential biomarker. On the other hand, although the pancreas is at the center of many metabolic pathways, the metabolic rewiring of pancreatic cancer is an underestimated topic since the number of metabolomics studies are not as numerous as some of the other omics investigations.

Although many novel markers have been discovered through omics studies of PDAC in the past decade, none of those novel biomarkers have yet been brought into routine clinical practice. However, there is a hope that various combinations of these biomarkers as a biomarker panel may result in a clinical output, and this fact makes the integration of multi-omics data more challenging on the way to translating omics markers into the clinic.

Another point that has a crucial role in translation to the clinic is sampling, where body fluids are favorable for the detection of the biomarkers. Later, these biomarkers also assist oncologists in deciding optimal therapeutic management by defining the way for precision treatment.

In conclusion, there is great attention focusing on multi-omics biomarkers in terms of their diagnostic, predictive, and prognostic potentials to fight against pancreatic cancer as well as other cancer types. One of the major medical concerns raised by oncologists is the identification of robust, reasonable, and reliable diagnostic biomarkers since early detection of pancreatic cancer is crucial for personalized therapy options and improved survival outcomes. This strategy can be accomplished by a systems biology approach that aims to organize multi-omics data despite the challenges. Successfully accomplishing multi-omics data integration by systems biology approaches will fulfill future expectations and the need for robust, accurate, and feasible biomarker panels for pancreatic cancer.

Table 1. A summary of methodology and sampling used in biomarker studies for pancreatic cancer.

"Omic" Level Description		Sample Origin	Altered Molecule/Microorganism	Expression Pattern	Detection Method *	Reference Study
Genomics	Mutation	Pancreatic tissue	CDKN2A, CDKN2B, TP53, SMAD4, KRAS	-	WES/WGS	[28]
Transcriptomics	Coding RNAs	T cell	EGLN3, PLAU	Downregulated	scRNA-seq	[72]
		T cell	MMP9	Dysregulated	scRNA-seq	[72]
		Mouse pancreatic tissue	ONECUT2, FOXQ1, MARCKSL1, MMP7, IGFBP7	Upregulated	scRNA-seq	[164]
		Tumor tissue	hsa_circ_100782	Upregulated	Microarray/qRT-PCR	[71]
	circRNAs	Tumor tissue/plasma/cell lines	hsa_circ_0006988	Upregulated	qRT-PCR	[165]
		Tumor tissue/cell lines	hsa_circ_0099999 (circZMYM2)	Upregulated	circRNA overexpression	[166]
		Tumor tissue	hsa_circ_0006215	Upregulated	circRNA overexpression	[167]
		Tumor tissue, plasma exosome	circ-IARS	Upregulated	circRNA overexpression	[168]
		Tumor tissue	circ-PDE8A	Upregulated	circRNA overexpression	[169]
		Tumor tissue/cell	hsa_circ_0001649	Downregulated	Microarray/qRT-PCR	[170]
		Tumor tissue/cell	hsa_circ_0005397 (circ-RHOT1)	Upregulated	Microarray/qRT-PCR	[171]
		Tumor tissue/cell lines	hsa_circ_0030235	Upregulated	circRNA overexpression	[172]
		Tumor tissue/cell lines	hsa_circ_0007534	Upregulated	circRNA overexpression	[173]
		Tumor tissue/cell lines	ciRS-7 (Cdr1as)	Upregulated	qRT-PCR	[174]

Table 1. *Cont.*

"Omic" Level Description	Sample Origin	Altered Molecule/Microorganism	Expression Pattern	Detection Method *	Reference Study
	Tumor tissue	hsa_circ_0007334	Upregulated	Microarray/qRT-PCR	[175]
	Tumor tissue	circLDLRAD3	Upregulated	circRNA knockdown	[176]
	Tumor tissue/cell	circASH2L	Upregulated	Microarray/qRT-PCR	[177]
	Tumor tissue/cell lines	circADAM9	Upregulated	circRNA knockdown	[178]
	Tumor tissue/cell	hsa_circ_001653	Upregulated	circRNA knockdown	[179]
	Tumor tissue/cell	circHIPK3	Upregulated	circRNA knockdown	[180]
	Tumor tissue/cell	circFOXK2	Upregulated	circRNA knockdown	[181]
	Tumor tissue	hsa_circ_0009065 (circBFAR)	Upregulated	circRNA overexpression	[70]
	Tumor tissue	hsa_circ_0086375 (circNFIB1)	Downregulated	circRNA knockdown	[182]
	Tumor tissue/cell	hsa_circ_0013912	Upregulated	circRNA overexpression	[183]
	Tumor tissue/cell lines	hsa_circ_001587	Downregulated	circRNA knockdown	[184]
	Tumor tissue	hsa_circ_0001946, hsa_circ_0005397	Upregulated	Microarray/qRT-PCR	[67]
	Tumor tissue	hsa_circ_0005785, hsa_circ_0006913, hsa_circ_0000257, hsa_circ_0041150, hsa_circ_0008719	Downregulated	Microarray/qRT-PCR	[67]
	Plasma	miR-21	Upregulated	Microarray/qRT-PCR	[49]
	Pancreatic juice	miR-155	Upregulated	qRT-PCR	[49]
	Tumor tissue/cell lines	miR-196a	Upregulated	Microarray/qRT-PCR	[185]
	Tumor tissue	miR-210	Upregulated	qRT-PCR	[186]
	Tumor tissue/cell line/serum	miR-1290	Upregulated	Microarray/qRT-PCR	[50]
	Tumor tissue/cell lines	miR-200a/miR-200b	Upregulated	Microarray/qRT-PCR	[51]
	Tumor tissue/plasma/serum	miR-18a	Upregulated	qRT-PCR	[55]
	Tumor tissue	miR-192	Upregulated	Microarray/qRT-PCR	[187]
	Blood	miR-22-3p/miR-642b/miR-885-5p	Upregulated	qRT-PCR	[188]
miRNAs	Tumor tissue	miR-23a/miR-31/miR-100/miR-143/miR-221	Upregulated	qRT-PCR	[43]
	Tumor tissue	miR-148a/miR-375/miR-217	Downregulated	qRT-PCR	[43]
	Plasma	miR-16 and miR-16 and miR-196a and CA 19-9 combination	Upregulated	qRT-PCR	[56]
	Peripheral Blood Mononuclear Cells	miR-27a-3p with CA 19-9	Upregulated	RNA-seq/qRT-PCR	[57]
	Tumor tissue/cell lines	miR-221/miR-222	Upregulated	qRT-PCR	[185]
	Tumor tissue/plasma	miR-744	Upregulated	Microarray/qRT-PCR	[62]
	Tumor tissue	miR-218	Downregulated	Microarray/qRT-PCR	[189]
	Tumor tissue	miR-494	Downregulated	Microarray/qRT-PCR	[46]

Table 1. *Cont.*

"Omic" Level	Description	Sample Origin	Altered Molecule/Microorganism	Expression Pattern	Detection Method *	Reference Study
		Tumor tissue	HOTAIR	Upregulated	qRT-PCR	[35]
		Tumor tissue	PVT1	Upregulated	qRT-PCR	[190]
		Tumor tissue	MALAT-1	Upregulated	qRT-PCR	[191]
		Tumor tissue	Gas5	Upregulated	qRT-PCR	[192]
		Tumor tissue	MEG3	Upregulated	qRT-PCR	[193]
		Tumor tissue	HULC	Upregulated	qRT-PCR	[194]
	Other ncRNAs	Tumor tissue	BC008363	Upregulated	Microarray/qRT-PCR	[195]
		Tumor tissue	HSATII	Upregulated	RNA-seq	[196]
		Serum/plasma	U2snRNA	Upregulated	Microarray/qRT-PCR	[197]
		Pancreatic juice and cell line	REG1B/SYCN	Upregulated	ELISA	[84]
		Serum	C4BPA	Upregulated	TMT labeling	[92]
Proteomics	Proteins	Plasma	IGFBP2/IGFBP3	Upregulated	Antibody-based and LC-MS/MS-based	[93]
		Serum	DTNBP1	Upregulated	MS	[94]
		Plasma	ctDNA with CA19-9, CEA, HGF, and osteopontin	Upregulated	Luminex bead-based immunoassays	[95]
		Plasma	Combination of CA19-9, TFPI, and TNC- FNIII-B	Upregulated	ELISA	[96]
		Plasma	Combination of TIMP1, LRG1, and CA19-9	Upregulated	ELISA	[97]
		Plasma	THBS-2 and CA19-9	Upregulated	ELISA	[98]
		Serum	Survivin	Upregulated	ELISA	[99]
		Pancreatic ductal fluid	Mucins and S100A8 or S100A9	Upregulated	MS	[100]
		Tumor tissue	FLT3, PCBP3	Upregulated	HDMS	[101]
		Tumor tissue	Combination of hENT1 and lactic acid		GC/TOF-MS	[116]
		Tumor tissue	Glucose, ascorbate, ethanolamine, and taurine	Upregulated	HRMAS-NMR	[117]
		Tumor tissue	Choline, ethanolamine, glycerophosphocholine, phenylalanine, tyrosine, aspartate, threonine, succinate, glycerol, lactate, glycine, glutamate, glutamine, and creatine	Downregulated	HRMAS-NMR	[117]
Metabolomics	Metabolites	Rat tumor tissue	Kynurenate and methionine	Downregulated	NMR	[116]
		Tumor tissue	N-glycosylation of MUC5AC, CEACAM5, IGFBP3, and LGALS3BP	Upregulated	HPLC, MS	[133]
		Serum	α-linked mannose and glycan involved the Thomsen–Friedenreich antigen, fucose, and Lewis antigens affected MUC1 and MUC5AC	Upregulated	Microarray, WB	[130]
		Serum	Asn-88 N-glycosylation and differential RNase-1 expression	Upregulated	ELISA, WB	[131]

Table 1. *Cont.*

"Omic" Level Description		Sample Origin	Altered Molecule/Microorganism	Expression Pattern	Detection Method *	Reference Study
Glycomics	Glycan alterations	Serum	α1-3 fucosylation in α-1-acid glycoprotein	Upregulated	ELLA, HILIC-MS, CZE	[134]
		Serum	CA19-9	Downregulated	Immunoassay	[139]
		Tumor biopsy	CD44 antigen (CD44)	Upregulated	WB	[138]
		Plasma	Glypican-1 (GPC1)	Upregulated	Flow cytometry	[137]
	Glycoproteins	Serum	Mucin-5AC, MUC1, and MUC16	Upregulated	Antibody-lectin sandwich array	[130]
Metagenomics	Microbiota	Oral microbiota	*Porphyromonas gingivali*, *Fusobacterium*, *Neisseria elongata*, and *Streptococcus mitis*	High amount	plasma antibody analysis, 16S rRNA sequencing	[145]
		Murine fecal microbiota	Proteobacteria, Actinobacteria, Fusobacteria, and Verrucomicrobia	High amount	qPCR, FISH, 16S rRNA gene sequencing	[143]
		Murine gut microbiota	Proteobacteria, Bacteroidetes, and Firmicutes	High amount	qPCR, FISH, 16S rRNA gene sequencing	[143]

* CZE: capillary zone electrophoresis, ELISA: enzyme-linked immunosorbent assay, ELLA: Enzyme-linked lectin assay, FISH: fluorescence in situ hybridization, GC: gas chromatography, HILIC: Hydrophilic interaction chromatography, HRMAS: high-resolution magic angle spinning, LC: liquid chromatography, MS: mass spectrometry, NMR: nuclear magnetic resonance, qRT-PCR: quantitative reverse transcription polymerase chain reaction, TOF: time of flight, WB: Western blot, WES: Whole exome sequencing, WGS: whole genome sequencing.

Author Contributions: Conceptualization, B.T., K.Y.A., and R.S.; writing—original draft preparation, B.T., E.Y., and G.G.; review and editing, B.T., K.Y.A., and R.S.; visualization, G.G.; supervision, B.T., K.Y.A., and R.S. All authors have read and agreed to the published version of the manuscript.

Funding: This research received no external funding.

Conflicts of Interest: The authors declare no conflict of interest.

References

1. Bray, F.; Ferlay, J.; Soerjomataram, I.; Siegel, R.L.; Torre, L.A.; Jemal, A. Global cancer statistics 2018: GLOBOCAN estimates of incidence and mortality worldwide for 36 cancers in 185 countries. *CA Cancer J. Clin.* **2018**, *68*, 394–424. [CrossRef]
2. Siegel, R.L.; Miller, K.D.; Jemal, A. Cancer statistics, 2020. *CA Cancer J. Clin.* **2020**, *70*, 7–30. [CrossRef] [PubMed]
3. Zhang, L.; Sanagapalli, S.; Stoita, A. Challenges in diagnosis of pancreatic cancer. *World J. Gastroenterol.* **2018**, *24*, 2047–2060. [CrossRef]
4. Maeda, S.; Unno, M.; Yu, J. Adjuvant and Neoadjuvant Therapy for Pancreatic Cancer. *J. Pancreatol.* **2019**, *2*, 100–106. [CrossRef]
5. Srivastava, S.; Koay, E.J.; Borowsky, A.D.; De Marzo, A.M.; Ghosh, S.; Wagner, P.D.; Kramer, B.S. Cancer overdiagnosis: A biological challenge and clinical dilemma. *Nat. Rev. Cancer* **2019**, *19*, 349–358. [CrossRef] [PubMed]
6. Wu, W.; Jin, G.; Wang, C.; Miao, Y.; Wang, H. The current surgical treatment of pancreatic cancer in China: A national wide cross-sectional study. *J. Pancreatol.* **2019**, *2*, 1. [CrossRef]
7. Zhang, Q.; Zeng, L.; Chen, Y.; Lian, G.; Qian, C.; Chen, S.; Li, J.; Huang, K. Pancreatic Cancer Epidemiology, Detection, and Management. *Gastroenterol. Res. Pract.* **2016**, *2016*. [CrossRef] [PubMed]
8. Wolfgang, C.L.; Herman, J.M.; Laheru, D.A.; Al, E. Recent progress in pancreatic cancer. *CA Cancer J. Clin.* **2013**, *63*, 318–348. [CrossRef] [PubMed]
9. Goral, V. Pancreatic cancer: Pathogenesis and diagnosis. *Asian Pac. J. Cancer Prev.* **2015**, *16*, 5619–5624. [CrossRef]
10. Klimstra, D.S.; Pitman, M.B.; Hruban, R.H. An algorithmic approach to the diagnosis of pancreatic neoplasms. *Arch. Pathol. Lab. Med.* **2009**, *133*, 454–464. [CrossRef]
11. McGuigan, A.; Kelly, P.; Turkington, R.C.; Jones, C.; Coleman, H.G.; McCain, R.S. Pancreatic cancer: A review of clinical diagnosis, epidemiology, treatment and outcomes. *World J. Gastroenterol.* **2018**, *24*, 4846–4861. [CrossRef] [PubMed]
12. Decker, G.A.; Batheja, M.J.; Collins, J.M.; Silva, A.C.; Mekeel, K.L.; Moss, A.A.; Nguyen, C.C.; Lake, D.F.; Miller, L.J. Risk factors for pancreatic adenocarcinoma and prospects for screening. *Gastroenterol. Hepatol.* **2010**, *6*, 246–254.
13. Becker, A.E.; Hernandez, Y.G.; Frucht, H.; Lucas, A.L. Pancreatic ductal adenocarcinoma: Risk factors, screening, and early detection. *World J. Gastroenterol.* **2014**. [CrossRef]
14. Costache, M.I.; Costache, C.A.; Dumitrescu, C.I.; Tica, A.A.; Popescu, M.; Baluta, E.A.; Anghel, A.C.; Saftoiu, A.; Dumitrescu, D. Which is the Best Imaging Method in Pancreatic Adenocarcinoma Diagnosis and Staging—CT, MRI or EUS? *Curr. Health Sci. J.* **2017**, *43*, 132–136. [CrossRef]

15. Zhang, X.; Shi, S.; Zhang, B.; Ni, Q.; Yu, X.; Xu, J. Circulating biomarkers for early diagnosis of pancreatic cancer: Facts and hopes. *Am. J. Cancer Res.* **2018**, *8*, 332–353.

16. Tjensvoll, K.; Nordgård, O.; Smaaland, R. Circulating tumor cells in pancreatic cancer patients: Methods of detection and clinical implications. *Int. J. Cancer* **2014**, *134*, 1–8. [CrossRef]

17. Martini, V.; Timme-Bronsert, S.; Fichtner-Feigl, S.; Hoeppner, J.; Kulemann, B. Circulating tumor cells in pancreatic cancer: Current perspectives. *Cancers* **2019**, *11*, 1659. [CrossRef] [PubMed]

18. Amantini, C.; Morelli, M.B.; Nabissi, M.; Piva, F.; Marinelli, O.; Maggi, F.; Bianchi, F.; Bittoni, A.; Berardi, R.; Giampieri, R.; et al. Expression Profiling of Circulating Tumor Cells in Pancreatic Ductal Adenocarcinoma Patients: Biomarkers Predicting Overall Survival. *Front. Oncol.* **2019**, *9*, 1–13. [CrossRef]

19. Bhat, K.; Wang, F.; Ma, Q.; Li, Q.; Mallik, S.; Hsieh, T.; Wu, E. Advances in Biomarker Research for Pancreatic Cancer. *Curr. Pharm. Des.* **2012**, *18*, 2439–2451. [CrossRef] [PubMed]

20. Zhang, W.H.; Wang, W.Q.; Han, X.; Gao, H.L.; Li, T.J.; Xu, S.S.; Li, S.; Xu, H.X.; Li, H.; Ye, L.Y.; et al. Advances on diagnostic biomarkers of pancreatic ductal adenocarcinoma: A systems biology perspective. *Comput. Struct. Biotechnol. J.* **2020**, *18*, 3606–3614. [CrossRef]

21. Young, M.R.; Wagner, P.D.; Grosh, S.; Al, E. Validation of Biomarkers for Early Detection of Pancreatic Cancer: Summary of the Alliance of Pancreatic Cancer Consortia for Biomarkers for Early Detection Workshop. *Pancreas* **2019**, *47*, 135–141. [CrossRef]

22. Adamska, A.; Domenichini, A.; Falasca, M. Pancreatic ductal adenocarcinoma: Current and evolving therapies. *Int. J. Mol. Sci.* **2017**, *18*, 1338. [CrossRef]

23. Tian, Q.; Price, N.D.; Hood, L. Systems Cancer Medicine: Towards Realization of Predictive, Preventive, Personalized, and Participatory (P4) Medicine. *J. Intern. Med.* **2012**, *271*, 111–121. [CrossRef]

24. Westerhoff, H.V.; Palsson, B.O. The evolution of molecular biology into systems biology. *Nat. Biotechnol.* **2004**, *22*, 1249–1252. [CrossRef]

25. Sousa, C.M.; Kimmelman, A.C. The complex landscape of pancreatic cancer metabolism. *Carcinogenesis* **2014**, *35*, 1441–1450. [CrossRef] [PubMed]

26. Liang, C.; Qin, Y.; Zhang, B.; Ji, S.; Shi, S.; Xu, W.; Liu, J.; Xiang, J.; Liang, D.; Hu, Q.; et al. Metabolic plasticity in heterogeneous pancreatic ductal adenocarcinoma. *Bioch. Biophys. Acta Rev. Cancer* **2016**, *1866*, 177–188. [CrossRef]

27. Löhr, M.; Klöppel, G.; Maisonneuve, P.; Lowenfels, A.B.; Lüttges, J. Frequency of K-ras mutations in pancreatic intraductal neoplasias associated with pancreatic ductal adenocarcinoma and chronic pancreatitis: A meta-analysis. *Neoplasia* **2005**, *7*, 17–23. [CrossRef] [PubMed]

28. Bardeesy, N.; DePinho, R.A. Pancreatic cancer biology and genetics. *Nat. Rev. Cancer* **2002**, *2*, 897–909. [CrossRef] [PubMed]

29. Perera, R.M.; Bardeesy, N. Pancreatic Cancer Metabolism—Breaking it down to build it back up. *Cancer Discov.* **2015**, *5*, 1247–1261. [CrossRef] [PubMed]

30. Son, J.; Lyssiotis, C.A.; Ying, H.; Wang, X.; Hua, S.; Ligorio, M.; Perera, R.M.; Ferrone, C.R.; Mullarky, E.; Shyh-, N.; et al. Glutamine supports pancreatic cancer growth through a Kras- regulated metabolic pathway. *Nature* **2013**, *496*, 101–105. [CrossRef]

31. Ying, H.; Kimmelman, A.C.; Lyssiotis, C.A.; Hua, S.; Chu, G.C.; Fletcher-sananikone, E.; Locasale, J.W.; Son, J.; Zhang, H.; Coloff, J.L.; et al. Oncogenic Kras Maintains Pancreatic Tumors through Regulation of Anabolic Glucose Metabolism. *Cell* **2012**, *149*, 656–670. [CrossRef]

32. Le, A.; Cooper, C.R.; Gouw, A.M.; Dinavahi, R.; Maitra, A.; Deck, L.M.; Royer, R.E.; Vander Jagt, D.L.; Semenza, G.L.; Dang, C.V. Inhibition of lactate dehydrogenase A induces oxidative stress and inhibits tumor progression. *Proc. Natl. Acad. Sci. USA* **2010**, *107*, 2037–2042. [CrossRef]

33. Alian, O.; Philip, P.; Sarkar, F.; Azmi, A. Systems Biology Approaches to Pancreatic Cancer Detection, Prevention and Treatment. *Curr. Pharm. Des.* **2014**. [CrossRef]

34. Ballehaninna, U.K.; Chamberlain, R.S. Biomarkers for pancreatic cancer: Promising new markers and options beyond CA 19-9. *Tumor Biol.* **2013**, *34*, 3279–3292. [CrossRef]

35. Kishikawa, T.; Otsuka, M.; Ohno, M.; Yoshikawa, T.; Takata, A.; Koike, K. Circulating RNAs as new biomarkers for detecting pancreatic cancer. *World J. Gastroenterol.* **2015**, *21*, 8527–8540. [CrossRef] [PubMed]

36. Hasan, S.; Jacob, R.; Manne, U.; Paluri, R. Advances in pancreatic cancer biomarkers. *Oncol. Rev.* **2019**, *13*, 69–76. [CrossRef] [PubMed]

37. Zhang, B.; Pan, X.; Cobb, G.P.; Anderson, T.A. microRNAs as oncogenes and tumor suppressors. *Dev. Biol.* **2007**, *302*, 1–12. [CrossRef] [PubMed]

38. Volinia, S.; Calin, G.A.; Liu, C.-G.; Al, E. A microRNA expression signature of human solid tumors defines cancer gene targets. *Proc. Natl. Acad. Sci. USA* **2006**, *103*, 2257–2261. [CrossRef]

39. Li, Y.; Sarkar, F.H. MicroRNA targeted therapeutic approach for pancreatic cancer. *Int. J. Biol. Sci.* **2016**, *12*, 326–337. [CrossRef]

40. Lu, J.; Getz, G.; Miska, E.A.; Alvarez-Saavedra, E.; Lamb, J.; Peck, D.; Sweet-Cordero, A.; Ebert, B.L.; Mak, R.H.; Ferrando, A.A.; et al. MicroRNA expression profiles classify human cancers. *Nature* **2005**, *435*, 834–838. [CrossRef]

41. Rosenfeld, N.; Aharonov, R.; Meiri, E.; Rosenwald, S.; Spector, Y.; Zepeniuk, M.; Benjamin, H.; Shabes, N.; Tabak, S.; Levy, A.; et al. MicroRNAs accurately identify cancer tissue origin. *Nat. Biotechnol.* **2008**, *26*, 462–469. [CrossRef]

42. Hernandez, Y.G. MicroRNA in pancreatic ductal adenocarcinoma and its precursor lesions. *World J. Gastrointest. Oncol.* **2016**, *8*, 18. [CrossRef]

43. Frampton, A.E.; Giovannetti, E.; Jamieson, N.B.; Krell, J.; Gall, T.M.; Stebbing, J.; Jiao, L.R.; Castellano, L. A microRNA meta-signature for pancreatic ductal adenocarcinoma. *Expert Rev. Mol. Diagn.* **2014**, *14*, 267–271. [CrossRef]
44. Bloomston, M.; Frankel, W.L.; Petrocca, F.; Volinia, S.; Alder, H.; Hagan, J.P.; Liu, C.G.; Bhatt, D.; Taccioli, C.; Croce, C.M. MicroRNA expression patterns to differentiate pancreatic adenocarcinoma from normal pancreas and chronic pancreatitis. *J. Am. Med. Assoc.* **2007**, *297*, 1901–1908. [CrossRef]
45. Caponi, S.; Funel, N.; Frampton, A.E.; Mosca, F.; Santarpia, L.; Van der Velde, A.G.; Jiao, L.R.; De Lio, N.; Falcone, A.; Kazemier, G.; et al. The good, the bad and the ugly: A tale of miR-101, miR-21 and miR-155 in pancreatic intraductal papillary mucinous neoplasms. *Ann. Oncol.* **2013**, *24*, 734–741. [CrossRef]
46. Szafranska, A.E.; Davison, T.S.; John, J.; Cannon, T.; Sipos, B.; Maghnouj, A.; Labourier, E.; Hahn, S.A. MicroRNA expression alterations are linked to tumorigenesis and non-neoplastic processes in pancreatic ductal adenocarcinoma. *Oncogene* **2007**, *26*, 4442–4452. [CrossRef]
47. Ho, A.S.; Huang, X.; Cao, H.; Christman-Skieller, C.; Bennewith, K.; Le, Q.T.; Koong, A.C. Circulating miR-210 as a novel hypoxia marker in pancreatic cancer. *Transl. Oncol.* **2010**, *3*, 109–113. [CrossRef]
48. Habbe, N.; Koorstra, J.M.; Mendell, J.T.; Offerhaus, G.J.; Kon, J.; Feldmann, G.; Mullendore, M.E.; Goggins, M.G. MicroRNA miR-155 is a biomarker of early pancreatic neoplasia. *Cancer Biol. Ther.* **2010**, *8*, 340–346. [CrossRef] [PubMed]
49. Daoud, A.Z.; Mulholland, E.J.; Cole, G.; McCarthy, H.O. MicroRNAs in Pancreatic Cancer: Biomarkers, prognostic, and therapeutic modulators. *BMC Cancer* **2019**, *19*, 1–13. [CrossRef] [PubMed]
50. Li, A.; Yu, J.; Kim, H.; Wolfgang, C.L.; Al, E. MicroRNA array analysis finds elevated serum miR-1290 accurately distinguishes patients with low-stage pancreatic cancer from healthy and disease controls. *Clin. Cancer Res.* **2014**, *19*, 3600–3610. [CrossRef] [PubMed]
51. Li, A.; Omura, N.; Hong, S.M.; Vincent, A.; Walter, K.; Griffith, M.; Borges, M.; Goggins, M. Pancreatic cancers epigenetically silence SIP1 and hypomethylate and overexpress miR-200a/200b in association with elevated circulating miR-200a and miR-200b levels. *Cancer Res.* **2010**, *70*, 5226–5237. [CrossRef] [PubMed]
52. Liu, M.; Du, Y.; Gao, J.; Liu, J.; Kong, X.; Gong, Y.; Li, Z.; Wu, H.; Chen, H. Aberrant expression mir-196a is associated with abnormal apoptosis, invasion, and proliferation of pancreatic cancer cells. *Pancreas* **2013**, *42*, 1169–1181. [CrossRef] [PubMed]
53. Carlsen, A.L.; Joergensen, M.T.; Knudsen, S.; De Muckadell, O.B.S.; Heegaard, N.H.H. Cell-free plasma microRNA in pancreatic ductal adenocarcinoma and disease controls. *Pancreas* **2013**, *42*, 1107–1113. [CrossRef]
54. Ganepola, G.A. Novel blood-based microRNA biomarker panel for early diagnosis of pancreatic cancer. *World J. Gastrointest. Oncol.* **2014**, *6*, 22. [CrossRef] [PubMed]
55. Morimura, R.; Komatsu, S.; Ichikawa, D.; Takeshita, H.; Tsujiura, M.; Nagata, H.; Konishi, H.; Shiozaki, A.; Ikoma, H.; Okamoto, K.; et al. Novel diagnostic value of circulating miR-18a in plasma of patients with pancreatic cancer. *Br. J. Cancer* **2011**, *105*, 1733–1740. [CrossRef] [PubMed]
56. Liu, J.; Gao, J.; Du, Y.; Li, Z.; Ren, Y.; Gu, J.; Wang, X.; Gong, Y.; Wang, W.; Kong, X. Combination of plasma microRNAs with serum CA19-9 for early detection of pancreatic cancer. *Int. J. Cancer* **2012**, *131*, 683–691. [CrossRef]
57. Wang, W.S.; Liu, L.X.; Li, G.P.; Chen, Y.; Li, C.Y.; Jin, D.Y.; Wang, X.L. Combined serum CA19-9 and miR-27a-3p in peripheral blood mononuclear cells to diagnose pancreatic cancer. *Cancer Prev. Res.* **2013**, *6*, 331–338. [CrossRef] [PubMed]
58. Halkova, T.; Cuperkova, R.; Minarik, M.; Benesova, L. MicroRNAs in pancreatic cancer: Involvement in carcinogenesis and potential use for diagnosis and prognosis. *Gastroenterol. Res. Pract.* **2015**, *2015*. [CrossRef] [PubMed]
59. Li, B.S.; Liu, H.; Yang, W.L. Reduced miRNA-218 expression in pancreatic cancer patients as a predictor of poor prognosis. *Genet. Mol. Res.* **2015**, *14*, 16372–16378. [CrossRef]
60. Ma, Y.B.; Li, G.X.; Hu, J.X.; Liu, X.; Shi, B.M. Correlation of miR-494 expression with tumor progression and patient survival in pancreatic cancer. *Genet. Mol. Res.* **2015**, *14*, 18153–18159. [CrossRef]
61. Passadouro, M.; Pedroso de Lima, M.C.; Faneca, H. MicroRNA modulation combined with sunitinib as a novel therapeutic strategy for pancreatic cancer. *Int. J. Nanomed.* **2014**, *9*, 3203–3217. [CrossRef]
62. Miyamae, M.; Komatsu, S.; Ichikawa, D.; Kawaguchi, T.; Hirajima, S.; Okajima, W.; Ohashi, T.; Imamura, T.; Konishi, H.; Shiozaki, A.; et al. Plasma microRNA profiles: Identification of miR-744 as a novel diagnostic and prognostic biomarker in pancreatic cancer. *Br. J. Cancer* **2015**, *113*, 1467–1476. [CrossRef]
63. Kung, J.T.Y.; Colognori, D.; Lee, J.T. Long noncoding RNAs: Past, present, and future. *Genetics* **2013**, *193*, 651–669. [CrossRef] [PubMed]
64. Baraniskin, A.; Nöpel-Dünnebacke, S.; Ahrens, M.; Jensen, S.G.; Zöllner, H.; Maghnouj, A.; Wos, A.; Mayerle, J.; Munding, J.; Kost, D.; et al. Circulating U2 small nuclear RNA fragments as a novel diagnostic biomarker for pancreatic and colorectal adenocarcinoma. *Int. J. Cancer* **2013**, *132*, 1–10. [CrossRef]
65. Li, J.; Yang, J.; Zhou, P.; Le, Y.; Zhou, C.; Wang, S.; Xu, D.; Lin, H.-K.; Gong, Z. Circular RNAs in cancer: Novel insights into origins, properties, functions and implications. *Am. J. Cancer Res.* **2015**, *5*, 472–480. [PubMed]
66. Liu, J.; Xu, D.; Wang, Q.; Zheng, D.; Jiang, X.; Xu, L. LPS induced miR-181a promotes pancreatic cancer cell migration via targeting PTEN and MAP2K4. *Dig. Dis. Sci.* **2014**. [CrossRef] [PubMed]
67. Li, H.; Hao, X.; Wang, H.; Liu, Z.; He, Y.; Pu, M.; Zhang, H.; Yu, H.; Duan, J.; Qu, S. Circular RNA Expression Profile of Pancreatic Ductal Adenocarcinoma Revealed by Microarray. *Cell. Physiol. Biochem.* **2016**, *40*, 1334–1344. [CrossRef]

68. Zhang, Q.; Wang, J.Y.; Zhou, S.Y.; Yang, S.J.; Zhong, S.L. Circular RNA expression in pancreatic ductal adenocarcinoma. *Oncol. Lett.* **2019**, *18*, 2923–2930. [CrossRef]

69. Seimiya, T.; Otsuka, M.; Iwata, T.; Tanaka, E.; Sekiba, K.; Shibata, C.; Moriyama, M.; Nakagawa, R.; Maruyama, R.; Koike, K. Aberrant expression of a novel circular RNA in pancreatic cancer. *J. Human Genet.* **2020**. [CrossRef]

70. Guo, X.; Zhou, Q.; Su, D.; Luo, Y.; Fu, Z.; Huang, L.; Li, Z.; Jiang, D.; Kong, Y.; Li, Z.; et al. Circular RNA circBFAR promotes the progression of pancreatic ductal adenocarcinoma via the miR-34b-5p/MET/Akt axis. *Mol. Cancer* **2020**, *19*. [CrossRef]

71. Limb, C.; Liu, D.S.K.; Veno, M.T.; Rees, E.; Krell, J.; Bagwan, I.N.; Giovannetti, E.; Pandha, H.; Strobel, O.; Rockall, T.A.; et al. The role of circular rnas in pancreatic ductal adenocarcinoma and biliary-tract cancers. *Cancers* **2020**, *12*, 3250. [CrossRef]

72. Peng, J.; Sun, B.F.; Chen, C.Y.; Zhou, J.Y.; Chen, Y.S.; Chen, H.; Liu, L.; Huang, D.; Jiang, J.; Cui, G.S.; et al. Single-cell RNA-seq highlights intra-tumoral heterogeneity and malignant progression in pancreatic ductal adenocarcinoma. *Cell Res.* **2019**, *29*, 725–738. [CrossRef]

73. Aslam, B.; Basit, M.; Nisar, M.A.; Khurshid, M.; Rasool, M.H. Proteomics: Technologies and their applications. *J. Chromatogr. Sci.* **2017**, *55*, 182–196. [CrossRef]

74. Alharbi, R.A. Proteomics approach and techniques in identification of reliable biomarkers for diseases. *Saudi J. Biol. Sci.* **2020**, *27*, 968–974. [CrossRef] [PubMed]

75. Domon, B.; Aebersold, R. Challenges and opportunities in proteomics data analysis. *Mol. Cell. Proteom.* **2006**, *5*, 1921–1926. [CrossRef]

76. Iuga, C.; Seicean, A.; Iancu, C.; Buiga, R.; Sappa, P.K.; Völker, U.; Hammer, E. Proteomic identification of potential prognostic biomarkers in resectable pancreatic ductal adenocarcinoma. *Proteomics* **2014**, *14*, 945–955. [CrossRef] [PubMed]

77. Chen, R.; Pan, S.; Brentnall, T.A.; Aebersold, R. Proteomic profiling of pancreatic cancer for biomarker discovery. *Mol. Cell. Proteom.* **2005**, *4*, 523–533. [CrossRef] [PubMed]

78. Jimenez-Luna, C.; Torres, C.; Ortiz, R.; Dieguez, C.; Martinez-Galan, J.; Melguizo, C.; Prados, J.C.; Caba, O. Proteomic biomarkers in body fluids associated with pancreatic cancer. *Oncotarget* **2018**, *9*, 16573–16587. [CrossRef] [PubMed]

79. Sileikis, A.; Petrulionis, M.; Kurlinkus, B.; Ger, M.; Kaupinis, A.; Cicenas, J.; Valius, M.; Strupas, K. Current Role of Proteomics in Pancreatic Cancer Biomarkers Research. *Curr. Proteom.* **2016**, *13*, 68–75. [CrossRef]

80. Chen, R.; Pan, S.; Yi, E.C.; Donohoe, S.; Bronner, M.P.; Potter, J.D.; Goodlett, D.R.; Aebersold, R.; Brentnall, T.A. Quantitative proteomic profiling of pancreatic cancer juice. *Proteomics* **2006**, *6*, 3871–3879. [CrossRef]

81. Chen, R.; Pan, S.; Cooke, K.; Moyes, K.W.; Bronner, M.P.; Goodlett, D.R.; Aebersold, R.; Brentnall, T.A. Comparison of pancreas juice proteins from cancer versus pancreatitis using quantitative proteomic analysis. *Pancreas* **2007**, *34*, 70–79. [CrossRef] [PubMed]

82. Grønborg, M.; Bunkenborg, J.; Kristiansen, T.Z.; Jensen, O.N.; Yeo, C.J.; Hruban, R.H.; Maitra, A.; Goggins, M.G.; Pandey, A. Comprehensive proteomic analysis of human pancreatic juice. *J. Proteome Res.* **2004**, *3*, 1042–1055. [CrossRef]

83. Lv, S.; Gao, J.; Al, E. Transthyretin, Identified by Proteomics, is Overabundant in Pancreatic Juice From Pancreatic Carcinoma and Originates From Pancreatic Islets. *Diagn. Cytopathol.* **2011**, *39*, 875–881. [CrossRef] [PubMed]

84. Makawita, S.; Smith, C.; Batruch, I.; Zheng, Y.; Rückert, F.; Grützmann, R.; Pilarsky, C.; Gallinger, S.; Diamandis, E.P. Integrated proteomic profiling of cell line conditioned media and pancreatic juice for the identification of pancreatic cancer biomarkers. *Mol. Cell. Proteom.* **2011**, *10*, 1–20. [CrossRef]

85. Park, J.Y.; Kim, S.A.; Chung, J.W.; Bang, S.; Park, S.W.; Paik, Y.K.; Song, S.Y. Proteomic analysis of pancreatic juice for the identification of biomarkers of pancreatic cancer. *J. Cancer Res. Clin. Oncol.* **2011**, *137*, 1229–1238. [CrossRef]

86. Tian, M.; Cui, Y.Z.; Song, G.H.; Zong, M.J.; Zhou, X.Y.; Chen, Y.; Han, J.X. Proteomic analysis identifies MMP-9, DJ-1 and A1BG as overexpressed proteins in pancreatic juice from pancreatic ductal adenocarcinoma patients. *BMC Cancer* **2008**, *8*, 1–11. [CrossRef]

87. Kim, H.; Park, J.; Wang, J.I.; Kim, Y. Recent advances in proteomic profiling of pancreatic Ductal Adenocarcinoma and the road ahead. *Expert Rev. Proteom.* **2017**, *14*, 963–971. [CrossRef]

88. Stark, A.; Donahue, T.R.; Reber, H.A.; Joe Hines, O. Pancreatic cyst disease a review. *JAMA J. Am. Med. Assoc.* **2016**, *315*, 1882–1893. [CrossRef] [PubMed]

89. Eric Thomas, C.; Sexton, W.; Benson, K.; Sutphen, R.; Koomen, J. Urine collection and processing for protein biomarker discovery and quantification. *Cancer Epidemiol. Biomark. Prev.* **2010**, *19*, 953–959. [CrossRef] [PubMed]

90. Thongboonkerd, V.; Malasit, P. Renal and urinary proteomics: Current applications and challenges. *Proteomics* **2005**, *5*, 1033–1042. [CrossRef]

91. Makawita, S.; Dimitromanolakis, A.; Soosaipillai, A.; Soleas, I.; Chan, A.; Gallinger, S.; Haun, R.S.; Blasutig, I.M.; Diamandis, E.P. Validation of four candidate pancreatic cancer serological biomarkers that improve the performance of CA19.9. *BMC Cancer* **2013**, *13*, 1. [CrossRef]

92. Sogawa, K.; Takano, S.; Iida, F.; Satoh, M.; Tsuchida, S.; Kawashima, Y.; Yoshitomi, H.; Sanda, A.; Kodera, Y.; Takizawa, H.; et al. Identification of a novel serum biomarker for pancreatic cancer, C4b-binding protein α-chain (C4BPA) by quantitative proteomic analysis using tandem mass tags. *Br. J. Cancer* **2016**, *115*, 949–956. [CrossRef] [PubMed]

93. Yoneyama, T.; Ohtsuki, S.; Honda, K.; Kobayashi, M.; Iwasaki, M.; Uchida, Y.; Okusaka, T.; Nakamori, S.; Shimahara, M.; Ueno, T.; et al. Identification of IGFBP2 and IGFBP3 as compensatory biomarkers for CA19-9 in early-stage pancreatic cancer using a combination of antibody-based and LC-MS/MS-based proteomics. *PLoS ONE* **2016**, *11*, 1–23. [CrossRef]

94. Guo, X.; Lv, X.; Fang, C.; Al, E. Dysbindin as a novel biomarker for pancreatic ductal adenocarcinoma identified by proteomic profiling. *Int. J. Cancer* **2006**, *139*, 1821–1829. [CrossRef] [PubMed]

95. Cohen, J.D.; Javed, A.A.; Thoburn, C.; Wong, F.; Tie, J.; Gibbs, P.; Schmidt, C.M.; Yip-Schneider, M.T.; Allen, P.J.; Schattner, M.; et al. Combined circulating tumor DNA and protein biomarker-based liquid biopsy for the earlier detection of pancreatic cancers. *Proc. Natl. Acad. Sci. USA* **2017**, *114*, 10202–10207. [CrossRef]

96. Balasenthil, S.; Huang, Y.; Liu, S.; Marsh, T.; Chen, J.; Stass, S.A.; KuKuruga, D.; Brand, R.; Chen, N.; Frazier, M.L.; et al. A Plasma Biomarker Panel to Identify Surgically Resectable Early-Stage Pancreatic Cancer. *J. Natl. Cancer Inst.* **2017**, *109*, 1–10. [CrossRef] [PubMed]

97. Capello, M.; Bantis, L.E.; Scelo, G.; Zhao, Y.; Li, P.; Dhillon, D.S.; Patel, N.J.; Kundnani, D.L.; Wang, H.; Abbruzzese, J.L.; et al. Sequential validation of blood-based protein biomarker candidates for early-stage pancreatic cancer. *J. Natl. Cancer Inst.* **2017**, *109*, 1–9. [CrossRef] [PubMed]

98. Kim, J.; Bamlet, W.R.; Oberg, A.L.; Chaffee, K.G.; Donahue, G.; Cao, X.J.; Chari, S.; Garcia, B.A.; Petersen, G.M.; Zaret, K.S. Detection of early pancreatic ductal adenocarcinoma with thrombospondin-2 & CA19-9 blood markers. *Sci. Transl. Med.* **2017**, *9*, 1–14. [CrossRef]

99. Ren, Y.Q.; Zhang, H.Y.; Su, T.; Wang, X.H.; Zhang, L. Clinical significance of serum survivin in patients with pancreatic ductal adenocarcinoma. *Eur. Rev. Med. Pharmacol. Sci.* **2014**, *18*, 3063–3068. [PubMed]

100. Chen, K.; Kim, P.D.; Jones, K.A.; Al, E. Potential Prognostic Biomarkers of Pancreatic Cancer. *Pancreas* **2014**, *43*, 1–7. [CrossRef]

101. Ger, M.; Kaupinis, A.; Petrulionis, M.; Kurlinkus, B.; Cicenas, J.; Sileikis, A.; Valius, M.; Strupas, K. Proteomic identification of FLT3 and PCBP3 as potential prognostic biomarkers for pancreatic cancer. *Anticancer Res.* **2018**, *38*, 5759–5765. [CrossRef]

102. Pan, S.; Chen, R.; Brand, R.E.; Hawley, S.; Tamura, Y.; Gafken, P.R.; Milless, B.P.; Goodlett, D.R.; Rush, J.; Brentnall, T.A. Multiplex targeted proteomic assay for biomarker detection in plasma: A pancreatic cancer biomarker case study. *J. Proteome Res.* **2012**, *11*, 1937–1948. [CrossRef]

103. Yoneyama, T.; Ohtsuki, S.; Ono, M.; Ohmine, K.; Uchida, Y.; Yamada, T.; Tachikawa, M.; Terasaki, T. Quantitative targeted absolute proteomics-based large-scale quantification of proline-hydroxylated α-fibrinogen in plasma for pancreatic cancer diagnosis. *J. Proteome Res.* **2013**, *12*, 753–762. [CrossRef]

104. Ansari, D.; Aronsson, L.; Sasor, A.; Welinder, C.; Rezeli, M.; Marko-Varga, G.; Andersson, R. The role of quantitative mass spectrometry in the discovery of pancreatic cancer biomarkers for translational science. *J. Transl. Med.* **2014**, *12*, 1–15. [CrossRef]

105. Bailey, P.; Chang, D.K.; Nones, K.; Johns, A.L.; Patch, A.M.; Gingras, M.C.; Miller, D.K.; Christ, A.N.; Bruxner, T.J.C.; Quinn, M.C.; et al. Genomic analyses identify molecular subtypes of pancreatic cancer. *Nature* **2016**, *531*, 47–52. [CrossRef]

106. Hristova, V.A.; Chan, D.W. Cancer biomarker discovery and translation: Proteomics and beyond. *Expert Rev. Proteom.* **2019**, *16*, 93–103. [CrossRef] [PubMed]

107. Ritchie, S.A.; Akita, H.; Takemasa, I.; Eguchi, H.; Pastural, E.; Nagano, H.; Monden, M.; Doki, Y.; Mori, M.; Jin, W.; et al. Metabolic system alterations in pancreatic cancer patient serum: Potential for early detection. *BMC Cancer* **2013**, *13*, 1. [CrossRef] [PubMed]

108. Gupta, R.; Amanam, I.; Chung, V. Current and future therapies for advanced pancreatic cancer. *J. Surg. Oncol.* **2017**, *116*, 25–34. [CrossRef]

109. Fontana, A.; Copetti, M.; Di Gangi, I.M.; Mazza, T.; Tavano, F.; Gioffreda, D.; Mattivi, F.; Andriulli, A.; Vrhovsek, U.; Pazienza, V. Development of a metabolites risk score for one-year mortality risk prediction in pancreatic adenocarcinoma patients. *Oncotarget* **2016**, *7*, 8968–8978. [CrossRef] [PubMed]

110. LaConti, J.J.; Laiakis, E.C.; Mays, D.D.; Peran, I.; Kim, S.E.; Shay, J.W.; Riegel, A.T.; Fornace, A.J.; Wellstein, A. Distinct serum metabolomics profiles associated with malignant progression in the KrasG12D mouse model of pancreatic ductal adenocarcinoma. *BMC Genom.* **2015**, *16*, S1. [CrossRef] [PubMed]

111. Di Gangi, I.M.; Mazza, T.; Fontana, A.; Copetti, M.; Fusilli, C.; Ippolito, A.; Mattivi, F.; Latiano, A.; Andriulli, A.; Vrhovsek, U.; et al. Metabolomic profile in pancreatic cancer patients: A consensusbased approach to identify highly discriminating metabolites. *Oncotarget* **2016**, *7*, 5815–5829. [CrossRef] [PubMed]

112. Tumas, J.; Baskirova, I.; Petrenas, T.; Norkuniene, J.; Strupas, K.; Sileikis, A. Towards a personalized approach in pancreatic cancer diagnostics through plasma amino acid analysis. *Anticancer Res.* **2019**, *39*, 2035–2042. [CrossRef] [PubMed]

113. Moore, H.B.; Culp-Hill, R.; Reisz, J.A.; Lawson, P.J.; Sauaia, A.; Schulick, R.D.; Del Chiaro, M.; Nydam, T.L.; Moore, E.E.; Hansen, K.C.; et al. The metabolic time line of pancreatic cancer: Opportunities to improve early detection of adenocarcinoma. *Am. J. Surg.* **2019**, *218*, 1206–1212. [CrossRef] [PubMed]

114. Mayerle, J.; Kalthoff, H.; Reszka, R.; Kamlage, B.; Peter, E.; Schniewind, B.; González Maldonado, S.; Pilarsky, C.; Heidecke, C.D.; Schatz, P.; et al. Metabolic biomarker signature to differentiate pancreatic ductal adenocarcinoma from chronic pancreatitis. *Gut* **2018**, *67*, 128–137. [CrossRef]

115. Urayama, S. Pancreatic cancer early detection: Expanding higher-risk group with clinical and metabolomics parameters. *World J. Gastroenterol.* **2015**, *21*, 1707–1717. [CrossRef] [PubMed]

116. Phua, L.C.; Goh, S.; Tai, D.W.M.; Leow, W.Q.; Alkaff, S.M.F.; Chan, C.Y.; Kam, J.H.; Lim, T.K.H.; Chan, E.C.Y. Metabolomic prediction of treatment outcome in pancreatic ductal adenocarcinoma patients receiving gemcitabine. *Cancer Chemother. Pharmacol.* **2018**, *81*, 277–289. [CrossRef]

117. Battini, S.; Faitot, F.; Imperiale, A.; Cicek, A.E.; Heimburger, C.; Averous, G.; Bachellier, P.; Namer, I.J. Metabolomics approaches in pancreatic adenocarcinoma: Tumor metabolism profiling predicts clinical outcome of patients. *BMC Med.* **2017**, *15*, 1–16. [CrossRef]

118. Wen, S.; Li, Z.; Feng, J.; Bai, J.; Lin, X.; Huang, H. Metabonomic changes from pancreatic intraepithelial neoplasia to pancreatic ductal adenocarcinoma in tissues from rats. *Cancer Sci.* **2016**, *107*, 836–845. [CrossRef]

119. Eidelman, E.; Twum-Ampofo, J.; Ansari, J.; Siddiqui, M.M. The metabolic phenotype of prostate cancer. *Front. Oncol.* **2017**, *7*, 1–6. [CrossRef] [PubMed]

120. Zhang, C.; Hua, Q. Applications of genome-scale metabolic models in biotechnology and systems medicine. *Front. Physiol.* **2016**, *6*, 1–8. [CrossRef]

121. Mardinoglu, A.; Nielsen, J. Systems medicine and metabolic modelling. *J. Intern. Med.* **2012**, *271*, 142–154. [CrossRef] [PubMed]

122. Lee, S.; Zhang, C.; Arif, M.; Liu, Z.; Benfeitas, R.; Bidkhori, G.; Deshmukh, S.; Al Shobky, M.; Lovric, A.; Boren, J.; et al. TCSBN: A database of tissue and cancer specific biological networks. *Nucleic Acids Res.* **2018**, *46*, D595–D600. [CrossRef] [PubMed]

123. Ghaffari, P.; Mardinoglu, A.; Asplund, A.; Shoaie, S.; Kampf, C.; Uhlen, M.; Nielsen, J. Identifying anti-growth factors for human cancer cell lines through genome-scale metabolic modeling. *Sci. Rep.* **2015**, *5*, 1–10. [CrossRef]

124. Yizhak, K.; Chaneton, B.; Gottlieb, E.; Ruppin, E. Modeling cancer metabolism on a genome scale. *Mol. Syst. Biol.* **2015**, *11*, 817. [CrossRef]

125. López-López, Á.; López-Gonzálvez, Á.; Barker-Tejeda, T.C.; Barbas, C. A review of validated biomarkers obtained through metabolomics. *Expert Rev. Mol. Diagn.* **2018**, *18*, 557–575. [CrossRef]

126. Armitage, E.G.; Ciborowski, M. Applications of Metobolomics in Cancer Studies. In *Metabolomics: From Fundamentals to Clinical Applications. Advances in Experimental Medicine and Biology*; Springer: Cham, Switzeland, 2017; Volume 965. [CrossRef]

127. Cheung, P.K.; Ma, M.H.; Tse, H.F.; Yeung, K.F.; Tsang, H.F.; Chu, M.K.M.; Kan, C.M.; Cho, W.C.S.; Ng, L.B.W.; Chan, L.W.C.; et al. The applications of metabolomics in the molecular diagnostics of cancer. *Expert Rev. Mol. Diagn.* **2019**, *19*, 785–793. [CrossRef]

128. Özdemir, V.; Arga, K.Y.; Aziz, R.K.; Bayram, M.; Conley, S.N.; Dandara, C.; Endrenyi, L.; Fisher, E.; Garvey, C.K.; Hekim, N.; et al. Digging deeper into precision/personalized medicine: Cracking the sugar code, the third alphabet of life, and sociomateriality of the cell. *OMICS J. Integr. Biol.* **2020**, *24*, 62–80. [CrossRef] [PubMed]

129. Wang, W. Validation and Development of N-glycan as Biomarker in Cancer Diagnosis. *Curr. Pharmacogenom. Personal. Med.* **2013**, *11*, 53–58. [CrossRef]

130. Yue, T.; Goldstein, I.J.; Hollingsworth, M.A.; Kaul, K.; Brand, R.E.; Haab, B.B. The Prevalence and Nature of Glycan Alterations on Specific Proteins in Pancreatic Cancer Patients Revealed Using Antibody-Lectin Sandwich Arrays. *Mol. Cell. Proteom.* **2009**, *8*, 1697–1707. [CrossRef]

131. Nakata, D. Increased N-glycosylation of Asn88 in serum pancreatic ribonuclease 1 is a novel diagnostic marker for pancreatic cancer. *Sci. Rep.* **2015**, *4*, 6715. [CrossRef]

132. Kori, M.; Aydin, B.; Gulfidan, G.; Beklen, H.; Kelesoglu, N.; İscan, A.C.; Turanli, B.; Erzik, C.; Karademir, B.; Arga, K.Y. The Repertoire of Glycan Alterations and Glycoproteins in Human Cancers. *OMICS J. Integr. Biol.* **2021**. [CrossRef]

133. Pan, S.; Chen, R.; Tamura, Y.; Crispin, D.A.; Lai, L.A.; May, D.H.; McIntosh, M.W.; Goodlett, D.R.; Brentnall, T.A. Quantitative Glycoproteomics Analysis Reveals Changes in N-Glycosylation Level Associated with Pancreatic Ductal Adenocarcinoma. *J. Proteome Res.* **2014**, *13*, 1293–1306. [CrossRef] [PubMed]

134. Balmaña, M.; Giménez, E.; Puerta, A.; Llop, E.; Figueras, J.; Fort, E.; Sanz-Nebot, V.; de Bolós, C.; Rizzi, A.; Barrabés, S.; et al. Increased α1-3 fucosylation of α-1-acid glycoprotein (AGP) in pancreatic cancer. *J. Proteom.* **2016**, *132*, 144–154. [CrossRef]

135. Chaturvedi, P.; Singh, A.P.; Chakraborty, S.; Chauhan, S.C.; Bafna, S.; Meza, J.L.; Singh, P.K.; Hollingsworth, M.A.; Mehta, P.P.; Batra, S.K. MUC4 Mucin Interacts with and Stabilizes the HER2 Oncoprotein in Human Pancreatic Cancer Cells. *Cancer Res.* **2008**, *68*, 2065–2070. [CrossRef] [PubMed]

136. Simeone, D.M.; Ji, B.; Banerjee, M.; Arumugam, T.; Li, D.; Anderson, M.A.; Bamberger, A.M.; Greenson, J.; Brand, R.E.; Ramachandran, V.; et al. CEACAM1, a Novel Serum Biomarker for Pancreatic Cancer. *Pancreas* **2007**, *34*, 436–443. [CrossRef] [PubMed]

137. Qian, J.; Tan, Y.; Zhang, Y.; Yang, Y.; Li, X. Prognostic value of glypican-1 for patients with advanced pancreatic cancer following regional intra-arterial chemotherapy. *Oncol. Lett.* **2018**. [CrossRef]

138. Li, X.P.; Zhang, X.W.; Zheng, L.Z.; Guo, W.J. Expression of CD44 in pancreatic cancer and its significance. *Int. J. Clin. Exp. Pathol.* **2015**, *8*, 6724–6731.

139. Aoki, S.; Motoi, F.; Murakami, Y.; Sho, M.; Satoi, S.; Honda, G.; Uemura, K.; Okada, K.; Matsumoto, I.; Nagai, M.; et al. Decreased serum carbohydrate antigen 19–9 levels after neoadjuvant therapy predict a better prognosis for patients with pancreatic adenocarcinoma: A multicenter case-control study of 240 patients. *BMC Cancer* **2019**, *19*, 252. [CrossRef]

140. Picardo, S.L.; Coburn, B.; Hansen, A.R. The microbiome and cancer for clinicians. *Critical Rev. Oncol. Hematol.* **2019**, *141*, 1–12. [CrossRef]

141. Michaud, D.; Izard, J. Microbiota, Oral Microbiome, and Pancreatic Cancer. *Cancer J.* **2014**, *20*, 203–206. [CrossRef]

142. Zambirinis, C.; Res, M.; Al, E. Pancreatic Cancer, Inflammation and Microbiome. *Cancer J.* **2014**, *20*, 195–202. [CrossRef] [PubMed]

143. Pushalkar, S.; Mautin, H.; Al, E. The Pancreatic Cancer Microbiome Promotes Oncogenesis by Induction of Innate and Adaptive Immune Suppression. *Cancer Discov.* **2018**, *8*, 403–416. [CrossRef] [PubMed]

144. Jacob, J.A. Study links periodontal disease bacteria to pancreatic cancer risk. *JAMA J. Am. Med. Assoc.* **2016**, *315*, 2653–2654. [CrossRef] [PubMed]
145. Wei, M.Y.; Shi, S.; Liang, C.; Meng, Q.C.; Hua, J.; Zhang, Y.Y.; Liu, J.; Zhang, B.; Xu, J.; Yu, X.J. The microbiota and microbiome in pancreatic cancer: More influential than expected. *Mol. Cancer* **2019**, *18*, 1–15. [CrossRef]
146. Fan, X.; Alekseyenko, A.V.; Wu, J.; Jacobs, E.J.; Gapstur, S.M.; Purdue, M.P.; Abnet, C.C.; Stolzenberg-Solomon, R.; Miller, G.; Ravel, J.; et al. Human oral microbiome and prospective risk for pancreatic cancer: A population-based nested case-control study. *Cancer Res.* **2018**, *67*, 120–127. [CrossRef]
147. Mitsuhashi, K.; Nosho, K.; Sukawa, Y.; Matsunaga, Y.; Ito, M.; Kurihara, H.; Kanno, S.; Igarashi, H.; Naito, T.; Adachi, Y.; et al. Association of Fusobacterium species in pancreatic cancer tissues with molecular features and prognosis. *Oncotarget* **2015**, *6*, 7209–7220. [CrossRef] [PubMed]
148. Michaud, D.S.; Izard, J.; Wilhelm-benartzi, C.S.; Grote, V.A.; Tjønneland, A.; Dahm, C.C.; Overvad, K. Plasma antibodies to oral bacteria and risk of pancreatic cancer in a large European prospective cohort study. *Gut* **2014**, *62*, 1–18. [CrossRef] [PubMed]
149. Pitt, J.M.; Vétizou, M.; Boneca, I.G.; Lepage, P.; Chamaillard, M.; Zitvogel, L. Enhancing the clinical coverage and anticancer efficacy of immune checkpoint blockade through manipulation of the gut microbiota. *OncoImmunology* **2017**, *6*, 10–12. [CrossRef]
150. Riquelme, E.; Zhang, Y.; Zhang, L.; Montiel, M.; Zoltan, M.; Dong, W.; Quesada, P.; Sahin, I.; Chandra, V.; San Lucas, A.; et al. Tumor Microbiome Diversity and Composition Influence Pancreatic Cancer Outcomes. *Cell* **2019**, *178*, 795–806.e12. [CrossRef] [PubMed]
151. Mendez, R.; Kesh, K.; Arora, N.; Di Martino, L.; McAllister, F.; Merchant, N.; Banerjee, S.; Banerjee, S. Microbial dysbiosis and polyamine metabolism as predictive markers for early detection of pancreatic cancer. *Carcinogenesis* **2020**, *41*, 561–570. [CrossRef]
152. Geller, L.T.; Barzily-Rokni, M.; Danino, T.; Jonas, O.H.; Shental, N.; Nejman, D.; Gavert, N.; Zwang, Y.; Cooper, Z.A.; Shee, K.; et al. Potential role of intratumor bacteria in mediating tumor resistance to the chemotherapeutic drug gemcitabine. *Science* **2017**, *15*, 1156–1160. [CrossRef] [PubMed]
153. Farrel, J.; Zhang, L. Variations of oral microbiota are associated with pancreatic diseases including pancreatic cancer. *Gut* **2013**, *61*, 582–588. [CrossRef]
154. Ren, Z.; Jiang, J.; Xie, H.; Li, A.; Lu, H.; Xu, S.; Zhou, L.; Zhang, H.; Cui, G.; Chen, X.; et al. Gut microbial profile analysis by MiSeq sequencing of pancreatic carcinoma patients in China. *Oncotarget* **2017**, *8*, 95176–95191. [CrossRef] [PubMed]
155. Thomas, R.M.; Gharaibeh, R.Z.; Gauthier, J.; Beveridge, M.; Pope, J.L.; Guijarro, M.V.; Yu, Q.; He, Z.; Ohland, C.; Newsome, R.; et al. Intestinal microbiota enhances pancreatic carcinogenesis in preclinical models. *Carcinogenesis* **2018**, *39*, 1068–1078. [CrossRef] [PubMed]
156. Regel, I.; Mayerle, J.; Mukund Mahajan, U. Current strategies and future perspectives for precision medicine in pancreatic cancer. *Cancers* **2020**, *12*, 1024. [CrossRef]
157. Fang, Y.; Yao, Q.; Chen, Z.; Xiang, J.; William, F.E.; Gibbs, R.A.; Chen, C. Genetic and molecular alterations in pancreatic cancer: Implications for personalized medicine. *Med. Sci. Monit.* **2013**, *19*, 916–926. [CrossRef]
158. Pishvaian, M.J.; Bender, R.J.; Halverson, D.; Rahib, L.; Hendifar, A.E.; Mikhail, S.; Chung, V.; Picozzi, V.J.; Sohal, D.; Blais, E.M.; et al. Molecular Profiling of Patients with Pancreatic Cancer: Initial Results from the Know Your Tumor Initiative. *Clin. Cancer Res.* **2018**, *24*, 5018–5027. [CrossRef]
159. Dreyer, S.B.; Jamieson, N.B.; Cooke, S.L.; Valle, J.W.; McKay, C.J.; Biankin, A.V.; Chang, D.K. PRECISION-Panc: The Next Generation Therapeutic Development Platform for Pancreatic Cancer. *Clin. Oncol.* **2020**, *32*, 1–4. [CrossRef]
160. Dreyer, S.B.; Jamieson, N.B.; Morton, J.P.; Sansom, O.J.; Biankin, A.V.; Chang, D.K. Pancreatic Cancer: From Genome Discovery to PRECISION-Panc. *Clin. Oncol.* **2020**, *32*, 5–8. [CrossRef]
161. Lomberk, G.; Dusetti, N.; Iovanna, J.; Urrutia, R. Emerging epigenomic landscapes of pancreatic cancer in the era of precision medicine. *Nat. Commun.* **2019**, *10*, 1–10. [CrossRef]
162. Aung, K.L.; Fischer, S.E.; Denroche, R.E.; Jang, G.-H.; Dodd, A.; Creighton, S.; Southwood, B.; Liang, S.-B.; Chadwick, D.; Zhang, A.; et al. Genomics-Driven Precision Medicine for Advanced Pancreatic Cancer: Early Results from the COMPASS Trial. *Clin. Cancer Res.* **2018**, *24*, 1344–1354. [CrossRef] [PubMed]
163. Goonetilleke, K.S.; Siriwardena, A.K. Systematic review of carbohydrate antigen (CA 19-9) as a biochemical marker in the diagnosis of pancreatic cancer. *Eur. J. Surg. Oncol. EJSO* **2007**, *33*, 266–270. [CrossRef]
164. Schlesinger, Y.; Yosefov-Levi, O.; Kolodkin-Gal, D.; Granit, R.Z.; Peters, L.; Kalifa, R.; Xia, L.; Nasereddin, A.; Shiff, I.; Amran, O.; et al. Single-cell transcriptomes of pancreatic preinvasive lesions and cancer reveal acinar metaplastic cells' heterogeneity. *Nat. Commun.* **2020**, *11*. [CrossRef]
165. Yang, F.; Liu, D.Y.; Guo, J.T.; Ge, N.; Zhu, P.; Liu, X.; Wang, S.; Wang, G.X.; Sun, S.Y. Circular RNA circ-LDLRAD3 as a biomarker in diagnosis of pancreatic cancer. *World J. Gastroenterol.* **2017**, *23*, 8345–8354. [CrossRef]
166. An, Y.; Cai, H.; Zhang, Y.; Liu, S.; Duan, Y.; Sun, D.; Chen, X.; He, X. CircZMYM2 competed endogenously with miR-335-5p to regulate JMJD2C in pancreatic cancer. *Cell. Physiol. Biochem.* **2018**, *51*, 2224–2236. [CrossRef]
167. Zhu, P.; Ge, N.; Liu, D.; Yang, F.; Zhang, K.; Guo, J.; Liu, X.; Wang, S.; Wang, G.; Sun, S. Preliminary investigation of the function of hsa_circ_0006215 in pancreatic cancer. *Oncol. Lett.* **2018**, *16*, 603–611. [CrossRef] [PubMed]
168. Li, J.; Li, Z.; Jiang, P.; Peng, M.; Zhang, X.; Chen, K.; Liu, H.; Bi, H.; Liu, X.; Li, X. Circular RNA IARS (circ-IARS) secreted by pancreatic cancer cells and located within exosomes regulates endothelial monolayer permeability to promote tumor metastasis. *J. Exp. Clin. Cancer Res.* **2018**, *37*, 1–16. [CrossRef] [PubMed]

169. Li, Z.; Yanfang, W.; Li, J.; Jiang, P.; Peng, T.; Chen, K.; Zhao, X.; Zhang, Y.; Zhen, P.; Zhu, J.; et al. Tumor-released exosomal circular RNA PDE8A promotes invasive growth via the miR-338/MACC1/MET pathway in pancreatic cancer. *Cancer Lett.* **2018**, *432*, 237–250. [CrossRef]
170. Jiang, Y.; Wang, T.; Yan, L.; Qu, L. A novel prognostic biomarker for pancreatic ductal adenocarcinoma: Hsa_circ_0001649. *Gene* **2018**, *675*, 88–93. [CrossRef] [PubMed]
171. Qu, S.; Hao, X.; Song, W.; Niu, K.; Yang, X.; Zhang, X.; Shang, R.; Wang, Q.; Li, H.; Liu, Z. Circular RNA circRHOT1 is upregulated and promotes cell proliferation and invasion in pancreatic cancer. *Epigenomics* **2019**, *11*, 53–63. [CrossRef]
172. Xu, Y.; Yao, Y.; Gao, P.; Cui, Y. Upregulated circular RNA circ_0030235 predicts unfavorable prognosis in pancreatic ductal adenocarcinoma and facilitates cell progression by sponging miR-1253 and miR-1294. *Biochem. Biophys. Res. Commun.* **2019**, *509*, 138–142. [CrossRef]
173. Hao, L.; Rong, W.; Bai, L.; Cui, H.; Zhang, S.; Li, Y.; Chen, D.; Meng, X. Upregulated circular RNA circ_0007534 indicates an unfavorable prognosis in pancreatic ductal adenocarcinoma and regulates cell proliferation, apoptosis, and invasion by sponging miR-625 and miR-892b. *J. Cell. Biochem.* **2019**, *120*, 3780–3789. [CrossRef]
174. Liu, L.; Liu, F.B.; Huang, M.; Xie, K.; Xie, Q.S.; Liu, C.H.; Shen, M.J.; Huang, Q. Circular RNA ciRS-7 promotes the proliferation and metastasis of pancreatic cancer by regulating miR-7-mediated EGFR/STAT3 signaling pathway. *Hepatobiliary Pancr. Dis. Int.* **2019**, *18*, 580–586. [CrossRef] [PubMed]
175. Yang, J.; Cong, X.; Ren, M.; Sun, H.; Liu, T.; Chen, G.; Wang, Q.; Li, Z.; Yu, S.; Yang, Q. Circular RNA hsa-circRNA-0007334 is Predicted to Promote MMP7 and COL1A1 Expression by Functioning as a miRNA Sponge in Pancreatic Ductal Adenocarcinoma. *J. Oncol.* **2019**, *2019*. [CrossRef] [PubMed]
176. Yao, J.; Zhang, C.; Chen, Y.; Gao, S. Downregulation of circular RNA circ-LDLRAD3 suppresses pancreatic cancer progression through miR-137-3p/PTN axis. *Life Sci.* **2019**, *239*, 116871. [CrossRef]
177. Chen, Y.; Li, Z.; Zhang, M.; Wang, B.; Ye, J.; Zhang, Y.; Tang, D.; Ma, D.; Jin, W.; Li, X.; et al. Circ-ASH2L promotes tumor progression by sponging miR-34a to regulate Notch1 in pancreatic ductal adenocarcinoma. *J. Exp. Clin. Cancer Res.* **2019**, *38*, 1–15. [CrossRef] [PubMed]
178. Xing, C.; Ye, H.; Wang, W.; Sun, M.; Zhang, J.; Zhao, Z.; Jiang, G. Circular RNA ADAM9 facilitates the malignant behaviours of pancreatic cancer by sponging miR-217 and upregulating PRSS3 expression. *Artif. Cells NanoMed. Biotechnol.* **2019**, *47*, 3920–3928. [CrossRef] [PubMed]
179. Zhang, X.; Li, H.; Zhen, T.; Dong, Y.; Pei, X.; Shi, H. hsa_circ_001653 Implicates in the Development of Pancreatic Ductal Adenocarcinoma by Regulating MicroRNA-377-Mediated HOXC6 Axis. *Mol. Ther. Nucleic Acids* **2020**, *20*, 252–264. [CrossRef]
180. Liu, Y.; Xia, L.; Dong, L.; Wang, J.; Xiao, Q.; Yu, X.; Zhu, H. CircHIPK3 promotes gemcitabine (GEM) resistance in pancreatic cancer cells by sponging miR-330-5p and targets RASSF1. *Cancer Manag. Res.* **2020**, *12*, 921–929. [CrossRef]
181. Wong, C.H.; Lou, U.K.; Li, Y.; Chan, S.L.; Tong, J.H.; To, K.F.; Chen, Y. CircFOXK2 Promotes Growth and Metastasis of Pancreatic Ductal Adenocarcinoma by Complexing with RNA-Binding Proteins and Sponging MiR-942. *Cancer Res.* **2020**, *80*, 2138–2149. [CrossRef]
182. Kong, Y.; Li, Y.; Luo, Y.; Zhu, J.; Zheng, H.; Gao, B.; Guo, X.; Li, Z.; Chen, R.; Chen, C. CircNFIB1 inhibits lymphangiogenesis and lymphatic metastasis via the miR-486-5p/PIK3R1/VEGF-C axis in pancreatic cancer. *Mol. Cancer* **2020**, *19*, 1–17. [CrossRef]
183. Guo, W.; Zhao, L.; Wei, G.; Liu, P.; Zhang, Y.; Fu, L. Blocking circ_0013912 suppressed cell growth, migration and invasion of pancreatic ductal adenocarcinoma cells in vitro and in vivo partially through sponging miR-7-5p. *Cancer Manag. Res.* **2020**, *12*, 7291–7303. [CrossRef]
184. Zhang, X.; Tan, P.; Zhuang, Y.; Du, L. hsa_circRNA_001587 upregulates SLC4A4 expression to inhibit migration, invasion, and angiogenesis of pancreatic cancer cells via binding to microRNA-223. *Am. J. Physiol. Gastrointest. Liver Physiol.* **2020**, *319*, G703–G717. [CrossRef] [PubMed]
185. Zhang, Y.; Li, M.; Wang, H.; Fisher, W.E.; Lin, P.H.; Yao, Q.; Chen, C. Profiling of 95 MicroRNAs in pancreatic cancer cell lines and surgical specimens by real-time PCR analysis. *World J. Surg.* **2009**, *33*, 698–709. [CrossRef] [PubMed]
186. Greither, T.; Grochola, L.F.; Udelnow, A.; Lautenschläger, C.; Würl, P.; Taubert, H. Elevated expression of microRNAs 155, 203, 210 and 222 in pancreatic tumors is associated with poorer survival. *Int. J. Cancer* **2010**, *126*, 73–80. [CrossRef]
187. Zhao, C.; Zhang, J.; Zhang, S.; Yu, D.; Chen, Y.; Liu, Q.; Shi, M.; Ni, C.; Zhu, M. Diagnostic and biological significance of microRNA-192 in pancreatic ductal adenocarcinoma. *Oncol. Rep.* **2013**, *30*, 276–284. [CrossRef]
188. Hussein, N.A.E.M.; Kholy, Z.A.E.; Anwar, M.M.; Ahmad, M.A.; Ahmad, S.M. Plasma miR-22-3p, miR-642b-3p and miR-885-5p as diagnostic biomarkers for pancreatic cancer. *J. Cancer Res. Clin. Oncol.* **2017**, *143*, 83–93. [CrossRef] [PubMed]
189. Zhu, Z.; Xu, Y.; Du, J.; Tan, J.; Jiao, H. Expression of microRNA-218 in human pancreatic ductal adenocarcinoma and its correlation with tumor progression and patient survival. *J. Surg. Oncol.* **2014**, *109*, 89–94. [CrossRef]
190. Sun, J.; Zhang, P.; Yin, T.; Zhang, F.; Wang, W. Upregulation of LncRNA PVT1 facilitates pancreatic ductal adenocarcinoma cell progression and glycolysis by regulating MiR-519d-3p and HIF-1A. *J. Cancer* **2020**, *11*, 2572–2579. [CrossRef]
191. Pang, E.J.; Yang, R.; Fu, X.B.; Liu, Y.F. Overexpression of long non-coding RNA MALAT1 is correlated with clinical progression and unfavorable prognosis in pancreatic cancer. *Tumor Biol.* **2015**, *36*, 2403–2407. [CrossRef] [PubMed]
192. Lu, X.; Fang, Y.; Wang, Z.; Xie, J.; Zhan, Q.; Deng, X.; Chen, H.; Jin, J.; Peng, C.; Li, H.; et al. Downregulation of gas5 increases pancreatic cancer cell proliferation by regulating CDK6. *Cell Tissue Res.* **2013**, *354*, 891–896. [CrossRef] [PubMed]

193. Modali, S.D.; Parekh, V.I.; Kebebew, E.; Agarwal, S.K. Epigenetic regulation of the lncRNA MEG3 and its target c-MET in pancreatic neuroendocrine tumors. *Mol. Endocrinol.* **2015**, *29*, 224–237. [CrossRef]
194. Peng, W.; Gao, W.; Feng, J. Long noncoding RNA HULC is a novel biomarker of poor prognosis in patients with pancreatic cancer. *Med. Oncol.* **2014**, *31*, 1–7. [CrossRef]
195. Li, J.; Liu, D.; Hua, R.; Zhang, J.; Liu, W.; Huo, Y.; Cheng, Y.; Hong, J.; Sun, Y. Long non-coding RNAs expressed in pancreatic ductal adenocarcinoma and lncRNA BC008363 an independent prognostic factor in PDAC. *Pancreatology* **2014**, *14*, 385–390. [CrossRef] [PubMed]
196. Ting, D.T.; Lipson, D. Aberrant Overexpression of Satellite Repeats in Pancreatic and Other Epithelial Cancers. *Science* **2011**, *331*, 593–596. [CrossRef] [PubMed]
197. Kuhlmann, J.D.; Baraniskin, A.; Hahn, S.A.; Mosel, F.; Bredemeier, M.; Wimberger, P.; Kimmig, R.; Kasimir-Bauer, S. Circulating U2 small nuclear RNA fragments as a novel diagnostic tool for patients with epithelial ovarian cancer. *Clin. Chem.* **2014**, *60*, 206–213. [CrossRef]

Journal of
Personalized
Medicine

Review

Drug Repurposing for Triple-Negative Breast Cancer

Marta Ávalos-Moreno [1,†], Araceli López-Tejada [1,2,†], Jose L. Blaya-Cánovas [1,2],
Francisca E. Cara-Lupiañez [1,2], Adrián González-González [1,2], Jose A. Lorente [1,3],
Pedro Sánchez-Rovira [2] and Sergio Granados-Principal [1,2,*]

1   GENYO, Centre for Genomics and Oncological Research, Pfizer/University of Granada/Andalusian Regional
    Government, PTS Granada, Avenida de la Ilustración, 18016 Granada, Spain;
    martaam97@gmail.com (M.Á.-M.); aracelilopeztejada@gmail.com (A.L.-T.); jose.blaya@genyo.es (J.L.B.-C.);
    francisca.cara@genyo.es (F.E.C.-L.); adrikangus@gmail.com (A.G.-G.); jose.lorente@genyo.es (J.A.L.)
2   UGC de Oncología Médica, Complejo Hospitalario de Jaén, 23007 Jaén, Spain; oncopsr@yahoo.es
3   Department of Legal Medicine, School of Medicine—PTS—University of Granada, 18016 Granada, Spain
*   Correspondence: sgranados@fibao.es or sergio.granados@genyo.es; Tel.: +34-651-55-79-21
†   These authors contributed equally to this study.

Received: 24 September 2020; Accepted: 28 October 2020; Published: 29 October 2020

Abstract: Triple-negative breast cancer (TNBC) is the most aggressive type of breast cancer which
presents a high rate of relapse, metastasis, and mortality. Nowadays, the absence of approved
specific targeted therapies to eradicate TNBC remains one of the main challenges in clinical practice.
Drug discovery is a long and costly process that can be dramatically improved by drug repurposing,
which identifies new uses for existing drugs, both approved and investigational. Drug repositioning
benefits from improvements in computational methods related to chemoinformatics, genomics,
and systems biology. To the best of our knowledge, we propose a novel and inclusive classification of
those approaches whereby drug repurposing can be achieved in silico: structure-based, transcriptional
signatures-based, biological networks-based, and data-mining-based drug repositioning. This review
specially emphasizes the most relevant research, both at preclinical and clinical settings, aimed at
repurposing pre-existing drugs to treat TNBC on the basis of molecular mechanisms and signaling
pathways such as androgen receptor, adrenergic receptor, STAT3, nitric oxide synthase, or AXL.
Finally, because of the ability and relevance of cancer stem cells (CSCs) to drive tumor aggressiveness
and poor clinical outcome, we also focus on those molecules repurposed to specifically target this cell
population to tackle recurrence and metastases associated with the progression of TNBC.

Keywords: triple-negative breast cancer; personalized medicine; computational methods; drug
repurposing; clinical trials; cancer stem cells

---

1. Introduction

Breast cancer is the second most common cancer and the second cause of cancer death among
US women, after lung cancer [1]. In 2020, it is estimated that 279,100 new cases will be diagnosed in
the United States and more than 42,000 deaths will be a consequence of this type of cancer [2]. It is
a heterogeneous disease that has been classified using immunohistochemical techniques to measure
the presence of three receptors: estrogen receptor (ER), progesterone receptor (PR), and overexpression
of human epidermal growth factor receptor 2 (HER2). Triple-negative breast cancer (TNBC) is
characterized by the lack of expression of these receptors and, consequently, there are no approved
targeted therapies [3]. Approximately 10% to 20% of new cases of breast cancer would be included in
this subtype, which presents poor prognosis with high risk of relapse compared to other breast cancer
subtypes [4]. TNBC is the breast cancer subtype with the poorest overall survival (OS) and the highest
rates of metastases [5], most commonly in lungs and brain [6]. Furthermore, it is more frequent in

women in younger ages and black race, presenting an incidence rate about twice as high compared with white race [1].

Histopathologically, TNBC is a heterogeneous group that mostly presents features of ductal invasive carcinomas, but also metaplastic, medullary, or apocrine characteristics. Based on the gene expression profile, TNBC is divided into four subtypes: basal-like 1 (BL1), basal-like 2 (BL2), luminal androgen receptor (LAR), and mesenchymal (M) [6]. As a result of the variety and the lack of receptors of TNBC, there are not targeted therapies, making it necessary the application of personalized medicine. Whereas TNBC has a higher sensitivity to chemotherapeutics in comparison to other breast cancers, this subtype presents a higher risk of recurrence, which makes the unraveling of new treatments important [5]. Nevertheless, the process of creating and testing a new drug for TNBC is a cost- and time-consuming challenge that requires a huge investment and comprises high failure rates. For this reason, drug repurposing has been considered an increasingly successful approach for developing new therapies [7].

2. Current Treatments for TNBC

Besides surgery, nowadays, chemotherapy is the only treatment approved by the Food and Drug Administration (FDA) for non-metastatic TNBC [8], which includes microtubule inhibitors, anthracyclines, alkylating agents, antimetabolites, and platinum (Table 1) [7,9]. The current standard of treatment is based on a combination of anthracyclines and taxane agents [10]. In spite of initial chemosensitivity of tumors and the use of different drug combinations to potentiate treatments, later chemoresistance is frequently developed and it is related to the high presence of cancer stem cells (CSC) [9]. All of these compounds are repurposed drugs as they have been previously approved for diseases other than TNBC [7,11,12].

Table 1. Summarized approved agents for non-metastatic triple-negative breast cancer (TNBC).

Class	Agent	Mechanism	Original Indication
Microtubule inhibitors	Paclitaxel Docetaxel	Disruption of microtubule dynamics leading to the end of cell division.	Ovarian cancer, atrial restenosis hormone-refractory prostate cancer
Anthracyclines	Doxorubicin, Epirubicin	Inhibition of DNA, RNA synthesis forming an anthracycline-DNA-topoisomerase II ternary complex. Harm of mitochondrial function. Generation of oxygen-free radicals. Activation of apoptosis and matrix metalloproteinase. Immune reactions.	Antibiotics from *Streptomyces peucetius* bacterium
Alkylating agents	Cyclophosphamide	Inhibition of DNA replication.	Immuno-modulator in autoimmune diseases. Immunosuppressant
Antimetabolites	Methotrexate	Antagonist of dihydrofolate reductase. Decrease the synthesis of purines and pyrimidines.	Leukemia
	Capecitabine	5-fluorouracil pro-drug. Inhibition of thymidylate synthetase.	Colon cancer
	Gemcitabine	Analogue of cytidine. Irreparable errors that inhibit DNA replication.	Anti-viral drug
Platinum	Carboplatin, Cisplatin	Damage of genetic material.	Testicular, ovarian, and bladder cancers

Additional therapeutic options have been recently approved by the FDA for metastatic TNBC, when patients do not respond to traditional treatments (Table 2) [13]. For instance, olaparib and talazoparib, two PARP (poly[adenosine diphosphate-ribose] polymerase) inhibitors of enzymes were approved for patients harboring germline mutations in *BRCA1/2* [8,13–15].

Table 2. Novel approved agents for metastatic TNBC.

Class	Agent	Mechanism	Original Indication
PARP inhibitors	Olaparib Talazoparib	Inhibition of PARP. Cell death due to accumulation of irreparable DNA damage.	Ovarian cancer with *BRCA* mutation
PD-L1 inhibitor	Atezolizumab	Block interaction with receptors PD-1 and reverse T-cell suppression.	Non-small cell lung cancer Bladder cancer
ADC	Sacituzumab govitecan	Targeted to Trop-2 and conjugated with SN-38, a DNA damaging agent.	-

Furthermore, the use of patient's immune system as an approach for cancer treatment, or immunotherapy, has strongly emerged as the fifth pillar of cancer therapy [16]. Immune escape is hallmark of tumor cells that promotes their development and progression, by decreasing immune recognition, for example, through the expression of immune suppressive molecules, or immune checkpoints, like cytotoxic T-lymphocyte-associated antigen-4 (CTLA-4) or programmed cell death-1 and their ligands (PD-1, PD-L1/2)(19–21). Ligand-receptor binding inhibits T-lymphocytes activity through their exhaustion. Physiologically, these molecules are checkpoint regulators of strength and last of LT-mediated immune response [16]. Interaction of PD-1/PD-L1 represents a mechanism of resistance to adaptive immune system by tumor cells in response to the endogenous antitumor response [16]. Nowadays, several checkpoint inhibitors (CPIs) (antibodies anti-CTLA-4, anti-PD-1, and anti-PD-L1) are under clinical use in cancer. In TNBC, combination of CPIs with targeted therapies and/or chemotherapy have been shown to be more effective than monotherapy, which showed a modest effectivity and durability [17]. Recently, atezolizumab, an inhibitor that targets PD-L1, has been approved in combination with paclitaxel for the treatment of patients with previously untreated metastatic TNBC (IMpassion130 study, NCT02425891) [18,19]. Despite of the great expectative on this new and expensive therapy, a small percentage of patients respond to it [16] because of several reasons such as the low tumor infiltration of lymphocytes (TILs, tumor infiltrating lymphocytes), presence of which is associated with a higher survival and good prognosis in early stage TNBC patients [17], low expression of PD-L1 on tumor cells, or the expression of other inhibitor molecules of immune system (IDO, CD73, TIGIT, or VISTA) [20].

Lastly, antibody-drug conjugates (ADC) represent a big potential to improve cancer treatment as they allow to target toxic drugs directly into cancer cells by using specific receptors. Sacituzumab govitecan is the newest therapeutic option available only after the failure of at least two other treatments [13]. This FDA-approved drug is an anti-trophoblast cell-surface antigen 2 (Trop-2) antibody conjugated with SN-38, a DNA damaging agent [21].

3. Drug Repurposing

The discovery and development of a new drug is a time-consuming process which requires great investments, being estimated to take between 10 and 17 years and a cost of US$2–3 billion [22,23]. Moreover, it comprises high failure rates in clinical trials, where almost 90% of the drugs are rejected because of unexpected properties [7]. Drug repurposing (also known as drug repositioning or drug reprofiling) is a strategy for identifying new uses for existing drugs, both approved and under investigation (Figure 1). This relatively new strategy allows to significantly shorten the time and reduce the costs of drug development, especially in the case of FDA-approved repurposed drugs, which would likely go through accelerated clinical trials owing to their previous safety and toxicological clinical studies [24]. It has been estimated that repurposing a drug would cost, on average, US$300 million [23]. Several methodologies can be considered for drug repurposing, from non-computational approaches including high-throughput screening [25] and methods based on experimental findings and previous literature, e.g., target-based, to computational strategies. Indeed, drug repurposing process can be highly improved via computational methods related to chemoinformatics, genomics, and systems biology. These methods allow to select, prior to in vitro experiments, drug candidates for repositioning in a rational manner [24,26,27].

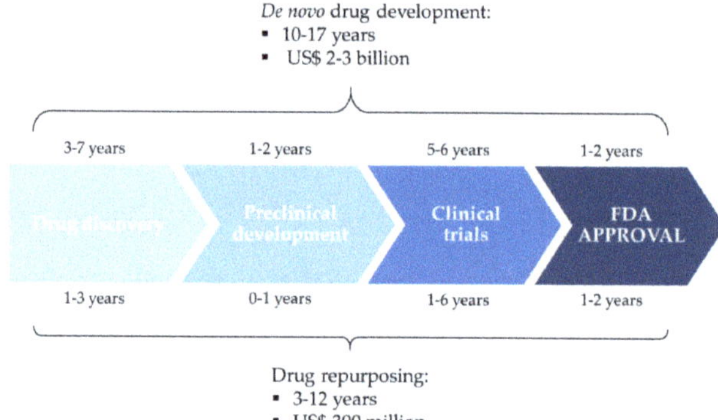

Figure 1. Comparison between de novo drug development and drug repurposing. Adapted from Ashburn and Thor [22].

3.1. Common Computational Approaches for Drug Repurposing

There are many different computational approaches for drug repurposing based on different types of data, including drug and target structures, drug–target interactions, or transcriptomes. Accordingly, several classifications have been suggested [24,28,29]. To date, it has not been determined which approach would be the best option for in silico drug repositioning, and no standardized method has been adopted. Hence, analyzing the retrieved literature, it was considered of interest reviewing and summarizing the most accessible, commonly used approaches (Figure 2), so as to provide a fuller view of the current strategies and the possibilities that in silico analysis has to offer. Thus, these various computational approaches have been categorized in: (1) structure-based, (2) transcriptional signatures-based, (3) biological networks-based, and (4) data-mining-based drug repurposing.

3.1.1. Structure-Based Drug Repurposing

Structure-based methods, which rely on both drug and receptor structure, are mainly based on virtual high-throughput screening (VHTS) of small chemical compounds from different databases such as PubChem (https://pubchem.ncbi.nlm.nih.gov/), DrugBank (www.drugbank.ca/), ChemSpider (www.chemspider.com/) or CheEMBL (www.ebi.ac.uk/chembl/). It allows the user to find, in silico, multiple drugs that will potentially interact with the target's binding site [24]. The 3D structure of the target, which is usually a protein, can be found in the Protein Data Bank (PDB, www.rcsb.org/). VHTS comprises a computational modelling technique known as molecular docking, which enables to predict ligand-receptor biding affinity via different scoring functions. There are several molecular docking programs, such as Glide (www.schrodinger.com/glide), GOLD (www.ccdc. cam.ac.uk/solutions/csd-discovery/components/gold/), UCSF DOCK (http://dock.compbio.ucsf.edu/), AutoDock Vina (http://vina.scripps.edu/), or Ledock software [30]. VHTS can also be inversely approached by finding a variety of biological targets that may have affinity for a particular ligand. Apart from molecular docking, the user can also perform pharmacophore mapping, which consists of searching of ligands that can be matched to a pharmacophore, i.e., a set of molecular features such as hydrogen bonds, hydrophobic groups, or chemical substructures, that enable the recognition of a ligand by a receptor and their biological activity. Pharmacophore features can be derived from protein-binding site or protein–ligand complexes structures, and software packages such as Catalyst (www.3dsbiovia.com/), Unity (Tripos, www.tripos.com), or PharmMapper can be used for pharmacophore searching [24,26]. Structure-based methods also encompass ligand/receptor profiling, based on a guilt-by-association principle. Ligand profiling consists of finding compounds that are chemically similar to a given drug, and consequently may have similar functional and biological properties. Likewise, receptor profiling consists of finding proteins that have similar binding sites to a particular receptor, therefore being likely to bind with the same ligands [24,26].

Figure 2. Diagram of the main computational approaches and software for drug repurposing.

3.1.2. Transcriptional Signature-Based Drug Repurposing

Transcriptional signatures related to a disease or transcriptional responses associated to a specific treatment can be used for drug repurposing. Potential drug candidates can be identified via negative correlation between the gene expression profile from a disease and the transcriptional signature induced by a small compound, with the aim of finding a drug that would reverse the disease state toward the normal one. Similarly, positive correlation can be used to identify small compounds that have similar transcriptional signatures to a genetically or chemically induced perturbation, so as to induce a similar gene expression [31]. Signature-based drug repurposing is also known as connectivity mapping, a concept first introduced with the creation of the Connectivity Map (CMap) database [32,33], which comprises a genome-wide dataset of transcriptional expression responses of human cell lines to perturbagens, e.g., chemical treatments or genetic perturbations [34]. Transcriptional data can be found in different public databases such as Gene Expression Omnibus (GEO; www.ncbi.nlm.nih.gov/geo/), Ensembl (www.ensembl.org/), or The Cancer Genome Atlas (TCGA; https://portal.gdc.cancer.gov/), and several tools are available for analyzing and comparing drug and disease transcriptional profiles. Examples of tools for signature-based repurposing are CMap (https://clue.io/), L1000CDS^2 (http://amp.pharm.mssm.edu/L1000CDS2/), and ksRepo free source [24].

3.1.3. Network-Based Drug Repurposing

Biological networks are data representations used to model biological interactions of any kind, where nodes represent various biological components, such as genes or proteins, and whereas edges represent the associations between them [28]. Network-based drug repositioning methods help inferring unknown disease-associated signaling pathways and therefore new therapeutic targets. There are different biological networks depending on the main source of biological data. Some interesting examples are protein–protein interaction (PPI) networks and drug–target interaction (DTI) networks. In PPI networks, nodes represent proteins. Most proteins are associated with other proteins, but only a limited number of them interact with multiple others. PPI networks allow to identify the most highly connected central proteins, generally known as hubs or hub proteins [35]. Alterations of hubs may affect the structure of the biological network, leading to dysfunction and disease [36]. Accordingly, PPI networking methods help predicting new disease-related targets for drug repurposing. PPI analysis can be performed with PRISM (Protein Interactions by Structural Matching; http://gordion.hpc.eng.ku.edu.tr/prism) server [36], or OmicsNet (https://omicsnet.ca/). Regarding DTIs, they are considered bipartite networks, where nodes represent both drugs and targets. There are several tools for predicting potential DTIs, such as DT-web (https://alpha.dmi.unict.it/dtweb/) or STITCH (http://stitch.embl.de/). Moreover, systems biology combines different network models with quantitative mathematical network models to infer the dynamics of biological systems, providing a more complete perspective for drug repurposing [24]. Complex biological networks can be found in the Causal Biological Networks (CBN, http://causalbionet.com/) database, and complex biological pathways can be found in KEGG database (www.kegg.jp/).

3.1.4. Data-Mining-Based Drug Repurposing

All the previously described methods are based on drug–target interactions. However, meta-analysis of data from clinical trials is another interesting approach for drug repurposing. Su et al. [37] described a novel method for drug repositioning using ClinicalTrials.gov (https://clinicaltrials.gov/) public database and two text mining tools, I2E (Linguamatics) and PolyAnalyst (Megaputer). It consists of, first, the extraction of Serious Adverse Event (SAE) data to identify drugs with fewer SAEs on the test arm than on the control arm and, second, the ranking of said drugs. Therefore, it allows to discover potential drug candidates for diseases different from those in the testing conditions.

4. Drug Repurposing for TNBC

The urgent necessity to find effective molecularly targeted treatments for TNBC has been translated into efforts by the research community to characterize and divide it into different subtypes with a more approachable profile. One of the first transcriptomic-based breast cancer classifications was performed by Perou et al., using cDNA microarrays and hierarchical clustering analysis to distinguish variations in gene expression patterns [38]. It gave a different approach to the commonly immunohistochemical characterization of breast cancers. Afterwards, several studies conducted similar genome-wide analyses [39–41], up until 2009 when Bernard et al. developed a qRT-PCR-based assay using only fifty genes (PAM50) to classify tumors into four intrinsic subtypes of breast cancer: luminal A, luminal B, HER2-enriched, and basal-like [42]. In 2007, Kreike et al. performed the first gene-expression-based classification of TNBC. After gene profiling, they identified all triple-negative breast tumors as basal-like, and classified them in five different subgroups [43]. In opposition, Prat et al. proved that basal-like cancers were not interchangeable with TNBCs [44], similarly to the findings of the study conducted by Lehman et al. in 2014 [45]. While the majority of TNBCs are basal-like, and vice versa, they should not be considered synonymous. These studies highlighted the necessity to further classify

TNBC in well-defined subtypes in order to successfully develop personalized therapies. The first transcriptomic-based TNBC classification which differentiated between basal-like and non-basal like TNBC subtypes was performed by Lehman et al. in 2011. They identified six TNBC subtypes with representative gene expression signatures and signaling pathways, including two basal-like (BL1, BL2), an immunomodulatory (IM), a mesenchymal (M), a mesenchymal stem-like (MSL), and a luminal androgen receptor (LAR) subtype [46]. A web-based tool (TNBCtype) was also developed for the classification of TNBC samples into the six mentioned subtypes [47]. Later in 2016, Lehman et al. refined their own classification algorithm and developed a new one (TNBCtype-4), which scaled down the number of subtypes to four: BL1, BL2, M, and LAR [48]. While several other TNBC classifications followed different approaches and described varying number of subtypes, they all broadly concurred in those four main subgroups [49–51]. Recently, based on both Lehman et al. and Ring et al. algorithms [48,52], Espinosa et al. identified various TNBC cell lines whose signatures remained stable between cell lines and xenografts for each of the four subtypes: HCC2157 for BL1 subtype; HCC70, SUM149PT and HCC1806 for BL2 subtype; BT-549 for M subtype; and MDA-MB-453 for LAR subtype [53]. Thus, those cell lines, representative of each subtype, should be considered for in vitro studies on the effectiveness of targeted therapies in all different subtypes. Among the previously mentioned TNBC subtypes, the dependency on androgen receptor (AR) signaling of the LAR subtype provides a feasible target for directed therapies, which makes it an excellent candidate for drug repurposing. Whereas patients with AR-dependent TNBCs, which have a better prognosis than those with other TNBC subtypes [54], would benefit from AR inhibition therapy, it has been suggested that this may also be beneficial for non-LAR patients with relatively lower AR expression [50,55,56]. However, not all TNBCs express AR, so a quadruple negative breast cancer subtype has also been addressed [57,58]. This subtype would not benefit from AR antagonist repurposing treatments, and so forth different molecular pathways would need to be targeted. Accordingly, we offer an insight on the main repurposed therapies which are currently being investigated for the treatment of TNBC based on their molecular targets, including both AR-directed and non-AR-directed therapies, as shown in Figure 3. We have also summarized drugs in preclinical phase for TNBC in Table 3 and those under clinical trials in Table 4.

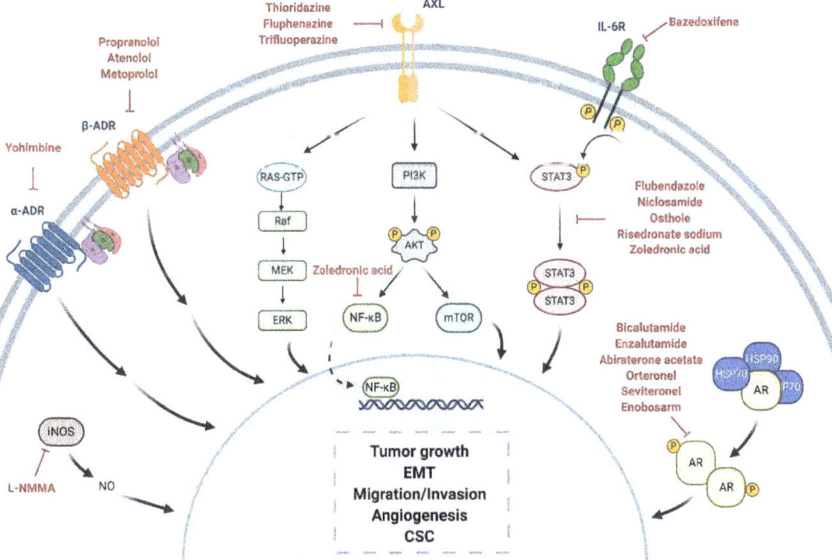

Figure 3. Overview of the different pathways investigated by drug repurposing. Repurposed inhibitors under investigation are shown in red. Created with BioRender.com.

Table 3. Summarized repurposed drugs to treat TNBC that are under investigation in the preclinical phase.

Mechanism	Compound	Pre-Clinical Effects	Original Indication	Repurposing Method	References
α-ADR antagonist	α -yohimbine	Reduction of tumor growth in vitro. Development of resistance to paclitaxel when treated in combination with catecholamines and/or cortisol in vitro. Reversion of tumor growth after stimulation with clonidine in vivo.	Impotence	Non computational: target-based	[59–61]
Non-selective β1/β2-blocker	Propranolol	Inhibition of cell proliferation, arrest of the cell cycle at G0/G1 and S, and induction of cell apoptosis in vitro. Inhibition of tumor growth in vivo. Combination of propranolol with paclitaxel increased the anti-tumor efficacy of paclitaxel in vivo. Associated with less advanced disease at diagnosis and decreased risk of metastasis and mortality. Reverted isoproterenol-induced cell inhibition.	Hypertension	Non computational: target-based	[61–65]
Selective β1-blocker	Atenolol	Reduction of norepinephrine-induced cell migration in vitro. Inhibition of cell proliferation in vitro. Combination with metformin enhanced reduction of angiogenesis and metastasis in vivo. Not associated with differences tumor incidence, risk of metastasis and mortality rates. Associated with significantly lower recurrence but no significant OS.	Hypertension	Non computational: target-based	[63,66–70]
	Metoprolol	Associated with significantly lower recurrence but no significant OS.	Hypertension	Non computational: target-based	[68]
STAT3 inhibitor	Bazedoxifene	Decrease of cell viability, migration, colony formation. Increase cell apoptosis. Improvement of sensitivity to paclitaxel if combination.	Osteoporosis	Computational: structure-based	[71,72]
	Flubendazole	Inhibition of cell proliferation in vitro and tumor growth in vivo. Reduction of CD44^{high}/CD24^{low} CSC population, mammosphere-forming ability and expression of stemness genes. Improvement of sensitivity to fluorouracil and doxorubicin if combination.	Anthelmintic	Non computational: target-based	[73]

Table 3. *Cont.*

Mechanism	Compound	Pre-Clinical Effects	Original Indication	Repurposing Method	References
	Niclosamide	Inhibition of cell proliferation in vitro and tumor growth in vivo. Reversion of EMT and inhibition of stem-like phenotype in cancer cells. Radiosensitizer in vitro and in vivo.	Anthelmintic	Non computational: screening	[74–76]
	Osthole	Induction of apoptosis in vitro. Reduction of tumor growth in vivo.	Osteoporosis	Non computational: literature-based	[77,78]
	Risedronate Sodium	Toxicity in TNBC cells in vitro.	Osteoporosis	Computational: structure-based	[79]
AXL pathway modulator	Thioridazine Fluphenazine Trifluoperazine	Decrease of cell invasion, proliferation, and viability and increase of apoptosis in vitro. Reduction of tumor growth and metastasis in vivo.	Anti-psychotics	Computational: transcriptional signature-based	[80]

Table 4. Summarized repurposed drugs for TNBC under current investigation in clinical trials.

Mechanism	Compound	Preclinical and Clinical Effects	Clinical Trials [1]	Original Indication	Repurposing Method	References
AR antagonist	Bicalutamide	Reduction of cellular proliferation and colony formation, and induction cell apoptosis in vitro. Decreased cellular viability and induced apoptosis in vivo. CBR at 6 months of 19% and median PFS of 12 weeks (n = 26; AR expression higher than 10% by IHC). Grade 1–3 AEs included fatigue, limb edema, or hot flashes.	Phase II—completed (NCT00468715) Phase II—recruiting (NCT02605486) Phase III—recruiting (NCT03055312)	Prostate cancer	Non computational: target-based	[81,82]
	Enzalutamide	Reduction of cell proliferation, migration and invasion and increased apoptosis in vitro. Inhibition of tumor viability by inducing cell apoptosis in vivo. CBR at 16 weeks of 25%, median PFS of 2.9 months and median OS of 12.7 months (n = 118) AR expression higher than 0% by IHC. CBR at 16 weeks of 33%, median PFS of 3.3 months and median OS of 17.6 months (n = 78; AR expression higher than 10% by IHC). Grade 3 AEs included fatigue.	Phase II—completed (NCT01889238) Phase II—recruiting (NCT02689427) Phase Ib/II—active (NCT02457910)	Prostate cancer	Non computational: target-based	[55,56,83,84]

bicalutamide, results in a lack of agonistic activity [54,55,58]. Different in vitro studies demonstrated that enzalutamide reduced cell proliferation, migration, and invasion and increased apoptosis [55,56,84], and it was correlated with decreased AREG mRNA expression in SUM159 cells after treatment with enzalutamide [56]. In vivo studies showed that enzalutamide inhibited tumor viability in TNBC xenografts by inducing cell apoptosis [56,84]. A single-arm, non-randomized phase II clinical trial evaluated the efficacy of enzalutamide in advanced AR-positive TNBC (NCT01889238). In this study, AR positivity was defined as AR expression higher than 0% by IHC (intent-to-treat population, ITT) or higher than 10% by IHC (evaluable subgroup). The ITT population (*n* = 118) and the evaluable subgroup (*n* = 78) showed a CBR at 16 weeks of 25 and 33%, respectively. Median PFS was 2.9 months in the ITT group and 3.3 in the evaluable group. Median OS was 12.7 and 17.6 in ITT and evaluable subgroup, respectively. The only treatment-related AE with grade 3 or higher was fatigue, meaning enzalutamide was well tolerated by AR-positive TNBC patients. This study supported further study of enzalutamide [83]. Moreover, other clinical studies are currently investigating the use of enzalutamide as an adjuvant in treating patients with AR-positive TNBC, including a phase II trial (NCT02689427) for enzalutamide in combination with paclitaxel and a phase Ib/II trial for enzalutamide in combination with taselisib (NCT02457910).

Abiraterone acetate. It was the first androgen-production inhibitor developed for the treatment of prostate cancer. It is a steroidal, non-selective inhibitor of 17α-hydroxylase/17,20-lyase (CYP17), a central, rate-limiting enzyme which plays a critical role in the androgen biosynthesis pathway [54,58,102]. The efficacy of abiraterone acetate was investigated in a phase II clinical trial in combination with prednisone in metastatic or locally advanced AR-positive TNBC patients (NCT01842321). AR positivity was defined as AR expression greater than 10% by IHC. Evaluable patients (*n* = 30) showed a CBR at 6 months of 20%, and the median PFS was 2.8 months. The most common treatment-related AEs were hypertension, fatigue, nausea, and hypokalemia, all grade 1–2 [85]. After this clinical trial, both in vitro and in vivo studies were performed to assess whether combining abiraterone acetate with a Chk1 inhibitor would enhance its efficacy. They showed that combination treatment with the inhibitor GDC-0575 had an additive effect on both MDA-MB-453 and SUM185PE cell lines in reducing cell proliferation. Whereas abiraterone acetate alone had a weak effect inducing apoptosis, Chk1 inhibitors doubled the effect, achieving statistical significance in MDA-MB-453 cells. Interestingly, a xenograft model with MDA-MB-453 cells injected orthotopically in the mammary gland ducts of NSG mice showed that abiraterone alone reduced tumor growth, and combination with GDC-0575 enhanced this effect [86].

Orteronel (TAK-700). It is a non-steroidal, selective, second-generation CYP17 inhibitor. Whereas clinical trials for the treatment of prostate cancer with orteronel were terminated in phase III because of a lack of significant effect on OS [54,58,103], it is currently being investigated in a phase II clinical study of women with AR-positive metastatic TNBC (NCT01990209).

Seviteronel (VT-464). It is another non-steroidal, selective, second-generation CYP17 inhibitor which, in contrast to orteronel, also inhibits AR activation [54,58]. It was demonstrated that seviteronel inhibited cellular growth and tumor volume in MDA-MB-453 cells and patient-derived xenografts (PDX), respectively [88,89]. Moreover, Michmerhuizen et al. proved that the AR inhibition with seviteronel induced radiosensitization, both in vitro and in vivo, whereas enzalutamide did not [104]. A phase I/II clinical study is investigating the activity of seviteronel in women with AR-positive TNBC (NCT02580448). Out of 16 patients with AR-positive TNBC, 6 were evaluable. Two patients (33%) had a 16-week CBR. The most common AEs were fatigue, nausea, and decreased appetite, all grade 1–2 [87]. A second phase II clinical trial is also currently investigating the effects of seviteronel in AR-positive TNBC patients (NCT02130700).

Enobosarm (MK-2866, ostarine, GTx-024). It is a non-steroidal SARM that achieves a tissue-selective modulation of AR action, hence minimizing the undesirable side-effects caused by antiandrogens [105]. In vitro studies showed that enobosarm inhibited cellular proliferation of MDA-MB-231 cells transiently expressing AR. Moreover, tumor growth was completely inhibited by enobosarm in a nude mice xenograft model with MDA-MB-231-AR cells [106]. There was a phase II clinical trial for enobosarm in AR-positive TNBC (NCT02368691), but it was terminated because of lack of efficacy.

4.2. Adrenergic Receptor

Adrenergic receptors (ADR), which can be classified as α or β receptors, belong to the G protein-coupled receptor (GPCR) superfamily. The activation of ADR, stimulated through the catecholamines, epinephrine and norepinephrine, derives in several stress response signaling pathways key in maintaining physiological homeostasis [107]. However, there is an increasing evidence that altered ADR stimulation may play a significant role in breast cancer progression, promoting cell proliferation, metastasis, tumor invasion, and angiogenesis [68,108,109]. Accordingly, it has been addressed that ADR-directed therapies, widely used for the treatment of hypertension and other pathologies, could be repurposed for TNBC. Several preclinical studies have investigated the effects of both α- and β-ADR antagonists in TNBC [61,64,66,67,110,111], and retrospective epidemiological studies have explored whether TNBC cancer patients under treatment with beta-blockers for hypertension had a significant better outcome that non-treated patients [63,68,108,112].

4.2.1. α-Adrenergic Receptor

α-adrenergic receptors can be subclassified as α1 (α1a, α1b, α1c) and α2 (α2a, α2b, α2c). Their ligands activate GPCRs and initiate a signaling cascade that, in the case of α1 receptors, increases intracellular calcium levels and is involved in blood pressure regulation, whereas α2 receptors signaling cascade decreases intracellular cyclic AMP (cAMP) levels and regulates neurotransmitters release [107]. Interestingly, activation of α-ADR has been associated with both tumor growth and chemoresistance in TNBC cell lines. Vazquez et al. showed that both epinephrine and norepinephrine, the natural ADR agonists, as well as clonidine, a synthetic α(2)-ADR agonist used in the treatment of hypertension [113], promoted cell proliferation in MDA-MB-231 cells [110]. Similarly, Bruzzone et al. demonstrated that clonidine increased tumor growth, whereas α(2)-ADR antagonist α-yohimbine reversed clonidine stimulation in breast cancer [114].

α-yohimbine (rauwolscine). It is an alkaloid and α(2)-ADR antagonist used as a mydriatic and in the treatment of impotence [115]. Piñero et al. found that yohimbine diminished tumor growth in vitro, and it was associated with inhibition of ERK1/2 phosphorylation in vivo [61]. It was also proved that α-yohimbine could reverse tumor growth after stimulation with clonidine in vivo [59]. Additionally, Flint et al. demonstrated that MDA-MB-231 cells developed resistance to paclitaxel when treated in combination with catecholamines and/or cortisol [60]. In the light of these results, we suggest the investigation of α-ADR antagonists for the treatment of TNBC and prevention of drug resistance.

4.2.2. β-Adrenergic Receptor

β-adrenergic receptors can also be subclassified as β1, β2, and β3. Activation of β1- and β2-ADR increases intracellular cAMP levels, as opposed to α2-ADR, regulating the sympathetic nervous system's stress response in several different tissues [107]. The signaling cascade induced by higher cAMP levels includes two main pathways. First, cAMP activation of protein kinase A (PKA) induces phosphorylation of several transcription factors, such as GATA family, and β-ADR kinase (BARK). The latter inhibits β-ADR signaling and activates Src kinase, leading to the activation of different transcription factors, including STAT3, and several kinases like focal adhesion kinase (FAK). Conversely, cAMP also leads to Rap1A activation, which induces B-Raf/mitogen-activated protein kinase (MAPK) signaling pathway and activation of multiple genes with effects on several cellular events [116]. It has been addressed that, in breast cancer, β-ADR signaling in β-ADR-expressing tumor cells activates metastatic-associated genes involved in inflammation, angiogenesis, and EMT processes, whereas it downregulates the expression of antitumoral response genes. Moreover, activation of β-ADR pathway in tumor stromal cells and tumor-associated macrophages seem to promote tumor growth and metastasis [109,116]. Several in vitro studies with different TNBC cell lines showed that β-ADR agonists stimulated cell migration, whereas β-ADR antagonists, such as atenolol and ICI118551, reverted this process [66,67,111]. Moreover, it was also demonstrated that β-blockers propranolol and ICI118551 decreased cell proliferation in TNBC, arresting the cell cycle and inducing cell apoptosis [62]. Oppositely, Slotkin et al. showed that treatment with β-ADR agonist isoproterenol lowered DNA synthesis and decreased cell proliferation, and that these effects were reverted by propranolol [64]. Similarly, in an experimental mouse model of breast cancer, β-ADR agonists isoprenaline and salbutamol inhibited breast cancer cell proliferation and tumor growth [61]. There seems to be conflicting results in the role of β-ADR signaling in breast cancer, indicating that it might be dependent on the cancer subtype. Accordingly, different retrospective observational cohort studies have been developed to further study the effects of different non selective β1/β2-blockers (propranolol, timolol) and selective β1-blockers (atenolol, bisoprolol, metoprolol) in breast cancer, more precisely in TNBC, so as to determine their effects in the cancer biology of each subtype [63,68,108,112]. The first observational study was performed by Powe et al. [108], in which breast cancer patients were divided into three subgroups: non-hypertensive control group ($n = 374$), hypertensive patients treated, prior to cancer diagnosis, either with β-blockers ($n = 43$) or with other antihypertensives ($n = 49$). Most β-blocker users had received selective blockers (25 with atenolol, 7 bisoprolol), but several had received non-selective ones (7 propranolol, 4 timolol). β-blocker users group suggested a significant lower risk of metastasis development, tumor recurrence, and breast cancer mortality. However, differences in β-ADR antagonists used by patients, and the lack of information in their cancer subtype made it necessary to perform further studies to assess the efficacy of non-selective β1/β2-blockers versus selective β1-blockers in TNBC.

Non-selective β1/β2-blockers (propranolol). Different studies showed that propranolol inhibited cell proliferation, arrested the cell cycle at G0/G1 and S, and induced cell apoptosis in vitro, and inhibited tumor growth in vivo [61,62,65]. Moreover, the anti-tumorigenic effects of this β-blocker were associated with a decrease in phosphorylation levels of ERK1/2 and the expression levels of cyclooxygenase 2 (COX-2) [62]. Interestingly, Pasquier et al. reported that, whereas combination of propranolol with chemotherapeutic drug paclitaxel seemed to have no additive effects in cellular cytotoxic effects in vitro, propranolol increased the anti-tumor efficacy of paclitaxel in an orthotopic xenograft model of TNBC, significantly increasing the median survival [65]. Barron et al. performed a study on women treated with propranolol for hypertension ($n = 70$) in the year before breast cancer diagnosis, in comparison with matching (1:2) non-users ($n = 4738$), and suggested that the use of propranolol was significantly associated with less advanced disease at diagnosis and decreased risk of metastasis and mortality [63]. However, like Ganz et al. pointed out, the limited size of the β-blocker users' group may be insufficient to prove propranolol benefits in breast cancer [117]. Moreover, the patient population was not subclassified based on cancer subtype or receptor status, so no conclusions can be drawn for TNBC subtype.

Selective β1-blockers (atenolol, metoprolol). In vitro studies demonstrated that atenolol inhibited cell proliferation in MDA-MB-435 cells [69], and enhanced metformin activity in vivo by reducing angiogenesis and metastasis [70]. In the same study mentioned above, Barron et al. also evaluated breast cancer patients treated with selective β1-blocker atenolol ($n = 525$) in the year before cancer diagnosis. However, they found no significant difference in between atenolol users and matched non-users in tumor incidence, risk of metastasis and mortality rates. These results indicated that the effects of propranolol in breast cancer were mediated by β2-ADR [63]. Melhem-Bertrandt et al. performed another retrospective study comparing breast cancer patients treated with β-blockers ($n = 102$), who received neoadjuvant chemotherapy, with non β-blockers users ($n = 1311$), as well as TNBC patients taking β-blockers ($n = 29$) compared to non-users ($n = 348$) [68]. The most commonly prescribed β-blockers were selective β1-blockers, first metoprolol (42%) followed by atenolol (37%). Interestingly, after age, race, stage, and receptor status adjustment, among some other parameters, users of β-blockers proved to have significantly lower recurrence but no significant OS among both breast cancer and TNBC patients, which seemed to contradict the findings of Barron et al. However, a subset analysis demonstrated that the subgroup of ER-positive breast cancer patients had no significant differences in tumor recurrence. Consequently, these results suggested that, whereas patients with any breast cancer subtype could benefit from a treatment with non-selective β-blockers via β2-ADR antagonism, only TNBC patients could benefit from a treatment with non-selective β-ADR inhibitors. Nevertheless, it has to be noted that not statistically significant results in the ER-positive subgroup may have been due to the relatively short follow-up time in the study of Melhem-Bertrandt et al. Additionally, in a retrospective study on TNBC patients taking β-blockers ($n = 74$), compared to non-users ($n = 726$), Botteri et al. also demonstrated that a treatment with β-blockers was associated with a decreased risk of recurrence, metastasis, and mortality, supporting previous findings [112]. Nevertheless, new prospective studies will be required to clarify whether the efficacy of β-blockers depends on breast cancer subtype and/or receptor status.

4.3. STAT3

Signal transducer and activator of transcription 3 (STAT3) is a tumor marker for early diagnosis and the activation of its pathway is related to breast cancer aggressiveness, as it plays an important role in progression, proliferation, apoptosis, metastasis, and chemoresistance [118]. The activation of this pathway involves several cytokines such as, interleukin 6 (IL-6) and interleukin 10 (IL-10), and growth factors, including epidermal growth factor (EGF), fibroblast growth factor (FGF), and insulin-like growth factor (IGF), which bind their receptors and activate Janus kinases (JAKs). JAKs phosphorylate themselves in a tyrosine domain included in their cytoplasmic fractions and they subsequently activate STAT3 via tyrosine phosphorylation. Once STAT homodimers are produced, they are translocated to the nucleus in order to create a complex with coactivators (e.g., p68) and ending up into the activation of transcription [118]. The upregulation of IL-6/STAT3/ROS can lead to the transcription of genes involved in breast cancer progression, as well as an augmentation in inflammation and generation of breast cancer stem cells (BCSCs). Furthermore, the activation of JAK2/STAT3 favors proliferation and motility of breast cancer cells by different mechanisms, including the suppression of apoptosis by upregulation of cyclin D-1, c-Myc, and Bcl-2, and promotion of EMT. Finally, resistance to several drugs like paclitaxel may be a consequence of this pathway. Because of its complexity and wide regulation of breast cancer cells, STAT3 is an interesting target candidate to treat in TNBC. As a matter of fact, several compounds that inhibit different mechanisms are being investigated. We will highlight some of them: bazedoxifene, flubendazole, niclosamide, osthole, and zoledronic acid [118].

Bazedoxifene. It is a selective ER modulator approved in 2013 by the FDA to treat and prevent osteoporosis in postmenopausal women [71]. Using a structure-based study for repurposing drugs, bazedoxifene was discovered as a novel inhibitor of IL-6 receptor by blocking signals of glycoprotein 130 [119]. Hence, in TNBC, its mechanism involves the upstreaming disruption of STAT3 pathway as ER is not expressed. Studies in in vitro and in vivo models of TNBC confirmed the decrease of cell viability, migration, colony formation, and increase of apoptosis. Furthermore, when this compound was administered in combination with paclitaxel, a synergistic effect as well as an improvement of sensitivity to paclitaxel was found, probably because of the inhibition of the resistance effect induced by IL-6 [71,72]. Those doses were administered in safety ranges that are registered in other indication trials of bazedoxifene. Subsequently, safe effects can be assured in endometrial, ovarian, and breast tissues, but it would be necessary to study possible secondary effects in other tissues that express ER [72]. Considering the association between STAT3 and EMT, their interplay in CSCs, and the in vitro effects of bazedoxifene, we suggest that this compound could act as an inhibitor of tumor-initiating cells, although this hypothesis must be further investigated.

Flubendazole. It is an FDA-approved anthelmintic agent to treat intestinal parasites whose mechanism of action is the disruption of tubulin polymerization. For this reason, it was considered as a repurposed candidate to treat breast cancer [120]. Even though flubendazole causes cell cycle arrest at G2/M phase and, consequently, inhibits cell proliferation in vitro and tumor growth in vivo at clinical doses, it also presents additional properties. As an STAT3 inhibitor, it also causes a reduction of CD44^{high}/CD24^{low} CSC population, mammosphere-forming ability, and the expression of stemness genes [73]. This fact is a positive characteristic as CSCs might have an essential role in metastasis and aggressiveness of TNBC [120]. Furthermore, in some studies flubendazole is shown to increase cytotoxicity activity of fluorouracil and doxorubicin, meaning it could reduce tumor chemoresistance [73].

Niclosamide. It is a FDA-approved anthelmintic agent to treat tapeworms, which is known to inhibit cell growth in vitro and tumor growth in vivo in TNBC studies [74]. Niclosamide was identified as an inhibitor of BCSCs owing to a high-throughput drug screening [76]. It reverses EMT and inhibits the stem-like phenotype in cancer cells suggesting that it may reverse cisplatin resistance [74]. Furthermore, Lu et al. proved that niclosamide is a radiosensitizer both in vitro and in vivo models of TNBC as it reversed radioresistance generated by activation of STAT and Bcl-2 and reduction of reactive oxygen species (ROS) [75].

Osthole (7-methoxy-8-isopentenoxycoumarin). It is a coumarin-derivative extract isolated from *C. monnieri* that presents interesting properties, such as anti-inflammatory and vasorelaxant [121]. Osthole has successful results in vivo treating osteoporosis as it stimulates osteoblast proliferation and differentiation and bone formation [77]. It also possesses anti-tumoral characteristics and, hence it can be a candidate for repositioning in TNBC. Dai et al. elucidated that osthole inhibits STAT3 phosphorylation, induced by IL-6, in a dose-dependent manner by avoiding the translocation of STAT3 to the nucleus, what causes cell cycle arrest and induction of apoptosis in TNBC cell lines. Moreover, in vivo assays with osthole confirmed the suppression of STAT3 phosphorylation as well as reduction of tumor growth in TNBC xenograft mice [78].

Risedronate sodium and zoledronic acid. They are two oral bisphosphonates to treat osteoporosis that were found to be possible candidates as STAT3 inhibitors by a comparative docking study in silico. Svranthi et al. also proved their toxicity in TNBC cells in vitro [79]. Furthermore, zoledronic acid has been largely analyzed for TNBC. Schech et al. proved that it inhibited cell viability, induced cell cycle arrest, reduced proliferative capacity, inhibited self-renewal capability, and decreased the expression of EMT markers (N-cadherin, Twist, and Snail). Mechanistically, they discovered that zoledronic acid inhibited phosphorylation of RelA, an active subunit of nuclear factor κB (NF-κB). Consequently, direct inactivation of NF-κB induced the loss of EMT transcription factor gene expression [91]. In vivo studies in mice also support the antitumor potential of zoledronic acid in combination with doxorubicin [92]. In a randomized phase II clinical trial (UMIN000003261), the combination of zoledronic acid and neoadjuvant chemotherapy was evaluated in TNBC patients. The pathologic complete response rate (pCR) was ameliorated in the combination group (35.3%) ($n = 17$) compared to patients treated with chemotherapy alone (11.8%) ($n = 17$). Such an improvement of pCR rate was translated into a higher disease-free survival in the combination group (70.6%) versus the chemotherapy group (94.1%) [90]. In contrast, a phase II clinical trial studying the application of pre-operative zoledronate prematurely ended because of a low accrual rate (NCT02347163). Further trials to assess the anti-tumor activity of zoledronic acid are currently ongoing in combination with atorvastatin and neoadjuvant standard chemotherapy (NCT03358017), as well as to evaluate the potential of zoledronic acid as an adjuvant therapy (NCT02595138, NCT04045522).

4.4. Nitric Oxide Synthase

Nitric oxide (NO) is a small molecule that is involved in several functions in the organism. It can be synthesized by three isoforms of nitric oxide synthase (NOS): neuronal (NOS1/nNOS), inducible (NOS2/iNOS), and endothelial (NOS3/eNOS). NO has a short half-life and interacts with different targets, which produces nitrites, nitrates, S-nitrosothiols, and nitrosamines, these being compounds that induce DNA damage and, therefore, gene mutations [122]. Glynn et al. proved that an increased expression of iNOS in ER^{-} breast cancer is correlated with poor survival of patients [123]. We later proved that iNOS is a biomarker of poor prognosis and a good therapeutic target in a cohort of 73 TNBC patients [93]. In a previous report, we identified two genes, *RPL39* (ribosomal protein

L39) and *MLF2* (myeloid leukemia factor 2), that are commonly mutated in lung metastases from breast cancer patients, and their inhibition significantly reduced BCSC self-renewal and number, tumor cell migration, invasion and generation of lung metastases, and tumor growth in in vitro and patient-derived xenografts (PDX) models of TNBC. Mechanistically, RPL39 and MLF2 expression was associated with iNOS signaling, and their mutations were associated with shorter median time to relapse in a cohort of 53 breast cancer patients, which suggests that iNOS inhibition represents a promising strategy for the treatment of TNBC [124]. In this regard, we reported that iNOS inhibitors diminish cancer cell proliferation and migration, CSC self-renewal and EMT by a targeting HIF1α and endoplasmic reticulum stress-transforming growth factor (TGFβ)-ATF4/ATF3 crosstalk [93]. Furthermore, we later confirmed that ATF4 is a transcriptional target of TGFβ-Smad2/3, is a biomarker of poor prognosis in TNBC patients, and promotes tumor progression by modulating CSCs, metastasis, relapse, and growth in PDX of TNBC [125]. Among the inhibitors tested, we reported that the pan-NOS inhibitor L-NMMA (NG-monomethyl-L-arginine) decreased cell proliferation, migration, and CSC self-renewal in vitro, and tumor growth (associated with less expression of Ki67), CSC self-renewal and tumor initiation in xenograft models of TNBC. Accordingly, we designed a safe and effective targeted therapy in TNBC by repurposing L-NMMA, previously studied in septic shock, with a dose regimen in combination with docetaxel that restrained tumor growth and prolonged mice survival [93]. Moreover, in combination with docetaxel, iNOS inhibition with L-NMMA enhanced the response to chemotherapy in PDX models of TNBC [94]. The translation of this therapeutic approach into clinic is under investigation in a phase Ib/II study in refractory locally advanced or metastatic TNBC patients (NCT02834403) [93,94]. Finally, iNOS has been associated with different signal transduction pathways such as vascular endothelial growth factor (VEGF). Increased levels of VEGF have been found in TNBC and it is known that NO can be responsible for it. Both iNOS and eNOS can induce VEGF and promote angiogenesis, thus L-NMMA (pan-NOS inhibitor) may be a good option to target this pathway [126].

4.5. Anexelekto (AXL)

AXL, named from the Greek word *anexelekto* which means "uncontrolled," is one of the TAM (Tyro3, AXL, and Mer) family of receptors tyrosine kinase (RTK) [127]. Structurally, in the extracellular part, it is composed of two immunoglobulin-like domains and two fibronectin III domains. The intracellular part presents an RTK domain that contains a KWIAIES motif of TAM family. Its activation results in the autophosphorylation at the cytoplasmic domain that unleashes different cascades and downstream targets that are highly context dependent. Some of these pathways are PI3K/protein kinase B (Akt), extracellular-signal-regulated kinase (ERK), and STAT, which can stimulate tumorigenic processes such as cell motility, invasion, or proliferation [128]. In TNBC patients, the high expression of AXL is a predictor of poor prognosis, produces mesenquimal phenotypes, by promoting EMT through the expression of Vimentin, Twist, Snail, and Slug, higher chemoresistance, tumorigenesis, metastases, and CSCs, which make it a potential candidate to treat TNBC [80,128,129]. AXL can be activated by mechanisms dependent and independent of the ligand GAS6. If it is mediated by GAS6, AXL activates signaling pathways like PI3K/Akt, MAPK, NF-κB, and JAK/STAT, which can stimulate tumorigenic processes. On the other hand, the GAS6-independent pathway involves EGFR that activates AXL, which finally unleashes Akt transcription and produces an increase of tumor cell proliferation and migration [128]. Targeted inhibition of EGFR may not be a good option in TNBC because AXL can be activated thought other pathways and the response to EGFR inhibitors is limited [130]. Because of drug repositioning three drugs included in the same family are considered as a possible CSC-targeted therapy.

Phenotiazines. Goyette et al. carried out a research of drug repurposing based on AXL knockdown gene signature. Using CMap, they found that three phenothiazines (thioridazine, fluphenazine, trifluoperazine) could produce a similar gene signature. These dopamine receptor antagonists are used as anti-psychotics and were tested in TNBC, obtaining good results both in vitro and in vivo. In vitro, decrease of cell invasion, proliferation and viability, and increase of apoptosis were seen in TNBC cell lines. Interestingly, an increased sensitivity to standard chemotherapy was also observed in combination with paclitaxel. In vivo, a significant reduction of tumor growth and metastasis were observed. Furthermore, mechanistic insights revealed that these compounds did not exert their activities by antagonizing with dopamine receptor. AXL activity was not decreased but a reduction of PI3K/Akt/mammalian target of rapamycin (mTOR) and ERK signaling was produced, unravelling that repurposed drugs generate the same consequences as AXL knockdown [80].

5. Drug Repositioning to Target Cancer Stem Cells in TNBC

The CSC model for tumor propagation underlines that solid tumors are hierarchically organized, and contain a subset of cancer cells with stem-cell-like characteristics known as CSCs or tumor-initiating cells, which are able to sustain tumor growth, progression, and recurrence, as well as metastasis. Consequently, this model would explain intra-tumor heterogeneity and dormant behavior of several types of cancer [131–133]. CSCs phenotype varies according to the type of cancer. BCSC are characterized by surface markers CD44$^+$/CD24$^{-/low}$ and aldehyde dehydrogenase 1 (ALDH1) enzyme activity. Interestingly, it has been suggested that the acquisition of a stemness phenotype in CD44$^+$/CD24$^{-/low}$ subpopulation is connected to EMT [134], key event in metastatic spread [131,135,136]. EMT is known to be regulated by different pathways, including the TGFβ, PI3K/Akt/mTOR, MAPK, or Wnt/β-catenin, which can be abnormally regulated during malignant processes in TNBC [131]. In fact, several studies have demonstrated that activation of EMT induced by TGFβ increases the subpopulation of CSCs in breast cancers [137,138]. Interestingly, CSCs have been proved to be more abundant in TNBC than in other breast cancer subtypes, which could explain its higher aggressiveness [139,140]. Therefore, efforts are being focused on the development of CSC-targeted therapies [141]. Additionally, several studies have shown that CSCs are intrinsically resistant to chemotherapy and radiotherapy, therefore, targeting CSCs in combination with conventional chemotherapy might decrease the aggressiveness of TNBC and prevent cancer relapse and improve survival [131–133]. It has been suggested that EMT inhibitors could be potential CSC-targeted therapies in breast cancer. In fact, activation of Wnt/β-catenin signaling has been correlated with the expression of CD44$^+$/CD24$^{-/low}$ CSC subpopulation. Whereas different Wnt inhibitors are currently under development for the treatment of cancer, several FDA-approved drugs, such as salinomycin, vitamin D3, or pyrvinium pamoate, have proven to inhibit this pathway, being possible candidates for repurposing [50,142]. Some other FDA-approved drugs have also been demonstrated to regulate EMT and/or affect CSCs via different molecular pathways, such as all-trans retinoic acid (ATRA) [143], benztropine mesylate [144,145], and chloroquine [146]. Moreover, some of the previously mentioned TNBC-directed repurposed drugs were shown to target EMT or CSCs as well, including flubendazole, niclosamide, zoledronic acid, and L-NMMA. All breast CSCs-targeted drugs that are being investigated are summarized in Table 5.

Table 5. Summary of drug candidates to target cancer stem cells (CSCs) under investigation by drug repurposing.

Mechanism	Compound	Cellular and Molecular Effects	Original Indication	Repurposing Method	References
Wnt, LRP6	Salinomycin	Decreased CD44$^+$/CD24$^{-/low}$ population both in vitro and in vivo. Inhibition of tumor growth and expression of CSC genes in vivo. Combination with LBH589 induced apoptosis and cell cycle arrest and regulates EMT in BCSCs.	Antibiotic	Non computational: high-throughput screening	[147–149]
Wnt/β-catenin, PI3K dependent pathway, lipid anabolism	Pyrvinium pamoate	Reduction of CSC self-renewal. Reduction of CD44$^+$/CD24$^{-/low}$ and ALDH+ populations. Reduction of expression of EMT markers (N-cadherin, Vimentin and Snail). Reduction of tumor growth in vivo.	Anthelmintic	Non computational: high-throughput screening	[142,130,131]
Notch-1, NF-κB1	Vitamin D3	Reduction of cell proliferation, CD44$^+$/CD24$^{-/low}$ population and mammosphere formation in vitro. Relative insensitivity to vitamin D3 treatment, but combination therapy with DETA NONOate achieved a significant decrease in mammosphere formation in vitro and tumor growth in vivo.	Vitamin supplement	Non computational: target-based	[152–154]
Notch-1, TGF-β	ATRA	Inhibition of mammospheres formation and reduction of CSC self-renewal. Reduction of ALDH1 CSC subpopulation.	Dermatologic diseases, acute promyelocytic leukemia	Computational: transcriptional signature-based	[155,156]
STAT3, NF-κB, and β-catenin	Benztropine mesylate	Inhibition of mammospheres formation and reduction of CSC self-renewal. Reduction of ALDH and CD44$^+$/CD24$^{-/low}$ populations.	Parkinson's disease	Computational: cell-based phenotypic screening	[144]
Jak2, DNMT1	Chloroquine	Inhibition of autophagy. Reduction of mammosphere formation efficiency and CD44$^+$/CD24$^{-/low}$ population in vitro. Sensitization to paclitaxel through the inhibition of autophagy in vitro. Combination of paclitaxel significantly reduced tumor growth and CD44$^+$/CD24$^{-/low}$ population in vivo. Phase II clinical trial for chloroquine in combination with taxanes: ORR of 45.16%, median PFS of 12.4 months and median OS of 25.4 months. 13.15% of patients experienced Grade ≥ 3 adverse events.	Antimalarial	Computational: transcriptional signature-based	[146,157] NCT01446016

Table 5. *Cont.*

Mechanism	Compound	Cellular and Molecular Effects	Original Indication	Repurposing Method	References
STAT3	Flubendazole	Loss of CD44$^+$/CD24$^{-/low}$ population. Decrease of mammosphere-forming ability. Suppression of stem cell genes expression.	Anthelmintic	Non computational: target-based	[73]
	Niclosamide	Reversion of EMT. Inhibition of stem-like phenotype.	Anthelmintic	Non computational: high-throughput screening	[74]
STAT3, NF-κB	Zoledronic acid	Induction of cell cycle arrest, decrease of cell viability, cell proliferation, self-renewal and expression of EMT markers in vitro.	Osteoporosis	Computational: structure-based. Non computational: literature-based	[91]
iNOS	L-NMMA	Decrease of mammosphere-forming ability.	Septic shock	Non computational: target-based	[93]

Salinomycin. It has been shown that LRP6, a co-receptor in the Wnt/β-catenin signaling pathway, is upregulated in TNBC, [158], and transcriptional knockdown decreased Wnt/β-catenin signaling, suppressing tumor growth in vivo [159]. Interestingly, the antibiotic salinomycin was demonstrated to induced the degradation of LRP6, inhibiting the Wnt pathway [147]. Gupta et al. studied the effects of salinomycin both in vitro and in vivo in comparison with paclitaxel. Salinomycin was found to decrease CD44$^+$/CD24$^{-/low}$ population both in cell culture and tumorspheres, whereas paclitaxel induced an increase of this cell population, showing that CSCs were resistant to paclitaxel but sensitive to salinomycin. This effect was later confirmed in mice orthotopically injected with SUM159 cells; it was shown that, compared to paclitaxel, salinomycin was able to inhibit tumor growth and the expression of CSC genes [149]. Moreover, a study investigating the efficacy of salinomycin in combination with LBH589 was proven to be a potential BCSCs-targeted therapy in TNBC by inducing apoptosis, arresting the cell cycle, and regulating EMT in breast CSCs [148].

Pyrvinium pamoate. This FDA-approved anthelmintic was discovered to inhibit the Wnt/β-catenin signaling pathway using a high-throughput screen in a *Xeropus* egg extract [160]. As a consequence of this inhibition, this drug is able to suppress self-renewal of CSC, it reduces both CD44$^+$/CD24$^{-/low}$ and ALDH$^+$ BCSCs and expression of EMT markers such as N-cadherin, vimentin, and Snail [142]. Furthermore, pyrvinium pamoate inhibits PI3K-dependent pathway via suppression of Akt/P70S6K signaling axis [151], as well as mitochondrial respiration function [161] and fatty acids and cholesterol anabolism, lipids that are crucial to Wnt/β-catenin pathways [150]. Reduction of tumor growth was observed in in vivo assays [142,151]. Xu et al. suggested that pyrvinium pamoate's effect on chemoresistance should be assessed in combination with traditional treatments based on the known association between BCSCs and Wnt pathways and the development of drug resistance [142].

Vitamin D3. Upon binding to its ligand, the vitamin D3 nuclear receptor (VDR) heterodimerizes with the retinoid X receptors (RXRs) and regulates the transcription of several genes involved in Wnt, TGFβ and Notch pathways in different types of cancer [143]. In breast cancer, vitamin D3 has been proved to decrease transcriptional levels of the Notch ligands, resulting in the inhibition of Notch-1 signaling, and levels of NF-κB1 [152,153]. Moreover, vitamin D3 has been shown to induce the downregulation of BRCA-1 expression, a commonly mutated gene in breast cancer, including TNBC [162]. In addition, Vitamin D3 was shown to reduce cell proliferation, CD44$^+$/CD24$^{-/low}$ population, and mammosphere formation [153]. Interestingly, Pervin et al. reported that, in breast cancer, VDR silencing was associated with EMT and a higher ability to form mammospheres, whereas its over-expression was followed by a decrease in mammosphere-forming ability. Moreover, in accordance with the inherent aggressiveness of TNBC, they reported that VDR was significantly downregulated in TNBC cells, which resulted in a relative insensitivity to vitamin D3 treatment. Accordingly, these authors showed that a combination therapy with DETA NONOate achieved a significant decrease in mammosphere formation in vitro and tumor growth in vivo [154]. Accordingly, vitamin D3 has been suggested to be a potential inhibitor of breast CSCs.

All-trans retinoic acid (ATRA). Also called tretinoin, is a retinoid used in dermatology which was approved to treat acute promyelocytic leukemia and has been investigated for the treatment of other cancers like lymphoma, leukemia, melanoma, lung cancer, or cervix [143]. In a HER2+ breast cancer cell line, Zanetti et al. proved that treatment of both ATRA and EGF suppressed tumorigenic effects of EGF. While EGF-treated cells developed an increase of Notch1 transcription and TGFβ pathway stimulation via SMAD3, ATRA+EGF-treated cells did not enhance levels of Notch1, and SMAD3 active form was also decreased as phosphorylation did not ensue. Hence, ATRA modulated and reduced EMT by inhibiting transcription of Notch1 and switching TGFβ pathway from a pro-migratory to anti-migratory program. In TNBC, further studies are needed to be done to verify these mechanisms [163]. Using CMap and introducing six analyses of up and down-regulated genes related to CSCs, Bhat-Nakshatri et al. found ATRA to be a good candidate for a CSC targeted therapy in breast cancer, although its effectiveness depends on tumor type. These gene signatures were obtained by comparison of gene expression in two opposite contexts: one associated with CSC versus a non-CSC conditioned control. In TNBC, it was more interesting in those subtypes having mesenchymal properties, as they are enriched for CD44$^+$/CD24$^{-/low}$ subpopulations. In vitro, ATRA produced a decrease in CSC self-renewal, determined by a mammosphere assay, and its effectiveness was augmented in cell lines with higher SOX2 expression. In addition, ATRA reduced levels of EGFR, SERPINE1, and Slug in a cell-line-type-dependent manner. MDA-MB-231 cell line was less sensitive to ATRA because of SOX2-independent characterization and KRAS mutation, which was responsible for resistance to ATRA. Thus, better results in mammosphere assays were obtained after the inhibition of KRAS pathway [155]. Furthermore, Ginestier et al. proved that treatment of ATRA reduced breast ALDH1$^+$ CSC population [156].

Benztropine mesylate. It is used for the treatment of Parkinson's disease. It acts as a central anticholinergic agent, as well as an antihistamine and a dopamine re-uptake inhibitor. Cell-based phenotypic screening and functional assays showed that benztropine mesylate inhibited mammosphere formation and self-renewal, reduced CSC subpopulations (both ALDH1$^+$ and CD44$^+$/CD24$^{-/low}$), and improved chemotherapy in vitro. In vivo, it impaired CSC frequency and their tumor-initiating potential [144]. In addition, Sogawa et al. studied that benztropine could modulate EMT via STAT3, NF-κβ, and β-catenin in colorectal cancer [145].

Chloroquine. It is an autophagy inhibitor primarily used as an antimalarial drug. Interestingly, autophagy has been associated with drug resistance and maintenance of CSC population. In accordance with this mechanism, Choi et al. identified chloroquine as a potential repurposed BCSC inhibitor after in silico gene expression signature analysis of CD44$^+$/CD24$^{-/low}$ population. In vitro assays showed that chloroquine alone reduced the mammosphere formation efficiency and CD44$^+$/CD24$^{-/low}$ population in SUM159 and MDA-MB-231 cells, which was associated with a decrease in the expression of Jak2 and DNA methyltransferase 1 (DNMT1). Moreover, chloroquine sensitized TNBC cells to paclitaxel through the inhibition of autophagy. In vivo assays with an orthotopic xenograft model proved that chloroquine plus paclitaxel significantly reduced tumor growth and CD44$^+$/CD24$^{-/low}$ population, as opposed to paclitaxel alone, which had no effect on tumor growth and increased the CD44$^+$/CD24$^{-/low}$ population, compared to controls, in accordance with previous in vitro assays [146]. A phase II clinical trial demonstrated the efficacy of chloroquine in combination with taxanes in the treatment of patients with advanced or metastatic anthracycline-refractory breast cancer (NCT01446016). Among their results, objective response rate (ORR) was 45.16%, patients showed a median PFS of 12.4 months and a median OS of 25.4 months. The combination was well tolerated, with only up to 13.15% of patients experiencing Grade ≥ 3 adverse events. These results suggest that chloroquine, in combination with taxanes, could be used for the treatment of TNBC patients [157].

Several of the previously mentioned target pathways in TNBC have been associated with EMT mechanisms, maintenance of tumor-initiating cells and/or tumor invasion, and drug resistance, including AR, ADR, STAT3, and AXL pathways. Correspondingly, we hypothesize that AR antagonists [56,58], the β-blocker propranolol [65] and atenolol [66,67,111], the STAT3 inhibitor bazedoxifene [71,72,118] and zoledronic acid [91], and phenothiazines (thioridazine, fluphenazine, trifluoperazine) [80] could act as potential inhibitors of BCSCs. Nevertheless, further investigations would still need to be performed. The pathways altered by these drug candidates to be potentially repurposed, as well as those included in Table 5, have been summarized in Figure 4.

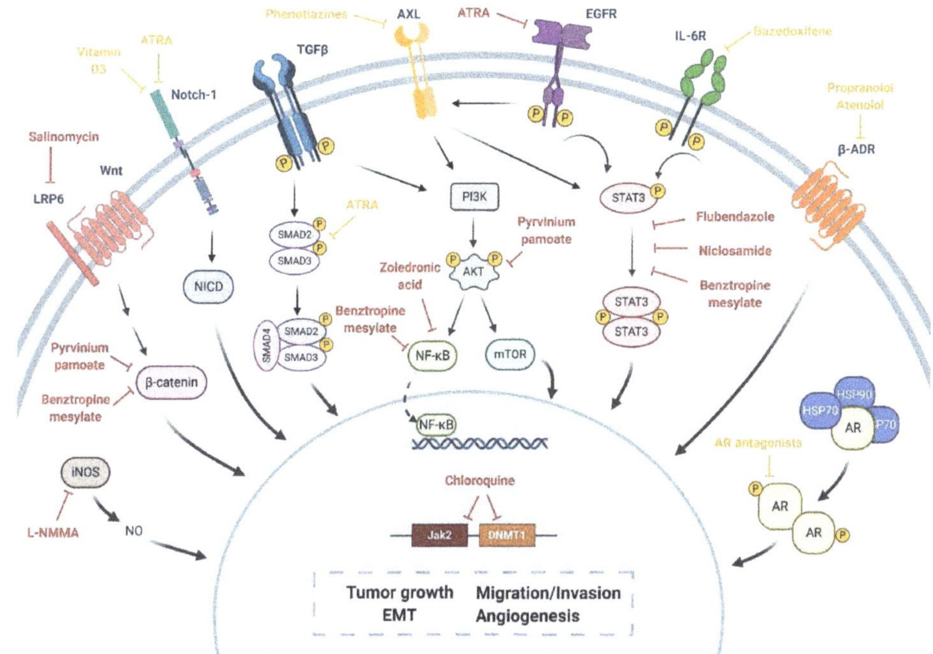

Figure 4. Overview of the different pathways investigated by drug repurposing to target breast cancer stem cells (BCSCs) and their potential inhibitors/modulators. Repurposed inhibitors under investigation are shown in red. Hypothesized inhibitors are shown in yellow. Created with BioRender.com.

6. Conclusions

The absence of targeted therapies for the treatment of TNBC, besides its inherent molecular and histopathologic complexity, strongly reduces the chance of patient recovery and life expectancy. It has therefore become imperative to find effective molecularly targeted treatments to overcome the aggressive progression of this breast cancer subtype. Whereas de novo research is a costly and long-term process, drug repurposing provides the possibility to reduce the time and investment needed to translate a drug from bench to bedside for a specific therapeutic purpose. Drug repositioning is achieved by means of different strategies, especially those including computational methods. Accordingly, several therapies with different molecular targets are currently being investigated for repurposing in TNBC, including androgen receptor, adrenergic receptor, STAT3, nitric oxide synthase, or AXL-directed therapies. However, because of the importance of CSCs in the progression and aggressiveness of this subtype of cancer, current efforts are also being directed to the search of compounds targeting this subset of tumor-initiating cells in TNBC. Herein, according to all repurposed drugs that are currently being studied for the treatment of TNBC, a few of them can be highlighted. AR antagonists bicalutamide, enzalutamide, and seviteronel, currently under clinical trials, seem to be particularly promising drugs in light of their association with the Wnt pathway, reduction of drug resistance, and induction of radiosensitization, respectively. However, clinical trials are evaluating the efficacy of these antiandrogens only in patients with a LAR subtype and, as a consequence, these drugs might not be successful in treating the rest TNBC patients. Other drugs that are currently in the clinical stage are also highlighted, including zoledronic acid, L-NMMA, and chloroquine. They decrease tumor viability, reduce CSC population and their capacity of self-renewal both in vitro and in vivo. Furthermore, they seem to sensitize these cells to chemotherapeutics, hence diminishing drug resistance. Finally, there are other drugs at preclinical stage that must be highlighted because they target CSCs or have been associated with a reduction of drug resistance, such as salinomycin, pyrvinium, vitamin D3, ATRA, benztropine, flubendazole, niclosamide, or propanolol. These drugs could be used as a monotherapy or in combination with chemotherapy to enhance the therapeutic response.

At the core of precision oncology, the high heterogeneity and molecular subtypes of TNBC should drive the diversity of approaches to tackle it, however, most studies do not discriminate between different subtypes. To date, only LAR subtype has really been addressed as an example of successful personalized drug repurposing. Besides the variety of molecular targets, a plethora of computational strategies hinder the ability to efficiently find potential repurposed drugs for TNBC patients. While having different tools for drug repositioning offers indeed a wide range of possibilities for personalized medicine, lack of a standardized protocol and a resolution of the most effective approach in the search of new uses for old drugs, raises the question: can computational drug repurposing actually be implemented as an improved method for drug discovery in personalized medicine and, more particularly, for TNBC? Factually, it is noticeable that some of the reviewed studies date from some years ago but none of those repurposed compounds have been yet approved for TNBC. While drug repurposing might increase the chances to help find new molecularly targeted candidates, hence improving the development of a more personalized medicine, the results suggest that not all candidates were as adequate as they might have seemed during in silico analysis, meaning that computational drug repurposing could not be as efficient as expected. It is therefore necessary for computational approaches to be validated and standardized, so as to reduce the chances of failure and allow drug repurposing to become an improved and attainable alternative with guarantees for personalized medicine. Be that as it may, drug repositioning has allowed to find new candidates that would not have been considered otherwise, making it still a powerful alternative for the search of a personalized treatment for TNBC patients.

Author Contributions: Conceptualization, S.G.-P.; writing, review and editing, M.Á.-M., A.L.-T., S.G.-P.; preparation of tables and figures, M.Á.-M., A.L.-T., F.E.C.-L., J.L.B.-C.; revision, S.G.-P., F.E.C.-L., J.L.B.-C., A.G.-G., J.A.L., P.S.-R.; funding acquisition, S.G.-P., P.S.-R., J.A.L. All authors have read and agreed to the published version of the manuscript.

Funding: This work and S.G.P. was funded by Instituto de Salud Carlos III (CP19/00029, CP14/00197, PI19/01533, PI15/00336), Ministerio de Economía y Competitividad (RTC-2016-5674-1) and European Regional Development Fund (European Union). A.L.T. is funded by the Ministerio de Ciencia, Innovación y Universidades (FPU19/04450) and J.B.C. is supported by Fundación Científica Asociación Española Contra el Cáncer, Junta Provincial de Jaén (AECC) (PRDJA19001BLAY).

Conflicts of Interest: The authors declare no conflict of interest.

Abbreviations

ADC	Antibody drug conjugate
ADR	Adrenergic receptor
AE	Adverse events
Akt	Protein kinase B
ALDH1	Aldehyde dehydrogenase 1
AR	Androgen receptor
AREG	Amphiregulin
ATRA	All-trans retinoic acid
AXL	Anexelekto
BARK	β-adrenergic receptor kinase
BCSC	Breast cancer stem cells
BL1	Basal-like 1
BL2	Basal-like 2
cAMP	Cyclic AMP
CBN	Causal biological networks
CBR	Clinical benefit rate
CMap	Connectivity Map

CPIs	Checkpoint inhibitors
CSC	Cancer stem cells
CTLA-4	Cytotoxic T-lymphocyte-associated antigen-4
DFS	Disease free survival
DTI	Drug-target interaction
EGF	Epidermal growth factor
EGFR	Epidermal growth factor receptor
EMT	Epithelial-to-mesenchymal transition
ER	Estrogen receptor
FAK	Focal adhesion kinase
FDA	Food and Drug Administration
FGF	Fibroblast growth factor
GEO	Gene Expression Omnibus
GPCR	G protein-coupled receptor
HER2	Human epidermal growth factor receptor 2
IGF	Insulin-like growth factor
IHC	Immunohistochemistry
IL-10	Interleukin 10
IL-6	Interleukin 6
IM	Immunomodulatory
ITT	Intent-to-treat population
JAKs	Janus kinases
LAR	Luminal androgen receptor
L-NMMA	NG-monomethyl-L-arginine
M	Mesenchymal
MAPK	Mitogen-activated protein kinase
mCRPC	Metastatic castration-resistant prostate cancer
MLF2	Myeloid leukemia factor 2
MSL	Mesenchymal stem–like
mTOR	Mammalian target of rapamycin
NO	Nitric oxide
NOS	Nitric oxide synthase
NOS1/nNOS	Neuronal nitric oxide synthase
NOS2/iNOS	Inducible nitric oxide synthase
NOS3/eNOS	Endothelial nitric oxide synthase
OS	Overall survival
PARP	Poly[adenosine diphosphate-ribose] polymerase
PDB	Protein Data Bank
pCR	Pathologic complete response
PD-1	Programmed cell death 1
PD-L1	Programmed cell death-ligand 1
PDX	Patient-derived xenografts
PFS	Progression free survival
PI3K	Phosphatidylinositol-3 kinase
PKA	Protein kinase A
PPI	Protein-protein interaction
PR	Progesterone receptor
PRISM	Protein Interactions by Structural Matching
ROS	Reactive oxygen species
RPL39	Ribosomal protein L39
RTK	Receptors tyrosine kinase
RXR	Retinoid X receptors
SARMs	Selective androgen receptor modulators
SAEs	Serious adverse events

STAT3	Signal transducer and activator of transcription 3
TAM	Tyro3, AXL and Mer
TCGA	The Cancer Genome Atlas
TGFβ	Transforming growth factor β
TILs	Tumor infiltrating lymphocytes
TNBC	Triple-negative breast cancer
Trop-2	Trophoblast cell-surface antigen 2
VDR	Vitamin D3 nuclear receptor
VEGF	Vascular endothelial growth factor
VHTS	Virtual high-throughput screening

References

1. DeSantis, C.E.; Ma, J.; Gaudet, M.M.; Newman, L.A.; Miller, K.D.; Goding Sauer, A.; Jemal, A.; Siegel, R.L. Breast cancer statistics, 2019. *CA Cancer J. Clin.* **2019**, *69*, 438–451. [CrossRef]
2. Siegel, R.L.; Miller, K.D.; Jemal, A. Cancer statistics, 2020. *CA Cancer J. Clin.* **2020**, *70*, 7–30. [CrossRef]
3. Hon, J.D.C.; Singh, B.; Sahin, A.; Du, G.; Wang, J.; Wang, V.Y.; Deng, F.M.; Zhang, D.Y.; Monaco, M.E.; Lee, P. Breast cancer molecular subtypes: From TNBC to QNBC. *Am. J. Cancer Res.* **2016**, *6*, 1864–1872.
4. Costa, R.L.B.; Gradishar, W.J. Triple-negative breast cancer: Current practice and future directions. *J. Oncol. Pract.* **2017**, *13*, 301–303. [CrossRef]
5. Lee, A.; Djamgoz, M.B.A. Triple negative breast cancer: Emerging therapeutic modalities and novel combination therapies. *Cancer Treat. Rev.* **2018**, *62*, 110–122. [CrossRef]
6. Jitariu, A.A.; Cîmpean, A.M.; Ribatti, D.; Raica, M. Triple negative breast cancer: The kiss of death. *Oncotarget* **2017**, *8*, 46652–46662. [CrossRef]
7. Aggarwal, S.; Verma, S.S.; Aggarwal, S.; Gupta, S.C. Drug repurposing for breast cancer therapy: Old weapon for new battle. *Semin. Cancer Biol.* **2019**. [CrossRef]
8. Waks, A.G.; Winer, E.P. Breast cancer treatment: A review. *J. Am. Med. Assoc.* **2019**, *321*, 288–300. [CrossRef]
9. Cruz-Lozano, M.; González-González, A.; Muñoz-Muela, E.; Cara, F.E.; Granados-Principal, S. Targeted therapies in triple negative breast cancer: Current knowledge and perspectives. *Recent Stud. Adv. Breast Cancer* **2019**, *2*, 1–17.
10. Park, J.H.; Ahn, J.-H.; Kim, S.-B. How shall we treat early triple-negative breast cancer (TNBC): From the current standard to upcoming immuno-molecular strategies. *ESMO Open* **2018**, *3*, e000357. [CrossRef]
11. Al-Mahmood, S.; Sapiezynski, J.; Garbuzenko, O.B.; Minko, T. Metastatic and triple-negative breast cancer: Challenges and treatment options. *Drug Deliv. Transl. Res.* **2018**, *8*, 1483–1507. [CrossRef]
12. Timothy, C.J.; Kogularaman, S.; Stephen, J.L. The next generation of platinum drugs: Targeted Pt(II) agents, nanoparticle delivery, and Pt(IV) prodrugs timothy. *Chem. Rev.* **2016**, *116*, 3436–3486. [CrossRef]
13. Treatment of Triple-Negative Breast Cancer. Available online: https://www.cancer.org/cancer/breast-cancer/treatment/treatment-of-triple-negative.html (accessed on 14 October 2020).
14. Robson, M.; Im, S.A.; Senkus, E.; Xu, B.; Domchek, S.M.; Masuda, N.; Delaloge, S.; Li, W.; Tung, N.; Armstrong, A.; et al. Olaparib for metastatic breast cancer in patients with a germline BRCA mutation. *N. Engl. J. Med.* **2017**, *377*, 523–533. [CrossRef]
15. Litton, J.K.; Rugo, H.S.; Ettl, J.; Hurvitz, S.A.; Gonçalves, A.; Lee, K.H.; Fehrenbacher, L.; Yerushalmi, R.; Mina, L.A.; Martin, M.; et al. Talazoparib in patients with advanced breast cancer and a germline BRCA mutation. *N. Engl. J. Med.* **2018**, *379*, 753–763. [CrossRef]
16. Oiseth, S.J.; Aziz, M.S. Cancer immunotherapy: A brief review of the history, possibilities, and challenges ahead. *J. Cancer Metastasis Treat.* **2017**, *3*, 250–261. [CrossRef]
17. Kwa, M.J.; Adams, S. Checkpoint inhibitors in triple-negative breast cancer (TNBC): Where to go from here. *Cancer* **2018**, *124*, 2086–2103. [CrossRef]
18. Soare, G.R.; Soare, C.A. Immunotherapy for breast cancer: First FDA approved regimen. *Discoveries* **2019**, *7*, e91. [CrossRef]
19. Schmid, P.; Adams, S.; Rugo, H.S.; Schneeweiss, A.; Barrios, C.H.; Iwata, H.; Dieras, V.; Hegg, R.; Im, S.A.; Shaw Wright, G.; et al. Atezolizumab and nab-paclitaxel in advanced triple-negative breast cancer. *N. Engl. J. Med.* **2018**, *379*, 2108–2121. [CrossRef]

20. Vonderheide, R.H.; Domchek, S.M.; Clark, A.S. Immunotherapy for breast cancer: What are we missing? *Clin. Cancer Res.* **2017**, *23*, 2640–2646. [CrossRef]

21. Nagayama, A.; Vidula, N.; Ellisen, L.; Bardia, A. Novel antibody–drug conjugates for triple negative breast cancer. *Ther. Adv. Med. Oncol.* **2020**, *12*, 1–12. [CrossRef]

22. Ashburn, T.T.; Thor, K.B. Drug repositioning: Identifying and developing new uses for existing drugs. *Nat. Rev. Drug Discov.* **2004**, *3*, 673–683. [CrossRef]

23. Nosengo, N. Can you teach old drugs new tricks? *Nature* **2016**, *534*, 314–316. [CrossRef]

24. Akhoon, B.A.; Tiwari, H.; Nargotra, A. In silico drug design methods for drug repurposing. In *In Silico Drug Design*; Academic Press: Cambridge, MA, USA, 2019; pp. 47–84.

25. MacArron, R.; Banks, M.N.; Bojanic, D.; Burns, D.J.; Cirovic, D.A.; Garyantes, T.; Green, D.V.S.; Hertzberg, R.P.; Janzen, W.P.; Paslay, J.W.; et al. Impact of high-throughput screening in biomedical research. *Nat. Rev. Drug Discov.* **2011**, *10*, 188–195. [CrossRef]

26. Hodos, R.A.; Kidd, B.A.; Shameer, K.; Readhead, B.P.; Dudley, J.T. In silico methods for drug repurposing and pharmacology. *Wiley Interdiscip. Rev. Syst. Biol. Med.* **2016**, *8*, 186–210. [CrossRef]

27. Karaman, B.; Sippl, W. Computational drug repurposing: Current trends. *Curr. Med. Chem.* **2018**, *26*, 5389–5409. [CrossRef]

28. Alaimo, S.; Pulvirenti, A. Network-based drug repositioning: Approaches, resources, and research directions. In *Methods in Molecular Biology*; Humana Press Inc.: New York, NY, USA, 2019; Volume 1903, pp. 97–113.

29. Gns, H.S.; Gr, S.; Murahari, M.; Krishnamurthy, M. An update on drug repurposing: Re-written saga of the drug's fate. *Biomed. Pharmacother.* **2019**, *110*, 700–716. [CrossRef]

30. Sohraby, F.; Bagheri, M.; Aryapour, H. Performing an in silico repurposing of existing drugs by combining virtual screening and molecular dynamics simulation. In *Methods in Molecular Biology*; Humana Press Inc.: New York, NY, USA, 2019; Volume 1903, pp. 23–43.

31. Wang, Y.; Yella, J.; Jegga, A.G. Transcriptomic data mining and repurposing for computational drug discovery. *Methods Mol. Biol.* **2019**, *1903*, 73–95. [CrossRef]

32. Lamb, J.; Crawford, E.D.; Peck, D.; Modell, J.W.; Blat, I.C.; Wrobel, M.J.; Lerner, J.; Brunet, J.P.; Subramanian, A.; Ross, K.N.; et al. The connectivity map: Using gene-expression signatures to connect small molecules, genes, and disease. *Science* **2006**, *313*, 1929–1935. [CrossRef]

33. Subramanian, A.; Narayan, R.; Corsello, S.M.; Peck, D.D.; Natoli, T.E.; Lu, X.; Gould, J.; Davis, J.F.; Tubelli, A.A.; Asiedu, J.K.; et al. A next generation connectivity map: L1000 platform and the first 1,000,000 profiles. *Cell* **2017**, *171*, 1437–1452.e17. [CrossRef]

34. CLUE Connectopedia. Available online: https://clue.io/connectopedia/ (accessed on 30 July 2020).

35. Higurashi, M.; Ishida, T.; Kinoshita, K. Identification of transient hub proteins and the possible structural basis for their multiple interactions. *Protein Sci.* **2008**, *17*, 72–78. [CrossRef]

36. Ozdemir, E.S.; Halakou, F.; Nussinov, R.; Gursoy, A.; Keskin, O. Methods for discovering and targeting druggable protein-protein interfaces and their application to repurposing. *Methods Mol. Biol.* **2019**, *1903*, 1–21. [CrossRef] [PubMed]

37. Su, E.W.; Sanger, T.M. Systematic drug repositioning through mining adverse event data in ClinicalTrials.gov. *PeerJ* **2017**, *2017*. [CrossRef]

38. Perou, C.M.; Sørile, T.; Eisen, M.B.; Van De Rijn, M.; Jeffrey, S.S.; Ress, C.A.; Pollack, J.R.; Ross, D.T.; Johnsen, H.; Akslen, L.A.; et al. Molecular portraits of human breast tumours. *Nature* **2000**, *406*, 747–752. [CrossRef]

39. Sørlie, T.; Perou, C.M.; Tibshirani, R.; Aas, T.; Geisler, S.; Johnsen, H.; Hastie, T.; Eisen, M.B.; Van De Rijn, M.; Jeffrey, S.S.; et al. Gene expression patterns of breast carcinomas distinguish tumor subclasses with clinical implications. *Proc. Natl. Acad. Sci. USA* **2001**, *98*, 10869–10874. [CrossRef]

40. Sørlie, T.; Tibshirani, R.; Parker, J.; Hastie, T.; Marron, J.S.; Nobel, A.; Deng, S.; Johnsen, H.; Pesich, R.; Geisler, S.; et al. Repeated observation of breast tumor subtypes in independent gene expression data sets. *Proc. Natl. Acad. Sci. USA* **2003**, *100*, 8418–8423. [CrossRef] [PubMed]

41. Hu, Z.; Fan, C.; Oh, D.S.; Marron, J.S.; He, X.; Qaqish, B.F.; Livasy, C.; Carey, L.A.; Reynolds, E.; Dressler, L.; et al. The molecular portraits of breast tumors are conserved across microarray platforms. *BMC Genom.* **2006**, *7*, 96. [CrossRef] [PubMed]

42. Bernard, P.S.; Parker, J.S.; Mullins, M.; Cheung, M.C.U.; Leung, S.; Voduc, D.; Vickery, T.; Davies, S.; Fauron, C.; He, X.; et al. Supervised risk predictor of breast cancer based on intrinsic subtypes. *J. Clin. Oncol.* **2009**, *27*, 1160–1167. [CrossRef]

43. Kreike, B.; van Kouwenhove, M.; Horlings, H.; Weigelt, B.; Peterse, H.; Bartelink, H.; van de Vijver, M.J. Gene expression profiling and histopathological characterization of triple-negative/basal-like breast carcinomas. *Breast Cancer Res.* **2007**, *9*, R65. [CrossRef]

44. Prat, A.; Adamo, B.; Cheang, M.C.U.; Anders, C.K.; Carey, L.A.; Perou, C.M. Molecular characterization of basal-like and non-basal-like triple-negative breast cancer. *Oncologist* **2013**, *18*, 123–133. [CrossRef]

45. Lehmann, B.D.; Pietenpol, J.A. Identification and use of biomarkers in treatment strategies for triple-negative breast cancer subtypes. *J. Pathol.* **2014**, *232*, 142–150. [CrossRef]

46. Lehmann, B.D.; Bauer, J.A.; Chen, X.; Sanders, M.E.; Chakravarthy, A.B.; Shyr, Y.; Pietenpol, J.A. Identification of human triple-negative breast cancer subtypes and preclinical models for selection of targeted therapies. *J. Clin. Investig.* **2011**, *121*, 2750–2767. [CrossRef] [PubMed]

47. Chen, X.; Li, J.; Gray, W.H.; Lehmann, B.D.; Bauer, J.A.; Shyr, Y.; Pietenpol, J.A. TNBCtype: A subtyping tool for triple-negative breast cancer. *Cancer Inform.* **2012**, *11*, 147–156. [CrossRef] [PubMed]

48. Lehmann, B.D.; Jovanović, B.; Chen, X.; Estrada, M.V.; Johnson, K.N.; Shyr, Y.; Moses, H.L.; Sanders, M.E.; Pietenpol, J.A. Refinement of triple-negative breast cancer molecular subtypes: Implications for neoadjuvant chemotherapy selection. *PLoS ONE* **2016**, *11*, e0157368. [CrossRef]

49. Burstein, M.D.; Tsimelzon, A.; Poage, G.M.; Covington, K.R.; Contreras, A.; Fuqua, S.A.W.; Savage, M.I.; Osborne, C.K.; Hilsenbeck, S.G.; Chang, J.C.; et al. Comprehensive genomic analysis identifies novel subtypes and targets of triple-negative breast cancer. *Clin. Cancer Res.* **2015**, *21*, 1688–1698. [CrossRef] [PubMed]

50. Le Du, F.; Eckhardt, B.L.; Lim, B.; Litton, J.K.; Moulder, S.; Meric-Bernstam, F.; Gonzalez-Angulo, A.M.; Ueno, N.T. Is the future of personalized therapy in triple-negative breast cancer based on molecular subtype? *Oncotarget* **2015**, *6*, 12890–12908. [CrossRef] [PubMed]

51. Liu, Y.R.; Jiang, Y.Z.; Xu, X.E.; Yu, K.D.; Jin, X.; Hu, X.; Zuo, W.J.; Hao, S.; Wu, J.; Liu, G.Y.; et al. Comprehensive transcriptome analysis identifies novel molecular subtypes and subtype-specific RNAs of triple-negative breast cancer. *Breast Cancer Res.* **2016**, *18*. [CrossRef] [PubMed]

52. Ring, B.Z.; Hout, D.R.; Morris, S.W.; Lawrence, K.; Schweitzer, B.L.; Bailey, D.B.; Lehmann, B.D.; Pietenpol, J.A.; Seitz, R.S. Generation of an algorithm based on minimal gene sets to clinically subtype triple negative breast cancer patients. *BMC Cancer* **2016**, *16*, 143. [CrossRef]

53. Espinosa Fernandez, J.R.; Eckhardt, B.L.; Lee, J.; Lim, B.; Pearson, T.; Seitz, R.S.; Hout, D.R.; Schweitzer, B.L.; Nielsen, T.J.; Lawrence, O.R.; et al. Identification of triple-negative breast cancer cell lines classified under the same molecular subtype using different molecular characterization techniques: Implications for translational research. *PLoS ONE* **2020**, *15*, e0231953. [CrossRef]

54. Rampurwala, M.; Wisinski, K.B.; O'Regan, R. Role of the androgen receptor in triple-negative breast cancer. *Clin. Adv. Hematol. Oncol.* **2016**, *14*, 186–193.

55. Caiazza, F.; Murray, A.; Madden, S.F.; Synnott, N.C.; Ryan, E.J.; O'Donovan, N.; Crown, J.; Duffy, M.J. Preclinical evaluation of the AR inhibitor enzalutamide in triple-negative breast cancer cells. *Endocr. Relat. Cancer* **2016**, *23*, 323–334. [CrossRef]

56. Barton, V.N.; D'Amato, N.C.; Gordon, M.A.; Lind, H.T.; Spoelstra, N.S.; Babbs, B.L.; Heinz, R.E.; Elias, A.; Jedlicka, P.; Jacobsen, B.M.; et al. Multiple molecular subtypes of triple-negative breast cancer critically rely on androgen receptor and respond to enzalutamide in vivo. *Mol. Cancer Ther.* **2015**, *14*, 769–778. [CrossRef]

57. Barton, V.N.; D'Amato, N.C.; Gordon, M.A.; Christenson, J.L.; Elias, A.; Richer, J.K. Androgen receptor biology in triple negative breast cancer: A case for classification as AR+ or quadruple negative disease. *Horm. Cancer* **2015**, *6*, 206–213. [CrossRef]

58. Christenson, J.L.; Trepel, J.B.; Ali, H.Y.; Lee, S.; Eisner, J.R.; Baskin-Bey, E.S.; Elias, A.D.; Richer, J.K. Harnessing a different dependency: How to identify and target androgen receptor-positive versus quadruple-negative breast cancer. *Horm. Cancer* **2018**, *9*, 82–94. [CrossRef] [PubMed]

59. Narayanan, R.; Mohler, M.L.; Bohl, C.E.; Miller, D.D.; Dalton, J.T. Selective androgen receptor modulators in preclinical and clinical development. *Nucl. Recept. Signal.* **2008**, *6*, e010. [CrossRef] [PubMed]

60. Bird, I.M.; Abbott, D.H. The hunt for a selective 17,20 lyase inhibitor; learning lessons from nature. *J. Steroid Biochem. Mol. Biol.* **2016**, *163*, 136–146. [CrossRef] [PubMed]

61. Crawford, E.D.; Schellhammer, P.F.; McLeod, D.G.; Moul, J.W.; Higano, C.S.; Shore, N.; Denis, L.; Iversen, P.; Eisenberger, M.A.; Labrie, F. Androgen receptor targeted treatments of prostate cancer: 35 years of progress with antiandrogens. *J. Urol.* **2018**, *200*, 956–966. [CrossRef]

62. Gucalp, A.; Traina, T.A. Targeting the androgen receptor in triple-negative breast cancer. *Curr. Probl. Cancer* **2016**, *40*, 141–150. [CrossRef]

63. Scher, H.I.; Fizazi, K.; Saad, F.; Taplin, M.E.; Sternberg, C.N.; Miller, K.; De Wit, R.; Mulders, P.; Chi, K.N.; Shore, N.D.; et al. Increased survival with enzalutamide in prostate cancer after chemotherapy. *N. Engl. J. Med.* **2012**, *367*, 1187–1197. [CrossRef]

64. Chen, J.; Kim, J.; Dalton, J.T. Discovery and therapeutic promise of selective androgen receptor modulators. *Mol. Interv.* **2005**, *5*, 173–188. [CrossRef]

65. Solomon, Z.J.; Mirabal, J.R.; Mazur, D.J.; Kohn, T.P.; Lipshultz, L.I.; Pastuszak, A.W. Selective androgen receptor modulators: Current knowledge and clinical applications. *Sex. Med. Rev.* **2019**, *7*, 84–94. [CrossRef]

66. Masiello, D.; Cheng, S.; Bubley, G.J.; Lu, M.L.; Balk, S.P. Bicalutamide functions as an androgen receptor antagonist by assembly of a transcriptionally inactive receptor. *J. Biol. Chem.* **2002**, *277*, 26321–26326. [CrossRef]

67. Huang, R.; Han, J.; Liang, X.; Sun, S.; Jiang, Y.; Xia, B.; Niu, M.; Li, D.; Zhang, J.; Wang, S.; et al. Androgen receptor expression and bicalutamide antagonize androgen receptor inhibit β-catenin transcription complex in estrogen receptor-negative breast cancer. *Cell. Physiol. Biochem.* **2017**, *43*, 2212–2225. [CrossRef]

68. Gucalp, A.; Tolaney, S.; Isakoff, S.J.; Ingle, J.N.; Liu, M.C.; Carey, L.A.; Blackwell, K.; Rugo, H.; Nabell, L.; Forero, A.; et al. Phase II trial of bicalutamide in patients with androgen receptor-positive, estrogen receptor-negative metastatic breast cancer. *Clin. Cancer Res.* **2013**, *19*, 5505–5512. [CrossRef] [PubMed]

69. Cochrane, D.R.; Bernales, S.; Jacobsen, B.M.; Cittelly, D.M.; Howe, E.N.; D'Amato, N.C.; Spoelstra, N.S.; Edgerton, S.M.; Jean, A.; Guerrero, J.; et al. Role of the androgen receptor in breast cancer and preclinical analysis of enzalutamide. *Breast Cancer Res.* **2014**, *16*. [CrossRef]

70. Traina, T.A.; Miller, K.; Yardley, D.A.; O'Shaughnessy, J.; Cortes, J.; Awada, A.; Kelly, C.M.; Trudeau, M.E.; Schmid, P.; Gianni, L.; et al. Results from a phase 2 study of enzalutamide (ENZA), an androgen receptor (AR) inhibitor, in advanced AR+ triple-negative breast cancer (TNBC). *J. Clin. Oncol.* **2015**, *33*, 1003. [CrossRef]

71. Yin, L.; Hu, Q. CYP17 inhibitors—Abiraterone, C17,20-lyase inhibitors and multi-targeting agents. *Nat. Rev. Urol.* **2014**, *11*, 32–42. [CrossRef] [PubMed]

72. Bonnefoi, H.; Grellety, T.; Tredan, O.; Saghatchian, M.; Dalenc, F.; Mailliez, A.; Cottu, P.; Abadie-Lacourtoisie, S.; You, B.; Mousseau, M.; et al. A phase II trial of abiraterone acetate plus prednisone in patients with triple-negative androgen receptor positive locally advanced or metastatic breast cancer (UCBG 12-1) original articles Annals of Oncology. *Ann. Oncol.* **2016**, *27*, 812–818. [CrossRef] [PubMed]

73. Grellety, T.; Callens, C.; Richard, E.; Briaux, A.; Velasco, V.; Pulido, M.; Gonçalves, A.; Gestraud, P.; MacGrogan, G.; Bonnefoi, H.; et al. Enhancing abiraterone acetate efficacy in androgen receptor–positive triple-negative breast cancer: Chk1 as a potential target. *Clin. Cancer Res.* **2019**, *25*, 856–867. [CrossRef]

74. Roviello, G.; Pacifico, C.; Chiriacò, G.; Generali, D. Is still there a place for orteronel in management of prostate cancer? Data from a literature based meta-analysis of randomized trials. *Crit. Rev. Oncol. Hematol.* **2017**, *113*, 18–21. [CrossRef]

75. Reese, J.; Babbs, B.; Christenson, J.; Spoelstra, N.; Elias, A.; Eisner, J.; Baskin-Bey, E.; Gertz, J.; Richer, J. Abstract P5-05-05: Targeting the androgen receptor with seviteronel, a CYP17 lyase and AR inhibitor, in triple negative breast cancer. *Cancer Res.* **2019**, *79*. [CrossRef]

76. Michmerhuizen, A.R.; Chandler, B.; Olsen, E.; Wilder-Romans, K.; Moubadder, L.; Liu, M.; Pesch, A.M.; Zhang, A.; Ritter, C.; Ward, S.T.; et al. Seviteronel, a novel CYP17 lyase inhibitor and androgen receptor antagonist, radiosensitizes ar-positive triple negative breast cancer cells. *Front. Endocrinol.* **2020**, *11*, 35. [CrossRef] [PubMed]

77. Gucalp, A.; Danso, M.A.; Elias, A.D.; Bardia, A.; Ali, H.Y.; Potter, D.; Gabrail, N.Y.; Haley, B.B.; Khong, H.T.; Riley, E.C.; et al. Phase (Ph) 2 stage 1 clinical activity of seviteronel, a selective CYP17-lyase and androgen receptor (AR) inhibitor, in women with advanced AR+ triple-negative breast cancer (TNBC) or estrogen receptor (ER)+ BC: CLARITY-01. *J. Clin. Oncol.* **2017**, *35*, 1102. [CrossRef]

78. Gao, W.; Dalton, J.T. Expanding the therapeutic use of androgens via selective androgen receptor modulators (SARMs). *Drug Discov. Today* **2007**, *12*, 241–248. [CrossRef] [PubMed]

79. Narayanan, R.; Ahn, S.; Cheney, M.D.; Yepuru, M.; Miller, D.D.; Steiner, M.S.; Dalton, J.T. Selective Androgen Receptor Modulators (SARMs) negatively regulate triple-negative breast cancer growth and epithelial: Mesenchymal stem cell signaling. *PLoS ONE* **2014**, *9*, e103202. [CrossRef]

80. Obeid, E.I.; Conzen, S.D. The role of adrenergic signaling in breast cancer biology. *Cancer Biomark.* **2013**, *13*, 161–169. [CrossRef]

81. Powe, D.G.; Voss, M.J.; Zänker, K.S.; Habashy, H.O.; Green, A.R.; Ellis, I.O.; Entschladen, F. Beta-blocker drug therapy reduces secondary cancer formation in breast cancer and improves cancer specific survival. *Oncotarget* **2010**, *1*, 628–638. [CrossRef]

82. Melhem-Bertrandt, A.; Chavez-MacGregor, M.; Lei, X.; Brown, E.N.; Lee, R.T.; Meric-Bernstam, F.; Sood, A.K.; Conzen, S.D.; Hortobagyi, G.N.; Gonzalez-Angulo, A.M. Beta-blocker use is associated with improved relapse-free survival in patients with triple-negative breast cancer. *J. Clin. Oncol.* **2011**, *29*, 2645–2652. [CrossRef]

83. Kafetzopoulou, L.E.; Boocock, D.J.; Dhondalay, G.K.R.; Powe, D.G.; Ball, G.R. Biomarker identification in breast cancer: Beta-adrenergic receptor signaling and pathways to therapeutic response. *Comput. Struct. Biotechnol. J.* **2013**, *6*, e201303003. [CrossRef]

84. Vázquez, S.M.; Mladovan, A.G.; Pérez, C.; Bruzzone, A.; Baldi, A.; Lüthy, I.A. Human breast cell lines exhibit functional α2-adrenoceptors. *Cancer Chemother. Pharmacol.* **2006**, *58*, 50–61. [CrossRef]

85. Pérez Piñero, C.; Bruzzone, A.; Sarappa, M.G.; Castillo, L.F.; Lüthy, I.A. Involvement of α2- and β2-adrenoceptors on breast cancer cell proliferation and tumour growth regulation. *Br. J. Pharmacol.* **2012**, *166*, 721–736. [CrossRef]

86. Drell, T.L., IV; Joseph, J.; Lang, K.; Niggemann, B.; Zaenker, K.S.; Entschladen, F. Effects of neurotransmitters on the chemokinesis and chemotaxis of MDA-MB-468 human breast carcinoma cells. *Breast Cancer Res. Treat.* **2003**, *80*, 63–70. [CrossRef] [PubMed]

87. Lang, K.; Drell, T.L.; Lindecke, A.; Niggemann, B.; Kaltschmidt, C.; Zaenker, K.S.; Entschladen, F. Induction of a metastatogenic tumor cell type by neurotransmitters and its pharmacological inhibition by established drugs. *Int. J. Cancer* **2004**, *112*, 231–238. [CrossRef]

88. Strell, C.; Niggemann, B.; Voss, M.J.; Powe, D.G.; Zänker, K.S.; Entschladen, F. Norepinephrine promotes the β1-integrin-mediated adhesion of MDA-MB-231 cells to vascular endothelium by the induction of a GROα release. *Mol. Cancer Res.* **2012**, *10*, 197–207. [CrossRef] [PubMed]

89. Slotkin, T.A.; Zhang, J.; Dancel, R.; Garcia, S.J.; Willis, C.; Seidler, F.J. β-adrenoceptor signaling and its control of cell replication in MDA-MB-231 human breast cancer cells. *Breast Cancer Res. Treat.* **2000**, *60*, 153–166. [CrossRef] [PubMed]

90. Barron, T.I.; Connolly, R.M.; Sharp, L.; Bennett, K.; Visvanathan, K. Beta blockers and breast cancer mortality: A population-based study. *J. Clin. Oncol.* **2011**, *29*, 2635–2644. [CrossRef] [PubMed]

91. Botteri, E.; Munzone, E.; Rotmensz, N.; Cipolla, C.; De Giorgi, V.; Santillo, B.; Zanelotti, A.; Adamoli, L.; Colleoni, M.; Viale, G.; et al. Therapeutic effect of β-blockers in triple-negative breast cancer postmenopausal women. *Breast Cancer Res. Treat.* **2013**, *140*, 567–575. [CrossRef]

92. Musini, V.M.; Pasha, P.; Gill, R.; Wright, J.M. Blood pressure lowering efficacy of clonidine for primary hypertension. *Cochrane Database Syst. Rev.* **2017**, *2017*. [CrossRef]

93. Bruzzone, A.; Pinero, P.C.; Rojas, P.; Romanato, M.; Gass, H.; Lanari, C.; Luthy, A.I. α(2)-Adrenoceptors Enhance Cell Proliferation and Mammary Tumor Growth Acting Through both the Stroma and the Tumor Cells. *Curr. Cancer Drug Targets* **2011**, *11*, 763–774. [CrossRef]

94. DrugBank Yohimbine. Available online: https://www.drugbank.ca/drugs/DB01392 (accessed on 2 July 2020).

95. Luthy, I.; Bruzzone, A.; Pinero, C.; Castillo, L.; Chiesa, I.; Vazquez, S.; Sarappa, M. Adrenoceptors: Non conventional target for breast cancer? *Curr. Med. Chem.* **2009**, *16*, 1850–1862. [CrossRef]

96. Flint, M.S.; Kim, G.; Hood, B.L.; Bateman, N.W.; Stewart, N.A.; Conrads, T.P. Stress hormones mediate drug resistance to paclitaxel in human breast cancer cells through a CDK-1-dependent pathway. *Psychoneuroendocrinology* **2009**, *34*, 1533–1541. [CrossRef]

97. Cole, S.W.; Sood, A.K. Molecular pathways: Beta-adrenergic signaling in cancer. *Clin. Cancer Res.* **2012**, *18*, 1201–1206. [CrossRef] [PubMed]

98. Xie, W.Y.; He, R.H.; Zhang, J.; He, Y.J.; Wan, Z.; Zhou, C.F.; Tang, Y.J.; Li, Z.; McLeod, H.L.; Liu, J. β-blockers inhibit the viability of breast cancer cells by regulating the ERK/COX-2 signaling pathway and the drug response is affected by ADRB2 single-nucleotide polymorphisms. *Oncol. Rep.* **2019**, *41*, 341–350. [CrossRef]

99. Pasquier, E.; Ciccolini, J.; Carre, M.; Giacometti, S.; Fanciullino, R.; Pouchy, C.; Montero, M.P.; Serdjebi, C.; Kavallaris, M.; André, N. Propranolol potentiates the anti-angiogenic effects and antitumor efficacy of chemotherapy agents: Implication in breast cancer treatment. *Oncotarget* **2011**, *2*, 797–809. [CrossRef]

100. Ganz, P.A.; Cole, S.W. Expanding our therapeutic options: Beta blockers for breast cancer? *J. Clin. Oncol.* **2011**, *29*, 2612–2616. [CrossRef] [PubMed]

101. Cakir, Y.; Plummer, H.K.; Tithof, P.K.; Schuller, H.M. Beta-adrenergic and arachidonic acid-mediated growth regulation of human breast cancer cell lines. *Int. J. Oncol.* **2002**, *21*, 153–157. [CrossRef]

102. Talarico, G.; Orecchioni, S.; Dallaglio, K.; Reggiani, F.; Mancuso, P.; Calleri, A.; Gregato, G.; Labanca, V.; Rossi, T.; Noonan, D.M.; et al. Aspirin and atenolol enhance metformin activity against breast cancer by targeting both neoplastic and microenvironment cells. *Sci. Rep.* **2016**, *6*. [CrossRef] [PubMed]

103. Ma, J.H.; Qin, L.; Li, X. Role of STAT3 signaling pathway in breast cancer. *Cell Commun. Signal.* **2020**, *18*, 33. [CrossRef] [PubMed]

104. Fu, S.; Chen, X.; Lo, H.W.; Lin, J. Combined bazedoxifene and paclitaxel treatments inhibit cell viability, cell migration, colony formation, and tumor growth and induce apoptosis in breast cancer. *Cancer Lett.* **2019**, *448*, 11–19. [CrossRef] [PubMed]

105. Li, H.; Xiao, H.; Lin, L.; Jou, D.; Kumari, V.; Lin, J.; Li, C. Drug design targeting protein-protein interactions (PPIs) using multiple ligand simultaneous docking (MLSD) and drug repositioning: Discovery of raloxifene and bazedoxifene as novel inhibitors of IL-6/GP130 interface. *J. Med. Chem.* **2014**, *57*, 632–641. [CrossRef]

106. Tian, J.; Chen, X.; Fu, S.; Zhang, R.; Pan, L.; Cao, Y.; Wu, X.; Xiao, H.; Lin, H.J.; Lo, H.W.; et al. Bazedoxifene is a novel IL-6/GP130 inhibitor for treating triple-negative breast cancer. *Breast Cancer Res. Treat.* **2019**, *175*, 553–566. [CrossRef]

107. Oh, E.; Kim, Y.J.; An, H.; Sung, D.; Cho, T.M.; Farrand, L.; Jang, S.; Seo, J.H.; Kim, J.Y. Flubendazole elicits anti-metastatic effects in triple-negative breast cancer via STAT3 inhibition. *Int. J. Cancer* **2018**, *143*, 1978–1993. [CrossRef]

108. Hou, Z.J.; Luo, X.; Zhang, W.; Peng, F.; Cui, B.; Wu, S.J.; Zheng, F.M.; Xu, J.; Xu, L.Z.; Long, Z.J.; et al. Flubendazole, FDA-approved anthelmintic, targets breast cancer stem-like cells. *Oncotarget* **2015**, *6*, 6326–6340. [CrossRef] [PubMed]

109. Liu, J.; Chen, X.; Ward, T.; Pegram, M.; Shen, K. Combined niclosamide with cisplatin inhibits epithelial-mesenchymal transition and tumor growth in cisplatin-resistant triple-negative breast cancer. *Tumor Biol.* **2016**, *37*, 9825–9835. [CrossRef]

110. Wang, Y.C.; Chao, T.K.; Chang, C.C.; Yo, Y.T.; Yu, M.H.; Lai, H.C. Drug screening identifies niclosamide as an inhibitor of breast cancer stem-like cells. *PLoS ONE* **2013**, *8*, e74538. [CrossRef]

111. Lu, L.; Dong, J.; Wang, L.; Xia, Q.; Zhang, D.; Kim, H.; Yin, T.; Fan, S.; Shen, Q. Activation of STAT3 and Bcl-2 and reduction of reactive oxygen species (ROS) promote radioresistance in breast cancer and overcome of radioresistance with niclosamide. *Oncogene* **2018**, *37*, 5292–5304. [CrossRef]

112. Wang, L.; Peng, Y.; Shi, K.; Wang, H.; Lu, J.; Li, Y.; Ma, C. Osthole inhibits proliferation of human breast cancer cells by inducing cell cycle arrest and apoptosis. *J. Biomed. Res.* **2015**, *29*, 132–138. [CrossRef]

113. Tang, D.Z.; Hou, W.; Zhou, Q.; Zhang, M.; Holz, J.; Sheu, T.J.; Li, T.F.; Cheng, S.D.; Shi, Q.; Harris, S.E.; et al. Osthole stimulates osteoblast differentiation and bone formation by activation of β-catenin-BMP signaling. *J. Bone Miner. Res.* **2010**, *25*, 1234–1245. [CrossRef] [PubMed]

114. Dai, X.; Yin, C.; Zhang, Y.; Guo, G.; Zhao, C.; Wang, O.; Xiang, Y.; Zhang, X.; Liang, G. Osthole inhibits triple negative breast cancer cells by suppressing STAT3. *J. Exp. Clin. Cancer Res.* **2018**, *37*, 1–11. [CrossRef] [PubMed]

115. Palaka, B.K.; Venkatesan, R.; Ampasala, D.R.; Periyasamy, L. Identification of novel inhibitors of signal transducer and activator of transcription 3 over signal transducer and activator of transcription 1 for the treatment of breast cancer by in-silico and in-vitro approach. *Process Biochem.* **2019**, *82*, 153–166. [CrossRef]

116. Schech, A.J.; Kazi, A.A.; Gilani, R.A.; Brodie, A.H. Zoledronic acid reverses the epithelial-mesenchymal transition and inhibits self-renewal of breast cancer cells through inactivation of NF- B. *Mol. Cancer Ther.* **2013**, *12*, 1356–1366. [CrossRef]

117. Ottewell, P.D.; Mönkkönen, H.; Jones, M.; Lefley, D.V.; Coleman, R.E.; Holen, I. Antitumor effects of doxorubicin followed by zoledronic acid in a mouse model of breast cancer. *J. Natl. Cancer Inst.* **2008**, *100*, 1167–1178. [CrossRef] [PubMed]

118. Ishikawa, T.; Akazawa, K.; Hasegawa, Y.; Tanino, H.; Horiguchi, J.; Miura, D.; Hayashi, M.; Kohno, N. Survival outcomes of neoadjuvant chemotherapy with zoledronic acid for HER2-negative breast cancer. *J. Surg. Res.* **2017**, *220*, 46–51. [CrossRef]

119. Burke, A.J.; Sullivan, F.J.; Giles, F.J.; Glynn, S.A. The yin and yang of nitric oxide in cancer progression. *Carcinogenesis* **2013**, *34*, 503–512. [CrossRef] [PubMed]

120. Glynn, S.A.; Boersma, B.J.; Dorsey, T.H.; Yi, M.; Yfantis, H.G.; Ridnour, L.A.; Martin, D.N.; Switzer, C.H.; Hudson, R.S.; Wink, D.A.; et al. Increased NOS2 predicts poor survival in estrogen receptor-negative breast cancer patients. *J. Clin. Investig.* **2010**, *120*, 3843–3854. [CrossRef] [PubMed]

121. Granados-Principal, S.; Liu, Y.; Guevara, M.L.; Blanco, E.; Choi, D.S.; Qian, W.; Patel, T.; Rodriguez, A.A.; Cusimano, J.; Weiss, H.L.; et al. Inhibition of iNOS as a novel effective targeted therapy against triple-negative breast cancer. *Breast Cancer Res.* **2015**, *17*, 1–16. [CrossRef] [PubMed]

122. Dave, B.; Granados-Principal, S.; Zhu, R.; Benz, S.; Rabizadeh, S.; Soon-Shiong, P.; Yu, K.D.; Shao, Z.; Li, X.; Gilcrease, M.; et al. Targeting RPL39 and MLF2 reduces tumor initiation and metastasis in breast cancer by inhibiting nitric oxide synthase signaling. *Proc. Natl. Acad. Sci. USA* **2014**, *111*, 8838–8843. [CrossRef] [PubMed]

123. González-González, A.; Muñoz-Muela, E.; Marchal, J.A.; Cara, F.E.; Molina, M.P.; Cruz-Lozano, M.; Jiménez, G.; Verma, A.; Ramírez, A.; Qian, W.; et al. Activating transcription factor 4 modulates TGFb-induced aggressiveness in triple-negative breast cancer via SMad2/3/4 and mTORC2 signaling. *Clin. Cancer Res.* **2018**, *24*, 5697–5709. [CrossRef]

124. Davila-Gonzalez, D.; Choi, D.S.; Rosato, R.R.; Granados-Principal, S.M.; Kuhn, J.G.; Li, W.F.; Qian, W.; Chen, W.; Kozielski, A.J.; Wong, H.; et al. Pharmacological inhibition of NOS activates ASK1/JNK pathway augmenting docetaxel-mediated apoptosis in triple-negative breast cancer. *Clin. Cancer Res.* **2018**, *24*, 1152–1162. [CrossRef]

125. Walsh, E.M.; Keane, M.M.; Wink, D.A.; Callagy, G.; Glynn, S.A. Review of triple negative breast cancer and the impact of inducible nitric oxide synthase on tumor biology and patient outcomes. *Crit. Rev. Oncog.* **2016**, *21*, 333–351. [CrossRef]

126. Lemke, G. Biology of the TAM receptors. *Cold Spring Harb. Perspect. Biol.* **2013**, *6*, a009076. [CrossRef]

127. Colavito, S.A. AXL as a target in breast cancer therapy. *J. Oncol.* **2020**, *2020*, 5291952. [CrossRef]

128. Asiedu, M.K.; Beauchamp-perez, F.D.; Ingle, J.N.; Behrens, M.D.; Radisky, D.C.; Knutson, K.L. AXL induces epithelial to mesenchymal transition and regulates the function of breast cancer stem cells. *Oncogene* **2014**, *33*, 1316–1324. [CrossRef] [PubMed]

129. Goyette, M.A.; Cusseddu, R.; Elkholi, I.; Abu-Thuraia, A.; El-Hachem, N.; Haibe-Kains, B.; Gratton, J.P.; Côté, J.F. AXL knockdown gene signature reveals a drug repurposing opportunity for a class of antipsychotics to reduce growth and metastasis of triple-negative breast cancer. *Oncotarget* **2019**, *10*, 2055–2067. [CrossRef]

130. Meyer, A.S.; Miller, M.A.; Gertler, F.B.; Lauffenburger, D.A. The receptor AXL diversifies EGFR signaling and limits the response to EGFR-targeted inhibitors in triple-negative breast cancer cells. *Sci. Signal.* **2013**, *6*, ra66. [CrossRef]

131. Visvader, J.E.; Lindeman, G.J. Cancer stem cells in solid tumours: Accumulating evidence and unresolved questions. *Nat. Rev. Cancer* **2008**, *8*, 755–768. [CrossRef] [PubMed]

132. O'Brien, C.A.; Kreso, A.; Dick, J.E. Cancer stem cells in solid tumors: An overview. *Semin. Radiat. Oncol.* **2009**, *19*, 71–77. [CrossRef] [PubMed]

133. Bomken, S.; Fišer, K.; Heidenreich, O.; Vormoor, J. Understanding the cancer stem cell. *Br. J. Cancer* **2010**, *103*, 439–445. [CrossRef] [PubMed]

134. Mani, S.A.; Guo, W.; Liao, M.J.; Eaton, E.N.; Ayyanan, A.; Zhou, A.Y.; Brooks, M.; Reinhard, F.; Zhang, C.C.; Shipitsin, M.; et al. The epithelial-mesenchymal transition generates cells with properties of stem cells. *Cell* **2008**, *133*, 704–715. [CrossRef] [PubMed]

135. Rosen, J.M.; Jordan, C.T. The increasing complexity of the cancer stem cell paradigm. *Science* **2009**, *324*, 1670–1673. [CrossRef]

136. Liu, S.; Cong, Y.; Wang, D.; Sun, Y.; Deng, L.; Liu, Y.; Martin-Trevino, R.; Shang, L.; McDermott, S.P.; Landis, M.D.; et al. Breast cancer stem cells transition between epithelial and mesenchymal states reflective of their normal counterparts. *Stem Cell Rep.* **2014**, *2*, 78–91. [CrossRef]

137. Asiedu, M.K.; Ingle, J.N.; Behrens, M.D.; Radisky, D.C.; Knutson, K.L. TGFβ/TNFα-mediated epithelial-mesenchymal transition generates breast cancer stem cells with a claudin-low phenotype. *Cancer Res.* **2011**, *71*, 4707–4719. [CrossRef] [PubMed]

138. Tan, E.J.; Olsson, A.K.; Moustakas, A. Reprogramming during epithelial to mesenchymal transition under the control of TGFβ. *Cell Adhes. Migr.* **2015**, *9*, 233–246. [CrossRef] [PubMed]

139. Idowu, M.O.; Kmieciak, M.; Dumur, C.; Burton, R.S.; Grimes, M.M.; Powers, C.N.; Manjili, M.H. CD44$^+$/CD24$^{-/low}$ cancer stem/progenitor cells are more abundant in triple-negative invasive breast carcinoma phenotype and are associated with poor outcome. *Hum. Pathol.* **2012**, *43*, 364–373. [CrossRef] [PubMed]

140. Li, H.; Ma, F.; Wang, H.; Lin, C.; Fan, Y.; Zhang, X.; Qian, H.; Xu, B. Stem cell marker aldehyde dehydrogenase 1 (ALDH1)-expressing cells are enriched in triple-negative breast cancer. *Int. J. Biol. Markers* **2013**, *28*, 357–364. [CrossRef] [PubMed]

141. Palomeras, S.; Ruiz-Martínez, S.; Puig, T. Targeting breast cancer stem cells to overcome treatment resistance. *Molecules* **2018**, *23*, 2193. [CrossRef]

142. Xu, L.; Zhang, L.; Hu, C.; Liang, S.; Fei, X.; Yan, N.; Zhang, Y.; Zhang, F. WNT pathway inhibitor pyrvinium pamoate inhibits the self-renewal and metastasis of breast cancer stem cells. *Int. J. Oncol.* **2016**, *48*, 1175–1186. [CrossRef]

143. Papi, A.; Orlandi, M. Role of nuclear receptors in breast cancer stem cells. *World J. Stem Cells* **2016**, *8*, 62–72. [CrossRef]

144. Cui, J.; Hollmén, M.; Li, L.; Chen, Y.; Proulx, S.T.; Reker, D.; Schneider, G.; Detmar, M. New use of an old drug: Inhibition of breast cancer stem cells by benztropine mesylate. *Oncotarget* **2017**, *8*, 1007–1022. [CrossRef]

145. Sogawa, C.; Eguchi, T.; Tran, M.T.; Ishige, M.; Trin, K.; Okusha, Y.; Taha, E.A.; Lu, Y.; Kawai, H.; Sogawa, N.; et al. Antiparkinson drug benztropine suppresses tumor growth, circulating tumor cells, and metastasis by acting on SLC6A3/dat and reducing STAT3. *Cancers* **2020**, *12*, 523. [CrossRef]

146. Choi, D.S.; Blanco, E.; Kim, Y.; Rodriguez, A.A.; Zhao, H.; Huang, T.H.; Chen, C.; Jin, G.; Landis, M.D.; Lacey, A.; et al. Chloroquine eliminates cancer stem cells through deregulation of Jak2 and DNMT1. *Stem Cells* **2014**, *32*, 2309–2323. [CrossRef]

147. Yang, L.; Wu, X.; Wang, Y.; Zhang, K.; Wu, J.; Yuan, Y.C.; Deng, X.; Chen, L.; Kim, C.C.H.; Lau, S.; et al. FZD7 has a critical role in cell proliferation in triple negative breast cancer. *Oncogene* **2011**, *30*, 4437–4446. [CrossRef] [PubMed]

148. Liu, C.C.; Prior, J.; Piwnica-Worms, D.; Bu, G. LRP6 overexpression defines a class of breast cancer subtype and is a target for therapy. *Proc. Natl. Acad. Sci. USA* **2010**, *107*, 5136–5141. [CrossRef]

149. King, T.D.; Suto, M.J.; Li, Y. The wnt/β-catenin signaling pathway: A potential therapeutic target in the treatment of triple negative breast cancer. *J. Cell. Biochem.* **2012**, *113*, 13–18. [CrossRef] [PubMed]

150. Gupta, P.B.; Onder, T.T.; Jiang, G.; Tao, K.; Kuperwasser, C.; Weinberg, R.A.; Lander, E.S. Identification of selective inhibitors of cancer stem cells by high-throughput screening. *Cell* **2009**, *138*, 645–659. [CrossRef] [PubMed]

151. Kai, M.; Kanaya, N.; Wu, S.V.; Mendez, C.; Nguyen, D.; Luu, T.; Chen, S. Targeting breast cancer stem cells in triple-negative breast cancer using a combination of LBH589 and salinomycin. *Breast Cancer Res. Treat.* **2015**, *151*, 281–294. [CrossRef] [PubMed]

152. Thorne, C.A.; Hanson, A.J.; Schneider, J.; Tahinci, E.; Orton, D.; Cselenyi, C.S.; Jernigan, K.K.; Meyers, K.C.; Hang, B.I.; Waterson, A.G.; et al. Small-molecule inhibition of Wnt signaling through activation of casein kinase 1α. *Nat. Chem. Biol.* **2010**, *6*, 829–836. [CrossRef]

153. Carrella, D.; Manni, I.; Tumaini, B.; Dattilo, R.; Papaccio, F.; Mutarelli, M.; Sirci, F.; Amoreo, C.A.; Mottolese, M.; Iezzi, M.; et al. Computational drugs repositioning identifies inhibitors of oncogenic PI3K/AKT/P70S6K-dependent pathways among FDAapproved compounds. *Oncotarget* **2016**, *7*, 58743–58758. [CrossRef]

154. Ishii, I.; Harada, Y.; Kasahara, T. Reprofiling a classical anthelmintic, pyrvinium pamoate, as an anti-cancer drug targeting mitochondrial respiration. *Front. Oncol.* **2012**, *2*, 1–4. [CrossRef]

155. Dattilo, R.; Mottini, C.; Camera, E.; Lamolinara, A.; Auslander, N.; Doglioni, G.; Muscolini, M.; Tang, W.; Planque, M.; Ercolani, C.; et al. Pyrvinium pamoate induces death of triple-negative breast cancer stem-like cells and reduces metastases through effects on lipid anabolism. *Cancer Res.* **2020**. [CrossRef]

156. So, J.Y.; Suh, N. Targeting cancer stem cells in solid tumors by vitamin D. *J. Steroid Biochem. Mol. Biol.* **2015**, *148*, 79–85. [CrossRef]

157. Wahler, J.; So, J.Y.; Cheng, L.C.; Maehr, H.; Uskokovic, M.; Suh, N. Vitamin D compounds reduce mammosphere formation and decrease expression of putative stem cell markers in breast cancer. *J. Steroid Biochem. Mol. Biol.* **2015**, *148*, 148–155. [CrossRef]

158. Pickholtz, I.; Saadyan, S.; Keshet, G.I.; Wang, V.S.; Cohen, R.; Bouwman, P.; Jonkers, J.; Byers, S.W.; Papa, M.Z.; Yarden, R.I. Cooperation between BRCA1 and vitamin D is critical for histone acetylation of the p21waf1 promoter and for growth inhibition of breast cancer cells and cancer stem-like cells. *Oncotarget* **2014**, *5*, 11827–11846. [CrossRef]

159. Pervin, S.; Hewison, M.; Braga, M.; Tran, L.; Chun, R.; Karam, A.; Chaudhuri, G.; Norris, K.; Singh, R. Correction: Down-regulation of vitamin D receptor in mammospheres: Implications for vitamin D resistance in breast cancer and potential for combination therapy. *PLoS ONE* **2013**, *8*, e53287. [CrossRef]

160. Zanetti, A.; Affatato, R.; Centritto, F.; Fratelli, M.; Kurosaki, M.; Barzago, M.M.; Bolis, M.; Terao, M.; Garattini, E.; Paroni, G. All-trans-retinoic acid modulates the plasticity and inhibits the motility of breast cancer cells role of notch1 and transforming growth factor (TGF β). *J. Biol. Chem.* **2015**, *290*, 17690–17709. [CrossRef]

161. Bhat-Nakshatri, P.; Goswami, C.P.; Badve, S.; Sledge, G.W.; Nakshatri, H. Identification of FDA-approved drugs targeting breast cancer stem cells along with biomarkers of sensitivity. *Sci. Rep.* **2013**, *3*, 2530. [CrossRef] [PubMed]

162. Ginestier, C.; Wicinski, J.; Cervera, N.; Monville, F.; Finetti, P.; Bertucci, F.; Wicha, M.S.; Birnbaum, D.; Charafe-Jauffret, E. Retinoid signaling regulates breast cancer stem cell differentiation. *Cell Cycle* **2009**, *8*, 3297–3302. [CrossRef] [PubMed]

163. Anand, K.; Niravath, P.; Patel, T.; Ensor, J.; Rodriguez, A.; Boone, T.; Wong, S.T.; Chang, J.C. A Phase II study of the efficacy and safety of Chloroquine in combination with Taxanes in the treatment of patients with advanced or metastatic anthracycline-refractory breast cancer. *Clin. Breast Cancer* **2020**. [CrossRef]

Publisher's Note: MDPI stays neutral with regard to jurisdictional claims in published maps and institutional affiliations.

Journal of
Personalized
Medicine

Review

Systems Biology and Experimental Model Systems of Cancer

Gizem Damla Yalcin [†] , **Nurseda Danisik** [†], **Rana Can Baygin** [†] **and Ahmet Acar** *

Department of Biological Sciences, Middle East Technical University, Universiteler Mah. Dumlupınar Bulvarı 1, Çankaya, Ankara 06800, Turkey; gizemdyalcin@gmail.com (G.D.Y.); nurseda.danisik@metu.edu.tr (N.D.); rana.baygin@metu.edu.tr (R.C.B.)
* Correspondence: acara@metu.edu.tr
† These authors contributed equally.

Received: 7 September 2020; Accepted: 16 October 2020; Published: 19 October 2020

Abstract: Over the past decade, we have witnessed an increasing number of large-scale studies that have provided multi-omics data by high-throughput sequencing approaches. This has particularly helped with identifying key (epi)genetic alterations in cancers. Importantly, aberrations that lead to the activation of signaling networks through the disruption of normal cellular homeostasis is seen both in cancer cells and also in the neighboring tumor microenvironment. Cancer systems biology approaches have enabled the efficient integration of experimental data with computational algorithms and the implementation of actionable targeted therapies, as the exceptions, for the treatment of cancer. Comprehensive multi-omics data obtained through the sequencing of tumor samples and experimental model systems will be important in implementing novel cancer systems biology approaches and increasing their efficacy for tailoring novel personalized treatment modalities in cancer. In this review, we discuss emerging cancer systems biology approaches based on multi-omics data derived from bulk and single-cell genomics studies in addition to existing experimental model systems that play a critical role in understanding (epi)genetic heterogeneity and therapy resistance in cancer.

Keywords: cancer systems biology; experimental model systems; next-generation sequencing; single-cell sequencing; patient-derived xenografts; patient-derived organoids

1. Introduction to Cancer Systems Biology

Cancer is an extremely complex disease with heterotypic interactions between cancer cells and neighboring stromal cells that support the proliferation, invasion, and the metastatic cascade of tumor cells [1,2]. Recently, multi-omics approaches empowered by next-generation technologies have enabled genomic characterization and evolutionary histories of both primary and metastatic cancer progression [3–6]. These technologies that shed light on the genome, transcriptome, metabolome, and proteome of cancer cells corroborate our understanding about systems biology-level approaches in cancer (Figure 1) [7]. Considering the challenges to unify high-throughput data obtained from multi-omics studies, system biology applications in cancer hold a key role to tackle this very problem. For example, cancer as a disease of numerous distinct cell types requires taking into consideration the combination of data derived from these different cell types together with the integration of various layers of genetic and non-genetic data that are forming the cellular systems. Thus, cancer systems biology can simplify the analysis of multi-layer data and offer effective and fast solutions for the development of novel drug technologies and the identification of predictive biomarkers in cancer therapies. Cancer systems biology is an emerging field with accumulating data obtained through network-driven and interdisciplinary science that ultimately aims to tailor better-personalized treatment modalities for patients based on their genetic and non-genetic profiles [8].

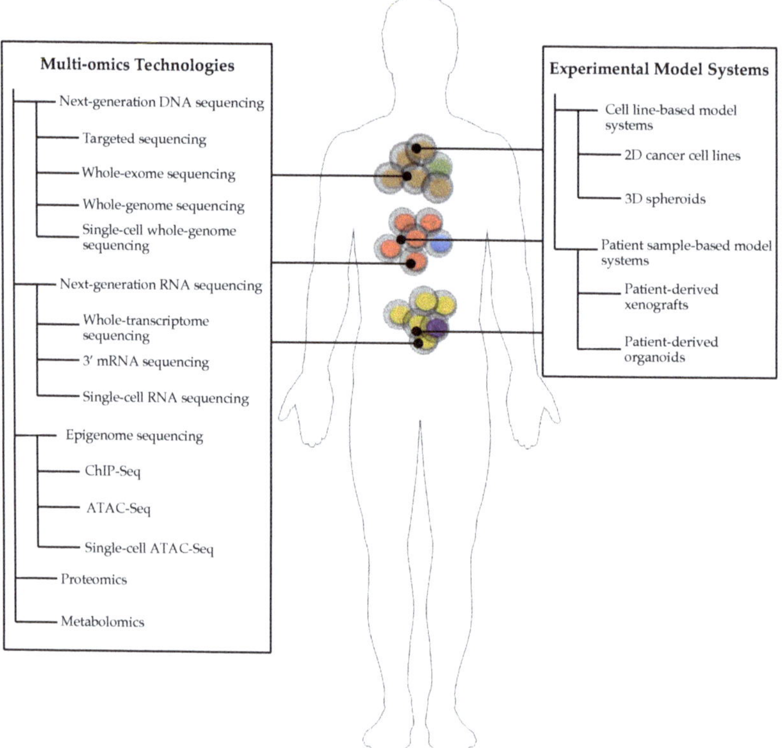

Figure 1. Comprehensive picture of systems biology approaches and experimental model systems constituting the core components of the biology of cancer.

The heterogeneous nature of cancers led to studies mapping the (epi)genomic alterations [9–11] both in primary [6,12] and metastatic cancers [5]. Through the high-throughput data obtained from cancer patients, it is now possible to combine this information and assess the genotype-to-phenotype link to further characterize the disease onset and clinical outcome. The combination of information derived from the genomic architecture and various gene networks from a single or a group of cells not only determines the fate of these cells during development but also a progression to cancer occurs as a result of the deregulation of these interactions. For example, while the regulation of Notch and Wnt signaling pathways are fine-tuned by each other in normal homeostasis [13], their aberrant expression and deregulation are commonly seen in cancers [14,15]. Therefore, understanding the genetic and epigenetic changes that cause persistent signaling activations and disrupting normal cellular homeostasis is still one of the biggest challenges to address in cancer systems biology.

2. Cancer Systems Biology for Precision Medicine

The vast majority of efforts focus on bridging the "big data" obtained from various multi-omics studies to new computational algorithms to ultimately offer more effective personalized cancer therapies. Despite the advancements in cancer therapy through systems biology approaches, treatment resistance is arguably one of the biggest challenges for better-personalized cancer treatments [16,17]. This is mainly due to the fact that cancer follows distinct evolutionary trajectories in patients compared to their genomic landscapes, not only during the initiation and metastasis cascade of cancer cells but also in response to the treatment in cancer therapies [18,19]. For this reason, the accurate identification of subclonal drivers holds great importance for the timing of the subclonal expansion and its diversity in cancer therapies [20]. This sophisticated subclonal identification tool, empowered by machine learning and population genetics, will potentially lead to developing more comprehensive computational methods by integrating with network-driven approaches for cancer systems biology in the future.

With the rapid developments in next-generation sequencing (NGS) technologies, previous microarray studies have been gradually replaced by massively parallel deep sequencing techniques such as whole-genome, whole-exome, targeted-panel, and RNA sequencing [21]. Initially, the microarray platforms have proved to be a very useful tool for genome-wide association studies (GWAS) in cancer systems biology; however, they demonstrated limitations such as covering only a the small fraction of the genome and failure to take into account more than common genetic risk factors [22]. Later, individual research groups started to apply NGS technologies to identify somatic alterations (single-nucleotide variations, copy-number alterations, structural variations) in cancer driver genes and to determine gene expression changes and open chromatin formations both in coding and non-coding regions of the genome [23–28]. Then, individual studies were followed by larger multigroup projects [4–6]. One of the remarkable efforts is The International Cancer Genome Consortium/The Cancer Genome Atlas (ICGC/TCGA) Pan-Cancer Analysis of Whole Genomes (PWAG) project, comprising a working group of 700 scientists, which recently reported their findings from 2600 whole-genome samples [4,29–31]. In addition to these studies that deciphered the evolutionary trajectories of tumors prospectively, recent technologies have also allowed the monitoring of clonal dynamics using "cellular barcodes" integrated into experimental model systems to map the tumor evolution at the single-cell resolution [32,33].

In addition, investigating cancer genomes at the single-cell resolution has taken a big step forward in the past few years [3]. Initial studies focused on the understanding of the transcriptome of single cells in a plate-based system wherein cells were required to be sorted individually, and thus the system lacked high-throughput capacity [34]. However, recent advances, especially with the use of droplet-based systems, have advanced our understanding about single-cell genomics through an increased capacity to profile thousands of single cells at the same time (single-cell RNA sequencing, scRNA-seq) [35,36]. The scRNA-seq technology provided a high-resolution picture not only of cellular states in developmental biology [37] but also in cancer biology where intratumoral heterogeneity and tumor cell plasticity are highly prevalent [38]. Furthermore, the information obtained from the analysis of single-cell WGS (scWGS) has proved to be informative for understanding intratumor heterogeneity and the evolutionary history of thousands of single cells comprising the bulk tumor population [39,40]. Recently, the high-throughput capacity for scWGS has improved significantly, and clonal/subclonal alterations at the single cell resolution were reported in thousands of cells [41]. To capture the epigenetic changes at the single-cell level, novel methods to map the single-cell epigenome have also been reported. For example, single-nuclei chromatin accessibility assays (ATAC-seq) inferring the chromatin open or closed states in single cells [42,43]. Lastly, the rapid developments in the single-cell biology have also resulted in novel methods such as parallel sequencing of single-cell genomes and transcriptomes [44] and joint profiling of single-cell chromatin accessibility and gene expression [45]. Various online databases containing cancer systems biology tools to document molecular profiles of cancer types are available and offered for the use of the cancer research community (Table 1). Importantly, various multi-omics data obtained using high-throughput sequencing methods enables the integration of these data into experimental model systems for the identification of the actionable targets in cancer. As such, these molecular data integrated with systems biology applications, for the function of transcriptional and proteomics networks, provide effective solutions for the treatment of cancer. Given that cancer is a systems biology disease, integration of the cellular information with the help of computational and mathematical modeling highlights the need to develop more advanced and sophisticated systems biology applications in cancer. This considerable challenge has especially become evident with a rapid increase in the accumulation of sequencing data over the past decade. Hence, to address this very challenge, systems biology approaches are timely positioned to offer novel solutions to better understand the underlying mechanisms of drug resistance and the identification of biomarkers that can predict the disease outcome and response to targeted therapies. Overall, integrating cellular networks with cancer (epi)genomes in both single and bulk cell populations has paved a way to advance our understanding for developing systems biology approaches for precision therapy to advance clinical decisions for patient benefits.

Table 1. A collection of databases.

Name	Description	Website	Reference
CaSNP	CaSNP performs quantitative analysis of copy number variation from SNP arrays in multiple cancer types	https://bioinformaticshome.com/tools/cnv/descriptions/CaSNP.html	[46]
OncoLand	OncoLand provides oncology data access in sample and gene directions.	https://omicsoftdocs.github.io/ArraySuiteDoc/tutorials/OncoLand/Introduction/	[47]
AGCOH	The Atlas of Genetics, Cytogenetics in Oncology and Hematology perform comprehensive genomic characterization and analysis of multiple cancer types	http://atlasgeneticsoncology.org/BackpageAbout.html	[48]
PCWAG	PCWAG—Pan-cancer Analysis of Whole Genomes provides common patterns of mutations from more than 2600 cancer whole genomes	http://dcc.icgc.org/pcawg	[4]
ChiTaRS	ChiTaRS contains chimeric transcripts and RNA-Seq data	http://chitars.bioinfo.cnio.es/	[49]
CanSAR	CanSAR provides information about translational research and drug discovery knowledgebase	https://cansarblack.icr.ac.uk/	[50]
OncoDB.HCC	Oncogenomics Database of Hepatocellular Carcinoma provides genomic, transcriptomic, and proteomic data	http://oncodb.hcc.ibms.sinica.edu.tw/index.htm	[51]
COSMIC	COSMIC performs a comprehensive database of somatic mutation in multiple cancer types	https://cancer.sanger.ac.uk/cosmic	[52]
canEvolve	canEvolve is a comprehensive database including genes, miRNA, and protein expression profiles; copy number changes for a variety of cancer types and protein–protein interactions	http://www.canevolve.org/AnalysisResults/AnalysisResults.html	[53]
CancerPPD	CancerPPD provides information about anticancer peptides and proteins in multiple cancer types	http://crdd.osdd.net/raghava/cancerppd/	[54]
PED	The Pancreatic Expression Database performs a comprehensive meta-analysis of pancreatic cancer	http://www.pancreasexpression.org/	[55]
CGP	Cancer Genome Project provides genotype and copy number changes information in tumors	https://www.sanger.ac.uk/group/cancer-genome-project	[56]
MethyCancer	MethyCancer provides information about DNA methylation and gene expression in a variety of cancer types	http://methycancer.psych.ac.cn/	[57]
CPTAC	Clinical Proteomic Tumor Analysis Consortium is a database containing an integration of genomic and proteomic data	https://proteomics.cancer.gov/	[58]
intOGen	Integrative Onco Genomics performs comprehensive genomic data of multiple cancer types	https://www.intogen.org/search	[59]
ArrayExpress	ArrayExpress focuses on microarray gene expression data	https://www.ebi.ac.uk/arrayexpress/	[60]
DriverDBv3	DriverDBv3 is a database of cancer omics	http://driverdb.tms.cmu.edu.tw/	[61]
PCDB	The Pancreatic Cancer Database provides genetic information in pancreatic cancer	http://www.pancreaticcancerdatabase.org	[62]
CancerDR	CancerDR contains anticancer drugs and their effectiveness against a variety of cell lines	http://crdd.osdd.net/raghava/cancerdr/	[63]
Platinum	Platinum provides knowledge about missense mutations on ligand–proteome interactions	http://biosig.unimelb.edu.au/platinum/	[64]

3. Experimental Model Systems of Cancer

Although cancer mortality rates are gradually diminishing, it is still one of the deadliest diseases in the world [65]. To develop more effective therapeutic solutions, cancer cell lines, 3D spheroids, in vivo patient-derived xenografts (PDXs), and ex vivo patient-derived organoids (PDOs) have been studied by various groups [66–68]. Due to the advances in the development of experimental model systems, there has been remarkable progress in understanding the underlying mechanisms of initiation, progression, and the metastatic cascade of cancer cells [69]. In addition to the advantages of each model system, traditional model systems have failed to recapitulate the response to drugs that are observed in the clinic. For instance, targeted therapies and chemotherapeutic agents that work well in preclinical model systems fail to proceed into clinical trials since specific model systems were unable to recapitulate the disease progression [70]. Therefore, in this section of the review, we sought to discuss current preclinical model systems used in cancer research and their role in predicting how cancer will progress and respond to the therapy when these model systems are integrated with system biology approaches.

4. Cell Line-Based Model Systems

Since the first human cancer cell line was established in 1951, 2D monolayer systems have provided major advantages in the understanding of tumor biology and cancer therapy [71]. Over the decades, 2D monolayer systems offered several advantages such as being easy to expand and hence allowing long-term culture times, being manipulated by gene insertions and deletions, and requiring inexpensive material for culturing [72]. On the other hand, this platform has many drawbacks, mainly its inability to mimic the 3D nature of tumor growth. The inadequacies of the 2D monolayer systems also include a lack of cell-to-extracellular matrix (ECM) contact that has been reported as responsible for the accurate detection of cell viability/death, drug metabolism, and expression of certain genes and protein in tumors [73]. Another major limitation of 2D monolayer systems is their inaccurate utility of oxygen and nutrients when compared to 3D culture systems that have proven to be more successful in mimicking real tumor masses [74]. Collectively, 2D monolayer systems have played a major role in understanding and designing cancer therapies for systems biology approaches; however, due to their insufficiency to predict real tumor outcomes in patients, more suitable model systems such as 3D culture systems have been developed.

The first 3D culture was performed using a soft agar solution by Hamburger and Salmon in 1977 [75]. Since that time, several 3D culture methods have been documented. Depending on the material used, the 3D culture systems can be divided into three categories: (i) cultured onto non-adherent plates, (ii) embedded into matrigel-like substances, and (iii) seeded into scaffold-based systems. The general approach for 3D culture systems is based on the formation of a spheroid structure in which cancer cells can form various layers. The 3D nature of spheroids has been demonstrated as a successful system in mimicking the features of the solid tumor mass [73]. Three-dimensional spheroids can also mimic tissue-specific functional characteristics in developmental processes. For example, cardiomyocyte spheroids can exhibit heart-like rhythms, and hepatocyte spheroids exhibit biochemical functions of the liver [76,77]. Three-dimensional culture systems have also been shown to mimic in vivo-like microenvironments via the establishment of complex cell-to-cell and cell-to-ECM communications. These interactions result in cellular signal transduction events similar to tumor tissues that can mediate their cell shape and proliferation [78]. In addition, drug response assays in 3D culture systems were shown to resemble in vivo studies more than 2D culture systems in terms of their success rates in preclinical studies [79,80]. In another study, sensitivities of the same cell line against different chemotherapeutic agents were reported as different in 2D vs. 3D culture systems [81]. For instance, in this study, HCT-116 cells grown in both 2D and 3D model systems and their sensitivities against four commonly used anticancer agents (melphalan, 5-Fluorouracil, oxaliplatin, and irinotecan) were tested. The efficacy of these inhibitors was higher in the 2D than the 3D culture system, suggesting that phenotypic differences and distinct cell-to-cell interactions between these model systems might be responsible for observing the differences in drug sensitivities.

5. Patient Sample-Based Model Systems

Patient-derived xenografts (PDXs) are preclinical models established by directly transplanting patient-derived tumor specimens into immunodeficient mice [82]. PDXs have been accepted as promising preclinical model systems that successfully mimic the testing of anticancer drugs [66]. This system provides several advantages, such as the preservation of tumor heterogeneity, molecular subtypes and the clinicopathological features of the tumors obtained from patients [83]. In addition, PDXs have been shown to successfully predict the drug response in the preclinical setting to test the effectiveness of therapeutic agents [84]. While PDXs offer several advantages as a preclinical model system, an increasing body of evidence suggests there are limitations [85]. Firstly, a significant proportion of tumor samples engrafted in mice may not successfully grow due to the host mouse environment causing a bottleneck. Secondly, engraftment times can be long so that the maintenance costs associated with each PDX prove prohibitive. Thirdly, there is still no standardized method for choosing the type of mouse or engraftment technique specific for each cancer type, which raises the possibility of obtaining non-reproducible results between different studies. Studies that overcome these limitations have shed light on the mechanisms of acquired drug resistance, especially in metastatic colorectal cancer (mCRC). For instance, a series of seminal studies published by the Bertotti Lab has demonstrated the use of a large PDX biobank to investigate the underlying mechanisms of drug resistance in mCRC [86–89]. Importantly, one of these studies played a critical role in assessing the genomic landscape of anti-EGFR antibody blockage in PDXs and functional consequences linked to clinical data in cancer patients [87]. Thus, PDXs have paved a way to develop a platform for the systematic analysis and evaluation of cancer therapies.

Patient-derived tumor organoids (PDOs) are ex vivo three-dimensional structures of tumors obtained from cancer patients and grown in the presence of an extracellular matrix [90]. Accumulating evidence suggests that PDOs can successfully predict the drug response in cancer patients in the clinic in addition to preserving the genetic and transcriptomic heterogeneity of the original tumor [67]. In addition, studies focused on comparing the histopathological features of tumors with PDOs revealed that the PDOs maintain similar morphological characteristics as the original tumor [90,91]. Importantly, PDOs also mimic the genomic and transcriptomic features of the tumors that they have derived from even after long ex vivo culture times [91–93]. To date, PDOs have been established from different cancer types including colorectal [93], gastrointestinal [91], pancreatic [94], prostate [95], bladder [96], breast [97], glioblastoma [98], and ovarian [99]. Three-dimensional cultures of PDOs that predict the outcome of drug treatment in cancer patients can be considered an important milestone for personalized medicine for the benefits of cancer patient [100]. When PDOs are established from individual patients in a short time, they can provide a window of opportunity to test therapeutic agents in parallel to the clinic, and thus the outcome of drug testing in the laboratory can prove informative for the decision making of treatment for patients.

Amongst the key studies about PDOs, van de Wetering et al. (2015) is the first study that reported a well-established and characterized PDO biobank from 20 primary CRC patients [93]. In this study, whole-exome sequencing (WES) and the RNA sequencing of samples resulted in preserved genetic heterogeneity and molecular cancer subtypes both in the primary tumor tissue and PDOs. In addition, the genetic heterogeneity of the primary tumor was mostly preserved during the establishment and long culture times of organoids in ex vivo. The histopathological assessment of samples suggested a very high similarity in terms of the phenotypic heterogeneity between PDOs and the parental tumor. In this important study, PDOs were treated with 58 chemotherapeutic agents, and those with TP53 loss of function mutation were resistant to MDM2 inhibitors and as a consequence acquired RAS mutations and therefore decreased sensitivity to an EGFR inhibitor. Importantly, in this study, colon tumor organoids carrying the RNF43 mutation were dramatically sensitive to Wnt inhibitors.

In another significant study, PDOs were examined for the first time to investigate whether PDOs as a preclinical model could predict the drug response seen in the gastrointestinal cancer patients in the clinic [91]. In this study, a living organoid biobank was established from metastatic gastrointestinal cancer patients who were previously recruited for phase I or II clinical trials. According to the phenotypic and genotypic profiling of organoid and patient tumor samples, both of them exhibited highly similar profiles. Then, this led the authors to assess drug responses of PDOs in the laboratory setting in parallel to the clinic. High-throughput drug screening of PDOs with Food and Drug Administration (FDA)-approved drugs was shown to be successful with a positive predictive value (predicting that a certain drug worked) of 88% and a negative predictive value (predicting that a certain drug did not work) of 100%. This suggests a promising forecasting potential for PDOs in terms of the treatment response.

6. Conclusions

Extensive (epi)genetic heterogeneity in cancer has been demonstrated in several studies. As a result of the aberrantly activated and sustained complex signaling networks both in cancer cells and neighboring tumor microenvironment, examples of the hallmarks of cancer were presented. To address genomic aberrations and signaling network complexity, there has been a growing need to develop more sophisticated approaches for cancer systems biology. Cancer systems biology can deliver solutions for the better understanding of intratumor heterogeneity and therapeutic opportunities. Specifically, improved cancer systems biology approaches integrated not only with multi-omics data from tumors but also with comprehensive patient-derived experimental model systems can guide clinicians for their decision-making to offer better therapeutic solutions with an ultimate aim to overcome treatment failure in cancer.

Author Contributions: Conceptualization, all authors; writing—original draft preparation, all authors; writing—review and editing, all authors. All authors have read and agreed to the published version the manuscript.

Funding: This research was funded by The Scientific and Technological Research Council of Turkey (TUBITAK) grant number 118C197. And the APC was funded by the same grant.

Acknowledgments: Ahmet Acar's laboratory is currently supported by the International Fellowship for Outstanding Researchers Program administered by The Scientific and Technological Research Council of Turkey (TUBITAK). We are grateful to Daniel Nichol for critical reading of the manuscript.

Conflicts of Interest: The authors declare no conflict of interest.

References

1. Hanahan, D.; Weinberg, R.A. Hallmarks of cancer: The next generation. *Cell* **2011**, *144*, 646–674. [CrossRef] [PubMed]
2. Pietras, K.; Östman, A. Hallmarks of cancer: Interactions with the tumor stroma. *Exp. Cell Res.* **2010**, *316*, 1324–1331. [CrossRef]
3. Han, X.; Zhou, Z.; Fei, L.; Sun, H.; Wang, R.; Chen, Y.; Chen, H.; Wang, J.; Tang, H.; Ge, W.; et al. Construction of a human cell landscape at single-cell level. *Nature* **2020**, *581*, 303–309. [CrossRef]
4. Campbell, P.J.; Getz, G.; Korbel, J.O.; Stuart, J.M.; Jennings, J.L.; Stein, L.D.; Perry, M.D.; Nahal-Bose, H.K.; Ouellette, B.F.F.; Li, C.H.; et al. Pan-cancer analysis of whole genomes. *Nature* **2020**, *578*, 82–93. [CrossRef]
5. Zehir, A.; Benayed, R.; Shah, R.H.; Syed, A.; Middha, S.; Kim, H.R.; Srinivasan, P.; Gao, J.; Chakravarty, D.; Devlin, S.M.; et al. Mutational landscape of metastatic cancer revealed from prospective clinical sequencing of 10,000 patients. *Nat. Med.* **2017**, *23*, 703–713. [CrossRef] [PubMed]
6. Jamal-Hanjani, M.; Wilson, G.A.; McGranahan, N.; Birkbak, N.J.; Watkins, T.B.K.; Veeriah, S.; Shafi, S.; Johnson, D.H.; Mitter, R.; Rosenthal, R.; et al. Tracking the evolution of non–small-cell lung cancer. *N. Eng. J. Med.* **2017**, *376*, 2109–2121. [CrossRef]
7. Chakraborty, S.; Hosen, M.I.; Ahmed, M.; Shekhar, H.U. Onco-Multi-OMICS Approach: A New Frontier in Cancer Research. *BioMed Res. Int.* **2018**, *2018*. [CrossRef]
8. Filipp, F.V. Precision medicine driven by cancer systems biology. *Cancer Metastasis Rev.* **2017**, *36*, 91–108. [CrossRef]
9. McGranahan, N.; Swanton, C. Clonal heterogeneity and tumor evolution: Past, present, and the future. *Cell* **2017**, *168*, 613–628. [CrossRef]

10. Hinohara, K.; Polyak, K. Intratumoral Heterogeneity: More Than Just Mutations. *Trends Cell Biol.* **2019**, *29*, 569–579. [CrossRef]

11. Guo, M.; Peng, Y.; Gao, A.; Du, C.; Herman, J.G. Epigenetic heterogeneity in cancer. *Biomark. Res.* **2019**, *7*, 23. [CrossRef] [PubMed]

12. Turajlic, S.; Xu, H.; Litchfield, K.; Rowan, A.; Chambers, T.; Lopez, J.I.; Nicol, D.; O'Brien, T.; Larkin, J.; Horswell, S.; et al. Tracking cancer evolution reveals constrained routes to metastases: TRACERx renal. *Cell* **2018**, *173*, 581–594.e12. [CrossRef] [PubMed]

13. Acar, A.; Hidalgo-Sastre, A.; Leverentz, M.K.; Mills, C.G.; Woodcock, S.; Baron, M.; Collu, G.M.; Brennan, K. Inhibition of Wnt signalling by Notch via two distinct mechanisms. *bioRxiv* **2020**. [CrossRef]

14. Collu, G.M.; Hidalgo-Sastre, A.; Brennan, K. Wnt-Notch signalling crosstalk in development and disease. *Cell. Mol. Life Sci.* **2014**, *71*, 3553–3567. [CrossRef]

15. Stylianou, S.; Clarke, R.B.; Brennan, K. Aberrant activation of Notch signaling in human breast cancer. *Cancer Res.* **2006**, *66*, 1517–1525. [CrossRef]

16. Marusyk, A.; Janiszewska, M.; Polyak, K. Intratumor heterogeneity: The Rosetta Stone of therapy resistance. *Cancer Cell* **2020**, *37*, 471–484. [CrossRef] [PubMed]

17. Werner, H.M.J.; Mills, G.B.; Ram, P.T. Cancer systems biology: A peek into the future of patient care? *Nat. Rev. Clin. Oncol.* **2014**, *11*, 167–176. [CrossRef]

18. Turajlic, S.; Sottoriva, A.; Graham, T.; Swanton, C. Resolving genetic heterogeneity in cancer. *Nat. Rev. Genet.* **2019**, *20*, 406–416. [CrossRef]

19. Greaves, M.; Maley, C.C. Clonal evolution in cancer. *Nature* **2012**, *481*, 306–313. [CrossRef]

20. Caravagna, G.; Heide, T.; Williams, M.J.; Zapata, L.; Nichol, D.; Chkhaidze, K.; Cross, W.; Cresswell, G.D.; Werner, B.; Acar, A.; et al. Subclonal reconstruction of tumors by using machine learning and population genetics. *Nat. Genet.* **2020**, *52*, 898–907. [CrossRef]

21. Levy, S.E.; Myers, R.M. Advancements in next-generation sequencing. *Annu. Rev. Genom. Hum. Genet.* **2016**, *17*, 95–115. [CrossRef] [PubMed]

22. Antman, E.; Weiss, S.; Loscalzo, J. Systems pharmacology, pharmacogenetics, and clinical trial design in network medicine. *Wiley Interdiscip. Rev. Syst. Biol. Med.* **2012**, *4*, 367–383. [CrossRef] [PubMed]

23. Barry, P.; Vatsiou, A.; Spiteri, I.; Nichol, D.; Cresswell, G.D.; Acar, A.; Trahearn, N.; Hrebien, S.; Garcia-Murillas, I.; Chkhaidze, K.; et al. The spatiotemporal evolution of lymph node spread in early breast cancer. *Clin. Cancer Res.* **2018**, *24*, 4763–4770. [CrossRef] [PubMed]

24. Spiteri, I.; Caravagna, G.; Cresswell, G.D.; Vatsiou, A.; Nichol, D.; Acar, A.; Ermini, L.; Chkhaidze, K.; Werner, B.; Mair, R.; et al. Evolutionary dynamics of residual disease in human glioblastoma. *Ann. Oncol.* **2019**, *30*, 456–463. [CrossRef]

25. Cross, W.; Kovac, M.; Mustonen, V.; Temko, D.; Davis, H.; Baker, A.M.; Biswas, S.; Arnold, R.; Chegwidden, L.; Gatenbee, C.; et al. The evolutionary landscape of colorectal tumorigenesis. *Nat. Ecol. Evol.* **2018**, *2*, 1661–1672. [CrossRef]

26. Kelso, T.W.R.; Porter, D.K.; Amaral, M.L.; Shokhirev, M.N.; Benner, C.; Hargreaves, D.C. Chromatin accessibility underlies synthetic lethality of SWI/SNF subunits in ARID1A-mutant cancers. *eLife* **2017**, *6*. [CrossRef]

27. Kim, H.; Zheng, S.; Amini, S.S.; Virk, S.M.; Mikkelsen, T.; Brat, D.J.; Grimsby, J.; Sougnez, C.; Muller, F.; Hu, J.; et al. Whole-genome and multisector exome sequencing of primary and post-treatment glioblastoma reveals patterns of tumor evolution. *Genome Res.* **2015**, *25*, 316–327. [CrossRef]

28. Zhang, X.; Choi, P.S.; Francis, J.M.; Imielinski, M.; Watanabe, H.; Cherniack, A.D.; Meyerson, M. Identification of focally amplified lineage-specific super-enhancers in human epithelial cancers. *Nat. Genet.* **2016**, *48*. [CrossRef]

29. Calabrese, C.; Davidson, N.R.; Demircioğlu, D.; Fonseca, N.A.; He, Y.; Kahles, A.; van Lehmann, K.; Liu, F.; Shiraishi, Y.; Soulette, C.M.; et al. Genomic basis for RNA alterations in cancer. *Nature* **2020**, *578*. [CrossRef]

30. Reyna, M.A.; Haan, D.; Paczkowska, M.; Verbeke, L.P.C.; Vazquez, M.; Kahraman, A.; Pulido-Tamayo, S.; Barenboim, J.; Wadi, L.; Dhingra, P.; et al. Pathway and network analysis of more than 2500 whole cancer genomes. *Nat. Commun.* **2020**, *11*. [CrossRef]

31. Rheinbay, E.; Nielsen, M.M.; Abascal, F.; Wala, J.A.; Shapira, O.; Tiao, G.; Hornshøj, H.; Hess, J.M.; Juul, R.I.; Lin, Z.; et al. Analyses of non-coding somatic drivers in 2658 cancer whole genomes. *Nature* **2020**, *578*. [CrossRef] [PubMed]

32. Bhang, H.E.C.; Ruddy, D.A.; Radhakrishna, V.K.; Caushi, J.X.; Zhao, R.; Hims, M.M.; Singh, A.P.; Kao, I.; Rakiec, D.; Shaw, P.; et al. Studying clonal dynamics in response to cancer therapy using high-complexity barcoding. *Nat. Med.* **2015**, *21*, 440–448. [CrossRef] [PubMed]

33. Acar, A.; Nichol, D.; Fernandez-Mateos, J.; Cresswell, G.D.; Barozzi, I.; Hong, S.P.; Trahearn, N.; Spiteri, I.; Stubbs, M.; Burke, R.; et al. Exploiting evolutionary steering to induce collateral drug sensitivity in cancer. *Nat. Commun.* **2020**, *11*, 1923. [CrossRef]

34. Tang, F.; Barbacioru, C.; Wang, Y.; Nordman, E.; Lee, C.; Xu, N.; Wang, X.; Bodeau, J.; Tuch, B.B.; Siddiqui, A.; et al. mRNA-Seq whole-transcriptome analysis of a single cell. *Nat. Methods* **2009**, *6*, 377–382. [CrossRef] [PubMed]

35. Macosko, E.Z.; Basu, A.; Satija, R.; Nemesh, J.; Shekhar, K.; Goldman, M.; Tirosh, I.; Bialas, A.R.; Kamitaki, N.; Martersteck, E.M.; et al. Highly parallel genome-wide expression profiling of individual cells using nanoliter droplets. *Cell* **2015**, *161*, 1202–1214. [CrossRef]

36. Klein, A.M.; Mazutis, L.; Akartuna, I.; Tallapragada, N.; Veres, A.; Li, V.; Peshkin, L.; Weitz, D.A.; Kirschner, M.W. Droplet barcoding for single-cell transcriptomics applied to embryonic stem cells. *Cell* **2015**, *161*, 1187–1201. [CrossRef]

37. Baron, C.S.; van Oudenaarden, A. Unravelling cellular relationships during development and regeneration using genetic lineage tracing. *Nat. Rev. Mol. Cell Biol.* **2019**, *20*, 753–765. [CrossRef]

38. González-Silva, L.; Quevedo, L.; Varela, I. Tumor Functional Heterogeneity Unraveled by scRNA-seq Technologies. *Trends Cancer* **2020**, *6*, 13–19. [CrossRef]

39. Navin, N.; Kendall, J.; Troge, J.; Andrews, P.; Rodgers, L.; McIndoo, J.; Cook, K.; Stepansky, A.; Levy, D.; Esposito, D.; et al. Tumour evolution inferred by single-cell sequencing. *Nature* **2011**, *472*, 90–94. [CrossRef]

40. Kim, C.; Gao, R.; Sei, E.; Brandt, R.; Hartman, J.; Hatschek, T.; Crosetto, N.; Foukakis, T.; Navin, N.E. Chemoresistance evolution in triple-negative breast cancer delineated by single-cell sequencing. *Cell* **2018**, *173*, 879–893.e13. [CrossRef]

41. Laks, E.; McPherson, A.; Zahn, H.; Lai, D.; Steif, A.; Brimhall, J.; Biele, J.; Wang, B.; Masud, T.; Ting, J.; et al. Clonal decomposition and DNA replication states defined by scaled single-cell genome sequencing. *Cell* **2019**, *179*, 1207–1207.e22. [CrossRef] [PubMed]

42. Cusanovich, D.A.; Daza, R.; Adey, A.; Pliner, H.A.; Christiansen, L.; Gunderson, K.L.; Steemers, F.J.; Trapnell, C.; Shendure, J. Multiplex single-cell profiling of chromatin accessibility by combinatorial cellular indexing. *Science* **2015**, *348*, 910–914. [CrossRef] [PubMed]

43. Granja, J.M.; Klemm, S.; McGinnis, L.M.; Kathiria, A.S.; Mezger, A.; Corces, M.R.; Parks, B.; Gars, E.; Liedtke, M.; Zheng, G.X.Y.; et al. Single-cell multiomic analysis identifies regulatory programs in mixed-phenotype acute leukemia. *Nat. Biotechnol.* **2019**, *37*, 1458–1465. [CrossRef] [PubMed]

44. Macaulay, I.C.; Haerty, W.; Kumar, P.; Li, Y.I.; Hu, T.X.; Teng, M.J.; Goolam, M.; Saurat, N.; Coupland, P.; Shirley, L.M.; et al. G&T-seq: Parallel sequencing of single-cell genomes and transcriptomes. *Nat. Methods* **2015**, *12*, 519–522. [CrossRef] [PubMed]

45. Cao, J.; Cusanovich, D.A.; Ramani, V.; Aghamirzaie, D.; Pliner, H.A.; Hill, A.J.; Daza, R.M.; McFaline-Figueroa, J.L.; Packer, J.S.; Christiansen, L.; et al. Joint profiling of chromatin accessibility and gene expression in thousands of single cells. *Science* **2018**, *361*, 1380–1385. [CrossRef]

46. Cao, Q.; Zhou, M.; Wang, X.; Meyer, C.A.; Zhang, Y.; Chen, Z.; Li, C.; Liu, X.S. CaSNP: A database for interrogating copy number alterations of cancer genome from SNP array data. *Nucleic Acids Res.* **2011**, *39*, D968–D974. [CrossRef] [PubMed]

47. de Anda-Jáuregui, G.; Hernández-Lemus, E. Computational oncology in the multi-omics era: State of the Art. *Front. Oncol.* **2020**, *10*, 423.

48. Huret, J.L.; Ahmad, M.; Arsaban, M.; Bernheim, A.; Cigna, J.; Desangles, F.; Guignard, J.C.; Jacquemot-Perbal, M.C.; Labarussias, M.; Leberre, V.; et al. Atlas of genetics and cytogenetics in oncology and haematology in 2013. *Nucleic Acids Res.* **2013**, *41*, D920–D924. [CrossRef]

49. Gorohovski, A.; Tagore, S.; Palande, V.; Malka, A.; Raviv-Shay, D.; Frenkel-Morgenstern, M. ChiTaRS-3.1-the enhanced chimeric transcripts and RNA-seq database matched with protein-protein interactions. *Nucleic Acids Res.* **2017**, *45*, D790–D795. [CrossRef]

50. Halling-Brown, M.D.; Bulusu, K.C.; Patel, M.; Tym, J.E.; Al-Lazikani, B. canSAR: An integrated cancer public translational research and drug discovery resource. *Nucleic Acids Res.* **2012**, *40*, D947–D956. [CrossRef]

51. Su, W.H.; Chao, C.C.; Yeh, S.H.; Chen, D.S.; Chen, P.J.; Jou, Y.S. OncoDB.HCC: An integrated oncogenomic database of hepatocellular carcinoma revealed aberrant cancer target genes and loci. *Nucleic Acids Res.* **2007**, *35*. [CrossRef] [PubMed]

52. Forbes, S.A.; Beare, D.; Boutselakis, H.; Bamford, S.; Bindal, N.; Tate, J.; Cole, C.G.; Ward, S.; Dawson, E.; Ponting, L.; et al. COSMIC: Somatic cancer genetics at high-resolution. *Nucleic Acids Res.* **2017**, *45*, D777–D783. [CrossRef]

53. Samur, M.K.; Yan, Z.; Wang, X.; Cao, Q.; Munshi, N.C.; Li, C.; Shah, P.K. canEvolve: A web portal for integrative oncogenomics. *PLoS ONE* **2013**, *8*, e56228. [CrossRef] [PubMed]

54. Tyagi, A.; Tuknait, A.; Anand, P.; Gupta, S.; Sharma, M.; Mathur, D.; Joshi, A.; Singh, S.; Gautam, A.; Raghava, G.P.S. CancerPPD: A database of anticancer peptides and proteins. *Nucleic Acids Res.* **2015**, *43*, D837–D843. [CrossRef]

55. Cutts, R.J.; Gadaleta, E.; Hahn, S.A.; Crnogorac-Jurcevic, T.; Lemoine, N.R.; Chelala, C. The pancreatic expression database: 2011 update. *Nucleic Acids Res.* **2011**, *39*, D1023–D1028. [CrossRef]

56. Hudson, T.J.; Anderson, W.; Aretz, A.; Barker, A.D.; Bell, C.; Bernabé, R.R.; Bhan, M.K.; Calvo, F.; Eerola, I.; Gerhard, D.S.; et al. International network of cancer genome projects. *Nature* **2010**, *464*, 993–998.

57. He, X.; Chang, S.; Zhang, J.; Zhao, Q.; Xiang, H.; Kusonmano, K.; Yang, L.; Sun, Z.S.; Yang, H.; Wang, J. MethyCancer: The database of human DNA methylation and cancer. *Nucleic Acids Res.* **2008**, *36*, D836–D841. [CrossRef] [PubMed]

58. Whiteaker, J.R.; Halusa, G.N.; Hoofnagle, A.N.; Sharma, V.; MacLean, B.; Yan, P.; Wrobel, J.A.; Kennedy, J.; Mani, D.R.; Zimmerman, L.J.; et al. CPTAC Assay Portal: A repository of targeted proteomic assays. *Nat. Methods* **2014**, *11*, 703–704. [CrossRef]

59. Perez-Llamas, C.; Gundem, G.; Lopez-Bigas, N. Integrative Cancer Genomics (IntOGen) in Biomart. *Database* **2011**, *2011*. [CrossRef]

60. Parkinson, H.; Kapushesky, M.; Shojatalab, M.; Abeygunawardena, N.; Coulson, R.; Farne, A.; Holloway, E.; Kolesnykov, N.; Lilja, P.; Lukk, M.; et al. ArrayExpress—A public database of microarray experiments and gene expression profiles. *Nucleic Acids Res.* **2007**, *35*. [CrossRef]

61. Liu, S.H.; Shen, P.C.; Chen, C.Y.; Hsu, A.N.; Cho, Y.C.; Lai, Y.L.; Chen, F.H.; Li, C.Y.; Wang, S.C.; Chen, M.; et al. DriverDBv3: A multi-omics database for cancer driver gene research. *Nucleic Acids Res.* **2020**, *48*. [CrossRef]

62. Thomas, J.K.; Kim, M.S.; Balakrishnan, L.; Nanjappa, V.; Raju, R.; Marimuthu, A.; Radhakrishnan, A.; Muthusamy, B.; Khan, A.A.; Sakamuri, S.; et al. Pancreatic Cancer Database: An integrative resource for pancreatic cancer. *Cancer Biol. Ther.* **2014**, *15*. [CrossRef]

63. Kumar, R.; Chaudhary, K.; Gupta, S.; Singh, H.; Kumar, S.; Gautam, A.; Kapoor, P.; Raghava, G.P.S. CancerDR: Cancer drug resistance database. *Sci. Rep.* **2013**, *3*. [CrossRef] [PubMed]

64. Pires, D.E.V.; Blundell, T.L.; Ascher, D.B. Platinum: A database of experimentally measured effects of mutations on structurally defined protein-ligand complexes. *Nucleic Acids Res.* **2015**, *43*. [CrossRef]

65. Jemal, A.; Siegel, R.; Ward, E.; Hao, Y.; Xu, J.; Thun, M.J. Cancer Statistics, 2009. *CA Cancer J. Clin.* **2009**, *59*. [CrossRef]

66. Hidalgo, M.; Amant, F.; Biankin, A.V.; Budinská, E.; Byrne, A.T.; Caldas, C.; Clarke, R.B.; de Jong, S.; Jonkers, J.; Mælandsmo, G.M.; et al. Patient-derived Xenograft models: An emerging platform for translational cancer research. *Cancer Discov.* **2014**, *4*. [CrossRef] [PubMed]

67. Sachs, N.; Clevers, H. Organoid cultures for the analysis of cancer phenotypes. *Curr. Opin. Genet. Dev.* **2014**, *24*, 68–73. [CrossRef]

68. Ben-David, U.; Beroukhim, R.; Golub, T.R. Genomic evolution of cancer models: Perils and opportunities. *Nat. Rev. Cancer* **2019**, *19*, 97–109. [CrossRef]

69. Dugger, S.A.; Platt, A.; Goldstein, D.B. Drug development in the era of precision medicine. *Nat. Rev. Drug Discov.* **2018**, *17*, 183–196. [CrossRef] [PubMed]

70. Dhandapani, M.; Goldman, A. Preclinical Cancer Models and Biomarkers for Drug Development: New Technologies and Emerging Tools. *J. Mol. Biomark. Diagn.* **2017**, *8*. [CrossRef] [PubMed]

71. Gey, G.O.; Coffmann, W.D.; Kubicek, M.T. Tissue culture studies of the proliferative capacity of cervical carcinoma and normal epithelium. *Cancer Res.* **1952**, *12*, 264–265.

72. Masters, J.R.W. Human cancer cell lines: Fact and fantasy. *Nat. Rev. Mol. Cell Biol.* **2000**, *1*, 233–236. [CrossRef]

73. Kapałczyńska, M.; Kolenda, T.; Przybyła, W.; Zajączkowska, M.; Teresiak, A.; Filas, V.; Ibbs, M.; Bliźniak, R.; Łuczewski, Ł.; Lamperska, K. 2D and 3D cell cultures—A comparison of different types of cancer cell cultures. *Arch. Med. Sci.* **2018**, *14*. [CrossRef] [PubMed]

74. Pampaloni, F.; Reynaud, E.G.; Stelzer, E.H.K. The third dimension bridges the gap between cell culture and live tissue. *Nat. Rev. Mol. Cell Biol.* **2007**, *8*, 839–845. [CrossRef] [PubMed]

75. Hamburger, A.W.; Salmon, S.E. Primary bioassay of human tumor stem cells. *Science* **1977**, *197*. [CrossRef] [PubMed]

76. Fukuda, J.; Nakazawa, K. Orderly arrangement of hepatocyte spheroids on a microfabricated chip. *Tissue Eng.* **2005**, *11*, 1254–1262. [CrossRef]

77. Desroches, B.R.; Zhang, P.; Choi, B.R.; King, M.E.; Maldonado, A.E.; Li, W.; Rago, A.; Liu, G.; Nath, N.; Hartmann, K.M.; et al. Functional scaffold-free 3-D cardiac microtissues: A novel model for the investigation of heart cells. *Am. J. Physiol. Heart Circ. Physiol.* **2012**, *302*. [CrossRef]

78. Achilli, T.M.; Meyer, J.; Morgan, J.R. Advances in the formation, use and understanding of multi-cellular spheroids. *Exp. Opin. Biol. Ther.* **2012**, *12*, 1347–1360. [CrossRef]

79. Lee, J.; Cuddihy, M.J.; Kotov, N.A. Three-dimensional cell culture matrices: State of the art. *Tissue Eng. Part B Rev.* **2008**, *14*, 61–68. [CrossRef] [PubMed]

80. Schilsky, R.L. Personalized medicine in oncology: The future is now. *Nat. Rev. Drug Discov.* **2010**, *9*, 363–366. [CrossRef]

81. Karlsson, H.; Fryknäs, M.; Larsson, R.; Nygren, P. Loss of cancer drug activity in colon cancer HCT-116 cells during spheroid formation in a new 3-D spheroid cell culture system. *Exp. Cell Res.* **2012**, *318*. [CrossRef]

82. Lai, Y.; Wei, X.; Lin, S.; Qin, L.; Cheng, L.; Li, P. Current status and perspectives of patient-derived xenograft models in cancer research. *J. Hematol. Oncol.* **2017**, *10*, 106. [CrossRef]

83. Chdiwa, T.; Kawai, K.; Noguchi, A.; Sato, H.; Hayashi, A.; Cho, H.; Shiozawa, M.; Kishida, T.; Morinaga, S.; Yokose, T.; et al. Establishment of patient-derived cancer xenografts in immunodeficient NOG mice. *Int. J. Oncol.* **2015**, *47*. [CrossRef]

84. Jhan, J.R.; Andrechek, E.R. Effective personalized therapy for breast cancer based on predictions of cell signaling pathway activation from gene expression analysis. *Oncogene* **2017**, *36*. [CrossRef] [PubMed]

85. Byrne, A.T.; Alférez, D.G.; Amant, F.; Annibali, D.; Arribas, J.; Biankin, A.V.; Bruna, A.; Budinská, E.; Caldas, C.; Chang, D.K.; et al. Interrogating open issues in cancer precision medicine with patient-derived xenografts. *Nat. Rev. Cancer* **2017**, *17*, 254–268. [CrossRef]

86. Lupo, B.; Sassi, F.; Pinnelli, M.; Galimi, F.; Zanella, E.R.; Vurchio, V.; Migliardi, G.; Gagliardi, P.A.; Puliafito, A.; Manganaro, D.; et al. Colorectal cancer residual disease at maximal response to EGFR blockade displays a druggable Paneth cell–like phenotype. *Sci. Transl. Med.* **2020**, *12*, eaax8313. [CrossRef]

87. Bertotti, A.; Papp, E.; Jones, S.; Adleff, V.; Anagnostou, V.; Lupo, B.; Sausen, M.; Phallen, J.; Hruban, C.A.; Tokheim, C.; et al. The genomic landscape of response to EGFR blockade in colorectal cancer. *Nature* **2015**, *526*. [CrossRef]

88. Bertotti, A.; Migliardi, G.; Galimi, F.; Sassi, F.; Torti, D.; Isella, C.; Corà, D.; di Nicolantonio, F.; Buscarino, M.; Petti, C.; et al. A molecularly annotated platform of patient- derived xenografts ("xenopatients") identifies HER2 as an effective therapeutic target in cetuximab-resistant colorectal cancer. *Cancer Discov.* **2011**, *1*. [CrossRef] [PubMed]

89. Lazzari, L.; Corti, G.; Picco, G.; Isella, C.; Montone, M.; Arcela, P.; Durinikova, E.; Zanella, E.R.; Novara, L.; Barbosa, F.; et al. Patient-derived xenografts and matched cell lines identify pharmacogenomic vulnerabilities in colorectal cancer. *Clin. Cancer Res.* **2019**, *25*. [CrossRef]

90. Yang, H.; Sun, L.; Liu, M.; Mao, Y. Patient-derived organoids: A promising model for personalized cancer treatment. *Gastroenterol. Rep.* **2018**, *6*, 243–245. [CrossRef] [PubMed]

91. Vlachogiannis, G.; Hedayat, S.; Vatsiou, A.; Jamin, Y.; Fernández-Mateos, J.; Khan, K.; Lampis, A.; Eason, K.; Huntingford, I.; Burke, R.; et al. Patient-derived organoids model treatment response of metastatic gastrointestinal cancers. *Science* **2018**, *359*. [CrossRef] [PubMed]

92. Weeber, F.; van de Wetering, M.; Hoogstraat, M.; Dijkstra, K.K.; Krijgsman, O.; Kuilman, T.; Gadellaa-Van Hooijdonk, C.G.M.; van der Velden, D.L.; Peeper, D.S.; Cuppen, E.P.J.G.; et al. Preserved genetic diversity in organoids cultured from biopsies of human colorectal cancer metastases. *Proc. Natl. Acad. Sci. USA* **2015**, *112*. [CrossRef] [PubMed]

93. van de Wetering, M.; Francies, H.E.; Francis, J.M.; Bounova, G.; Iorio, F.; Pronk, A.; van Houdt, W.; van Gorp, J.; Taylor-Weiner, A.; Kester, L.; et al. Prospective derivation of a living organoid biobank of colorectal cancer patients. *Cell* **2015**, *161*. [CrossRef] [PubMed]

94. Boj, S.F.; Hwang, C.I.; Baker, L.A.; Chio, I.I.C.; Engle, D.D.; Corbo, V.; Jager, M.; Ponz-Sarvise, M.; Tiriac, H.; Spector, M.S.; et al. Organoid models of human and mouse ductal pancreatic cancer. *Cell* **2015**, *160*. [CrossRef]

95. Gao, D.; Vela, I.; Sboner, A.; Iaquinta, P.J.; Karthaus, W.R.; Gopalan, A.; Dowling, C.; Wanjala, J.N.; Undvall, E.A.; Arora, V.K.; et al. Organoid cultures derived from patients with advanced prostate cancer. *Cell* **2014**, *159*. [CrossRef]

96. Lee, S.H.; Hu, W.; Matulay, J.T.; Silva, M.V.; Owczarek, T.B.; Kim, K.; Chua, C.W.; Barlow, L.M.J.; Kandoth, C.; Williams, A.B.; et al. Tumor evolution and drug response in patient-derived organoid models of bladder cancer. *Cell* **2018**, *173*. [CrossRef]

97. Sachs, N.; de Ligt, J.; Kopper, O.; Gogola, E.; Bounova, G.; Weeber, F.; Balgobind, A.V.; Wind, K.; Gracanin, A.; Begthel, H.; et al. A Living Biobank of Breast Cancer Organoids Captures Disease Heterogeneity. *Cell* **2018**, *172*. [CrossRef]

98. Hubert, C.G.; Rivera, M.; Spangler, L.C.; Wu, Q.; Mack, S.C.; Prager, B.C.; Couce, M.; McLendon, R.E.; Sloan, A.E.; Rich, J.N. A three-dimensional organoid culture system derived from human glioblastomas recapitulates the hypoxic gradients and cancer stem cell heterogeneity of tumors found in vivo. *Cancer Res.* **2016**, *76*. [CrossRef]

99. Nelson, L.; Tighe, A.; Golder, A.; Littler, S.; Bakker, B.; Moralli, D.; Murtuza Baker, S.; Donaldson, I.J.; Spierings, D.C.J.; Wardenaar, R.; et al. A living biobank of ovarian cancer ex vivo models reveals profound mitotic heterogeneity. *Nat. Commun.* **2020**, *11*. [CrossRef]

100. Aboulkheyr Es, H.; Montazeri, L.; Aref, A.R.; Vosough, M.; Baharvand, H. Personalized Cancer Medicine: An Organoid Approach. *Trends Biotechnol.* **2018**, *36*, 358–371. [CrossRef]

Publisher's Note: MDPI stays neutral with regard to jurisdictional claims in published maps and institutional affiliations.

Article

Circulating miR-1246 Targeting UBE2C, TNNI3, TRAIP, UCHL1 Genes and Key Pathways as a Potential Biomarker for Lung Adenocarcinoma: Integrated Biological Network Analysis

Siyuan Huang [1], Yong-Kai Wei [2], Satyavani Kaliamurthi [3,4], Yanghui Cao [5], Asma Sindhoo Nangraj [6], Xin Sui [1], Dan Chu [7], Huan Wang [7], Dong-Qing Wei [3,4,6], Gilles H. Peslherbe [3], Gurudeeban Selvaraj [3,4] and Jiang Shi [7,*]

[1] Department of Oncology, The First Affiliated Hospital of Zhengzhou University, No.1 Jianshe East Road, Zhengzhou 450052, China; huangsy1989822@163.com (S.H.); suixin0319@163.com (X.S.)

[2] College of Science, Henan University of Technology, Zhengzhou 450001, China; ykwei@haut.edu.cn

[3] Centre for Research in Molecular Modeling and Department of Chemistry and Biochemistry, Concordia University, 7141 Sherbrooke Street West, Montréal, QC H4B 1R6, Canada; satyavani.mkk@haut.edu.cn (S.K.); dqwei@sjtu.edu.cn (D.-Q.W.); Gilles.Peslherbe@concordia.ca (G.H.P.); gurudeeb99@haut.edu.cn (G.S.)

[4] Center of Interdisciplinary Science-Computational Life Sciences, College of Biological Engineering, Henan University of Technology, No.100, Lianhua Street, Hi-Tech Development Zone, Zhengzhou 450001, China

[5] Department of General Surgery, Henan Tumor Hospital, No.127 Dongming Road, Zhengzhou 450008, China; caoyanghui1010@163.com

[6] The State Key Laboratory of Microbial Metabolism, College of Life Sciences and Biotechnology, Shanghai Jiao Tong University, Shanghai 200240, China; asma_sindhoo@sjtu.edu.cn

[7] Department of Respiratory, The First Affiliated Hospital of Zhengzhou University, No.1 Jianshe East Road, Zhengzhou 450052, China; jiaoluodeguangmang@163.com (D.C.); wanghuanyisheng@163.com (H.W.)

* Correspondence: fccshij@zzu.edu.cn; Tel.: +86-15824836717

Received: 20 August 2020; Accepted: 28 September 2020; Published: 11 October 2020

Abstract: Analysis of circulating miRNAs (cmiRNAs) before surgical operation (BSO) and after the surgical operation (ASO) has been informative for lung adenocarcinoma (LUAD) diagnosis, progression, and outcomes of treatment. Thus, we performed a biological network analysis to identify the potential target genes (PTGs) of the overexpressed cmiRNA signatures from LUAD samples that had undergone surgical therapy. Differential expression (DE) analysis of microarray datasets, including cmiRNAs (GSE137140) and cmRNAs (GSE69732), was conducted using the Limma package. cmiR-1246 was predicted as a significantly upregulated cmiRNA of LUAD samples BSO and ASO. Then, 9802 miR-1246 target genes (TGs) were predicted using 12 TG prediction platforms (MiRWalk, miRDB, and TargetScan). Briefly, 425 highly expressed overlapping miRNA-1246 TGs were observed between the prediction platform and the cmiRNA dataset. ClueGO predicted cell projection morphogenesis, chemosensory behavior, and glycosaminoglycan binding, and the PI3K–Akt signaling pathways were enriched metabolic interactions regulating miRNA-1245 overlapping TGs in LUAD. Using 425 overlapping miR-1246 TGs, a protein–protein interaction network was constructed. Then, 12 PTGs of three different Walktrap modules were identified; among them, ubiquitin-conjugating enzyme E2C (UBE2C), troponin T1(TNNT1), T-cell receptor alpha locus interacting protein (TRAIP), and ubiquitin c-terminal hydrolase L1(UCHL1) were positively correlated with miR-1246, and the high expression of these genes was associated with better overall survival of LUAD. We conclude that PTGs of cmiRNA-1246 and key pathways, namely, ubiquitin-mediated proteolysis, glycosaminoglycan binding, the DNA metabolic process, and the PI3K–Akt–mTOR signaling pathway, the neurotrophin and cardiomyopathy signaling pathway, and the MAPK signaling pathway provide new insights on a noninvasive prognostic biomarker for LUAD.

Keywords: lung adenocarcinoma; circulating miR-1246; glycosaminoglycan binding; prognosis; PI3K–Akt signaling pathways; TargetScan; UBE2C

---

1. Introduction

Resection-based therapy is a key player that increases the patient's survival in nonsmall-cell lung cancer (NSCLC). However, long-term survival remains below 50% in NSCLC patients as there is a frequent recurrence of disease development following surgery and treatment [1]. It may increase many concerns associated with a favorable therapeutic strategy. Analysis of circulating microRNAs (cmiRNAs) before surgical operation (BSO) and after the surgical operation (ASO) has provided significant information for NSCLC diagnosis, progression, and outcomes of treatment [2]. One of the epigenetic biomarkers, known as cmiRNA, serves as a potential source for diagnosing NSCLC and its subtypes [3–6]. There are three important advantages in using cmiRNAs as a biomarker for NSCLC; they are as follows: (1) diagnostic feasibility from body fluids, (2) elevated stability and protection from endogenous enzymes (RNAase), and (3) accumulation of pathologic information from various tumorous sites, which overcomes the difficulty of tumor heterogenicity [7]. One of the common subtypes in NSCLC is lung adenocarcinoma (LUAD), which accounts for ~40% of all lung cancers. Stable and essential biomarkers for early diagnosis of LUAD are still insufficient [8]. Notably, the cmiRNAs are more stable in serum samples [9], it facilitates the augmentation of miRNA as a promising blood-based diagnostic biomarker. Aberrant changes in the level of miRNA, correlated with tumor growth, results in metastasis, invasion, drug resistance, and progression in LUAD patients. The expression level of miRNA-33a plays a vital role in the progression of LUAD; it could be an ideal biomarker for the diagnosis and prognosis of LUAD patients who have received adjuvant chemotherapy [10]. Upregulated oncogenic miRNAs (miR-130b, miR-182-5p, miRNA-17, and miRNA-222) were reported to cause the development and progression of LUAD [11–13]; moreover, downregulated miRNAs (miR-486-5p, miR-101, miR-133a), also called tumor-suppressive miRNAs, were reported to repress the development of NSCLC [14–16]. miR-21 and miR-24 were significantly lower in ASO serum samples of lung carcinoma patients when compared to the samples of BSO patients. The findings depicted that both miRNAs (21 and 24) could be employed as biomarkers for the prediction of cancerous growth reappearance ASO [2]. Besides that, Asakura et al. [17] compared the diagnostic indexes of miR (17-3p, -1268b, and -6075) BSO and ASO of serum samples of LC ($n = 180$) patients. After surgery, the level of miRNAs was significantly reduced; it enhances their potential in the screening of resectable lung cancer, including adenocarcinoma.

Gene signatures are an essential condition for potential clinical practice in cancer. It has many important inferences that are used to reclassify the disease [18]. The identification of gene signatures from patients who have undergone surgical therapy provides new insights on the diagnosis and therapeutic implications of LUAD. Thus, we performed integrated biological network analysis to identify the potential target genes (PTGs) of the shared cmiRNA signatures BSO and ASO of LUAD samples. Microarray analysis of gene expression profiles is a standard and well-known method to identify key hub genes and pathways [19,20]. Initially, we collected the cmiRNA and cmRNA datasets from the Gene Expression Omnibus (GEO) database. Then, we performed differential expression (DE) using the Limma algorithm and identified miR-1246 as a potential upregulated gene in BSO and ASO samples of LUAD. miR-1246 TGs were predicted from 12 different TG prediction platforms. Then, the overlapping genes of miR-1246 TGs and DE-cmRNAs were used to construct a protein–protein interaction (PPI) network. After that, different modules were extracted from the PPI network using the Walktrap algorithm. Moreover, 12 potential target genes (PTGs) were predicted from the modules based on degree centrality measures, and their functional and pathway enrichment terms were determined. Furthermore, the PTGs were validated using the expression of miR-1246 and the PTG correlation analysis survival curve and immune–histochemical analysis. This study will provide new insights into the underlying molecular mechanism in LUAD, which might contribute to the clinical therapy of LUAD patients.

2. Materials and Methods

2.1. Data Collection

Microarray datasets GSE137140 and GSE69732 of cmiRNAs and cmRNAs were extracted from the GEO database [17,21]. The workflow of the study is shown in Figure 1. The cmiRNA study was performed using a 3D-Gene Human miRNA V21_1.0.0 platform, which included cmiRNA profiles of 3924 samples consisting of 1566 BSO, 180 ASO of cancer, and 2178 noncancer controls. The histological types include adenocarcinoma (1217), squamous carcinoma (221), adenosquamous carcinoma (18), and small cell carcinoma (23) and other (87) subtype samples. However, in this study, we included 1217 BSO and 180 ASO LUAD samples and 1774 noncancer controls. The cmRNA study was performed using the Illumina HumanHT-12 WG-DASL V4.0 R2 expression bead chip platform, which included cmRNA profiles of six samples of lung cancer and noncancer controls.

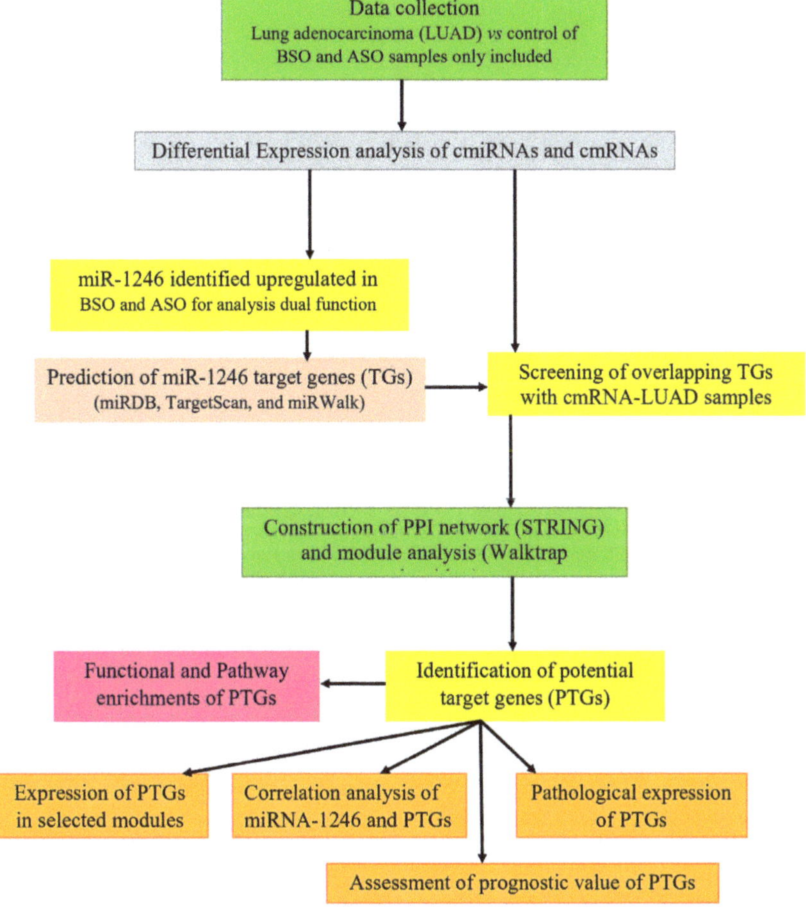

Figure 1. Schematic illustration of the study.

2.2. Differential Expression of cmiRNAs and cmRNAs

The R package "limma" (linear models for microarray analysis) of the Bioconductor project was used to retrieve, preprocess datasets, and perform differential expression of cmiRNAs [22]. Log2 transformation, Benjamini and Hochberg and *t*-test were used to perform normalization and calculate the false discovery rate (FDR; $p < 0.05$) of samples [23]. The total number of samples were divided into two groups, control versus BSO samples and control versus ASO samples, for cmiRNAs and cmRNAs as LUAD versus control. The analysis demonstrated that miR-1246 was upregulated in both samples, which was used for further analysis.

2.3. miR-1246 Target Gene Prediction

The web interface miRWalk 2.0 (http://zmf.umm.uni-heidelberg.de/apps/zmf/mirwalk2/) was employed to predict the target genes of miR-1246. The database contains comprehensive and experimentally verified information about miRNA-target gene interaction [24]. We extracted all the possible miR-1246 target genes from the database. It includes 11 other different miRNA-target prediction resources, namely, TargetScan, miRanda, miRDB, mirbridge, miRmap, miRNAMap, PITA, Pictar2, RNA22, and RNAhybrid. The target genes of miR-1246 were predicted from more than five databases and were used for further analysis.

2.4. Screening of Overlapping Target Genes

The overlapping miR-1246 target genes between the list of genes predicted from 12 different databases and DE-cmRNAs in LUAD samples were selected based on the standard log2FC >1. A Venny (https://bioinfogp.cnb.csic.es/tools/venny/index.html) [25] intersection diagram was used to facilitate more distinctively identified miR-1246 TGs from more than five databases and DE-cmRNAs in LUAD samples. These overlapping genes were employed to perform sequential bioinformatics analysis to discover the molecular mechanism of miR-1246 in LUAD.

2.5. Construction of PPI Network

The search tool for retrieval of interacting genes/proteins (STRING; https://string-db.org/) [26] is a database that is used to construct the PPI network. Currently, the database consists of 18,838 human proteins with a core of 25,914,693 network interactions. In this study, we constructed the PPI network from identified TGs using the STRING interactome. The highest confidence interaction score was set to 0.9, which reduces false-positive interactions [27].

2.6. Identification of Modules and Hub Genes

The R package "*igraph*" was used to extract modules based on the Walktrap algorithm from the PPI network. It runs several short random walks within a group of nodes that are highly connected to detect small modules. From the modules, the PTGs (nodes) were identified with two different centrality measures, "degree" and "betweenness" [28]. The degree of the gene is the number of maximum connections it has with the other genes. Genes with a high degree act as hubs within the network. The betweenness of a gene is the number of paths that pass through it when considering the pair-wise shortest paths between all genes in the network. A node that occurs between two dense clusters will have a high betweenness.

2.7. Functional Enrichment Analysis

We have used ClueGO v2.5.3, which is a Cytoscape v3.8.0 plugin for function and pathway enrichment analysis of PTGs [29,30]. A list of overlapping miR-1246 TGs or PTGs was provided as input into ClueGO with selected specific parameters, such as species (*Homo sapiens*), ID type (Entrez gene ID), and different enrichment functions (biological process or cellular component or molecular function or KEGG pathways), for the analysis. Each enrichment was calculated based on the Bonferroni method ($p < 0.005$) and a kappa score of 0.96. In ClueGO, the kappa score is employed to identify term–term interactions revealed as edges on the network and correlate terms and pathways into functional groups, depending on shared genes. The high kappa score indicates stronger network connectivity of PTGs to the GO terms.

$$k = p_o - p_e / 1 - p_e = 1 - (1 - p_o / 1 - p_e)$$

where p_o represents a relative observed agreement among raters, and p_e represents the hypothetical probability of chance agreement [31]. Functional enrichment analysis results were visualized using ImageGP (http://www.ehbio.com/ImageGP/index.php/Home/Reg/reg.html).

2.8. Validation of Potential Target Genes (PTGs)

2.8.1. Expression of PTGs in LUAD

The Gene Expression Profiling Interactive Analysis 2 (GEPIA2; http://gepia.cancer-pku.cn/) web interface was employed to validate the expression level of PTGs in LUAD [32]. GEPIA consists of comprehensive RNA sequencing information from TCGA and the Genotype–Tissue Expression (GTEx) project. The expression level of PTGs is illustrated in the box plot, and $p < 0.05$ was considered to be statistically important.

2.8.2. Correlation Analysis of miR-1246 and PTGs

The expression data of miR-1246 and PTGs in LUAD were obtained from the TCGA database. The expression data were transformed with log2 for normalization. Then, the relationship with miR-1246 and PTGs was elucidated using Spearman's correlation analysis. The linear regression plot was used as a visual representation of the trend of the relationships. It was performed with GraphPad Prism (USA).

2.8.3. Survival Analysis

The PTGs were identified from the Walktrap modules. The R package *"survival"* was employed to calculate the Kaplan–Meier (KM) survival plot with a hazard ratio (HR) and log-rank test of the hub, which was implemented in the KM plotter web interface [33]. The database retrieved the gene expression profiles, and clinical data include TNM (Stage I, II, III, and IV), gender (male and female), smoking history (smoker and nonsmoker), histology (adenocarcinoma and squamous cell carcinoma), and grade (G1, G2, G3, and GX) of 1925 patients from The Cancer Genome Atlas (TCGA), Cancer Biomedical Informatics Grid (caBIG), and GEO. We analyzed the overall survival rate of the PTGs as input and obtained the plot from the tool.

2.8.4. Protein Expression Analysis in LUAD

The Human Protein Atlas database (HPAD) was used to validate the immune-histochemistry of PTGs. The database facilitates system-level studies on the transcriptome of the coding genes and pathological expression of genes in different cancer types. The staining profiles for proteins of the PTGs in human LUAD tissue based on immunohistochemistry using tissue microarrays. Further, the name of the antibody, tissue type, staining levels (high, medium, low, and not detected), intensity, and quality of the IHC analysis data were retrieved from the database for interpreting results [34,35].

3. Results

3.1. Differentially Expressed cmiRNAs and cmRNAs

In total, 5132 DE-cmiRNAs, which included 2242 underexpressed and 324 overexpressed cmiRNAs, were obtained from BSO samples. Moreover, 1646 underexpressed and 920 overexpressed cmiRNAs were obtained from ASO samples. The top 10 overexpressed and underexpressed cmiRNAs are illustrated in Table 1. In both the samples, miR-1246 was highly upregulated, with log2FC as 7.09 in BSO samples and 6.28 in ASO samples. Therefore, further studies were carried out using miR-1246. There were 306 overexpressed cmRNAs, and 743 under-expressed cmRNAs identified from differential expression. The top 10 overexpressed and underexpressed cmRNAs are illustrated in Table 2.

Table 1. Top 10 overexpressed and underexpressed circulating miRNAs (cmiRNAs).

miRNA_ID	Log2FC	*p*-Value	miRNA_ID	Log2FC	*p*-Value
BSO overexpressed			**BSO underexpressed**		
hsa-miR-1246	6.28	2.79×10^{-110}	hsa-miR-373-5p	−5.92	0
hsa-miR-8060	5.69	6.62×10^{-189}	hsa-miR-1199-5p	−6.05	0
hsa-miR-920	5.46	0	hsa-miR-208b-5p	−6.07	0
hsa-miR-6131	5.31	9.32×10^{-187}	hsa-miR-6777-5p	−6.07	0
hsa-miR-4259	5.08	9.10×10^{-249}	hsa-miR-4648	−6.32	0
hsa-miR-6849-5p	4.61	2.22×10^{-172}	hsa-miR-4435	−6.38	0
hsa-miR-193a-5p	4.39	4.87×10^{-182}	hsa-miR-4276	−6.46	0
hsa-miR-6717-5p	4.24	2.02×10^{-226}	hsa-miR-6857-5p	−6.49	0
hsa-miR-3934-5p	4.11	2.63×10^{-128}	hsa-miR-92a-2-5p	−7.19	0
hsa-miR-1343-3p	3.96	0	hsa-miR-1203	−7.37	0
ASO overexpressed			**ASO underexpressed**		
hsa-miR-1246	7.09	0	hsa-miR-3184-5p	−8.41	0
hsa-miR-1290	6.17	0	hsa-miR-1203	−1.54	2.73×10^{-214}
hsa-miR-29b-1-5p	6.03	0	hsa-miR-4730	−1.60	0
hsa-miR-191-5p	5.75	0	hsa-miR-873-3p	−1.64	1.79×10^{-173}
hsa-miR-451a	5.64	0	hsa-miR-92a-2-5p	−1.74	0
hsa-miR-103a-3p	5.17	0	hsa-miR-4276	−1.89	2.65×10^{-242}
hsa-miR-4755-3p	5.09	0	hsa-miR-3184-5p	−2.01	0
hsa-miR-6131	4.99	0	hsa-miR-4648	−2.05	3.64×10^{-225}
hsa-miR-4771	4.96	0	hsa-miR-6857-5p	−2.36	4.82×10^{-302}
hsa-miR-4480	4.89	0	hsa-miR-4481	−2.55	1.76×10^{-312}

Table 2. Top 10 overexpressed and underexpressed cmRNAs.

Gene Symbol	Description	Log2FC	*p*-Value
Overexpressed genes			
BTBD11	BTB domain containing 11	3.108	4.69×10^{-4}
ZNF683	Zinc finger protein 683	1.991	6.82×10^{-3}
GPATCH4	G-patch domain containing 4	1.754	8.86×10^{-4}
EHMT1	Euchromatic histone lysine methyltransferase 1	1.652	3.61×10^{-3}
RAB6B	Ras-related protein Rab-6B	1.576	9.06×10^{-3}
C12orf5	TP53 induced glycolysis regulatory phosphatase	1.569	1.44×10^{-3}
GNLY	Granulysin	1.569	9.71×10^{-3}
RPGRIP1	X-linked retinitis pigmentosa GTPase regulator-interacting protein 1	1.542	3.44×10^{-4}
CPT1B	Carnitine palmitoyltransferase I	1.527	4.17×10^{-3}
SRI	Sorcin	1.525	1.38×10^{-3}
Underexpressed genes			
WISP3	WNT1-inducible-signaling pathway protein 3	−1.855	5.39×10^{-3}
HFE2	Hemojuvelin	−1.858	3.08×10^{-3}
LOR	Loricrin	−1.861	4.96×10^{-3}
SLC26A11	Sodium-independent sulfate anion transporter	−1.875	3.97×10^{-3}
DCAF12L2	DDB1- and CUL4-associated factor 12-like protein 2	−1.885	3.31×10^{-4}
DKFZp564N2472	POM121 transmembrane nucleoporin-like 12	−1.885	4.22×10^{-3}
FRG2C	FSHD region gene 2 family member C	−1.921	4.13×10^{-4}
PRM2	Protamine 2	−1.95	8.97×10^{-3}
PTCH2	Patched 2	−2.022	4.04×10^{-3}
NNAT	Neuronatin	−2.298	9.95×10^{-3}

3.2. Identification of Overlapping miR-1246 Target Genes

Briefly, 9802 miR-1246 TGs were predicted by 12 target gene prediction platforms. Additionally, the differential expression mRNA of LUAD predicted 1049 genes. The intersection of these gene sets using Venny demonstrated that 425 miR-1246 TGs were highly expressed in LUAD (Figure 2A).

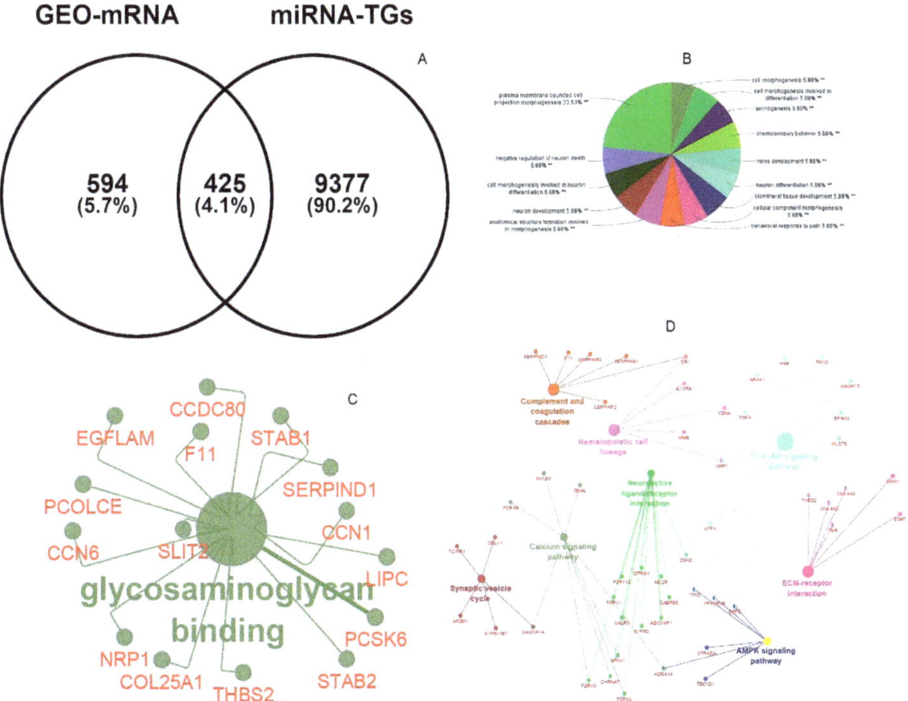

Figure 2. (**A**) Venny interactive diagram of overlapping miR-1246 genes and differential expression (DE)-cmRNAs. (**B**) Biological process of overlapping genes. (**C**) Molecular functions of overlapping genes. (**D**) KEGG pathway enrichment of overlapping genes.

3.3. Functional and Pathway Enrichment of Overlapping miR-1246 Target Genes

The functional and pathway enrichment terms have a great consequence in the regulatory mechanism of miR-1246 target genes. Plasma-membrane-bound cell projection morphogenesis, chemosensory behavior, and neuron development are important biological process terms (Figure 2B), and glycosaminoglycan binding (Figure 2C) is the major molecular functional term of miR-1245 overlapping target genes. KEGG pathways have demonstrated that nicotine addiction, neomycin, kanamycin and gentamicin biosynthesis, complement and coagulation cascades, ECM-receptor interaction, and PI3K–Akt signaling pathways are enriched pathway terms regulating miR-1245 overlapping target genes in LUAD (Figure 2D).

3.4. Modules and PTGs Identification

Using 425 overlapping miR-1246 TGs, the PPI network was constructed with 3133 nodes and 4228 interactions (Figure 3). Then, using the Walktrap algorithm, 21 modules with a minimum of three nodes were predicted from the PPI network. Among them, the top three highly interconnected modules, having more numbers of nodes, were selected for PTG analysis. Module 1 (33 nodes; $p = 1.98 \times 10^{-74}$); Module 2 (15 nodes; $p = 1.21 \times 10^{-08}$), and Module 3 (10 nodes; $p = 1.62 \times 10^{-14}$) were employed to identify the PTGs (Figure 4). Table 3 demonstrates the degree and betweenness centrality measures of 12 different PTGs, which include ubiquitin-conjugating enzyme E2C (UBE2C), tubulin folding cofactor E (TBCE), DnaJ heat shock protein family (Hsp40) member A3 (DNAJA3), paired like homeodomain 2 (PITX2), transforming growth factor-beta-induced factor 1 (TGIF1), T-cell receptor alpha locus interacting protein (TRAIP), ubiquitin c-terminal hydrolase L1 (UCHL1), troponin I3 (TNNI3), troponin T1 (TNNT1), neuroblastoma RAS (NRAS) viral oncogene, Rac family small GTPase 3 (RAC3), and the Ephrin-A4 (EFNA4) precursor.

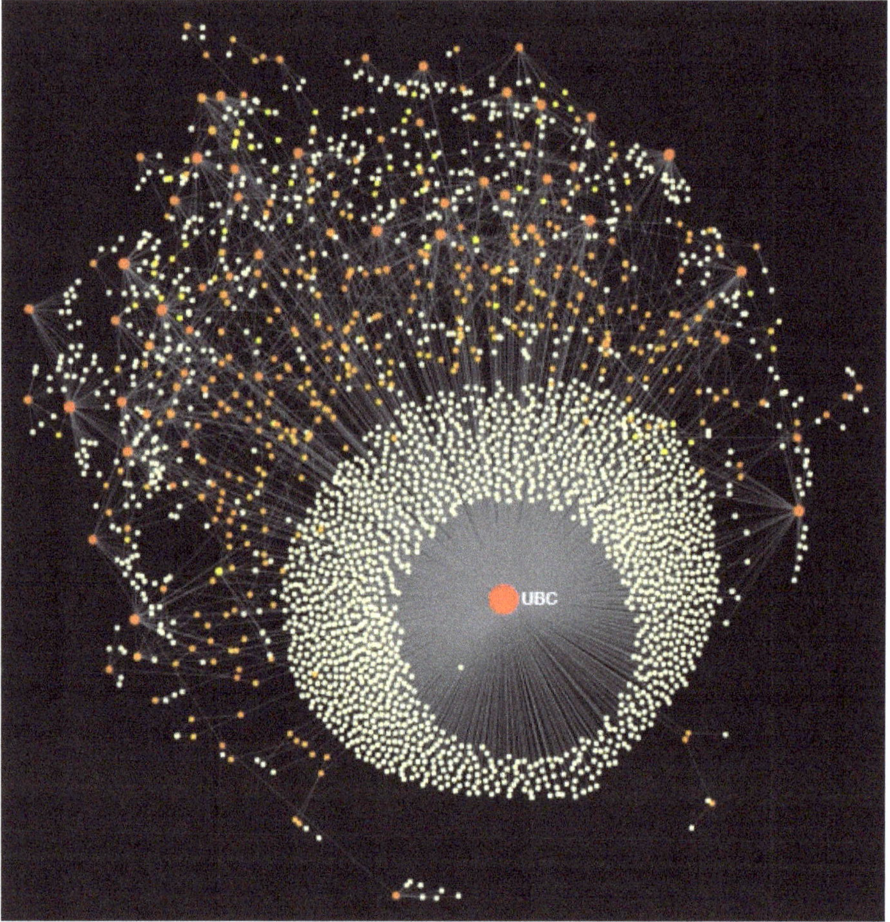

Figure 3. Protein–protein interaction (PPI) network of overlapping genes of miR-1246 targets and differentially expressed circulating mRNAs in force atlas layout (red color indicates downregulated genes, orange color indicates upregulated genes, and yellow color indicates interconnected genes). Change in the size of the nodes depends on degree centrality measures. UBC is the major node of the subnetwork that is enriched in ubiquitin-mediated proteolysis.

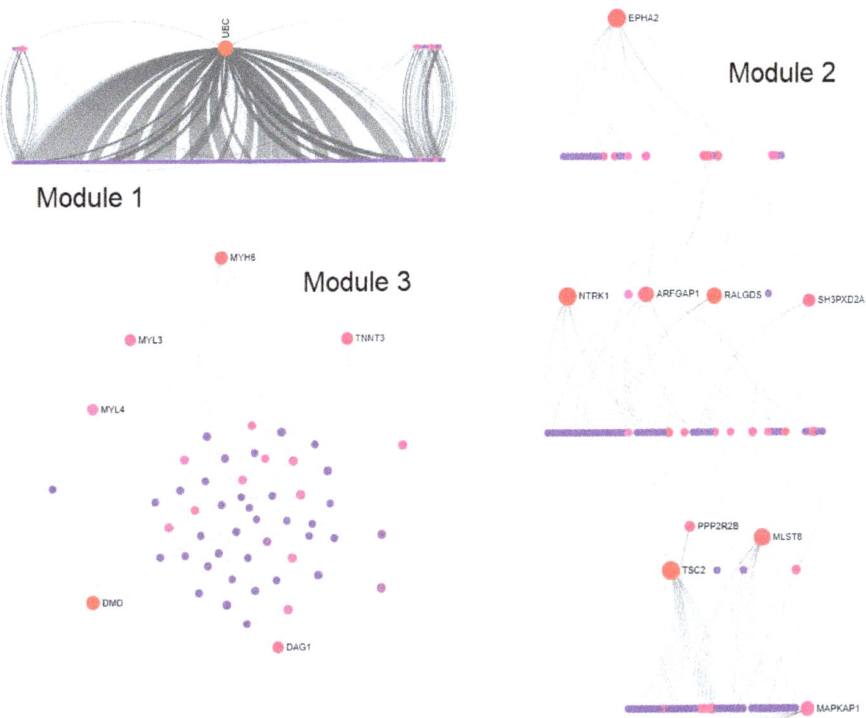

Figure 4. Walktrap modules of potential target genes extracted from the PPI network. Sugiyama layout of Module 1 (33 nodes; $p = 1.98 \times 10^{-74}$); linear bipartite/tripartite layout of Module 2 (15 nodes; $p = 1.21 \times 10^{-08}$); circular bipartite/tripartite layout of Module 3 (10 nodes; $p = 1.62 \times 10^{-14}$). Red color indicates downregulated, pink color indicates upregulated genes, and blue color indicates interconnected genes. Change in the size of the nodes depends on degree centrality measures.

Table 3. Degree and betweenness centrality measures of 12 different potential target genes (PTGs).

Official Symbol	Gene ID	Official Full Name	Chromosome Location	Exon Count	Degree	Betweenness
UBE2C	11,065	Ubiquitin conjugating enzyme E2 C	20q13.12	8	34	7811.25
TBCE	6905	Tubulin folding cofactor E	1q42.3	18	13	39.11
DNAJA3	9093	DNAJ heat shock protein family (Hsp40) member 3	16p13.3	12	12	6127.74
PITX2	5308	Paired-like homeodomain transcription factor 2	4q25	9	07	4584.14
TGIF1	7050	TGFB induced factor homeobox 1	18p11.31	12	07	22.32
TRAIP	10,293	TRAF interacting protein	3p21.31	16	06	1533.11
UCHL1	7345	Ubiquitin C-terminal hydrolase L1	4p13	9	06	1537.48
TNNI3	7137	Troponin I3	19q13.42	8	04	0.23
TNNT1	7138	Troponin T1	19q13.42	15	04	10.91
NRAS	4893	Neuroblastoma RAS viral oncogene homolog	1p13.2	7	03	247.07
RAC3	5881	Rac family small GTPase 3	17q25.3	6	03	630.68
EFNA4	1945	Ephrin A4	1q21.3	4	03	0

3.5. Function and Pathway Enrichments of PTGs

The following were enriched molecular function terms: In Module 1, the TNF signaling pathway, ubiquitin-mediated proteolysis, NOD-like receptor signaling pathways; in Module 2, regulation of MAP kinase activity, negative regulation of cysteine-type endopeptidase activity involved in the apoptotic processes, and transmembrane receptor kinase activity; in Module 3, channel inhibitor activity, calcium-dependent ATPase activity, and calmodulin-binding. The top three biological processes of the three different modules are microtubule cytoskeleton organization, the DNA metabolic process, regulation of the intrinsic apoptotic signaling pathway, activation of MAPK activity, positive regulation of endothelial cell proliferation, Rac protein signal transduction, positive regulation of cell–matrix adhesion, regulation of the force of heart contraction, and regulation of skeletal muscle contraction, respectively. The figure indicates the significant BP terms in the three modules (Figure 5A). PTGs of modules were enriched in many signaling pathways, including necroptosis, protein processing in the endoplasmic reticulum, ubiquitin-mediated proteolysis, Parkinson's disease, pathways in cancer, the MAPK signaling pathway, the mTOR signaling pathway, the Ras signaling pathway, the PI3K–Akt signaling pathway, Epstein–Barr virus infection, cardiac muscle contraction, adrenergic signaling in cardiomyocytes, hypertrophic cardiomyopathy, dilated cardiomyopathy, and the neurotrophin signaling pathway. Among the pathways, three pathways, namely, cancer, ubiquitin-mediated proteolysis, and Epstein–Barr virus infection, had a high number of gene counts (>75). Ten pathways had a moderate level of gene counts (>50), including the MAPK signaling pathway, the mTOR signaling pathway, the Ras signaling pathway, the PI3K–Akt signaling pathway, necroptosis, protein processing in the endoplasmic reticulum, and dilated cardiomyopathy; two pathways had fewer gene counts (>25) (Figure 5B).

3.6. Validation of PTGs

3.6.1. Expression of PTGs

There were 12 different PTGs identified from the modules, namely, UBE2C, TBCE, DNAJA3, PITX2, TGIF1, TRAIP, UCHL1, TNNI3, TNNT1, NRAS, RAC3, and EFNA4, and they demonstrated a high level of expression in LUAD tissues (Figure 6). As miR-1246 was upregulated in LUAD, the differentially expressed genes in LUAD have essential importance to act as potential target genes of miR-1246.

3.6.2. Spearman's Correlation Analysis of PTGs

Spearman's correlation analysis indicated that four of the 12 PTGs was significantly and positively correlated with miR-1246: UBE2C ($r = 0.32$, $p = 2.2 \times 10^{-08}$), TNNT1 ($r = 0.023$, $p = 0.07$), TRAIP ($r = 0.58$, $p = 8.7 \times 10^{-28}$), and UCHL1 ($r = 0.44$, $p = 6.5 \times 10^{-15}$) (Figure 7).

3.6.3. Prognostic Impact of PTGs

KM plots demonstrated the prognostic impact of the PTGs, which was identified from three different modules of the PPI network. The results explained that the high expression of UBE2C, UCHL1, TRAIP, TNNT1, TNNI3, and RAC3 were associated with poor overall survival of lung adenocarcinoma patients ($p < 0.05$; Figure 8). Moreover, the high expressions of PITX2, NRAS, ENFA4, DNAJA3, TBCE, and TGIF1 were correlated with longer overall survival of LUAD patients (Figure 9).

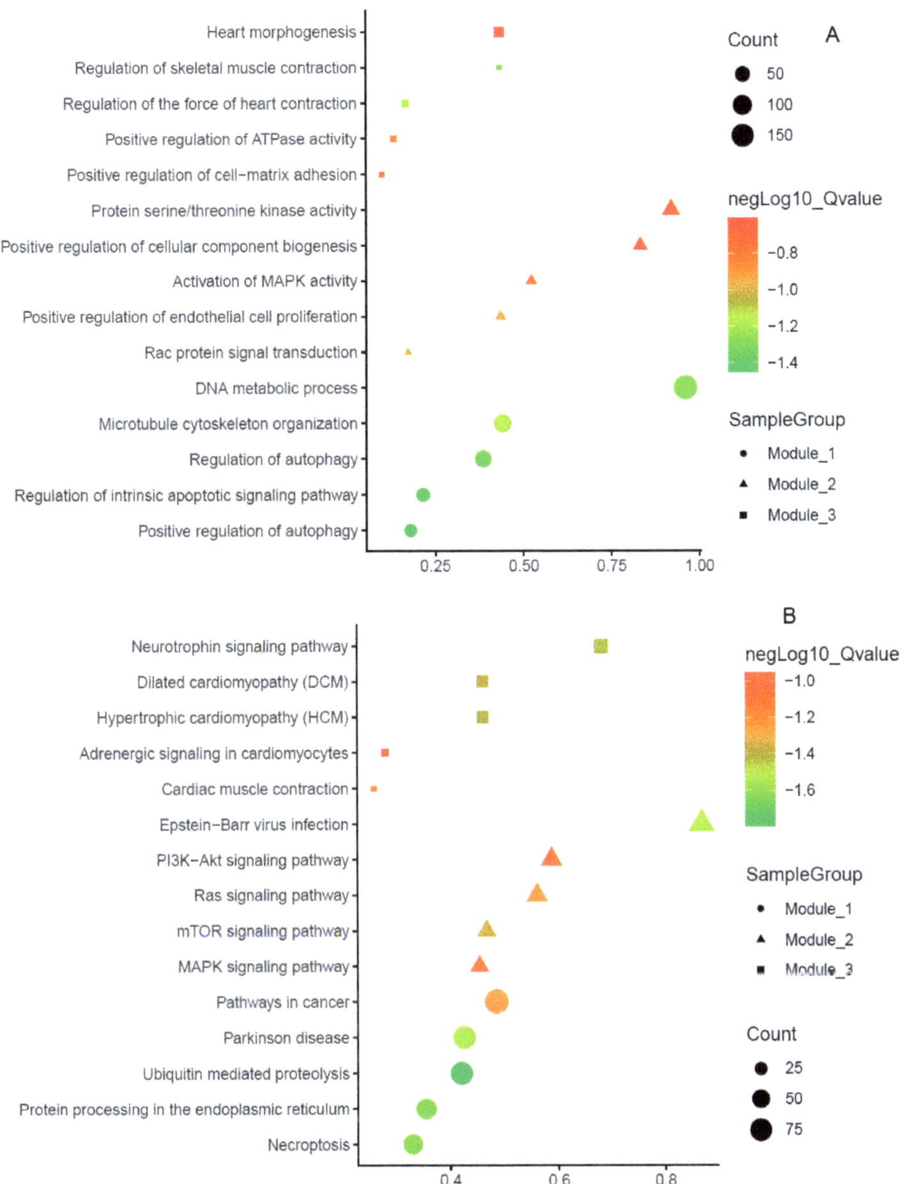

Figure 5. Functional enrichment terms of potential target genes. (**A**) Biological process; (**B**) KEGG pathways.

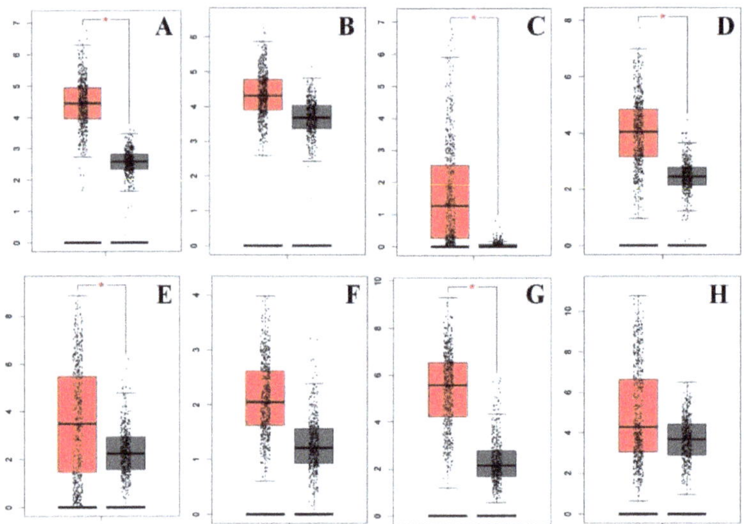

Figure 6. Expression of potential target genes in lung adenocarcinoma (LUAD) and control samples from the Gene Expression Profiling Interactive Analysis (GEPIA) database. Expression of the potential target genes was detected in 483 LUAD tissues (red in color) and 347 normal tissues (black in color). Eight potential target genes (**A**) EFNA4, (**B**) NRAS, (**C**) PITX2, (**D**) RAC3, (**E**) TNNT1, (**F**) TRAIP, (**G**) UBE2C, and (**H**) UCHL1 were upregulated in LUAD tissues compared to control.

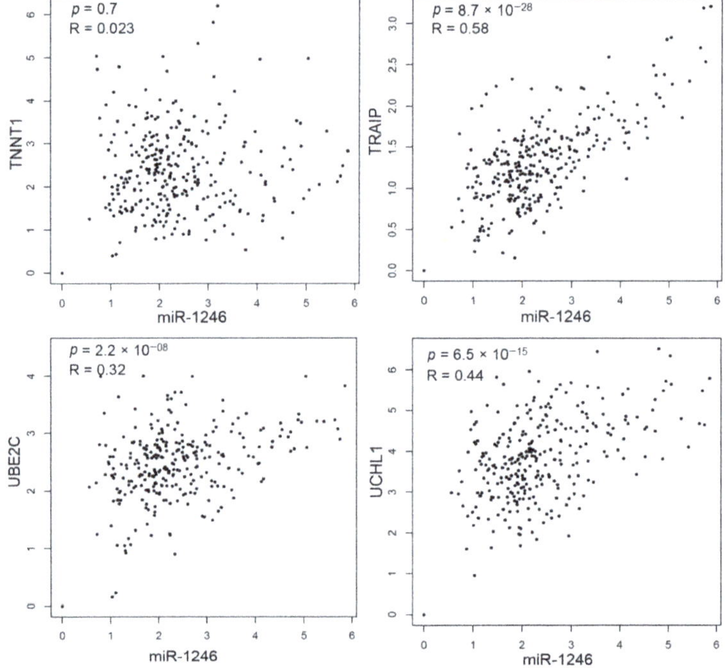

Figure 7. Correlation analysis between miR-1246 and 4 PTGs: UBE2C ($r = 0.32$, $p = 2.2 \times 10^{-08}$), TNNT1 ($r = 0.023$, $p = 0.07$), TRAIP ($r = 0.58$, $p = 8.7 \times 10^{-28}$), and UCHL1 ($r = 0.44$, $p = 6.5 \times 10^{-15}$).

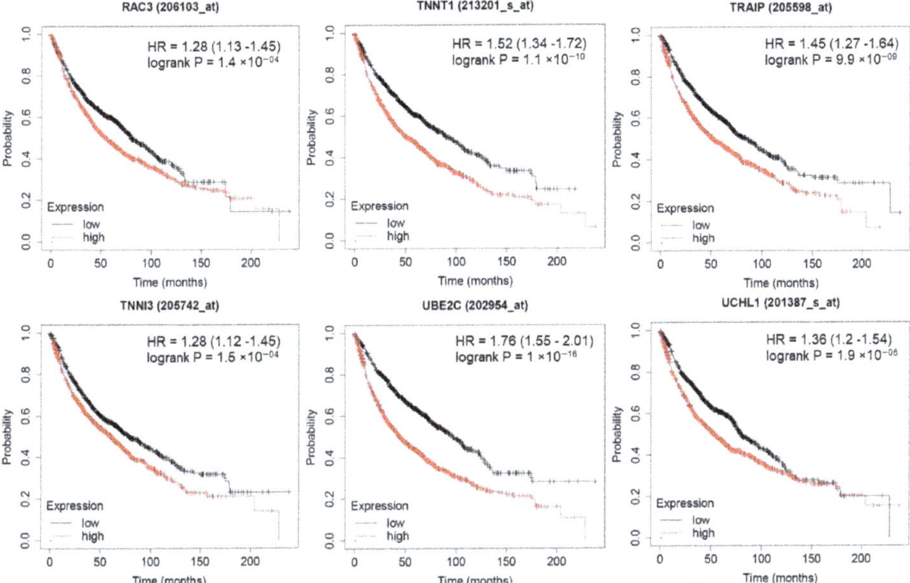

Figure 8. The prognostic value of PTGs in LUAD patients. High expression of UBE2C, UCHL1, TRAIP, TNNT1, TNNI3, and RAC3 was associated with poor overall survival of LUAD patients ($p < 0.05$).

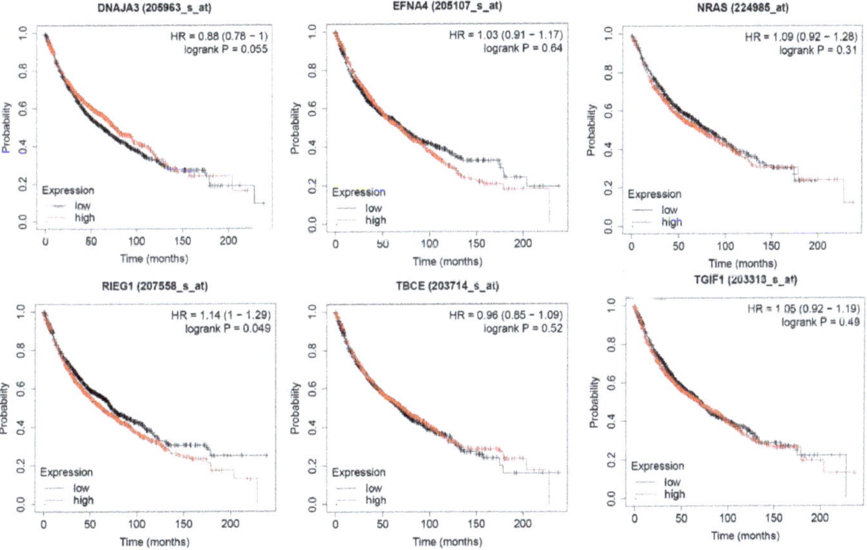

Figure 9. The prognostic value of PTGs in LUAD patients. High expression of PITX2, NRAS, ENFA4, DNAJA3, TBCE, and TGIF1 was correlated with longer overall survival of LUAD patients ($p < 0.05$).

3.6.4. Protein Expression of PTGs

The immune–histochemistry of pathological slides of the human protein atlas database (HPAD) indicated that the protein expressions of PTGs were drastically higher in LUAD tissues compared with adjacent normal tissues (Figure 10). The IHC data for UBE2C (https://www.proteinatlas.org/ENSG00000175063-UBE2C/pathology/lung+cancer#img), UCHL1 (https://

www.proteinatlas.org/ENSG00000154277-UCHL1/pathology/lung+cancer#img), TRAIP (https://www.proteinatlas.org/ENSG00000183763-TRAIP/pathology/lung+cancer#img), and RAC3 (https://www.proteinatlas.org/ENSG00000169750-RAC3/pathology/lung+cancer#img) have a strong intensity, and the intensity indicates that these PTGs played an initiative role and may be used as a biomarker. The other PTGs have weak or low intensity, which may be a reason for the availability of a limited number of samples in the database.

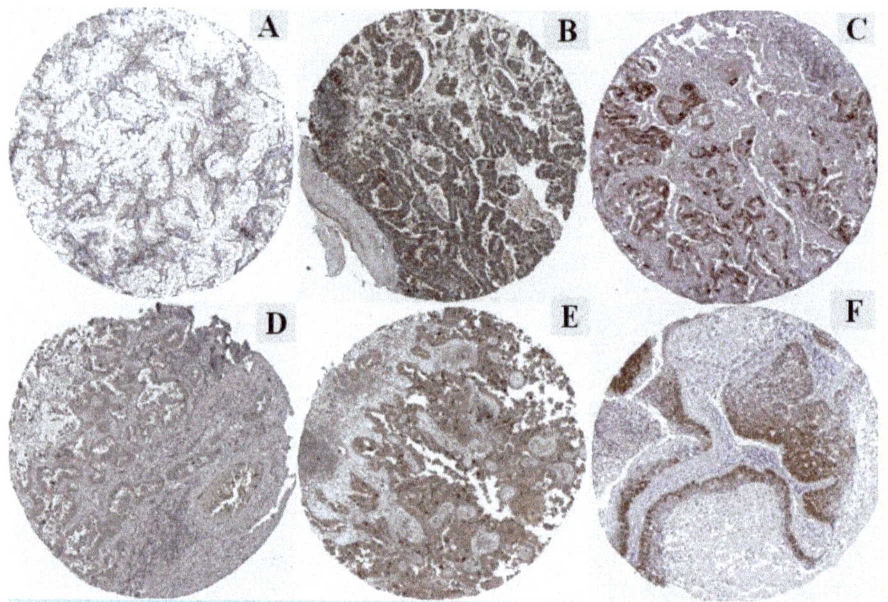

Figure 10. Immunohistochemistry of the PTGs based on the Human Protein Atlas database (HPAD). (**A**) Median staining of EFNA4 in LUAD (antibody: CAB021350; magnification of 4 × 10; substructures: cytoplasmic/membranous). (**B**) High staining of NRAS in LUAD (antibody: CAB010157; magnification of 4 × 10; substructures: cytoplasmic/membranous). (**C**) Medium staining of RAC3 in LUAD (antibody: HPA047820; magnification of 4 × 10; substructures: cytoplasmic/membranous). (**D**) Medium staining of TRAIP in LUAD (antibody: HPA036262; magnification of 4 × 10; substructures: cytoplasmic/membranous). (**E**) Medium staining of UBE2C in LUAD (antibody: CAB011464; magnification of 4 × 10; substructures: cytoplasmic/membranous. (**F**) Medium staining of UCHL1 in LUAD (antibody: CAB002580; magnification of 4 × 10; substructures: cytoplasmic/membranous). The staining intensity is strong and quantity: >75% to all the selected protein morphology.

4. Discussion

Continuous intricacy in earlier diagnosis is the main reason for the increased rate of LUAD individuals. Discovering potential and novel biomarkers and its interactive gene-level mechanism may lead to higher chances in the diagnosis and prognosis of LUAD [36]. Prognostic markers that include growth factor and hormone receptors, proliferation and angiogenesis markers, and proteases provide molecular characteristics and assist the course of therapy [37–39]. On this concern, the result of the study highlighted miR-1246 gene targets and key signaling pathways in LUAD.

MiR-1246 plays an imperative role in different cancers through their targets. For example, the expression of miR-1246 was significantly correlated with chemoresistance and cancer stem-cell-like characteristics and could identify a worse prognosis in cancer patients (pancreatic) by targeting cyclin G2 protein-coding genes such as CCNG2 [40]. Moreover, Li et al. [41] reported that miR-1246 enhances the proliferation and invasion of breast cancer cells by repressing the level of its CCNG2 target gene.

Du et al. [42] found that miR-1246 targeted thrombospondin-2 (THBS2) to inhibit cancerous growth and enhanced apoptosis in cervical cancer. Moreover, miR-1246 directly targeted death receptor 5 (DR5), which promotes proliferation and increases radioresistance in lung cancers [43]. However, to our knowledge, the specific role of miR-1246 in LUAD has been insufficiently investigated. Thus, in the present findings, we examined miR-1246 expression through miRNA sequencing data, which increased promisingly in LUAD patients. Hypothetically, target genes facilitate the functions of miR-1246. Therefore, we studied the probable target genes of miR-1246 and its enriched pathways through KEGG analysis.

For the biological process, we explored plasma-membrane-bound cell projection morphogenesis as a key function. The top five functional terms were chemosensory behavior, neuron development, negative regulation of neuron death, biomineral tissue development, and cellular component morphogenesis and axonogenesis. For molecular function, glycosaminoglycan binding is the major term. Glycosaminoglycan binds to different protein targets through electrostatic interactions between positively charged amino acids and negatively charged uronic acids. It is found to be concerned in multiple signaling cascades as it is mandatory for angiogenesis, cancer invasion, and metastasis. Similarly, it can also inhibit tumor progression and act as a drug target [44]. Salanti et al. [45] reported that parasite-derived protein could be exploited to target not only common but also complex malignancies like melanoma-associated glycosaminoglycan modification. Moreover, the targeting of glycosaminoglycans chains by tetrabranched peptide-like NT4 provides insights into the role of heparan sulfate proteoglycans in cancer cell adhesion and migration [46]. Based on those findings, we speculated that the regulation of miRNAs in glycosaminoglycan binding might improve the efficacy of LUAD therapeutics theoretically. Exclusively, we hypothesized that miR-1246 might be implicated in the regulation of glycosaminoglycans, which may persuade the treatment of LUAD patients.

Accordingly, we constructed a PPI interaction network of the TGs and selected the most densely connected modules based on the degree and between centrality measures. Moreover, we studied the gene expressions of UBE2C, TBCE, DNAJA3, PITX2, TGIF1, TRAIP, UCHL1, TNNI3, TNNT1, NRAS, RAC3, and EFNA4, which were high in LUAD tissue samples. Among these, UBE2C, TNNT1, TRAIP, and UCHL1 were positively correlated with miR-1246. The protein expression of ENFA4, NRAS, RAC3, TRAIP, UBE2C, and UCHL1 were upregulated in LUAD compared with that of control.

On survival analysis, we determined that UBE2C, UCHL1, TRAIP, TNNT1, TNNI3, and RAC3 were associated with poor overall survival of LUAD patients. Moreover, the high expression of PITX2, NRAS, ENFA4, DNAJA3, TBCE, and TGIF1 was correlated with longer overall survival of LUAD patients. The high expression of UBE2C is found in the advanced stage of cancer, which might point out its involvement in cancer progression and invasion. Additionally, patients with higher UBE2C levels showed a shorter overall survival (OS) time and worst OS prognosis. It indicated that UBE2C overexpression positively correlated in several cancers [47]. Accordingly, we hypothesized that post-translational modification of protein like UBE2C plays key roles in protein degradation, and protein interactions and their dysregulations in the earlier stage may lead to LUAD. Based on that, miR-1246 can be implied to target UBE2C, which may contribute to improving LUAD patients' prognosis and the survival of LUAD patients.

The ubiquitin-conjugating enzyme-2C (UBE-2C) is majorly responsible for the destructive cleavage of mitotic cyclin proteins for spindle assembly, which leads to the progression of the cell cycle. Moreover, the expressions of UBE-2C protein or mRNA are aberrantly expressed in various cancer types that lead to poor clinical results. Therefore, UBE-2C acts as a potential biomarker in cancer [48]. In the case of gastric cancer, Zhang et al. [49] reported that upregulated miR-17/20a significantly enhances the growth of gastric cancer cells by directly targeted UBE-2C. Jin et al. [50] reported that the miR-548e-5p, together with UBE-2C and zinc finger E-box binding homeobox (ZEB1/2), acts as a potential diagnostic biomarker and target for NSCLC. Moreover, another ubiquitin-protein UCHL1 was reported to promote uterine serous cancer cell proliferation, cell cycle progression [51], and TGFβ-induced breast cancer metastasis [52]. So far, no study has mentioned the relationship between miR-1246 and UBE-2C and

UCHL1 in LUAD. We are the first to report the targeting relationship between miR-1246 and UBE-2C. Based on the key enriched GO items, we supposed that miR-1246 targets UBE-2C and UCHL1 for the regulation of ubiquitin-mediated proteolysis and, therefore, persuades the prognosis of LUAD patients. However, more evidence is needed to validate this hypothesis.

TNNT1 is one of the isoforms of the troponin protein and is highly expressed in skeletal muscle. It plays key roles in muscle contraction and relaxation. In addition, reports have suggested that TNNT1 could contribute to cell proliferation in breast cancer [53]. Moreover, Hao et al. [54] reported that TNNT1 might promote the progression of colon adenocarcinoma by mediating the epithelial–mesenchymal transition process. In the present study, we found that TNNT1 was involved in cardiac muscle contraction, regulation of skeletal muscle contraction, troponin 1, and C binding molecular functions and associated with poor prognosis of LUAD. As a result, we assumed that miR-1246 might target TNNT1 and, thus, be associated with skeletal muscle regulation and poor prognosis of LUAD patients. However, future studies are required to supply more evidence.

TRAIP is a ring-type E3 ubiquitin ligase involved in many cellular functions, namely, NF-κB activation, DNA damage response, mitosis, and carcinogenesis [55]. Initially, it was considered a tumor suppressor in basal cell carcinomas and breast cancer [56]. Moreover, Guo et al. [57] reported that TRAIP exhibited as an oncogene in liver cancer. In the present study, we found that TRAIP was involved in pathways in cancer, regulation of autophagy, and DNA metabolic process enrichment terms and associated with poor prognosis of LUAD. As a result, we assumed that miR-1246 might target TRAIP as the metabolic function plays a key role in the prognosis of LUAD patients. However, future studies are required to supply more evidence.

Moreover, we also expected to determine the pathways that are associated with miR-1246 and its potential target genes in LUAD. We identified 12 important pathways that might have an essential role in the incidence and development of LUAD: pathways in cancer, ubiquitin-mediated proteolysis, protein processing in the endoplasmic reticulum, necroptosis, hypertrophic cardiomyopathy, cardiac muscle contraction, focal adhesion, the PI3K–Akt and Ras signaling pathways, the neurotrophin and mTOR signaling pathways, and the MAPK signaling pathway. Many studies have reported that the PI3K–Akt signaling pathway, the Ras signaling pathway, the neurotrophic-signaling pathway, the mTOR-signaling pathway, and the MAPK signaling pathway are regulated in the development of lung adenocarcinoma patients [58,59]. The PI3K/AKT/mTOR signal pathway is a key intracellular signal transduction pathway, with an essential function in cell proliferation, growth, survival, vesicle trafficking, glucose transport, and cytoskeletal organization. As we mentioned above, downregulation of PITX2, TGIF1, and TRAIP, and upregulation of TBCE may be involved in the different cellular processes (transcription, RNA splicing, cell cycle, and apoptosis) through the PI3K–Akt–mTOR signaling pathway and the MAPK signaling pathway, which might improve the survival of LUAD patients. Additionally, underexpression of TNNI3, TNNT1, and PITX2 might involve four independent pathways of cardiac muscle contraction, adrenergic signaling in cardiomyocytes, hypertrophic cardiomyopathy, and dilated cardiomyopathy, which are also incidentally associated through some somatic mutation and proto-oncogenic activities in LUAD. Besides, targeting Epstein–Barr virus infection and necroptosis pathways regulates the development of LUAD and the nonapoptotic form of regulated cell death, which may advance the prognosis of LUAD patients. Moreover, ubiquitin-mediated proteolysis and Parkinson's disease pathways are also indirectly associated through proteolysis and cell proliferation in LUAD. UBE-2C and UCHL1 are enriched in these pathways; we wondered if miR-1246 might target UBE-2C and UCHL1 and participate in the regulation of the cellular process, which may improve the prognosis of LUAD patients.

5. Conclusions

We have done a detailed and complete study about LUAD and miR-1246 using public datasets, with comprehensive biological network analysis useful for cancer research. Further experimental studies are still necessary to validate the results, which is a tough but promising task. We assumed that miR-1246 might target UBE2C, TNNT1, TRAIP, and UCHL1 during the regulation of ubiquitin-mediated proteolysis, glycosaminoglycan binding, DNA metabolism, the PI3K–Akt–mTOR signaling pathway, the neurotrophin and cardiomyopathy signaling pathway, and the MAPK signaling pathway. Upregulated UBE-2C, TNNT1, TRAIP, and UCHL1 may point out better survival of LUAD patients through the ubiquitin-mediated proteolysis, protein processing in the endoplasmic reticulum, and skeletal muscle contraction pathways. Moreover, similarly, this study had the limitations of other data-mining methods; the results of Limma and miRNA target prediction databases can be biased due to insufficient resources. To enhance the reliability of the results, immunohistochemical data from HPAD were employed for confirmation. Due to the constraint of HPAD, we could not get all the related IHCs of the tumor and adjacent normal samples of each potential target gene.

Author Contributions: J.S., and G.S. conceived and designed the experiment. S.H., G.S., Y.-K.W., and S.K. performed data collection and network analysis. Y.C., X.S., D.C., H.W., and A.S.N., performed the preparation of the figures. S.H., D.C., H.W., and A.S.N. performed the cytoscape analysis. S.H., S.K., G.S., and G.H.P. compiled the manuscript. J.S., G.S., G.H.P., and D.-Q.W. provided expert comments and performed the editing and proofreading of the manuscript. All authors have read and agreed to the published version of the manuscript.

Funding: Authors are thankful for the financial support from the China Postdoctoral Science Foundation (Grant No.: 2018M632766) to G.S., and the Henan Province Postdoctoral Science Foundation (Grant No.: 001802029 and 001803035) to S.K. and G.S.

Acknowledgments: The authors duly acknowledge all financial supporters mentioned in the funding section.

Conflicts of Interest: The authors declare no conflict of interest.

References

1. Subotic, D.D.; Van Schil, P.P.; Grigoriu, B. Radiation therapy for post-operative recurrence: Yes, but only for limited indications. *Eur. Respir. J.* **2016**, *48*, 278–279. [CrossRef]
2. Le, H.B.; Zhu, W.Y.Y.; Chen, D.D.; He, J.Y.; Huang, Y.Y.; Liu, X.G.; Zhang, Y.K. Evaluation of dynamic change of serum miR-21 and miR-24 in pre-and post-operative lung carcinoma patients. *Med. Oncolog.* **2012**, *29*, 3190–3197. [CrossRef]
3. Liloglou, T.; Bediaga, N.G.; Brown, B.R.; Field, J.K.; Davies, M.P. Epigenetic biomarkers in lung cancer. *Cancer Lett.* **2014**, *342*, 200–212. [CrossRef]
4. Rothschild, S.I. Epigenetic therapy in lung cancer–role of microRNAs. *Front. Oncolog.* **2013**, *3*, 158. [CrossRef]
5. Bhargava, A.; Bunkar, N.; Aglawe, A.; Pandey, K.C.; Tiwari, R.; Chaudhury, K.; Goryacheva, I.Y.; Mishra, P.K. Epigenetic biomarkers for risk assessment of particulate matter associated lung cancer. *Curr. Drug Targets* **2018**, *19*, 1127–1147. [CrossRef] [PubMed]
6. Afzali, F.; Salimi, M. Unearthing regulatory axes of breast cancer circRNAs networks to find novel targets and fathom pivotal mechanisms. *Interdiscip. Sci.* **2019**, *11*, 711–722. [CrossRef]
7. Han, H.; Zhang, Z.; Yang, X.; Yang, W.; Xue, C.; Cao, X. miR-23b suppresses lung carcinoma cell proliferation through CCNG1. *Oncolog. Lett.* **2018**, *16*, 4317–4324.
8. Feng, M.; Zhao, J.; Wang, L.; Liu, J. Upregulated expression of serum exosomal microRNAs as diagnostic biomarkers of lung adenocarcinoma. *Ann. Clin. Lab. Sci.* **2018**, *48*, 712–718. [PubMed]
9. Mitchell, P.S.; Parkin, R.K.; Kroh, E.M.; Fritz, B.R.; Wyman, S.K.; Pogosova-Agadjanyan, E.L.; Peterson, A.; Noteboom, J.; O'briant, K.C.; Allen, A.; et al. Circulating microRNAs as stable blood-based markers for cancer detection. *Proc. Natl. Acad. Sci. USA* **2008**, *105*, 10513–10518. [CrossRef]
10. Hou, L.K.; Ma, Y.S.; Han, Y.; Lu, G.X.; Luo, P.; Chang, Z.Y.; Xie, R.; Yang, H.; Chai, L.; Cai, M.; et al. Association of microRNA-33a molecular signature with non-small cell lung cancer diagnosis and prognosis after chemotherapy. *PLoS ONE* **2017**, *12*, e0170431. [CrossRef] [PubMed]

11. Hirono, T.; Jingushi, K.; Nagata, T.; Sato, M.; Minami, K.; Aoki, M.; Takeda, A.H.; Umehara, T.; Egawa, H.; Nakatsuji, Y.; et al. MicroRNA-130b functions as an oncomiRNA in non-small cell lung cancer by targeting tissue inhibitor of metalloproteinase-2. *Sci. Rep.* **2019**, *9*, 6956. [CrossRef] [PubMed]

12. Yang, L.; Dou, Y.; Sui, Z.; Cheng, H.; Liu, X.; Wang, Q.; Gao, P.; Qu, Y.; Xu, M. Upregulated miRNA-182-5p expression in tumor tissue and peripheral blood samples from patients with non-small cell lung cancer is associated with downregulated Caspase 2 expression. *Exp. Ther. Med.* **2020**, *19*, 603–610. [CrossRef] [PubMed]

13. Hetta, H.F.; Zahran, A.M.; El-Mahdy, R.I.; Nabil, E.E.; Esmaeel, H.M.; Alkady, O.A.; Elkady, A.; Mohareb, D.A.; Mostafa, M.M.; John, J. Assessment of Circulating miRNA-17 and miRNA-222 Expression Profiles as Non-Invasive Biomarkers in Egyptian Patients with Non-Small-Cell Lung Cancer. *Asian Pac. J. Cancer Prev.* **2019**, *20*, 1927. [CrossRef] [PubMed]

14. Yu, S.; Geng, S.; Hu, Y. miR-486–5p inhibits cell proliferation and invasion through repressing GAB2 in non-small cell lung cancer. *Oncolog. Lett.* **2018**, *16*, 3525–3530. [CrossRef]

15. Zhang, J.G.; Guo, J.F.; Liu, D.L.; Liu, Q.; Wang, J.J. MicroRNA-101 exerts tumor-suppressive functions in non-small cell lung cancer through directly targeting enhancer of zeste homolog 2. *J. Thorac. Oncolog.* **2011**, *6*, 671–678. [CrossRef]

16. Moriya, Y.; Nohata, N.; Kinoshita, T.; Mutallip, M.; Okamoto, T.; Yoshida, S.; Suzuki, M.; Yoshino, I.; Seki, N. Tumor suppressive microRNA-133a regulates novel molecular networks in lung squamous cell carcinoma. *J. Hum. Genet.* **2012**, *57*, 38–45. [CrossRef]

17. Asakura, K.; Kadota, T.; Matsuzaki, J.; Yoshida, Y.; Yamamoto, Y.; Nakagawa, K.; Takizawa, S.; Aoki, Y.; Nakamura, E.; Miura, J.; et al. A miRNA-based diagnostic model predicts resectable lung cancer in humans with high accuracy. *Commun. Biol.* **2020**, *3*, 134. [CrossRef]

18. Shi, X.; Yi, H.; Ma, S. Measures for the degree of overlap of gene signatures and applications to TCGA. *Brief. Bioinform.* **2015**, *16*, 735–744. [CrossRef]

19. Wei, D.Q.; Selvaraj, G.; Kaushik, A.C. Computational Perspective on the Current State of the Methods and New Challenges in Cancer Drug Discovery. *Curr. Pharm. Des.* **2018**, *24*, 3725. [CrossRef]

20. Selvaraj, G.; Kaliamurthi, S.; Kaushik, A.C.; Khan, A.; Wei, Y.K.; Cho, W.C.; Gu, K.; Wei, D.Q. Identification of target gene and prognostic evaluation for lung adenocarcinoma using gene expression meta-analysis, network analysis and neural network algorithms. *J. Biomed. Inform.* **2018**, *86*, 120–134. [CrossRef]

21. Shih, C.L.; Luo, J.D.; Chang, J.W.C.; Chen, T.L.; Chien, Y.T.; Yu, C.J.; Chiou, C.C. Circulating messenger RNA profiling with microarray and next-generation sequencing: Cross-platform comparison. *Cancer Genom. Proteom.* **2015**, *12*, 223–230.

22. Gentleman, R.; Carey, V.; Huber, W.; Irizarry, R.; Dudoit, S. *Bioinformatics and Computational Biology Solutions Using R and Bioconductor*; Springer Science and Business Media: New York, NY, USA, 2006.

23. Benjamini, Y.; Hochberg, Y. Controlling the false discovery rate: A practical and powerful approach to multiple testing. *J. R. Stat. Soc. Ser. B (Methodol.)* **1995**, *57*, 289–300. [CrossRef]

24. Dweep, H.; Gretz, N. miRWalk2. 0: A comprehensive atlas of microRNA-target interactions. *Nat. Methods* **2015**, *12*, 697. [CrossRef] [PubMed]

25. Oliveros, J.C. Venny. An Interactive Tool for Comparing Lists with Venn's Diagrams. 2007–2015. Available online: https://bioinfogp.cnb.csic.es/tools/venny/index.html (accessed on 25 July 2020).

26. Szklarczyk, D.; Morris, J.H.; Cook, H.; Kuhn, M.; Wyder, S.; Simonovic, M.; Santos, A.; Doncheva, N.T.; Roth, A.; Bork, P.; et al. The STRING database in 2017: Quality-controlled protein-protein association networks, made broadly accessible. *Nucleic Acids Res.* **2016**, *45*, D362–D368. [CrossRef] [PubMed]

27. Bozhilova, L.V.; Whitmore, A.V.; Wray, J.; Reinert, G.; Deane, C.M. Measuring rank robustness in scored protein interaction networks. *BMC Bioinform.* **2019**, *20*, 446. [CrossRef]

28. Pons, P.; Latapy, M. Computing Communities in Large Networks Using Random Walks. In *International Symposium on Computer and Information Sciences*; Springer: Berlin/Heidelberg, Germany, October 2005; pp. 284–293.

29. Kohl, P.; Sachs, F.; Franz, M.R. *Cardiac Mechano-Electric Coupling and Arrhythmias*; Oxford University Press: Oxford, UK, 2011.

30. Bindea, G.; Mlecnik, B.; Hackl, H.; Charoentong, P.; Tosolini, M.; Kirilovsky, A.; Fridman, W.H.; Pagès, F.; Trajanoski, Z.; Jérôme, G.; et al. ClueGO: A Cytoscape plug-in to decipher functionally grouped gene ontology and pathway annotation networks. *Bioinform. (Oxf. Engl.)* **2009**, *25*, 1091–1093. [CrossRef]

31. McHugh, M.L. Interrater reliability: The kappa statistic. *Biochem. Medica Biochem. Medica* **2012**, *22*, 276–282. [CrossRef]

32. Tang, Z.; Kang, B.; Li, C.; Chen, T.; Zhang, Z. GEPIA2: An enhanced web server for large-scale expression profiling and interactive analysis. *Nucleic Acids Res.* **2019**, *47*, W556–W560. [CrossRef]

33. Gyorffy, B.; Surowiak, P.; Budczies, J.; Lanczky, A. Online survival analysis software to assess the prognostic value of biomarkers using transcriptomic data in non-small-cell lung cancer. *PLoS ONE* **2013**, *8*, e82241. [CrossRef]

34. Uhlen, M.; Björling, E.; Agaton, C.; Al-KhaliliSzigyarto, C.; Amini, B.; Andersen, E.; Andersson, A.C.; Angelidou, P.; Asplund, A.; Asplund, C.; et al. A human protein atlas for normal and cancer tissues based onantibody proteomics. *Mol. Cell. Proteom.* **2005**, *4*, 1920–1932. [CrossRef]

35. Uhlen, M.; Zhang, C.; Lee, S.; Sjöstedt, E.; Fagerberg, L.; Bidkhori, G.; Benfeitas, R.; Arif, M.; Liu, Z.; Edfors, F.; et al. A pathology atlas of the human cancer transcriptome. *Science* **2017**, *357*, 6352. [CrossRef] [PubMed]

36. Selvaraj, G.; Selvaraj, C.; Wei, D.Q. Computational advances in chronic diseases diagnostics and therapy-I. *Curr. Drug Targets* **2020**, *21*, 1–2. [CrossRef] [PubMed]

37. Selvaraj, G.; Kaliamurthi, S.; Lin, S.; Gu, K.; Wei, D.Q. Prognostic impact of tissue inhibitor of metalloproteinase-1 in non-small cell lung cancer: Systematic review and meta-analysis. *Curr. Med. Chem.* **2019**, *26*, 7694. [CrossRef] [PubMed]

38. Kaliamurthi, S.; Selvaraj, G.; Junaid, M.; Khan, A.; Gu, K.; Wei, D.Q. Cancer immunoinformatics: A promising era in the development of peptide vaccines for human papillomavirus-induced cervical cancer. *Curr. Pharm. Des.* **2018**, *24*, 3791–3817. [CrossRef] [PubMed]

39. Nangraj, A.S.; Selvaraj, G.; Kaliamurthi, S.; Cho, W.C.S.; Aman, C.; Cho, W.C.; Wei, D.Q. Integrated PPI and WGCNA retrieving shared gene signatures between Barrett's esophagus and esophageal adenocarcinoma. *Front. Pharm.* **2020**, *11*, 881. [CrossRef] [PubMed]

40. Hoshino, I.; Yokota, H.; Ishige, F.; Iwatate, Y.; Takeshita, N.; Nagase, H.; Uno, T.; Matsubara, H. Radiogenomics predicts the expression of microRNA-1246 in the serum of esophageal cancer patients. *Sci. Rep.* **2020**, *10*, 2532. [CrossRef]

41. Li, X.J.; Ren, Z.J.; Tang, J.H.; Yu, Q. Exosomal MicroRNA MiR-1246 promotes cell proliferation, invasion and drug resistance by targeting CCNG2 in breast cancer. *Cell. Physiol. Biochem.* **2017**, *44*, 1741–1748. [CrossRef]

42. Du, P.; Lai, Y.H.; Yao, D.S.; Chen, J.Y.; Ding, N. Downregulation of microRNA-1246 inhibits tumor growth and promotes apoptosis of cervical cancer cells by targeting thrombospondin-2. *Oncolog. Lett.* **2019**, *18*, 2491–2499. [CrossRef] [PubMed]

43. Yuan, D.; Xu, J.; Wang, J.; Pan, Y.; Fu, J.; Bai, Y.; Zhang, J.; Shao, C. Extracellular miR-1246 promotes lung cancer cell proliferation and enhances radioresistance by directly targeting DR5. *Oncotarget* **2016**, *7*, 32707. [CrossRef]

44. Morla, S. Glycosaminoglycans and Glycosaminoglycan Mimetics in Cancer and Inflammation. *Int. J. Mol. Sci.* **2019**, *20*, 1963. [CrossRef]

45. Salanti, A.; Clausen, T.M.; Agerbæk, M.Ø.; Al Nakouzi, N.; Dahlbäck, M.; Oo, H.Z.; Lee, S.; Gustavsson, T.; Rich, J.R.; Hedberg, B.J.; et al. Targeting Human Cancer by a Glycosaminoglycan Binding Malaria Protein. *Cancer Cell* **2015**, *28*, 500–514. [CrossRef]

46. Brunetti, J.; Depau, L.; Falciani, C.; Gentile, M.; Mandarini, E.; Riolo, G.; Lupetti, P.; Pini, A.; Bracci, L. Insights into the role of sulfated glycans in cancer cell adhesion and migration through use of branched peptide probe. *Sci. Rep.* **2016**, *6*, 27174. [CrossRef] [PubMed]

47. Dastsooz, H.; Cereda, M.; Donna, D.; Oliviero, S. A Comprehensive Bioinformatics Analysis of UBE2C in Cancers. *Int. J. Mol. Sci.* **2019**, *20*, 2228. [CrossRef]

48. Xie, C.; Powell, C.; Yao, M.; Wu, J.; Dong, Q. Ubiquitin-conjugating enzyme E2C: A potential cancer biomarker. *Int. J. Biochem. Cell Biol.* **2014**, *47*, 113–117. [CrossRef]

49. Zhang, Y.; Han, T.; Wei, G.; Wang, Y. Inhibition of microRNA-17/20a suppresses cell proliferation in gastric cancer by modulating UBE2C expression. *Oncolog. Rep.* **2015**, *33*, 2529–2536. [CrossRef] [PubMed]

50. Jin, D.; Guo, J.; Wu, Y.; Du, J.; Wang, X.; An, J.; Hu, B.; Kong, L.; Di, W.; Wang, W. UBE2C, directly targeted by miR-548e-5p, increases the cellular growth and invasive abilities of cancer cells interacting with the EMT marker protein zinc finger E-box binding homeobox 1/2 in NSCLC. *Theranostics* **2019**, *9*, 2036. [CrossRef] [PubMed]

51. Kwan, S.Y.; Au-Yeung, C.L.; Yeung, T.L.; Rynne-Vidal, A.; Wong, K.K.; Risinger, J.I.; Lin, H.K.; Schmandt, R.E.; Yates, M.S.; Mok, S.C.; et al. Ubiquitin Carboxyl-Terminal Hydrolase L1 (UCHL1) Promotes Uterine Serous Cancer Cell Proliferation and Cell Cycle Progression. *Cancers* **2020**, *12*, 118. [CrossRef] [PubMed]

52. Liu, S.; González-Prieto, R.; Zhang, M.; Geurink, P.P.; Kooij, R.; Iyengar, P.V.; van Dinther, M.; Bos, E.; Zhang, X.; Sylvia, E.; et al. Deubiquitinase activity profiling identifies UCHL1 as a candidate oncoprotein that promotes TGFβ-induced breast cancer metastasis. *Clin. Cancer Res.* **2020**, *26*, 1460–1473. [CrossRef] [PubMed]

53. Shi, Y.; Zhao, Y.; Zhang, Y.; AiErken, N.; Shao, N.; Ye, R.; Lin, Y.; Wang, S. TNNT1 facilitates proliferation of breast cancer cells by promoting G1/S phase transition. *Life Sci.* **2018**, *208*, 161–166. [CrossRef] [PubMed]

54. Hao, Y.H.; Yu, S.Y.; Tu, R.S.; Cai, Y.Q. TNNT1, a prognostic indicator in colon adenocarcinoma, regulates cell behaviors and mediates EMT process. *Biosci. Biotechnol. Biochem.* **2020**, *84*, 111–117. [CrossRef]

55. Chapard, C.; Hohl, D.; Huber, M. The TRAF-interacting protein (TRAIP) is a novel E2F target with peak expression in mitosis. *Oncotarget* **2015**, *6*, 20933–20945. [CrossRef] [PubMed]

56. Zhou, Q.; Geahlen, R.L. The protein-tyrosine kinase Syk interacts with TRAF-interacting protein TRIP in breast epithelial cells. *Oncogene* **2009**, *28*, 1348–1356. [CrossRef] [PubMed]

57. Guo, Z.; Zeng, Y.; Chen, Y.; Liu, M.; Chen, S.; Yao, M.; Zhang, P.; Zhong, F.; Jiang, K.; He, S.; et al. TRAIP promotes malignant behaviors and correlates with poor prognosis in liver cancer. *Biomed. Pharm.* **2020**, *124*, 109857. [CrossRef] [PubMed]

58. Sanchez-Vega, F.; Mina, M.; Armenia, J.; Chatila, W.K.; Luna, A.; La, K.C.; Dimitriadoy, S.; Liu, D.L.; Kantheti, H.S.; Saghafinia, S.; et al. Oncogenic signaling pathways in the cancer genome atlas. *Cell* **2018**, *173*, 321–337. [CrossRef] [PubMed]

59. Harvey, A.J. Overview of Cell Signaling Pathways in Cancer. In *Predictive Biomarkers in Oncology*; Springer: Cham, Switzerland, 2019; pp. 167–182.

Publisher's Note: MDPI stays neutral with regard to jurisdictional claims in published maps and institutional affiliations.

Journal of
*Personalized
Medicine*

Article

Sestrin2 Expression Has Regulatory Properties and Prognostic Value in Lung Cancer

Hee Sung Chae [1,†], Minchan Gil [1,†], Subbroto Kumar Saha [1], Hee Jeung Kwak [1], Hwan-Woo Park [2], Balachandar Vellingiri [3] and Ssang-Goo Cho [1,*]

1   Department of Stem Cell and Regenerative Biotechnology, Incurable Disease Animal Model & Stem Cell Institute (IDASI), Konkuk University, 120 Neungdong-ro, Gwangjin-gu, Seoul 05029, Korea; gmltjdgk@konkuk.ac.kr (H.S.C.); minchangil@gmail.com (M.G.); subbroto@konkuk.ac.kr (S.K.S.); hjeong9581@konkuk.ac.kr (H.J.K.)
2   Department of Cell Biology, Konyang University College of Medicine, Daejeon 35365, Korea; hwanwoopark@konyang.ac.kr
3   Human Molecular Cytogenetics and Stem Cell Laboratory, Department of Human Genetics and Molecular Biology, Bharathiar University, Coimbatore 641-046, India; geneticbala@buc.edu.in
*   Correspondence: ssangoo@konkuk.ac.kr; Tel.: +82-2-450-4207; Fax: +82-2-444-4207
†   These authors contributed equally to this study.

Received: 25 July 2020; Accepted: 27 August 2020; Published: 1 September 2020

Abstract: Lung cancer remains the most dangerous type of cancer despite recent progress in therapeutic modalities. Development of prognostic markers and therapeutic targets is necessary to enhance lung cancer patient survival. Sestrin family genes (Sestrin1, Sestrin2, and Sestrin3) are involved in protecting cells from stress. In particular, Sestrin2, which mainly protects cells from oxidative stress and acts as a leucine sensor protein in mammalian target of rapamycin (mTOR) signaling, is thought to affect various cancers in different ways. To investigate the role of Sestrin2 expression in lung cancer cells, we knocked down Sestrin2 in A549, a non-small cell lung cancer cell line; this resulted in reduced cell proliferation, migration, sphere formation, and drug resistance, suggesting that Sestrin2 is closely related to lung cancer progression. We analyzed Sestrin2 expression in human tissue using various bioinformatic databases and confirmed higher expression of Sestrin2 in lung cancer cells than in normal lung cells using Oncomine and the Human Protein Atlas. Moreover, analyses using Prognoscan and KMplotter showed that Sestrin2 expression is negatively correlated with the survival of lung cancer patients in multiple datasets. Co-expressed gene analysis revealed Sestrin2-regulated genes and possible associated pathways. Overall, these data suggest that Sestrin2 expression has prognostic value and that it is a possible therapeutic target in lung cancer.

Keywords: Sestrin2; lung cancer; knockdown; cancer progression; bioinformatics; patient survival

---

1. Introduction

Cancer, one of the leading causes of death in modern society, poses a threat to human health worldwide. Among the various cancers, lung and bronchial cancer is the most dangerous cancer type, with 228,150 new patients and 142,670 deaths reported in 2019 in the United States alone [1]. Cancer occurrence is gradually increasing with population increase and aging, although there have been considerable advances in cancer therapy. Developments in the identification of novel cancer targets and markers are required to improve human health.

Sestrin family genes consist of Sestrin1, Sestrin2, and Sestrin3. Under conditions of stress, sestrins regulate stress-inducible metabolism and protect cells against various kinds of stressors such as hypoxia, DNA damage, and oxidative and metabolic stress [2,3]. Sestrin1, also called p53-activated gene 26 (PA26), is involved in the growth arrest and DNA damage response pathways [4]. Sestrin2, also known as hypoxia-inducible gene 95 (Hi95), is involved in mediating the response to hypoxia and is upregulated by other stressors, such as DNA damage and oxidative stress [5,6]. Sestrin3, as well as Sestrin2, is known to mediate the regulation of mammalian target of rapamycin 1 (mTORC1) and Akt activation [7,8]. Expression of these genes decreases the levels of intracellular reactive oxygen species (ROS) and promotes resistance against oxidative stress [9,10]. A recent study revealed that Sestrin1 and Sestrin2 activate Nrf2 and subsequently increase Srx, which is important for oxidative metabolism [11]. Sestrin has been shown to interact with p53 and forkhead box class O (FoxO) transcription factors and mediate antioxidant regulation [12]. Activation of 5′ adenosine monophosphate-activated protein kinase (AMPK) and inhibition of mTORC1 are important for cell cycle and cell lifespan [13]. Since sestrins can modulate pathways of cellular metabolism, sestrin expression seems to play an important role in prolonging life and inhibiting aging [2].

Sestrin2 is an intracellular leucine sensor protein that negatively regulates mTORC1 signaling by binding to GAP Activity Towards Rags 2 (GATOR2), a subunit of the GATOR complex, in the absence of leucine. In the presence of leucine, Sestrin2 detaches from GATOR2 and consequently activates mTORC1 [14–16]. Sestrin2 plays a variety of roles throughout the body and is responsible for mediating the cellular response against various environmental stressors [2]. Genotoxic stresses, such as UV or gamma irradiation, and genotoxic molecules promote the transcription of Sestrin1 and Sestrin2 through the p53 pathway, resulting in cell cycle inhibition and modulation of metabolism in the stressed cells [17]. Oxidative stress activates the Nuclear factor erythroid-2-related factor 2 (NRF2) and Jun N-terminal kinase-Activator protein 1 (JNK-AP1) pathways, which induce the expression of Sestrin2 [18,19]. Sestrin2 has also been shown to act as a tumor suppressor gene in various cancers [20–22]. In colorectal cancer, Sestrin2 suppresses mTORC1 signaling by activating AMPK/mTORC pathway, resulting in the suppression of tumor cell growth [21]. Sestrin2 knockdown accelerates colorectal carcinogenesis [22]. Moreover, Sestrin2 is known to be downregulated in bladder cancer, and when Sestrin2 is induced in response to mitogen-activated protein kinase 8 (MAPK8)-JUN-dependent transcription, it suppresses bladder cancer growth [23]. However, contrary to these results, Sestrin2 is still expressed in various cancers and may be necessary to increase cancer viability under certain conditions [24].

Based on previous reports, the present study aims to investigate the role of Sestrin2 in the survival, migration, and sphere formation of lung cancer cells. Further, this study also aims to conduct bioinformatic analyses for gene expression, prognostic value, and potential related pathways in human samples using various cancer gene expression databases. The outcome of this study with respect to Sestrin2 may indicate its potential role in prognostics and therapeutics for lung cancer.

2. Materials and Methods

2.1. Cell Culture

Human lung carcinoma cell line A549 was purchased from The Korea Cell Line Bank (Seoul, Korea). A549 was cultured in RPMI 1640 medium (Sigma-Aldrich, St. Louis, MO, USA) supplemented with 10% heat-inactivated fetal bovine serum (FBS, Gibco, Thermo Fisher Scientific Ltd., Waltham, MA, USA), 100 U/mL penicillin, and 100 mg/mL streptomycin (Gibco). Cells were seeded in cell culture plates (SPL Lifesciences, Pocheon-si, Korea) and maintained at 37 °C in a humidified atmosphere containing 5% CO_2.

2.2. Lentivirus Production and Infection

The short hairpin (shRNA) lentiviral plasmids for Sestrin2 knockdown (shSESN2-1 and shSESN2-2) and scramble control were purchased from VectorBuilder (Chicago, IL, USA). The sequence of shRNA was designed as previously reported [25,26]. pLSLPw-shLUC and shSESN2 were provided by Dr. Andrei Budanov [25]. To prepare the lentivirus, we cultured human embryonic kidney (HEK) 293T cells up to 80% confluence in 6-well plates and transfected with scramble and Sestrin2-targeted shRNA plasmids using lipofectamine 3000 (Thermo Fisher Scientific Ltd., Waltham, MA, USA). After 48 h, the virus-containing medium was collected and filtered with a 0.45 µm membrane filter. The scramble and shRNA targeting SESN2 (shSESN2) lentivirus supernatant were used to infect A549 cells, which were incubated overnight. Afterward, the media was replaced with fresh culture media and the cells were grown. Puromycin was used for treatment 24 h after media change, and the concentration was 4 µg/mL.

2.3. Total RNA Extraction and RT-PCR

Total RNA was extracted using Labozol (Labopass, Cosmogenetech, Seoul, Korea) with the experimental protocol provided. The concentration of total RNA was measured using Nanodrop (IMPLEN, CA, USA). Complementary DNA (cDNA) was synthesized using 2 µg of total RNA with Moloney Murine Leukemia Virus (MMLV) reverse transcriptase (Promega) as per the experimental protocol provided. RT-PCR was performed using r-Tap Plus Master Mix (Elpis Biotech, Daejeon, Korea), and PCR products were analyzed by electrophoresis on a ~1.5% agarose gel containing ethidium bromide (EtBr) and bands were observed under UV light. Relative expression was measured using ImageJ (https://imagej.net/). Primer sequences are listed in Supplementary Table S1.

2.4. Cell Survival Assay

For cell proliferation analysis, control and Sestrin2 knockdown cells were seeded in 6-well plates (5×10^4 cells/well) and cultured for 24, 48, 72, and 96 h. Cells were counted using a hemocytometer following trypan blue staining. Cell proliferation assay was also carried out using EZ-cytox reagent (DoGen, Seoul, Korea). Around 1×10^4 cells/well were seeded in 96-well plate and cultured. EZ-cytox was added at a ratio of 1 to 10 and held in an incubator for 2 h. Afterward, the absorbance was measured at 450 nm using a fluorescence microplate reader. To observe the drug sensitivity, the cells were seeded in 6-well plates (1×10^5 cells/well) and grown. After 24 h, doxorubicin and cisplatin were added to each well and mixed well. The final concentration was 1 µM for doxorubicin and 10 µM for cisplatin. After 24 h of treatments, the cells were counted using hemocytometer following trypan blue staining.

2.5. Wound Healing Assay

For the wound healing assay, 95% confluent cells were cultured in a 6-well plate. Cells were treated with mitomycin C (10 µg/mL) for 2 h, then the monolayer was scratched using a 200 µL tip and the media was replaced with fresh culture media with 10% fetal bovine serum (FBS). The wound area was marked, and photos were taken every 12 h. Pictures were analyzed, and closure percentage was measured using ImageJ.

2.6. Sphere-Forming Assay

Cells (1×10^5) were seeded in a non-coated 60 mm petri dish (SPL Lifesciences, catalogue number 11035) and cultured in the presence of sphere-forming media (serum-free DMEM/F12 media containing B27 supplement, 20 ng/mL epidermal growth factor (EGF) (Sigma-Aldrich, St. Louis, MO, USA), 10 µg/mL insulin (Sigma-Aldrich), and 1% bovine serum albumin (Sigma-Aldrich)). After 5 days, spheres were collected and stained with crystal violet (Sigma-Aldrich). Sphere sizes were measured using ImageJ.

2.7. Dichlorodihydrofluorescein Diacetate (DCFDA) Cellular ROS Assay

Cells (1×10^4) were seeded in 96-well plates. After 24 h, 2′,7′-dichlorodihydrofluorescein diacetate (DCFDA) (Invitrogen, Waltham, MA, USA) was added to the cells at a final concentration of 10 mM, followed by incubation at 37 °C for 30 min. After removing the DCFDA media, cells were washed with DPBS and fluorescence was measured immediately on a fluorescence microplate reader (SpectraMAX,

Molecular Devices). For flow cytometry, cells (5×10^5) were seeded and cultured in 6-well plates. After 24 h, DCFDA was added to cells and incubated for 30 min. DCFDA media was removed, and cells were washed with Dulbecco's Phosphate-Buffered Saline (DPBS), detached by trypsin, and analyzed by flow cytometry (FACSCalibur, Becton Dickinson).

2.8. Oncomine Database Analysis

The expression of Sestrin2 mRNA was analyzed using Oncomine with the Okayama Lung Statistics and Selamat Lung Statistics datasets (https://www.oncomine.org/resource/login.html) [27,28]. mRNA expression was compared between lung cancer and normal tissues using parameters with a threshold p-value of 1×10^{-4} and gene ranking in the top 10%. We also analyzed genes co-expressed with Sestrin2 in the Bass lung dataset.

2.9. The Human Protein Atlas

Expression of Sestrin2 protein in lung cancer tissue and normal tissue was analyzed using the Human Protein Atlas (https://www.proteinatlas.org/). The normal lung tissue from Patient 2268 was compared with lung cancer tissue from patient 3391 that stained with Sestrin2 antibody (HPA018191, Sigma-Aldrich, St. Louis, MO, USA).

2.10. Prognoscan and Kaplan-Meier Plotter

The correlation between Sestrin2 expression and survival rate in lung cancer patients was analyzed using Prognoscan and the Kaplan-Meier plotter database. Prognoscan is a database that includes prognostic data for various cancers (http://dna00.bio.kyutech.ac.jp/PrognoScan/) [29]. Gene Expression Omnibus (GSE)3141–overall survival (hazard ratio = 2.38) and GSE11117–overall survival (hazard ratio = 2.59) were analyzed using Prognoscan with a Cox p-value < 0.05. The Kaplan-Meier plotter database was used to analyze mRNA Affymetrix Genechip and RNA-sequencing datasets for lung squamous cell carcinoma patient.

2.11. cBioPortal Database Analysis

The mutation and alteration of Sestrin2 were analyzed using cBioportal, a free-access bioinformatic website (http://www.cbioportal.org/) [30]. cBioportal provides clinical characteristic data from 225 cancer studies in The Cancer Genome Atlas (TCGA) datasets. The mutations in 2704 cases of lung cancer and gene alteration in 2324 cases were analyzed. Copy number alteration analysis was performed using the GISTIC (Genomic Identification of Significant Targets in Cancer) algorithm and plotted using TCGA mRNA expression data.

2.12. Enrichr Gene Ontology (GO) Analysis

To analyze the ontology of Sestrin2 and co-expressed genes, the Enrichr database was used (https://amp.pharm.mssm.edu/Enrichr) [31]. GO and pathway analyses were visualized as a bar graph. GO biological process, molecular function, cellular component, and Kyoto Encyclopedia of Genes and Genomes information were also included in the analysis.

2.13. Statistical Analysis

Data were analyzed using GraphPad Prism 6 (Sandiego, CA, USA) and Excel 2006 (Microsoft Corporation, Redmond, WA, USA). Statistical analyses were conducted using a T-test and statistical significance was defined as * $p < 0.05$.

3. Results

3.1. Knockdown of Sestrin2 in a Lung Cancer Cell Line Leads to Reduced Cancer Cell Survival and Migration

We detected relatively high Sestrin2 expression in A549, a non-small cell lung cancer cell line compared to other cell lines tested (Supplementary Figure S1). To investigate the effect of Sestrin2 on lung cancer cells, we examined the effects of Sestrin2 knockdown in these cells. Knockdown was performed using Sestrin2-targeted shRNA cloned in a lentiviral vector. Reverse transcription-polymerase chain reaction (RT-PCR) analysis revealed that expression of Sestrin2 was reduced by shRNA in A549 cells (Figure 1A). Sestrin2 expression was decreased 72% by shSESN2-1 and 92% by shSESN2-2 compared to the scramble control. To observe the effect of Sestrin2 in cancer cells, we compared the viability of A549 cells treated with both shSESN2 and scramble control. The number of Sestrin2 knockdown cells with shSESN2-1 and SESN2-2 was significantly reduced compared to that in the scramble control (Figure 1B and Supplementary Figure S2). We performed a wound healing assay with A549 cells to examine the effect of Sestrin2 expression on cancer cell migration (Figure 1C). The results showed that the gap distance of the wound in scramble control cells was more closed than that in either Sestrin2 knockdown cultures. The expression of epithelial–mesenchymal transition (EMT) markers, which might contribute to cancer metastasis, was also observed (Figure 1D). RT-PCR revealed that the expression of EMT markers (Vimentin, Snail, *ZEB1*) was significantly reduced in Sestrin2 knockdown cells compared to that in scramble cells. Overall, we suggest that Sestrin2 expression is related to survival and migration in the A549 lung cancer cell line.

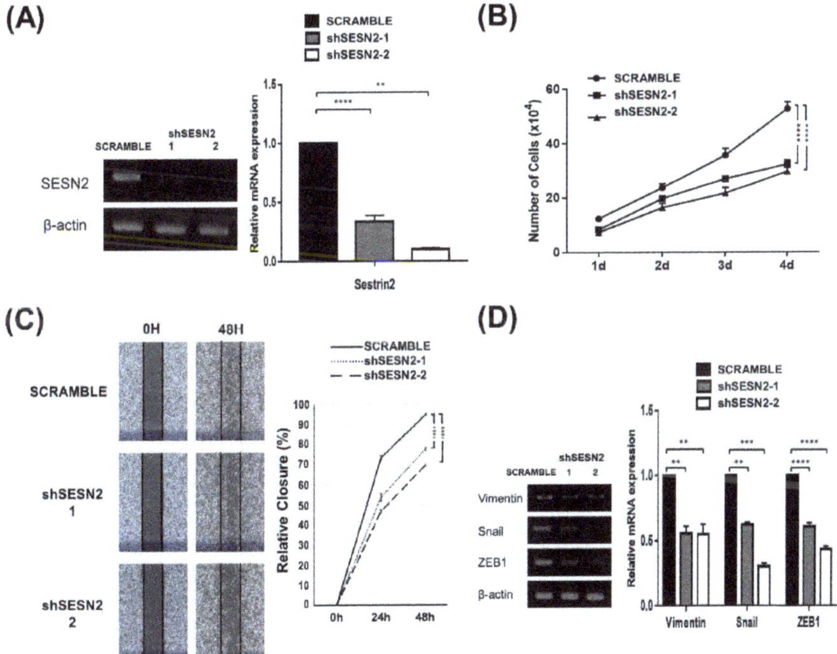

Figure 1. Survival and migration were decreased in response to Sestrin2 knockdown in A549 lung cancer cells. (**A**) Expression of Sestrin2 in shSESN2-1, shSESN2-2, and scramble A549 cells as measured by RT-PCR followed by subsequent agarose gel analysis. Band density was measured with ImageJ and

is plotted as the value relative to scramble control. (**B**) Survival of scramble, shSESN2-1, and shSESN2-2 cells was measured using trypan blue staining. The number of cells was counted using a hemocytometer. (**C**) The wound healing assay revealed cell movement capacity. Cells were observed at the indicated time, and closure percentage is plotted. The photo was taken under an inverted light microscope and closure percentage was measured using ImageJ. (**D**) The mRNA expression of epithelial–mesenchymal transition markers (Vimentin, Snail, ZEB1) was downregulated in Sestrin2 knockdown cells compared to that in scramble cells. Band density was measured using ImageJ and is plotted on the right (** $p < 0.01$; *** $p < 0.005$; **** $p < 0.0001$).

3.2. Knockdown of Sestrin2 in Lung Cancer Cells Decreases Cancer Cell Stemness and Drug Resistance

To investigate the role of Sestrin2 in cancer cell stemness, we determined the expression of stemness marker genes by RT-PCR (Figure 2A). Expression of stemness markers Oct4, Sox2, and Nanog was decreased in Sestrin2-knockdown A549 cells compared to that in the scramble control. The effect of Sestrin2 gene on cancer stemness by sphere-forming assay was also determined (Figure 2B). The size of the spheres formed by the Sestrin2 knockdown A549 cells was smaller than that formed by scramble A549 cells. This result showed that Sestrin2 knockdown reduced lung cancer stemness. To evaluate the effect of Sestrin2 on drug sensitivity, the expression of drug resistance marker genes (*ABCG*, *ABCA2*) was determined using RT-PCR (Figure 2C). Expression of the drug resistance marker genes was decreased in Sestrin2 knockdown A549 cells compared to that in cells with scramble (Figure 2C). In addition, cell survival assay was performed on knockdown and scramble cells treated with the anticancer drugs doxorubicin and cisplatin, which induce oxidative stress by increasing the ROS level [32] (Figure 2D). The survival rate of Sestrin2 knockdown cells was significantly decreased compared to that of scramble cells regardless of anticancer drug treatment. However, the percentage of cells recovered with doxorubicin treatment was 49.7% for scramble control, 41.6% for shSESN2-1, 41.1% for shSESN2-2 compared to no treatment control. The reduced recovery rate of Sestrin2 knockdown cells in doxorubicin treatment suggests that cells became more sensitive to doxorubicin treatment. These results suggest that the expression of Sestrin2 could be involved in mediating the development of cancer stemness and drug resistance in lung cancer cell lines.

3.3. Expression of Sestrin2 is Related to ROS Regulation in A549 Lung Cancer Cells

NF-E2-related factor 2 (*NRF2*) is a critical transcription factor regulating intracellular antioxidants and detoxification enzymes [33]. In cancers, the NRF2-mediated antioxidant pathways protect cells from drugs such as doxorubicin and cisplatin [34]. Because Sestrin2 activates the *NRF2* pathway in cancer cells [11], the effect of Sestrin2 knockdown on *NRF2* and oxidative status of A549 cells was investigated. For ROS measurement by DCFDA assay, Sestrin2 knockdown cells without GFP expression were generated, and the knockdown of Sestrin2 and downregulation of *NRF2* and heme oxygenase (*HO-1*) were confirmed in A549 cells (Figure 3A). Reduced expression of *NRF2* and *HO-1* were also observed in Sestrin2 knockdown A549 cells with the shRNA vectors used in Figures 1 and 2 (Supplementary Figure S3). The intracellular ROS level was then measured using the DCFDA assay. In the Sestrin2 knockdown cells, ROS levels were significantly increased by nearly threefold (Figure 3B). The increase in ROS levels was also indicated by flow cytometry (Figure 3C). These results suggest that Sestrin2 affects the regulation of the NRF2-HO-1 pathway and ROS level in A549 cancer cells.

3.4. Sestrin2 Expression and Correlation with Patient Survival in Lung Cancer

Knockdown of Sestrin2 in A549 cells suppressed cancer cell properties such as proliferation, migration, stemness, and drug resistance, which are critical to cancer progression. To examine the role of Sestrin2 in human lung cancer, we used a publicly available gene expression database of cancer tissues. In analysis using the Oncomine database, Sestrin2 mRNA expression was higher in lung cancer tissue than in normal lung tissue in the Okayama Lung Statistics dataset (fold change, 1.295; p-value: 3.34×10^{-7}) (Figure 4A). Histochemistry data for lung cancer using the Human Protein Atlas revealed that Sestrin2 protein was highly expressed in lung cancer (Figure 4B). The lung tumor was strongly stained by Sestrin2 antibody HPA018191 (patient ID = 3391), while pneumocytes of the normal lung were stained less strongly (patient ID = 2268). Based on these results, we suggest Sestrin2 is highly expressed in lung cancer than in the normal tissue.

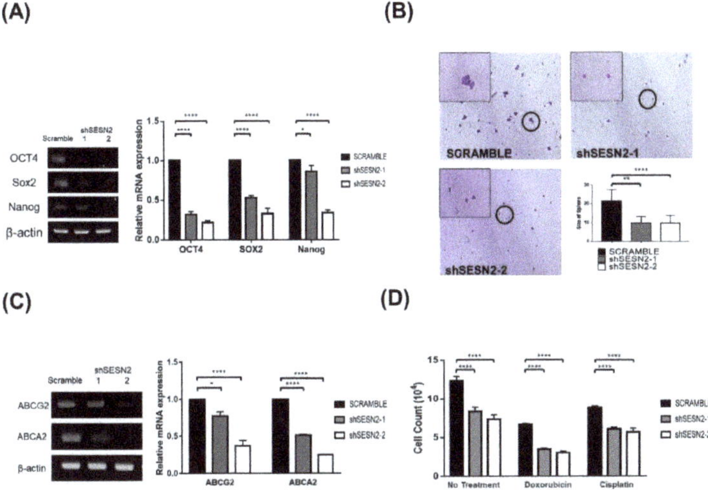

Figure 2. Effect of Sestrin2 expression on stemness and drug resistance in A549 cells. (**A**) Expression of stemness marker in scramble and shSESN2 A549 cells was analyzed using RT-PCR. mRNA expression relative to scramble control is shown in the graph. (**B**) Sphere-forming assay of scramble and Sestrin2 knockdown A549 cells. Cells were seeded in a petri dish and cultured in sphere-forming media. Spheres were evaluated after 5 days of culture using crystal violet, and then photos were taken. The size of the spheres was measured by ImageJ. The enlarged photo at the top left represents spheres in a circle. (**C**) Expression of the drug resistance marker genes (*ABCG2*, *ABCA2*) in scramble and shSESN2 A549 cells was analyzed using RT-PCR. (**D**) Drug resistance assay with doxorubicin and cisplatin, ROS-generating anticancer drugs. Scrambled control, shSESN2-1, and shSESN2-2 cells were treated with 1 µM doxorubicin or 10 µM cisplatin for 24 h and subjected to cell counting with trypan exclusion (* $p < 0.05$; ** $p < 0.01$; **** $p < 0.0001$).

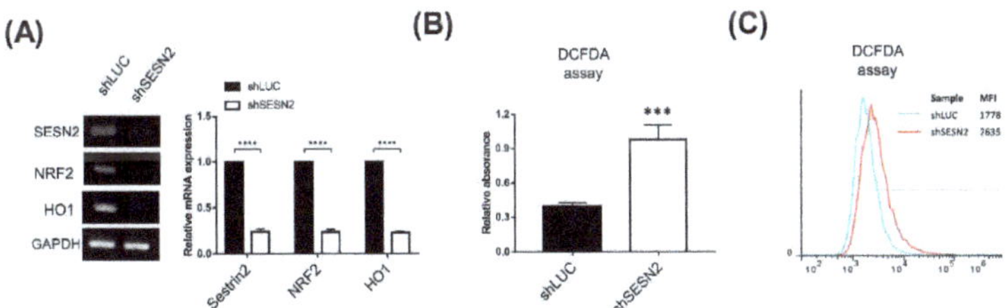

Figure 3. Sestrin2 knockdown leads to reactive oxygen species (ROS) overproduction by inhibiting the oxidative stress response. (**A**) Expression of *NRF2* and *HO-1* in control and shSESN2 A549 cells measured by RT-PCR. A549 cells were transduced with lentiviral pLSLPw-shLUC and shSESN2 plasmids. (**B**) 2′,7′-Dichlorodihydrofluorescein diacetate (DCFDA) cellular ROS assay. A549 cells were stained with DCFDA and washed. The emitted fluorescence was measured using a fluorescence microplate reader (**B**) and flow cytometer (**C**). MFI: mean fluorescence intensity (*** $p < 0.005$; **** $p < 0.0001$).

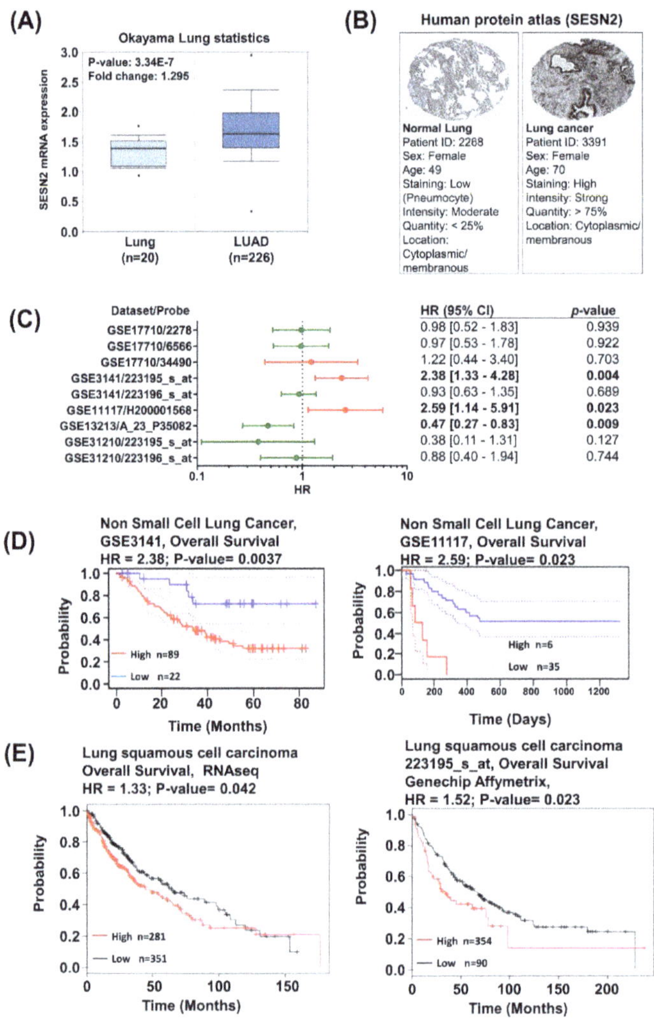

Figure 4. Sestrin2 mRNA expression analysis in lung cancer patients using various bioinformatic databases. (**A**) Oncomine database analysis of Okayama Lung statistics comparing Sestrin2 mRNA expression in normal lung with that in lung cancer. (**B**) Human Protein Atlas analysis for patient tissue staining. The normal lung tissue is stained with a low amount of Sestrin2 antibody (patient ID = 2268) while the lung cancer tissue is stained with a high amount of Sestrin2 antibody (patient ID = 3391) (**C**) Forest plots of GEO datasets evaluating association of sestrin2 expression and overall survival in lung cancer datasets in Prognoscan database. Hazard ratio (HR) with 95% confidential interval (CI) and p-values are labeled in the right column of each forest plot. (**D**) The survival rate graph compares high (red) and low (blue) Sestrin2 expression in non-small cell lung cancer patients. Prognoscan database analysis survival curve plotter using GSE3141–overall survival (hazard ratio = 2.38, *p*-value = 0.0037) (high *n* = 89, low *n* = 22) and GSE11117–overall survival (hazard ratio = 2.59, *p*-value = 0.023) (high *n* = 6, low *n* = 35) datasets. (**E**) The survival rate graph compares high (red) and low (black) Sestrin2 expression in lung squamous cell carcinoma patients. The 223195_s_at dataset was analyzed with RNAseq and Affymetrix Genechip using KM plotter.

The correlation between Sestrin2 expression and lung cancer patient survival was analyzed using KM plotter and Prognoscan. In meta-analysis using Prognoscan database, patient overall survival was significantly correlated with Sestrin2 expression in three lung cancer datasets (Figure 4C). The survival rate of the lung cancer patient group with high Sestrin2 expression was lower than that in the patient group with low Sestrin2 expression in the GSE3141 dataset (*p*-value: 0.0037) and in the GSE11117 (*p*-value: 0.023) (Figure 4D). However, Sestrin2 expression is positively correlated with patient overall survival in the GSE13213. In KM plotter analysis, the lung squamous carcinoma patient group with higher Sestrin2 expression had worse overall survival than the patient group with lower Sestrin2 expression in the RNAseq dataset (*p*-value: 0.042) and in the Affymetrix Genechip dataset with probe 223195_s_at (*p*-value: 0.023) (Figure 4D). Overall, Sestrin2 expression is negatively correlated with survival of patients with lung cancer in multiple lung cancer expression datasets.

3.5. Mutation and Alteration of Sestrin2 Gene in Lung Cancer

Mutations in lung cancer patients were analyzed using cBioportal web. The mutations of Sestrin2 in lung cancer patients were analyzed across 4510 samples from 4154 patients in 16 studies. Thirty-six mutations were analyzed in the Sestrin2 protein (Figure 5A). Mutation occurred in 2.19% of the samples, and deep deletion occurred in 0.55% of the samples, resulting in gene alteration in 2.73% of the lung adenocarcinoma Broad dataset (Figure 5B,C). Expression of Sestrin2 was also analyzed based on gene alteration (Figure 5D). Expression of Sestrin2 increased the following in order: deep deletion (DD), shallow deletion (SD), diploid (D), gain (G), and amplification (A) in lung adenocarcinoma and lung squamous cell carcinoma. We suggest that mRNA expression of Sestrin2 is associated with copy number alteration.

3.6. Genes co-Expressed with Sestrin2 in Lung Cancer

To explore Sestrin2-related pathways, the genes co-expressed with Sestrin2 were analyzed in the Bass lung dataset using Oncomine (Figure 6A). Highly co-expressed genes are listed by correlation rate. Ontology analysis was performed with Sestrin2 and its 11 co-expressed genes using Enrichr (Figure 6B). In GO analysis, Sestrin2 and its positively co-expressed genes were analyzed in biological processes, molecular function, and cellular components. In GO biological process, regulation of cell cycle process and protein tetramerization were highly related. Moreover, DNA-directed RNA polymerase 2 holoenzyme and RNA polymerase II transcription factor complex were associated in GO cellular component. In GO molecular function analysis, the most highly ranked terms were N-6 methyladenosine-containing RNA binding and leucine binding. In the Kyoto Encyclopedia of Genes and Genomes (KEGG) pathway database, the term 'basal transcription factors' was related to Sestrin2 and positively co-expressed genes. To check whether Sestrin2 expression regulates the expression of co-expressed genes, we analyzed the expression of the top 5 genes in Sestrin2-knockdown A549 cells by RT-PCR. Expression of all top 5 co-expressed genes was reduced in Sestrin2-knockdown cells (Figure 6C). These co-expressed gene profiles implied that Sestrin2 expression could regulate the expression of multiple co-expressed genes, possibly related to tumor progression.

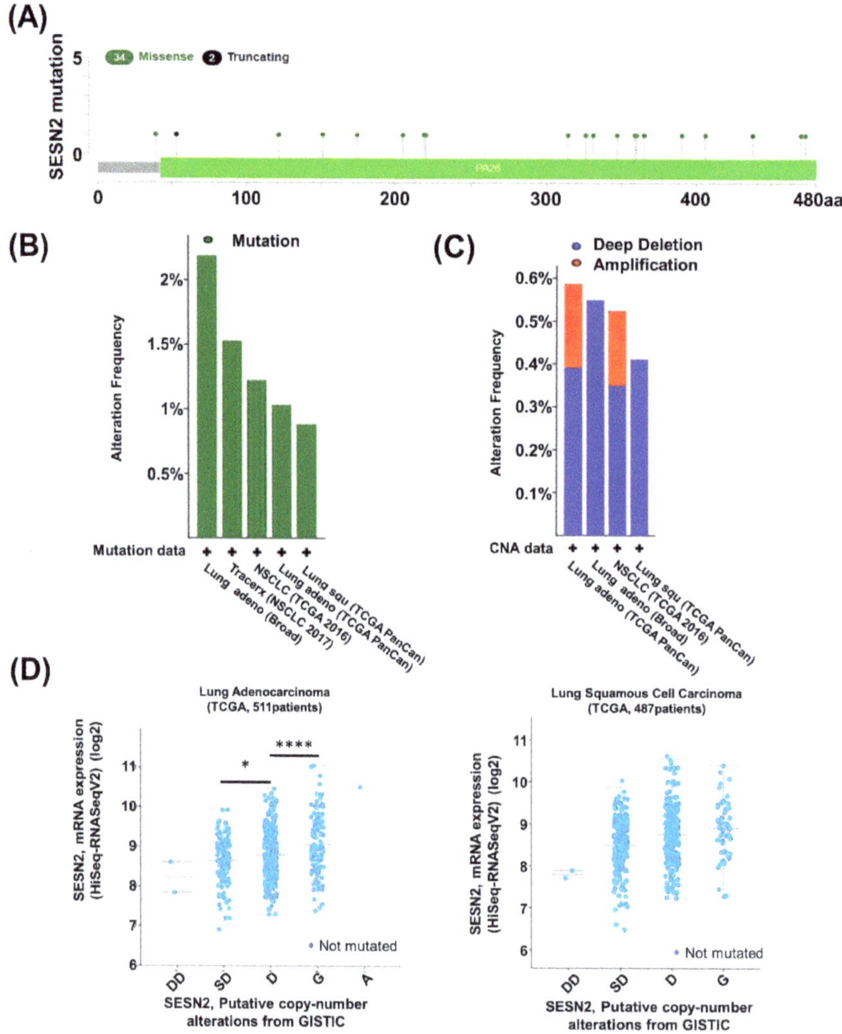

Figure 5. Sestrin2 mutation and alteration in TCGA lung cancer. (**A**) The mutation plot shows the location and type of mutation in Sestrin2. (**B**) Sestrin2 mutation analysis using cBioportal. Green: mutation. (**C**) Sestrin2 alteration analysis using cBioportal. Red: amplification; blue: deep deletion. (**D**) Copy number alteration of Sestrin2 mRNA expression in TCGA lung adenocarcinoma and TCGA lung squamous cell cancer datasets. Sestrin2 expression positively related to the copy number alteration status, deep deletion (DD), shadow deletion (SD), diploid (D), gain (G), and amplification (A). (* $p < 0.05$; **** $p < 0.0001$).

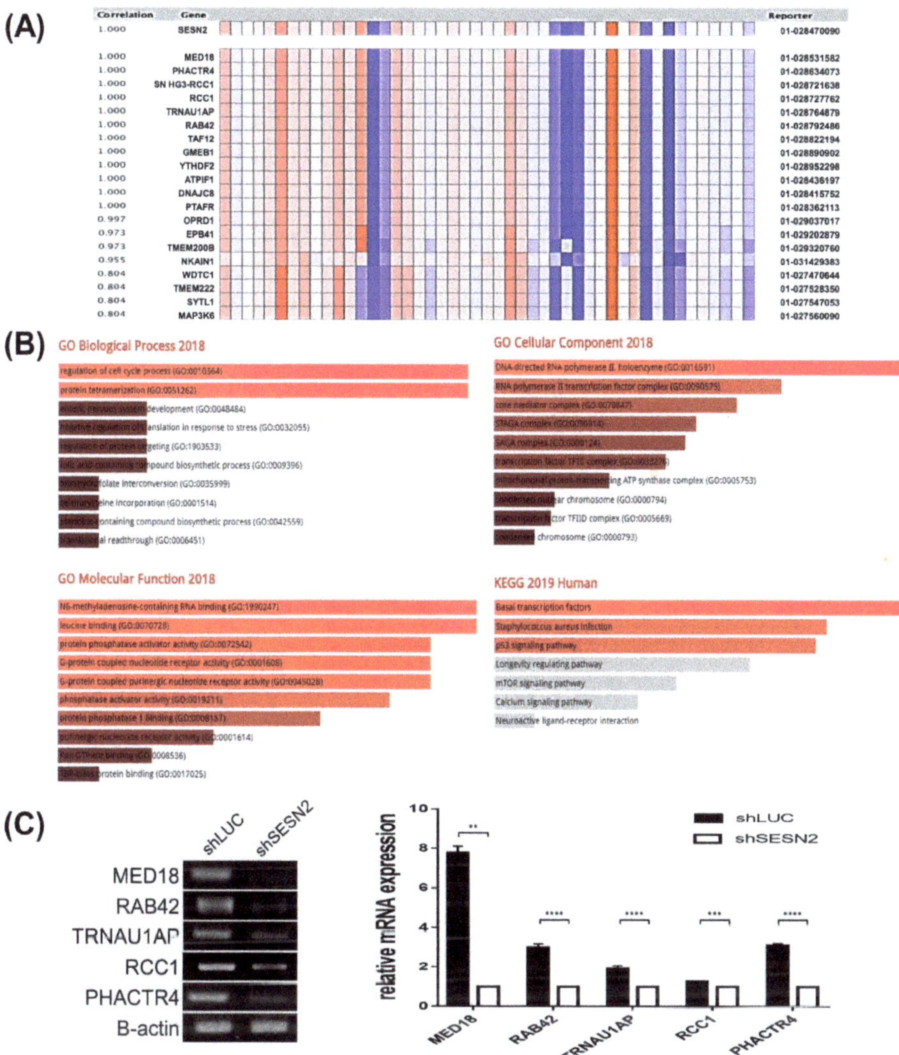

Figure 6. Sestrin2 co-expression gene analysis using RT-PCR (**A**) Co-expression gene analysis in the Bass lung dataset using the Oncomine database. (**B**) GO and KEGG analysis with Sestrin2 and co-expressed genes using Enrichr; bar graph listed by p-value. The brighter the bar color, the more significant the related pathway. (**C**) The expression of top 5 co-expression genes was downregulated in Sestrin2 knockdown A549 cell as shown by RT-PCR. Fold-change was measured by ImageJ; graph shown to the right. (** $p < 0.01$; *** $p < 0.005$; **** $p < 0.0001$).

4. Discussion

Sestrin2, a highly evolutionarily conserved protein, is involved in mediating cellular responses to various stressors. It has a protective effect against physiological and pathological conditions, mainly through regulating oxidative stress and inflammation [35]. Sestrin2 is a leucine sensor protein that regulates mTORC1 signaling that is related to cell proliferation and growth. Sestrin2 also plays an important role in cell protection and homeostasis, mainly by the downregulation of ROS and mTOR signaling [36]. It belongs to the family of stress-inducible proteins that has a pivotal role in regulating antioxidants, autophagy, and apoptosis, thereby enabling protection from any form of DNA damage, oxidative stress, hypoxia, or metabolic stress [37]. As a result, Sestrin2 keeps cells healthy, and it has been suggested that it may prevent cancer. In this paper, we investigated the effect of Sestrin2 knockdown in A549, a non-small cell lung cancer cell line, and analyzed the prognostic value of Sestrin2 expression in human lung cancer by employing various bioinformatic tools on various lung cancer datasets.

In the A549 lung cancer cell line, Sestrin2 knockdown led to downregulation of cancer properties, confirming the oncogenic function of Sestrin2. Our data showed that Sestrin2 knockdown resulted in reduced tumor cell proliferation and migration (Figure 1). Moreover, in the sphere-forming assay, the size of sphere was significantly decreased upon Sestrin2 knockdown in A549 cells (Figure 2). Based on these results, we suggest that Sestrin2 has oncogenic effects in lung cancer cells. Consistently, lung cancer cells from Sestrin2-deficient mice showed slower growth rates than did those from wild type mice [38]. However, Sestrin2 knockdown has also been reported to promote proliferation of cancer cells, inhibit apoptosis of cells [38–40], and enhance migration in the wound healing assay [40], which is in complete contrast to our results. Several previous studies have reported that Sestrin2 can work as a tumor suppressor gene in various cancers [20–22]. Sestrin2 was proposed to regulate AMPK/mTORC pathway activation and tumor cell growth in colorectal cancer [21], and Sestrin2 knockdown accelerated colorectal carcinogenesis [22]. Sestrin2 is also known to be downregulated in bladder cancer, and Sestrin2 expression upon MAPK-JUN-dependent transcription leads to the suppression of bladder cancer growth [23]. However, contrary to these results, Sestrin2 is still expressed in various cancers and may be necessary to increase cancer viability under certain conditions [24]. In lung cancers, there have also been conflicting reports about the role of Sestrin2. Downregulated Sestrin2 expression reduces death-receptor-induced apoptosis in lung cancer cell lines [25]. Sestrin2 expression is positively correlated with patient survival in 210 non-small cell lung cancer (NSCLC) tissue samples [41]. However, in glutamine-depleted lung cancer cells, upregulated Sestrin2 increases cell survival [42]. Based on contradictory reports, Sestrin2 expression studies on cancer progression could reveal opposite results, likely dependent on different cellular conditions, which need to be characterized in detail in future studies. Differences in the effects of Sestrin2 knockdown on cellular proliferation and migration might be due to differences in culture conditions between laboratories, originating from use of different reagents such as batches of fetal bovine serum, or different protocol details such as cell numbers used for each assay or subculture.

Sestrin2 knockdown increased the intracellular ROS concentration in A549 cells with reduced expression of antioxidant genes nrf-2 and HO-1 (Figure 3). In A549 cells, reduction in intracellular ROS concentration by the antioxidant molecule N-acetyl cysteine enhanced cellular proliferation [43], suggesting that reduced intracellular ROS level is a favorable condition for proliferation. Therefore, increased amount of ROS in Sestrin2 knockdown A549 cells may be a negative regulator of cellular proliferation and/or apoptosis induction. Treatment with doxorubicin or cisplatin induces cell death via the increase of ROS in A549 cells [44,45]. Sensitization to doxorubicin and cisplatin in Sestrin2 knockdown cell was not apparently detectable because impaired proliferation of Sestrin2 knockdown cells already reduced the recovered number of cells in no treatment control cells. However, the reduction in the rate of survival in Sestrin2 knockdown A549 cells in the doxorubicin-treated group is larger than that of no treatment control, suggesting sensitization to doxorubicin treatment in Sestrin2 knockdown cells (Figure 2D). Overall, our in vitro Sestrin2 knockdown experiment supports the assumption that reduced expression of Sestrin2 could be a favorable prognostic marker for survival of lung cancer patients.

In addition, we analyzed gene expression databases with various web tools to investigate the expression and prognostic value of Sestrin2 in lung cancers. In a dataset in the Oncomine database, mRNA expression of Sestrin2 was upregulated in lung cancer compared to that in normal tissue. Sestrin2 protein expression was upregulated in lung cancer patients in the Human Protein Atlas.

In addition, Sestrin2 mRNA expression was negatively correlated with the survival of lung cancer patients in multiple datasets. These results suggest that overexpressed Sestrin2 could have a poor prognostic value in lung cancer, which was in agreement with our in vitro data using A549 cells.

In the tumorigenesis processes, somatic loss-of-function or gain-of-function alterations in specific genes could have carcinogenic effects. However, mutations in the Sestrin2 gene have not been studied. Therefore, we used cBioPortal to determine mutations and CNAs in Sestrin2 gene. we a found several missense and truncating mutations within Sestrin2 protein-coding sequences (Figure 5A). The impact of each mutation in Sestrin2 has not been experimentally validated. We also found that expression of Sestrin2 was associated with the copy number alterations. This result implies that augmented Sestrin2 expression could be caused by the copy number alteration in lung cancer cells.

The co-expressed gene profile of Sestrin2 revealed pathways associated with Sestrin2 (Figure 6). The most highly rated gene ontology terms of GO biological process was regulation of cell cycle, which is closely related to cancer growth. Highly ranked terms in GO cellular component included RNA polymerase II (GO:0016591 and GO:0090575). Other terms including core mediator complex, STAGA complex, SAGA complex, transcription factor TFTC complex, condensation nuclear chromosome, and condensed chromosome are related to histone acetylation and chromosomal condensation. Most of the terms for GO cellular component suggested that Sestrin2 may be involved in transcriptional control through chromosomal condensation. The most highly ranked term in GO molecular function, N6-methyladenosine-containing RNA binding, may also be involved in transcriptional control; N6-methyladenosine is the most frequent mRNA modification significantly affecting gene expression and splicing [46]. KEGG pathway analysis includes p53 and mTOR signaling pathways, which were already known from previous studies [16]. Most importantly, knockdown of Sestrin2 also suppressed the expression of most highly correlated genes, which means that Sestrin2 is the upstream regulator of these associated pathways. This co-expressed gene analysis strongly suggests that Sestrin2 may be a key regulator of gene expression in lung cancer cells, which remains to be elucidated in further studies.

In this study, we investigated the impact of Sestrin2 expression in lung cancer with knockdown in a lung cancer cell line in vitro, and bioinformatic analysis using gene expression datasets of lung cancer. Further subsequent investigation using lung cancer cells including key cancer pathway analysis and in vivo study using animal model remains to be studied to elucidate the underlying mechanism of Sestrin2 in lung cancer.

5. Conclusions

In conclusion, Sestrin2 knockdown in lung cancer cells suppressed cancer cell properties, including proliferation, migration, stemness, and drug resistance. In human cancer expression datasets, increased expression of Sestrin2 and correlation of Sestrin2 expression with lung cancer patient survival was observed. Sestrin2 may be an upstream regulatory gene for its associated pathways. Thus, Sestrin2 may have prognostic value and serve as a therapeutic target in lung cancer.

Supplementary Materials: The following are available online at http://www.mdpi.com/2075-4426/10/3/109/s1, Supplementary Table S1. Primer list for RT-PCR; Supplementary Figure S1. RT-PCR for mRNA expression of *NRF2* and *HO-1* in A549 cells treated with scramble, shSESN2-1, or shSESN2-2 lentiviruses. Supplementary Figure S2. Cell Proliferation assay of scramble, shSESN2-1, and shSESN2-2 cells. Cell viability was measured using EZ cytox reagent at 1, 2, 3, and 4 day after seeing and incubating cells (1×10^4 per well in 96 well plate). The absorbance was measured at 450 nm using a fluorescence microplate reader. Supplementary Figure S3. Expression of NRF2 and HO1 by RT PCR in knockdown A549 lung cancer cell.

Author Contributions: Conceptualization, M.G., S.K.S. and S.-G.C.; methodology, S.K.S. and M.G.; validation, M.G.; formal analysis, H.S.C. and M.G.; investigation, H.S.C., M.G. and S-G.C.; data curation, M.G., S.K.S. and S.-G.C.; writing—original draft preparation, H.S.C. and M.G.; writing—review and editing, M.G., H.J.K., H.-W.P., B.V., and S.-G.C.; supervision, S.-G.C.; project administration, S.-G.C.; funding acquisition, S.-G.C. All authors have read and agreed to the published version of the manuscript.

Funding: This study was funded by grants from the National Research Foundation (NRF) of the Korean government (NRF-2019M3A9H1030682 and NRF-2015R1A5A1009701).

Acknowledgments: We thank Andrei Budanov for giving us the shLUC and shSESN2 plasmid set.

Conflicts of Interest: The authors declare no conflict of interest. The funders had no role in the design of the study; in the collection, analyses, or interpretation of data; in the writing of the manuscript, or in the decision to publish the results.

References

1. Siegel, R.L.; Miller, K.D.; Jemal, A. Cancer statistics, 2019. *CA Cancer J. Clin.* **2019**, *69*, 7–34. [CrossRef]
2. Lee, J.H.; Budanov, A.V.; Karin, M. Sestrins orchestrate cellular metabolism to attenuate aging. *Cell Metab.* **2013**, *18*, 792–801. [CrossRef]
3. Tsilioni, I.; Filippidis, A.S.; Kerenidi, T.; Budanov, A.V.; Zarogiannis, S.G.; Gourgoulianis, K.I. Sestrin-2 is significantly increased in malignant pleural effusions due to lung cancer and is potentially secreted by pleural mesothelial cells. *Clin. Biochem.* **2016**, *49*, 726–728. [CrossRef]
4. Budanov, A.V. Stress-responsive sestrins link p53 with redox regulation and mammalian target of rapamycin signaling. *Antioxid. Redox Signal.* **2011**, *15*, 1679–1690. [CrossRef]
5. Seo, K.; Seo, S.; Ki, S.H.; Shin, S.M. Sestrin2 inhibits hypoxia-inducible factor-1α accumulation via AMPK-mediated prolyl hydroxylase regulation. *Free Radic. Biol. Med.* **2016**, *101*, 511–523. [CrossRef]
6. Budanov, A.V.; Sablina, A.A.; Feinstein, E.; Koonin, E.V.; Chumakov, P.M. Regeneration of peroxiredoxins by p53-regulated sestrins, homologs of bacterial AhpD. *Science* **2004**, *304*, 596–600. [CrossRef]
7. Chen, C.-C.; Jeon, S.-M.; Bhaskar, P.T.; Nogueira, V.; Sundararajan, D.; Tonic, I.; Park, Y.; Hay, N. FoxOs inhibit mTORC1 and activate Akt by inducing the expression of Sestrin3 and Rictor. *Dev. Cell* **2010**, *18*, 592–604. [CrossRef]
8. Lee, J.H.; Budanov, A.V.; Talukdar, S.; Park, E.J.; Park, H.L.; Park, H.W.; Bandyopadhyay, G.; Li, N.; Aghajan, M.; Jang, I.; et al. Maintenance of metabolic homeostasis by Sestrin2 and Sestrin3. *Cell Metab.* **2012**, *16*, 311–321. [CrossRef]
9. Kopnin, P.B.; Agapova, L.S.; Kopnin, B.P.; Chumakov, P.M. Repression of sestrin family genes contributes to oncogenic Ras-induced reactive oxygen species up-regulation and genetic instability. *Cancer Res.* **2007**, *67*, 4671–4678. [CrossRef]
10. Nogueira, V.; Park, Y.; Chen, C.-C.; Xu, P.-Z.; Chen, M.-L.; Tonic, I.; Unterman, T.; Hay, N. Akt determines replicative senescence and oxidative or oncogenic premature senescence and sensitizes cells to oxidative apoptosis. *Cancer Cell* **2008**, *14*, 458–470. [CrossRef]
11. Bae, S.H.; Sung, S.H.; Oh, S.Y.; Lim, J.M.; Lee, S.K.; Park, Y.N.; Lee, H.E.; Kang, D.; Rhee, S.G. Sestrins activate Nrf2 by promoting p62-dependent autophagic degradation of Keap1 and prevent oxidative liver damage. *Cell Metab.* **2013**, *17*, 73–84. [CrossRef] [PubMed]
12. Hagenbuchner, J.; Kuznetsov, A.; Hermann, M.; Hausott, B.; Obexer, P.; Ausserlechner, M.J. FOXO3-induced reactive oxygen species are regulated by BCL2L11 (Bim) and SESN3. *J. Cell Sci.* **2012**, *125*, 1191–1203. [CrossRef] [PubMed]
13. Ganesan, H.; Balasubramanian, V.; Iyer, M.; Venugopal, A.; Subramaniam, M.D.; Cho, S.-G.; Vellingiri, B. mTOR signalling pathway-A root cause for idiopathic autism? *BMB Rep.* **2019**, *52*, 424. [CrossRef]
14. Parmigiani, A.; Nourbakhsh, A.; Ding, B.; Wang, W.; Kim, Y.C.; Akopiants, K.; Guan, K.-L.; Karin, M.; Budanov, A.V. Sestrins inhibit mTORC1 kinase activation through the GATOR complex. *Cell Rep.* **2014**, *9*, 1281–1291. [CrossRef]
15. Chantranupong, L.; Wolfson, R.L.; Orozco, J.M.; Saxton, R.A.; Scaria, S.M.; Bar-Peled, L.; Spooner, E.; Isasa, M.; Gygi, S.P.; Sabatini, D.M. The Sestrins interact with GATOR2 to negatively regulate the amino-acid-sensing pathway upstream of mTORC1. *Cell Rep.* **2014**, *9*, 1–8. [CrossRef]
16. Budanov, A.V.; Karin, M. p53 target genes sestrin1 and sestrin2 connect genotoxic stress and mTOR signaling. *Cell* **2008**, *134*, 451–460. [CrossRef]
17. Velasco-Miguel, S.; Buckbinder, L.; Jean, P.; Gelbert, L.; Talbott, R.; Laidlaw, J.; Seizinger, B.; Kley, N. PA26, a novel target of the p53 tumor suppressor and member of the GADD family of DNA damage and growth arrest inducible genes. *Oncogene* **1999**, *18*, 127. [CrossRef]
18. Shin, B.Y.; Jin, S.H.; Cho, I.J.; Ki, S.H. Nrf2-ARE pathway regulates induction of Sestrin-2 expression. *Free Radic. Biol. Med.* **2012**, *53*, 834–841. [CrossRef]

19. Zhang, X.-Y.; Wu, X.-Q.; Deng, R.; Sun, T.; Feng, G.-K.; Zhu, X.-F. Upregulation of sestrin 2 expression via JNK pathway activation contributes to autophagy induction in cancer cells. *Cell. Signal.* **2013**, *25*, 150–158. [CrossRef]

20. Pasha, M.; Eid, A.H.; Eid, A.A.; Gorin, Y.; Munusamy, S. Sestrin2 as a novel biomarker and therapeutic target for various diseases. *Oxidative Med. Cell. Longev.* **2017**, *2017*. [CrossRef]

21. Wei, J.-L.; Fang, M.; Fu, Z.-X.; Zhang, S.R.; Guo, J.-B.; Wang, R.; Lv, Z.-B.; Xiong, Y.-F. Sestrin 2 suppresses cells proliferation through AMPK/mTORC1 pathway activation in colorectal cancer. *Oncotarget* **2017**, *8*, 49318. [CrossRef] [PubMed]

22. Ro, S.-H.; Xue, X.; Ramakrishnan, S.K.; Cho, C.-S.; Namkoong, S.; Jang, I.; Semple, I.A.; Ho, A.; Park, H.-W.; Shah, Y.M. Tumor suppressive role of sestrin2 during colitis and colon carcinogenesis. *Elife* **2016**, *5*, e12204. [CrossRef]

23. Liang, Y.; Zhu, J.; Huang, H.; Xiang, D.; Li, Y.; Zhang, D.; Li, J.; Wang, Y.; Jin, H.; Jiang, G. SESN2/sestrin 2 induction-mediated autophagy and inhibitory effect of isorhapontigenin (ISO) on human bladder cancers. *Autophagy* **2016**, *12*, 1229–1239. [CrossRef]

24. Budanov, A.V.; Shoshani, T.; Faerman, A.; Zelin, E.; Kamer, I.; Kalinski, H.; Gorodin, S.; Fishman, A.; Chajut, A.; Einat, P. Identification of a novel stress-responsive gene Hi95 involved in regulation of cell viability. *Oncogene* **2002**, *21*, 6017. [CrossRef]

25. Ding, B.; Parmigiani, A.; Yang, C.; Budanov, A.V. Sestrin2 facilitates death receptor-induced apoptosis in lung adenocarcinoma cells through regulation of XIAP degradation. *Cell Cycle* **2015**, *14*, 3231–3241. [CrossRef]

26. Xia, J.; Zeng, W.; Xia, Y.; Wang, B.; Xu, D.; Liu, D.; Kong, M.G.; Dong, Y. Cold atmospheric plasma induces apoptosis of melanoma cells via Sestrin2-mediated nitric oxide synthase signaling. *J. Biophotonics* **2019**, *12*, e201800046. [CrossRef] [PubMed]

27. Rhodes, D.R.; Yu, J.; Shanker, K.; Deshpande, N.; Varambally, R.; Ghosh, D.; Barrette, T.; Pander, A.; Chinnaiyan, A.M. ONCOMINE: A cancer microarray database and integrated data-mining platform. *Neoplasia* **2004**, *6*, 1–6. [CrossRef]

28. Rhodes, D.R.; Kalyana-Sundaram, S.; Mahavisno, V.; Varambally, R.; Yu, J.; Briggs, B.B.; Barrette, T.R.; Anstet, M.J.; Kincead-Beal, C.; Kulkarni, P. Oncomine 3.0: Genes, pathways, and networks in a collection of 18,000 cancer gene expression profiles. *Neoplasia* **2007**, *9*, 166–180. [CrossRef]

29. Mizuno, H.; Kitada, K.; Nakai, K.; Sarai, A. PrognoScan: A new database for meta-analysis of the prognostic value of genes. *BMC Med Genom.* **2009**, *2*, 18. [CrossRef]

30. Gao, J.; Aksoy, B.A.; Dogrusoz, U.; Dresdner, G.; Gross, B.; Sumer, S.O.; Sun, Y.; Jacobsen, A.; Sinha, R.; Larsson, E. Integrative analysis of complex cancer genomics and clinical profiles using the cBioPortal. *Sci. Signal.* **2013**, *6*, pl1. [CrossRef]

31. Kuleshov, M.V.; Jones, M.R.; Rouillard, A.D.; Fernandez, N.F.; Duan, Q.; Wang, Z.; Koplev, S.; Jenkins, S.L.; Jagodnik, K.M.; Lachmann, A. Enrichr: A comprehensive gene set enrichment analysis web server 2016 update. *Nucleic Acids Res.* **2016**, *44*, W90–W97. [CrossRef]

32. Kim, S.J.; Kim, H.S.; Seo, Y.R. Understanding of ROS-Inducing Strategy in Anticancer Therapy. *Oxidative Med. Cell. Longev.* **2019**, *2019*. [CrossRef]

33. Zhang, D.D. Mechanistic studies of the Nrf2-Keap1 signaling pathway. *Drug Metab. Rev.* **2006**, *38*, 769–789. [CrossRef]

34. Wang, X.-J.; Sun, Z.; Villeneuve, N.F.; Zhang, S.; Zhao, F.; Li, Y.; Chen, W.; Yi, X.; Zheng, W.; Wondrak, G.T. Nrf2 enhances resistance of cancer cells to chemotherapeutic drugs, the dark side of Nrf2. *Carcinogenesis* **2008**, *29*, 1235–1243. [CrossRef] [PubMed]

35. Wang, L.-X.; Zhu, X.-M.; Yao, Y.-M. Sestrin2: Its potential role and regulatory mechanism in host immune response in diseases. *Front. Immunol.* **2019**, *10*, 2797. [CrossRef]

36. Fernandes, V.M.; McCormack, K.; Lewellyn, L.; Verheyen, E.M. Integrins regulate apical constriction via microtubule stabilization in the Drosophila eye disc epithelium. *Cell Rep.* **2014**, *9*, 2043–2055. [CrossRef]

37. Cordani, M.; Donadelli, M.; Strippoli, R.; Bazhin, A.V.; Sanchez-Alvarez, M. Interplay between ROS and Autophagy in Cancer and Aging: From Molecular Mechanisms to Novel Therapeutic Approaches. *Oxidative Med. Cell. Longev.* **2019**, *2019*. [CrossRef]

38. Ding, B.; Haidurov, A.; Chawla, A.; Parmigiani, A.; van de Kamp, G.; Dalina, A.; Yuan, F.; Lee, J.H.; Chumakov, P.M.; Grossman, S.R. p53-inducible SESTRINs might play opposite roles in the regulation of early and late stages of lung carcinogenesis. *Oncotarget* **2019**, *10*, 6997. [CrossRef] [PubMed]

39. Ben-Sahra, I.; Dirat, B.; Laurent, K.; Puissant, A.; Auberger, P.; Budanov, A.; Tanti, J.; Bost, F. Sestrin2 integrates Akt and mTOR signaling to protect cells against energetic stress-induced death. *Cell Death Differ.* **2013**, *20*, 611–619. [CrossRef]
40. Xu, H.; Sun, H.; Zhang, H.; Liu, J.; Fan, F.; Li, Y.; Ning, X.; Sun, Y.; Dai, S.; Liu, B. An shRNA based genetic screen identified Sesn2 as a potential tumor suppressor in lung cancer via suppression of Akt-mTOR-p70S6K signaling. *PLoS ONE* **2015**, *10*, e0124033. [CrossRef]
41. Chen, K.-B.; Xuan, Y.; Shi, W.-J.; Chi, F.; Xing, R.; Zeng, Y.-C. Sestrin2 expression is a favorable prognostic factor in patients with non-small cell lung cancer. *Am. J. Transl. Res.* **2016**, *8*, 1903.
42. Byun, J.-K.; Choi, Y.-K.; Kim, J.-H.; Jeong, J.Y.; Jeon, H.-J.; Kim, M.-K.; Hwang, I.; Lee, S.-Y.; Lee, Y.M.; Lee, I.-K. A positive feedback loop between Sestrin2 and mTORC2 is required for the survival of glutamine-depleted lung cancer cells. *Cell Rep.* **2017**, *20*, 586–599. [CrossRef]
43. Hann, S.S.; Zheng, F.; Zhao, S. Targeting 3-phosphoinositide-dependent protein kinase 1 by N-acetyl-cysteine through activation of peroxisome proliferators activated receptor alpha in human lung cancer cells, the role of p53 and p65. *J. Exp. Clin. Cancer Res.* **2013**, *32*, 43. [CrossRef]
44. Guo, J.; Xu, B.; Han, Q.; Zhou, H.; Xia, Y.; Gong, C.; Dai, X.; Li, Z.; Wu, G. Ferroptosis: A novel anti-tumor action for cisplatin. *Cancer Res. Treat. Off. J. Korean Cancer Assoc.* **2018**, *50*, 445. [CrossRef]
45. Poornima, P.; Kumar, V.B.; Weng, C.F.; Padma, V.V. Doxorubicin induced apoptosis was potentiated by neferine in human lung adenocarcima, A549 cells. *Food Chem. Toxicol.* **2014**, *68*, 87–98. [CrossRef]
46. Dominissini, D.; Moshitch-Moshkovitz, S.; Schwartz, S.; Salmon-Divon, M.; Ungar, L.; Osenberg, S.; Cesarkas, K.; Jacob-Hirsch, J.; Amariglio, N.; Kupiec, M.; et al. Topology of the human and mouse m6A RNA methylomes revealed by m6A-seq. *Nature* **2012**, *485*, 201–206. [CrossRef]

MDPI

St. Alban-Anlage 66

4052 Basel

Switzerland

Tel. +41 61 683 77 34

Fax +41 61 302 89 18

www.mdpi.com

Journal of Personalized Medicine Editorial Office

E-mail: jpm@mdpi.com

www.mdpi.com/journal/jpm